Calculus&M

VECTOR CALCULUS: Measuring in Two and Three Dimensions

Bill Davis
Ohio State University

Horacio Porta
University of Illinois, Urbana-Champaign

Jerry Uhl
University of Illinois, Urbana-Champaign

Addison-Wesley Publishing Company

Reading, Massachusetts • Menlo Park, California • New York
Don Mills, Ontario • Wokingham, England • Amsterdam • Bonn
Sydney • Singapore • Tokyo • Madrid • San Juan • Milan • Paris

None of Addison-Wesley, Wolfram Research, Inc., or the authors of Calculus&*Mathematica* makes any warranty or representation, either express or implied, with respect to the Calculus&*Mathematica* package, or its performance, merchantability or fitness for particular purposes.

None of Addison-Wesley, Wolfram Research, Inc., or the authors of Calculus&*Mathematica* shall be liable for any direct, indirect, special, incidental, or consequential damages resulting from any defect in the Calculus&*Mathematica* package, even if they have been advised of the possibility of such damages. Some states do not allow the exclusion or limitation of implied warranties or consequential damages, so the above exclusion may not apply to you.

Mathematica is a registered trademark, and *MathLink* and *MathReader* are trademarks of Wolfram Research, Inc.

Mathematica is not associated with Mathematica, Inc., Mathematica Policy Research, Inc., or MathTech, Inc.

Many of the designations used by manufacturers and sellers to distinguish their products are claimed as trademarks. Where those designations appear in this book, and Addison-Wesley was aware of a trademark claim, the designations have been printed in initial cap or all caps.

Copyright © 1994 by Addison-Wesley Publishing Company, Inc.

All rights reserved. No part of this publication may be reproduced, stored in a retrieval system, or transmitted, in any form or by any means, electronic, mechanical, photocopying, recording or otherwise, without the prior written permission of the publisher. Printed in the United States of America.

ISBN 0-201-58469-7

1 2 3 4 5 6 7 8 9 10-CRS-99 98 97 96 95 94

Preface

Now you get the chance to work measurements associated with functions of two and three variables. You'll learn about tangent vectors, velocity vectors, acceleration vectors, and curvature. You'll learn what the gradient is and what it means. You'll also learn how to plot paths of trajectories in fluid flow and paths of trajectories in a force field. You'll learn how to make measurements of the net flow of a fluid across a curve or a surface, both visually and with integrals. You'll measure the volume of a 3D solid and the area of a curved surface. You'll do plenty of intriguing plots and you'll learn from what you plot. And, to top it off, you'll learn some good calculational strategies for handling nasty 2D and 3D integrals.

Along the way, you'll have the opportunity to learn how to bounce light beams off curves and surfaces, to discover the reflecting properties of parabolas and ellipses, to plot the motion of the Sun-Earth-Mars system, to learn how to bore holes with a robotic router, to use the temperature gradient to program heat-seeking missiles and how to plot the path of a cork in a fluid flow. You'll work with electric fields and steady state heat and you'll have the opportunity to work with stealth technology and with some Star Wars window of vulnerability plots adapted from NASA work.

Students have reported that this part of Calculus&Mathematica has been especially useful in their physics and engineering courses. There's a lot of good stuff to soak up here. Jump in.

How to Use This Book

In Calculus&Mathematica, great care has been taken to put you in a position to learn visually. Instead of forcing you to attempt to learn by memorizing impenetrable jargon, you will be put in the position in which you will experience mathematics by seeing it happen and by making it happen. And you'll often be asked to describe what you see. When you do this, you'll be engaging in active mathematics as opposed to the passive mathematics you were probably asked to do in most other math courses. In order to take full advantage of this crucial aspect of Calculus&Mathematica, your first exposure to a new idea should be on the live computer screen where you can interact with the electronic text to your own satisfaction. This means that you should avoid "introductory lectures" and you should avoid reading this book at first. After you have some familarity with new ideas as found on the computer screen, you should seek out others for discussion and you can refer to this book to brush up on a point or two after you leave the computer. In the final analysis, this book is nothing more than a partial record of what happens on the screen.

Once you have participated in the mathematics and science of each lesson, you can sharpen your hand skills and check up on your calculus literacy by trying the questions in the Literacy Sheet associated with each lesson. The Literacy Sheets appear at the end of each book.

Those without previous Calculus&Mathematica experience should try to become familiar with Lesson 1.09 (Parametric Plotting) and with Lesson 2.05 (2D Integrals) before diving into the lessons in this book.

Significant Changes from the Traditional Course

The main change is the visual, conceptual approach. Students learn the basic ideas visually and then they learn about the integrals that make the associated measurements. The result: Instead of studying the symbols in a formula, the students learn they meaning of a given measurement formula and when to apply it.

One obvious change is the sudents' ability to plot whatever they want. In the traditional course, visualization of two dimensional regions and three dimensional curves and surfaces has always been a problem. The traditional response has been to look only at surfaces that have circular, elliptical, hyperbolic, or parabolic cross sections. There is no such restriction in this course.

Instead of delivering most surfaces implicitly or explictly, this course delivers most surfaces parametrically.

Vectors are used right at the beginning as strong motivators in experiments involving laser burns, pursuit models, and planet plotting—first with the sun at the origin and second with Earth at the origin.

The reflecting properties of parabolic and elliptical reflectors are discovered experimentally by the students and explained through student calculations.

Students learn about the principal unit normal and the unit binormals and they retain the idea because they use these two normal vectors to plot tubes, horns, and ribbons centered on curves. The value of this plotting experience as a strong driver for the acquisition and retention of the idea of moving frames cannot be underestimated.

The idea of curvature is developed cleanly with vector ideas so that the two dimensional version can be immediately transported to three dimensions. For the first time in a calculus course, students plot actual curves and osculating circles in two and three dimensions.

The meaning of velocity vectors and acceleration vectors, the tangential components of the acceleration vectors, and the normal components of acceleration vectors are set up with student-produced plots and write-ups.

Instead of looking at the dot product and cross product as isolated topics, the dot product and cross product are studied under the unifying theme of perpendicularity.

Students learn about ascent and descent paths by working with plots of the gradient field. The idea is reinforced by experiments like programming a heat-seeking missile on an ascent path through the gradient field of the temperature function.

Students get experience in optimization by working with Beale's valley function, Rosenbrock's banana function, optimation problems from metallurgy and least squares data fit in more than one dimension.

Local maximizers and local minimizers are deemphasized in favor of global maximizers and global minimizers.

Vector fields are visually introduced as flow models. Students drop floating corks into plots of vector fields with the goal of spotting sources and sinks. Students plot normal and tangential components of vector fields on curves with the goal of coming up with their own visual estimations of the net flow of a given vector field across a given curve (flux) and the net flow of a given vector field along a given curve (circulation or rotation). After the ideas are established in the students' minds visually, the associated measurements coming from path integrals (line integrals) and the Gauss-Green formula are introduced.

Students learn to use the divergence and curl of a vector field in a conceptual way as opposed to the traditional algebraic way. They are able to work with a formula and the meaning of the formula interchangeably. For instance, upon learning that the divergence of a given vector field Field$[x, y] = \{m[x, y], n[x, y]\}$ is positive at all points within a given closed curve C, the student will say that all the points inside C are sources of new fluid. They will go on to say that the net flow of this vector field across C is from inside to outside and consequently the net flow-across-C measurement $\int_C -n[x, y]\, dx + m[x, y]\, dy$ is positive.

The Laplacian is introduced as a measurement for detecting sources and sinks in the gradient field. Students explain why functions that measure steady state heat must have vanishing Laplacians. Similarly students explain, in two and three dimensions, why a function whose Laplacian vanishes cannot have maximum or minimum values.

Students get experience with electric fields in two and three dimensions by plotting them and making measurements on them. Students explain the reasoning behind

Gauss's law for the electric flux. Students have reported that this experience has been very helpful to them in their physics courses.

The two dimensional Jacobian is introduced as a local area conversion factor; the three dimensional Jacobian is introduced as a local volume conversion factor. Students study the action of transformations by plotting Jacobians and their gradient fields and learn why a Jacobian that vanishes at a point can cause big problems.

Students learn how to react to a given 2D or 3D integral and to come up with a custom transformation of a nasty 2D or 3D integral into an easily calculated integral. Traditional courses usually limit themselves to polar, cylindrical, or spherical transformations of integrals. There is no such restriction here. For traditional students, integrating on an ellipsoid is a big deal; for students in this course, integrating on an ellipsoid is routine. Reason: Students know that if they can plot a surface of a solid parametrically, then usually the plotting intructions set up the transformation they want to use to integrate on or within the same solid.

Contents

VECTOR CALCULUS: Measuring in Two and Three Dimensions

Prerequistes Those without previous Calculus&*Mathematica* experience should try to become familiar with Lesson 1.09 (Parametric Plotting) and with Lesson 2.05 (2D Integrals) before diving into these lessons.

■ 3.01) Vectors Point the Way

Mathematics Vectors: How they are plotted, how they are moved, how they are added and how they are multiplied by numbers. Tangent vectors, velocity vectors and tangent lines. Dot product of two vectors and the component (push) of one vector in the direction of another.

Science and math experience Bouncing light beams off curves in two dimensions. Experiments dealing with reflecting properties of parabolas and ellipses. Pursuit models. Parametric formulas for a line. Laser zapping along the tangent line. Velocity and acceleration vectors. Analyzing an object's motion and speed by looking at tangential components of its acceleration. Normal and tangential components of acceleration. Plotting the motion of the Earth-Mars-Sun system with Earth at the origin. Stealth technology.

■ 3.02) Perpendicularity

Mathematics Cross products and dot products. Planes in three dimensions. Normal vectors for surfaces in three dimensions.

Science and math experience Flatness and plotting. Using the main unit normal and binormal as moving frames to plot tubes, horns, ribbons and corrugations centered on a given curve. Experiments with linearizations (tangent plane approximations). Bouncing light beams off surfaces in three dimensions. Kissing circles and curvature in two and three dimensions. Boring holes with a robotic router.

■ 3.03) The Gradient

Mathematics The gradient and the chain rule. Level curves, level surfaces and the gradient as normal vector. The gradient as a vector pointing in the direction of greatest initial increase. Linearizations and the chain rule. Total differential. Lagrange multipliers.

Science and math experience Using the gradient for maximization and minimization. Ascent and descent paths through the gradient field with application to the idea behind *Mathematica*'s FindMinimum instruction. Data Fit: Vibrating string and one dimensional heat. Programming heat-seeking missiles. Cobb-Douglas manufacturing model. Experience with Beale's valley function and Rosenbrock's banana function. Optimation problems from metallurgy. Duffin's barge problem.

■ 3.04) 2D Vector Fields and Their Trajectories

Mathematics Vector fields as fluid flow. Trajectories in vector fields as the path a cork floats on. Solutions of differential equations $y'[x] = f[x, y[x]]$ as trajectories in a vector field.

Science and math experience Tangential and normal components of vector fields on given curves. Visual experiments dealing with the net flow of vector fields across given curves. Visual experiments dealing with the net flow of vector fields along given curves. The 2D electric field. Dipoles in 2D. The gradient field. Experiments with how gradient fields look near maximizers and minimizers. Where trajectories in the gradient field want to go. Where trajectories in the negative gradient field want to go. Looking for spigots and drains by following the path of a cork. Logistic harvesting model.

■ 3.05) Flow Measurements by Integrals

Mathematics Path integrals (line integrals) as measurements of the net flow of a given vector field across a given curve. Path integrals (line integrals) as measurements of the net flow of a given vector field along a given curve. Path independence and gradient fields. Recognition of gradient fields.

Science and math experience Sources and sinks of 2D vector fields. Sources and sinks at singularities in vector fields. Work as flow in a force field. Which way to go to make a force field do most of the work. Force fields and their trajectories. Models for water flow. Clockwise versus counterclockwise flow.

■ 3.06) Sources, Sinks, Swirls, and Singularities

Mathematics Divergence and rotation of a 2D vector field. Sources as points at which the divergence is positive; sinks as points at which the divergence is negative. Using the Gauss-Green formula to measure the flow of a 2D vector field across a closed curve by means of a 2D integral. Using the Gauss-Green formula to measure the flow of a 2D vector field along a closed curve by means of a 2D integral. Singularity sources, sinks and swirls.

Science and math experience Why it is that if all points inside a given closed curve C are sources of a given 2D vector field, then the net flow of the vector field across C must be from inside to outside. Encapsulating singularities with small circles centered on the singularity. Flow measurements in the presence or absence of singularities. 2D electric fields. Dipole fields. Gauss's law for calculating the flux of combined 2D electric fields. Parallel flow. The Laplacian as the divergence of the gradient field. Why harmonic functions cannot have local maxima or minima. Steady state heat and Laplace's equation in two dimensions. Spin fields.

■ 3.07) Transforming 2D Integrals

Mathematics uv-paper and xy-paper when $u = u[x,y]$ and $v = v[x,y]$. The uv grid on xy-paper. Linearizing the uv grid on xy-paper. The Jacobian as the area conversion factor for converting xy-paper area measurements into uv-paper area measurements. Transforming 2D integrals: How it is done and why it is done.

Science and math experience Semi-log and log-log paper. Flow measurements for 2D vector fields. If the boundary of a region can be plotted with *Mathematica*, then the plotting instructions usually carry enough information to make it possible to measure the area of the region. Why crazy things are likely to happen when the area conversion factor is 0. Analyzing a transformation by plotting its area conversion factor and by plotting the gradient field of its area conversion factor. Experiments relating linear equations and area measurements in two dimensions. Experiments with eigenvectors as stretching directions and eigenvalues as stretching factors in linear transformations in two dimensions.

■ 3.08) Transforming 3D Integrals

Mathematics The 3D integral $\iiint_R f[x, y, z]\, dx\, dy\, dz$ via slicing and accumulating. $\iiint_R dx\, dy\, dz$ as a volume measurement. Average value of a function on a region. The Jacobian as a local volume conversion factor in 3D. Transforming 3D integrals. Mass and density.

Science and math experience If the whole skin of a solid region can be plotted with *Mathematica*, then the plotting instructions usually carry enough information to make it possible to measure the volume of the solid region. Cylinders, spheres, and tubes: Plotting them and integrating on them. Integrating on solids bounded by sets of surfaces. Switching the order of integration. Tubes, horns, and squashed doughnuts. Drilling and slicing spheres. Experiments relating linear equations to volume measurements. Centroids and centers of mass. Bidding on rocket nose cones.

■ 3.09) Spherical Coordinates

Mathematics Meaning of each of the spherical parameters. Plotting and measuring volumes of spheres, ellipsoids, cones, and measuring volumes of each. Integration with spherical coordinates.

Science and math experience Earth-Moon animations. Snail shells. Star Wars window of vulnerability plots adapted from NASA work. Using spherical coordinates to design and paint flowers. Centering and aligning the general 3D ellipsoid. Ice cream cone and tops. Passing a plane between two disjoint solid disks. Measuring the volume inside 4D spheres. Experiments with eigenvectors as stretching directions and eigenvalues as stretching factors in linear transformations in three dimensions.

■ 3.10) 3D Surface Measurements

Mathematics Measuring area on surfaces. Surface integrals for measuring flow of 3D vector fields across surfaces. Sources, sinks, and Gauss's formula in 3D.

Science and math experience Measuring flow across surfaces: Gauss's formula versus calculation by a surface integral. Substitute surfaces to avoid calculational nightmares. Encapsulating singularities with small spheres centered on the singularity. 3D electric fields and Gauss's law for calculating the flux of combined 3D electric fields. The Laplacian as the divergence of the 3D gradient field. Sources and sinks in the 3D gradient field and their relation to max-min. Why harmonic functions of three variables cannot have local maxima or minima. Steady state heat in a solid. Morphing and Moebius strips.

■ **3.11) 3D Flow Along**

Mathematics Measuring flow along a 3D curve with a path integral. The curl of a 3D vector field. Orientation and Stokes's formula. Stokes's formula as an outgrowth of the 2D Gauss-Green formula.

Science and math experience Fingering a 3D vector field. The curl as the axis of the greatest counterclockwise swirl. Paddle wheels. Parallel flow and irrotational flow. Ideal fluid flow. Work done by 3D force fields. Recognition of 3D gradient fields. Path independence.

LESSON 3.01

Vectors Point the Way

Basics

■ **B.1)** Vectors: How you move them, how you add them, how you subtract them, and how you multiply them by numbers

A vector X in the plane is a stick with:

→ one end (its tail) at $\{0,0\}$ and

→ the other end (its tip) at a specified location $\{x[1], x[2]\}$.

Most folks like to draw a vector as an arrow with the arrowhead at its tip. Here's a look at the vector running from $\{0,0\}$ to $\{3,4\}$:

In[1]:=
```
X = {3,4};
Show[Arrow[X,Tail->{0,0},Blue],
Axes->True,
PlotRange->All,
AxesOrigin->{0,0}];
```

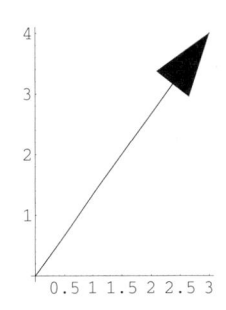

Most folks like to see the arrowhead at the tip so they can tell which end of the stick is the tip. Here is the vector running from $\{0,0\}$ (tail) to $\{-5,4\}$ (tip):

Lesson 3.01 Vectors Point the Way

```
In[2]:=
X = {-5,4};
Show[Arrow[X,Tail->{0,0},Red],
Axes->True,PlotRange->All,
AxesLabel->{"x","y"},
AxesOrigin->{0,0}];
```

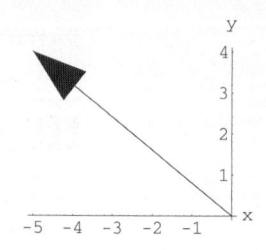

Here are some others:

```
In[3]:=
X = {-2,-4};
Show[Arrow[X,Tail->{0,0},Blue],
Axes->True,PlotRange->All,
AxesLabel->{"x","y"},
AxesOrigin->{0,0}];
```

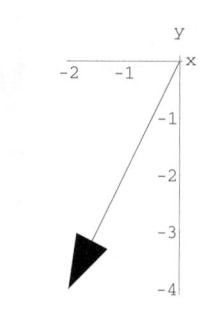

```
In[4]:=
X = {3,-1};
Show[Arrow[X,Tail->{0,0}],
Axes->True,PlotRange->All,
AxesLabel->{"x","y"},
AxesOrigin->{0,0}];
```

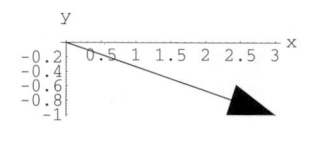

Play with these by rerunning with vectors of your own choice.

A vector $X = \{x[1], x[2]\}$ is written the same as the coordinates of its tip because everyone knows its tail is at $\{0, 0\}$.

You can also work with three-dimensional vectors by writing $X = \{x[1], x[2], x[3]\}$ for the vector whose tail is at $\{0, 0, 0\}$ and whose tip is at $\{x[1], x[2], x[3]\}$. Here's the vector $\{1.5, 5.1, 2.0\}$ shown along with the three-dimensional coordinate axes:

```
In[5]:=
X = {1.5,5.1,2.0}; spacer = 0.2; h = 1;
threedims = Graphics3D[{{Blue,Line[{{-h,0,0},{h,0,0}}]},
Text["x",{h + spacer,0,0}],{Blue,Line[{{0,-h,0},{0,h,0}}]},
Text["y",{0,h + spacer,0}],{Blue,Line[{{0,0,0},{0,0,h}}]},
Text["z",{0,0,h + spacer}]}];
```

```
In[6]:=
CMView = {2.7,1.6,1.2};
Show[Arrow[X,Tail->{0,0,0},Red],
threedims,PlotRange->All,
ViewPoint->CMView,Boxed->False];
```

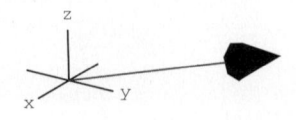

Basics (B.1)   3

Here's another:

In[7]:=
```
X = {1,-3,1};
Show[Arrow
[X,Tail->{0,0,0},Red],threedims,
PlotRange->All,
ViewPoint->CMView,Boxed->False];
```

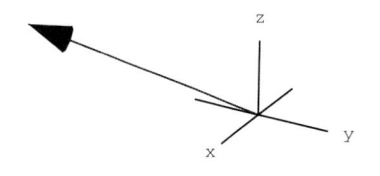

Here's a vector parallel to the yz-plane:

In[8]:=
```
X = {0,2,1};
Show[Arrow
[X,Tail->{0,0,0},Red],threedims,
PlotRange->All,
ViewPoint->CMView,Boxed->False];
```

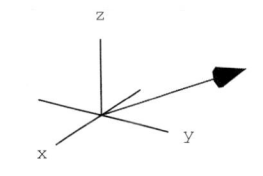

Play by rerunning with three-dimensional vectors of your own choice.

B.1.a.i) | How do you move vectors to new positions?

Answer: Very easily. Here is a vector in two dimensions with its tail at $\{0,0\}$:

In[9]:=
```
X = {3,4};
Show[Arrow[X,Tail->{0,0},Blue],
AxesLabel->{"x","y"},
Axes->True,AxesOrigin->{0,0}];
```

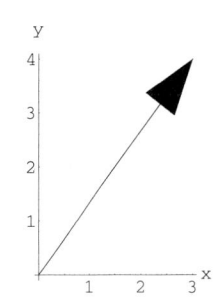

Here's the same vector X shown twice, once with its tail at $\{0,0\}$ and once with its tail at $\{2,1\}$:

In[10]:=
```
X = {3,4};
Show[Arrow[X,Tail->{0,0},Blue],
Arrow[X,Tail->{2,1},Blue],
Axes->True,AxesLabel->{"x","y"},
PlotRange->All,AxesOrigin->{0,0}];
```

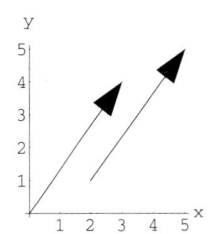

The two arrows have equal lengths and they both point in the same direction. They're parallel. Try it again for a different vector and a different tail:

Lesson 3.01 Vectors Point the Way

```
In[11]:=
X = {-1,4};
Show[Arrow[X,Tail->{0,0},Blue],
Arrow[X,Tail->{-2,3},Blue],
Axes->True,AxesLabel->{"x","y"},
PlotRange->All,AxesOrigin->{0,0}];
```

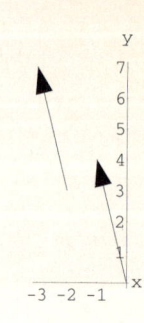

Same length; same direction. Here's a vector X shown with its tail at $\{0,0\}$ and shown with a whole squadron of its transplants:

```
In[12]:=
X = {2,3}; Clear[tail,k];
tail[1] = {0,0}; tail[2] = {6,4};
tail[3] = {6,6}; tail[4] = {6,8};
tail[5] = {4,8}; tail[6] = {2,8};
squadron = Table[Arrow[X,Tail->tail[k],Red],{k,1,6}];
Show[squadron,Axes->True,AxesLabel->{"x","y"},
PlotRange->All,AxesOrigin->{0,0}];
```

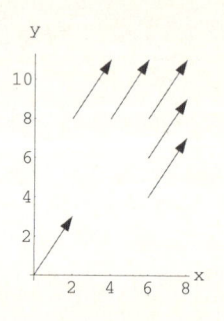

Check it out for a different X:

```
In[13]:=
X = {-3,-2}; Clear[tail,k];
tail[1] = {0,0}; tail[2] = {6,4};
tail[3] = {6,6}; tail[4] = {6,8};
tail[5] = {4,8}; tail[6] = {2,8};
squadron = Table[Arrow[X,Tail->tail[k],Red],{k,1,6}];
Show[squadron,Axes->True,AxesLabel->{"x","y"},
PlotRange->All,AxesOrigin->{0,0}];
```

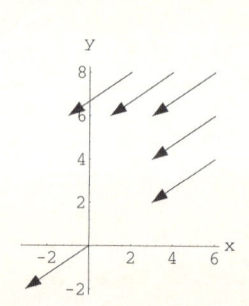

Check it out in three dimensions:

```
In[14]:=
X = {3,0,2}; Clear[tail,k];
tail[1] = {0,0,0}; tail[2] = {1,0,0};
tail[3] = {-1,0,0}; tail[4] = {0,1,0};
tail[5] = {0,-1,0}; tail[6] = {0,0,1};
tail[7] = {1,0,1};
squadron = Table[Arrow[X,Tail->tail[k],Red],{k,1,7}];
Show[squadron,threedims,PlotRange->All,
ViewPoint->CMView,Boxed->False];
```

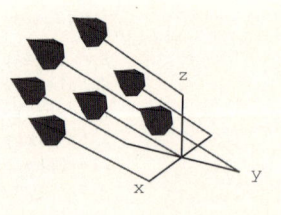

Same length; same direction.

Basics (B.1) 5

B.1.a.ii) Are there any rules about moving vectors to new positions?

Answer: You can put the tail anywhere you like, but you must be careful not to change its direction or its length.

B.1.b) How do you add vectors?

Answer: Very easily. For instance, if $X = \{3, 8\}$ and $Y = \{5, 4\}$ are vectors, then you add them to get
$$X + Y = \{3, 8\} + \{5, 4\} = \{8, 12\}$$
just by adding the corresponding components. *Mathematica* can do this too:

In[15]:=
```
X = {3,8}; Y = {5,4}; X + Y
```
Out[15]=
```
{8, 12}
```

You can add three-dimensional vectors:

In[16]:=
```
X = {3,-5,2}; Y = {3,4,-7}; X + Y
```
Out[16]=
```
{6, -1, -5}
```

You cannot add vectors from different dimensions:

In[17]:=
```
X = {3,-5}; Y = {3,4,-7}; X + Y
```
```
Thread::tdlen:
   Objects of unequal length in {3, -5} + <<1>> cannot be combined.
```
Out[17]=
```
{3, -5} + {3, 4, -7}
```

Here is a way of seeing what's happening in two dimensions: Look at a picture of $X = \{3, 10\}$ and $Y = \{8, 2\}$ and $X + Y$:

In[18]:=
```
X = {3,10}; Y = {8,2};
Show[Arrow[X,Tail->{0,0},Red],
Graphics[Text["X",X/2]],Arrow[Y,Tail->{0,0},Red],
Graphics[Text["Y",Y/2]],Arrow[X + Y,Tail->{0,0},Red],
Graphics[Text["X+Y",(X+Y)/2]],Axes->Automatic];
```

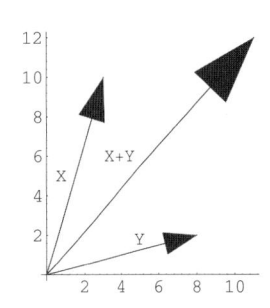

$X + Y$ represents the combined push of X and Y. See what happens when you move Y without changing its direction, so that its tail is at the tip of X:

Lesson 3.01 Vectors Point the Way

In[19]:=
```
Show[Arrow[X,Tail->{0,0},Red],
 Graphics[Text["X",X/2]],
 Arrow[Y,Tail->X,Red],
 Graphics[Text["Y",X + Y/2]],
 Arrow[X + Y,Tail->{0,0},Blue],
 Graphics[Text["X+Y",(X+Y)/2]],
 Axes->Automatic];
```

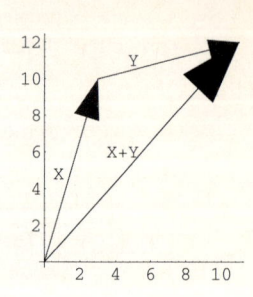

A triangle! This triangle shows that you have your choice of ways of getting from $\{0,0\}$ to the tip of $X + Y$.

→ Route 1: You can walk directly on the vector $X + Y$ from its tail to its tip.

→ Route 2: You can walk on X to the tip of X and then hook the tail of Y on the tip of X and finish the trip by walking along Y to its tip.

Lazy folks usually take Route 1. Rerun this for two-dimensional vectors of your own choice. Now check out what happens in three dimensions:

In[20]:=
```
X = {3,-4,7}; Y = {1,2,3}; spacer = 0.2; h = 1;
threedims = Graphics3D[{{Blue,Line[{{-h,0,0},
 {h,0,0}}]},Text["x",{h + spacer,0,0}],{Blue,
 Line[{{0,-h,0},{0,h,0}}]},Text["y",{0,h + spacer,0}],
 {Blue,Line[{{0,0,0},{0,0,h}}]},
 Text["z",{0,0,h + spacer}]}]; CMView = {2.7,1.6,1.2};
Show[Arrow[X,Tail->{0,0,0},Red],Graphics3D[Text["X",X/2]],
 Arrow[Y,Tail->{0,0,0},Red],Graphics3D[Text["Y",Y/2]],
 Arrow[X + Y,Tail->{0,0,0},Red],Graphics3D[Text["X+Y",(X+Y)/2]],
 threedims,PlotRange->All,
 ViewPoint->CMView,Boxed->False];
```

Again, in three dimensions, $X + Y$ represents the combined push of X and Y. See what happens when you move Y without changing its direction so that its tail is at the tip of X:

In[21]:=
```
Show[Arrow[X,Tail->{0,0,0},Red],
 Graphics3D[Text["X",X/2]],
 Arrow[Y,Tail->X,Red],
 Graphics3D[Text["Y",X + Y/2]],
 Arrow[X + Y,Tail->{0,0,0},Blue],
 Graphics3D[Text["X+Y",(X+Y)/2]],
 threedims,PlotRange->All,
 ViewPoint->CMView,Boxed->False];
```

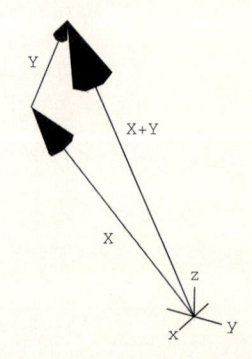

Another triangle. Just as in two dimensions, this triangle shows that you have your choice of ways of getting from $\{0,0\}$ to the tip of $X+Y$.

→ Route 1: You can walk directly on the vector $X+Y$ from its tail to its tip.

→ Route 2: You can walk on X to the tip of X and then hook the tail of Y on the tip of X and finish the trip by walking along Y to its tip.

Rerun the pictures above for different vectors X and Y until you get the hang of vector addition.

B.1.c) | How do you subtract vectors?

Answer: With no trouble. For instance, if $X = \{8,2\}$ and $Y = \{3,4\}$ are vectors, then you subtract Y from X to get
$$X - Y = \{8,2\} - \{3,4\} = \{5,-2\}.$$
Mathematica can do this too:

In[22]:=
```
X = {8,2}; Y = {3,4};
X - Y
```
Out[22]=
```
{5, -2}
```

Here is a way of seeing what's happening.

Look at a picture of $X = \{8,2\}, Y = \{3,4\}$ and $X - Y$:

In[23]:=
```
X = {8,2}; Y = {3,4};
Show[Arrow[X,Tail->{0,0},Red],
Graphics[Text["X",X/2]],
Arrow[Y,Tail->{0,0},Red],
Graphics[Text["Y",Y/2]],
Arrow[X - Y,Tail->{0,0},Red],
Graphics[Text["X-Y",(X-Y)/2]],
Axes->Automatic];
```

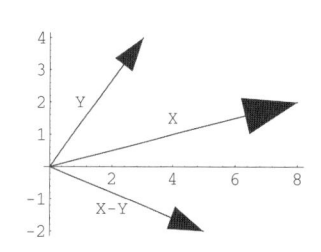

This time X is the combined push of Y and $X - Y$. See what happens when you move $X - Y$ without changing its direction so that its tail is at the tip of Y:

In[24]:=
```
Show[Arrow[X,Tail->{0,0},Blue],
Graphics[Text["X",X/2]],
Arrow[Y,Tail->{0,0},Red],
Graphics[Text["Y",Y/2]],
Arrow[X - Y,Tail->Y,Red],
Graphics[Text["X-Y",Y + (X-Y)/2]],
Axes->Automatic];
```

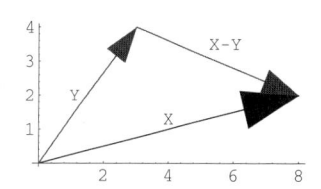

Lesson 3.01 Vectors Point the Way

Another triangle. You can get to the tip of X by going directly along X, or you can go from the tail of Y to the tip of Y and then ride on $X - Y$ to the tip of X. This is not a surprise because $X = (X - Y) + Y$. Rerun for some other vectors.

Here it is in three dimensions:

In[25]:=
```
X = {4,2,8}; Y = {-2,6,-3}; spacer = 0.2; h = 1;
threedims = Graphics3D[{{Blue,Line[{{-h,0,0},{h,0,0}}]},
Text["x",{h + spacer,0,0}],{Blue,Line[{{0,-h,0},{0,h,0}}]},
Text["y",{0,h + spacer,0}],{Blue,Line[{{0,0,0},{0,0,h}}]},
Text["z",{0,0,h + spacer}]}]; CMView = {2.7,1.6,1.2};
Show[Arrow[X,Tail->{0,0,0},Blue],Graphics3D[Text["X",X/2]],
Arrow[Y,Tail->{0,0,0},Red],Graphics3D[Text["Y",Y/2]],
Arrow[X - Y,Tail->Y,Red],
Graphics3D[Text["X-Y",Y + (X-Y)/2]],
threedims,PlotRange->All,
ViewPoint->CMView,Boxed->False];
```

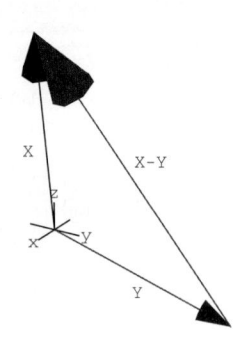

Rerun for other X and Y of your own choice until you get the picture down pat.

B.1.d) | How do you multiply vectors by numbers?

Answer: You just do it. For instance, if $X = \{3, 1\}$ then

$$2X = \{6, 2\}.$$

Mathematica can do this too:

In[26]:=
```
X = {3,1}; 2X
```
Out[26]=
{6, 2}

Here are X and $2X$ both shown with their tails at $\{0, 0\}$ for $x\{1, 2\}$:

In[27]:=
```
X = {1,2};
Show[Arrow[2X,Tail->{0,0},Red],
Graphics[Text["2X",3X/2]],
Arrow[X,Tail->{0,0},Blue],
Graphics[Text["X",X/2]],
Axes->Automatic];
```

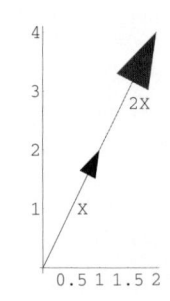

$2X$ points the same direction as X, but $2X$ is twice as long as X. That makes some sense because $2X$ is the same thing as $X + X$.

Here are X and $0.3\,X$ both shown with their tails at $\{0,0\}$:

In[28]:=
```
X = {1,2};
Show[Arrow[0.3 X,Tail->{0,0},Red],
Graphics[Text["0.3 X",0.3 X/2]],
Arrow[X,Tail->{0,0},Blue],
Graphics[Text["X",X/2]],Axes->Automatic];
```

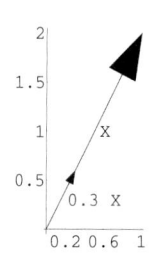

$0.3\,X$ points in the same direction as X points. The only difference is that the length of $0.3\,X$ is 0.3 times the length of X. The same idea carries over to three dimensions as well.

■ B.2) Tangent vectors, velocity vectors, and tangent lines

B.2.a.i) Here's a curve in two dimensions:

In[29]:=
```
Clear[x,y,t,P]; x[t_] = 3 Sin[t];
y[t_] = Cos[t]; P[t_] = {x[t], y[t]};
curveplot = ParametricPlot[{P[t]},{t,0,5},
AspectRatio->Automatic,
PlotStyle->{{Thickness[0.01],Blue}}];
```

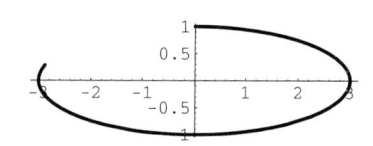

Here's what you get when you add plots of the vectors $P'[t] = \{x'[t], y'[t]\}$ with their tails at $P[t]$ for some choices of t:

In[30]:=
```
Show[curveplot,
Table[Arrow[P'[t],Tail->P[t],Red],
{t,0,5,5/6}]];
```

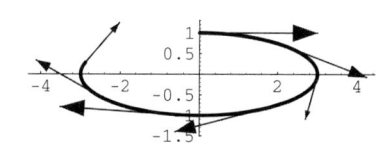

Hot plot.

> Describe what you see in terms of tangent vectors and velocity vectors.

Answer: The vectors you see are
$$P'[t] = \{x'[t], y'[t]\}$$
plotted with their tails at
$$P[t] = \{x[t], y[t]\}$$

for selected t's. The vectors are tangent to the curve. If you imagine t to be time, and you agree that you are at $P[t] = \{x[t], y[t]\}$ at time t, then the vector $P'[t]$ measures your velocity at time t. Take another look at the plot:

In[31]:=
```
Show[curveplot,
  Table[Arrow[P'[t],Tail->P[t],Red],
  {t,0,5,5/6}]];
```

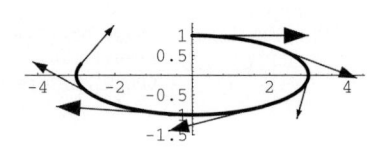

The direction of $P'[t]$ measures the instantaneous direction you are moving at time t. The length of $P'[t]$ measures your instantaneous speed at time t. The plot above shows that you start at the top at time $t = 0$, moving rather quickly to the right. As you enter the turn on the right, you slow down until the curve flattens out. Then you speed up, slowing down as you go into the turn on the left.

B.2.a.ii) Does this work in three dimensions as well?

Answer: Try it and see:

In[32]:=
```
Clear[x,y,z,t,P];
x[t_] = Sqrt[t + 1] Cos[t];
y[t_] = Sqrt[t + 1] Sin[t];
z[t_] = t; P[t_] = {x[t],y[t],z[t]};
curveplot = ParametricPlot3D[P[t],{t,0,10},DisplayFunction->Identity];
tangentvectors = Table[Arrow[P'[t],Tail->P[t],Red],{t,0,10}];
```

In[33]:=
```
threedims = ThreeAxes[2,0.2];
Show[threedims,curveplot,tangentvectors,
  ViewPoint->CMView,Boxed->False,
  DisplayFunction->$DisplayFunction];
```

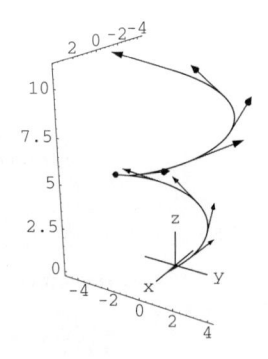

B.2.b.i) Here's a curve shown with a certain point on the curve:

In[34]:=
```
Clear[x,y,t,P]; x[t_] = t^2;
y[t_] = t + Cos[3 t]; P[t_] = {x[t], y[t]};
curveplot = ParametricPlot[{P[t]},{t,0,3},
  PlotStyle->{{Thickness[0.01],Blue}},
  AxesLabel->{"x","y"},Epilog->{PointSize[0.05],
  Red,Point[P[1.9]]}];
```

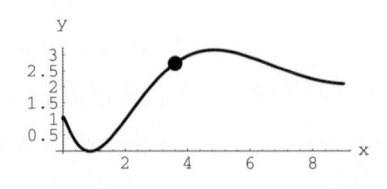

The special point is:

In[35]:=
    P[1.9]

Out[35]=
    {3.61, 2.73471}

> Plot the line tangent to this curve at this special point.

Answer: As a first step, show the plot above along with the tangent vector at $P[1.9]$:

In[36]:=
    Show[curveplot,Arrow[P'[1.9],
    Tail->P[1.9],Red]];

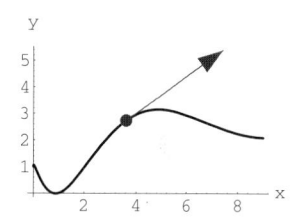

The line tangent to the curve at $P[1.9]$ runs right through the shaft of the arrow. Here's how you can use the tangent vector $P'[1.9]$ to come up with a plot of the tangent line.

In[37]:=
    Clear[tanline,s]
    tanline[s_] = P[1.9] + s P'[1.9];
    tanlineplot = ParametricPlot[tanline[s],{s,0,2},
    PlotStyle->Thickness[0.01],
    DisplayFunction->Identity];
    Show[curveplot,Arrow[P'[1.9],Tail->P[1.9],Red],
    tanlineplot];

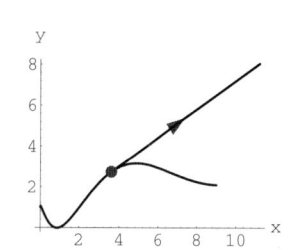

That's only part of the tangent line; you got it by plotting $P[1.9] + s\,P'[1.9]$ with s running from 0 to 2. To get the stuff on the left, run s through some negative numbers as well as positive numbers:

In[38]:=
    Clear[tanline,s]
    tanline[s_] = P[1.9] + s P'[1.9];
    tanlineplot = ParametricPlot[tanline[s],{s,-2,2},
    PlotStyle->Thickness[0.01],
    DisplayFunction->Identity];
    Show[curveplot,Arrow[P'[1.9],Tail->P[1.9],Red],
    tanlineplot];

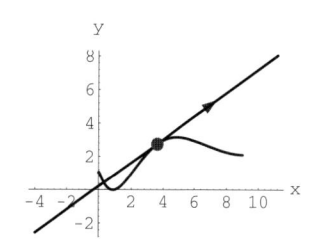

There's that tangent line in all its glory.

Lesson 3.01 Vectors Point the Way

The upshot of all this is: If you have a curve described by the plot of $P[t]$ for $a \le t \le b$ and you want to plot the tangent line through a point $P[c]$ on the curve, you plot $P[c] + s\, P'[c]$ and run s from negative to positive.

B.2.b.ii) Does this work in three dimensions too?

Answer: Get a life!

Of course it works in three dimensions.

■ B.3) Length of a vector, dot product, and distance between two points

B.3.a) Here are two cleared vectors $X = \{x[1], x[2]\}$ and $Y = \{y[1], y[2]\}$ in two dimensions:

In[39]:=
```
Clear[X,Y,x,y]; X = {x[1],x[2]}
```
Out[39]=
```
{x[1], x[2]}
```

In[40]:=
```
Y = {y[1],y[2]}
```
Out[40]=
```
{y[1], y[2]}
```

Here is the dot product, $X \bullet Y$, of these vectors:

In[41]:=
```
X.Y
```
Out[41]=
```
x[1] y[1] + x[2] y[2]
```

Here are two cleared vectors $X = \{x[1], x[2], x[3]\}$ and $Y = \{y[1], y[2], y[3]\}$ in three dimensions:

In[42]:=
```
Clear[X,Y,x,y]; X = {x[1],x[2],x[3]}
```
Out[42]=
```
{x[1], x[2], x[3]}
```

In[43]:=
```
Y = {y[1],y[2],y[3]}
```
Out[43]=
```
{y[1], y[2], y[3]}
```

Here is the dot product, $X \bullet Y$, of these vectors:

In[44]:=
```
X.Y
```
Out[44]=
```
x[1] y[1] + x[2] y[2] + x[3] y[3]
```

> Describe how dot products are calculated.

Answer: In two or three dimensions, the dot product $X \bullet Y$ just multiplies each slot in X by the corresponding slot in Y, and then adds them up. Check it out:

In[45]:=
```
{1,2}.{0,1}
```
Out[45]=
```
2
```
In[46]:=
```
{-1,1}.{8,2}
```
Out[46]=
```
-6
```
In[47]:=
```
{1,1,1}.{1,2,3}
```
Out[47]=
```
6
```

You cannot take the dot product of vectors from different dimensions:

In[48]:=
```
X = {1,0}; Y = {2,1,4};
X.Y
Dot::dotsh:
   Tensors {1, 0} and {2, 1, 4}
      have incompatible shapes.
```
Out[48]=
```
{1, 0} . {2, 1, 4}
```

That hacked off *Mathematica*.

B.3.b) Given a two-dimensional vector, $X = \{x[1], x[2]\}$, the length of X is measured by
$$\|X\| = \sqrt{x[1]^2 + x[2]^2} \ .$$
Given a three-dimensional vector, $X = \{x[1], x[2], x[3]\}$, the length of X is measured by
$$\|X\| = \sqrt{x[1]^2 + x[2]^2 + x[3]^2} \ .$$

> Explain why the formula
> $$\|X\| = \sqrt{X \bullet X}$$
> works in either dimension.

14   Lesson 3.01 Vectors Point the Way

Answer: Try it out in two dimensions:

In[49]:=
```
Clear[x]; X = {x[1],x[2]}; length = Sqrt[x[1]^2 + x[2]^2]
```

Out[49]=
$$\text{Sqrt}[x[1]^2 + x[2]^2]$$

$\sqrt{X \bullet X}$ is given by:

In[50]:=
```
Sqrt[X.X]
```

Out[50]=
$$\text{Sqrt}[x[1]^2 + x[2]^2]$$

This tells you that the formula $||X|| = \sqrt{X \bullet X}$ works in two dimensions. Try it out in three dimensions:

In[51]:=
```
Clear[x]; X = {x[1],x[2],x[3]}; length = Sqrt[x[1]^2 + x[2]^2 + x[3]^2]
```

Out[51]=
$$\text{Sqrt}[x[1]^2 + x[2]^2 + x[3]^2]$$

$\sqrt{X \bullet X}$ is given by:

In[52]:=
```
Sqrt[X.X]
```

Out[52]=
$$\text{Sqrt}[x[1]^2 + x[2]^2 + x[3]^2]$$

This tells you that the formula $||X|| = \sqrt{X \bullet X}$ works in three dimensions too.

Handy little formula, and it's nothing more or less than your old friend, the Pythagorean theorem, in action.

B.3.c.i) Why does

$$||X - Y|| = \sqrt{(X - Y) \bullet (X - Y)}$$

calculate the distance between the tip of X and the tip of Y when X and Y are positioned so that their tails are at the origin?

Answer: Look at a picture:

In[53]:=
```
X = {7,3}; Y = {-4,7};
Show[Arrow[X,Tail->{0,0},Blue],
Graphics[Text["X",X/2]],
Arrow[Y,Tail->{0,0},Blue],
Graphics[Text["Y",Y/2]],
Arrow[X - Y,Tail->Y,Red],Graphics[
Text["X - Y",Y + (X - Y)/2]],Axes->True];
```

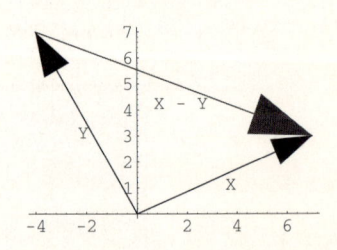

Rerun for different vectors X and Y and you will see that the distance between the tip of X and the tip of Y is the same as the length of $X - Y$.

The length of $X - Y$ is $\sqrt{(X-Y) \bullet (X-Y)}$, so the distance between the tip of X and the tip of Y is given by

$$||X - Y|| = \sqrt{(X-Y) \bullet (X-Y)}.$$

B.3.c.ii) Measure the distance between $\{2, 7\}$ and $\{3, -5\}$ in two dimensions.

Measure the distance between $\{2, 7, -6\}$ and $\{3, -5, 6\}$ in three dimensions.

Answer: The distance between the points $\{2, 7\}$ and $\{3, -5\}$ in two dimensions is measured by:

In[54]:=
```
X = {2,7}; Y = {3,-5};
N[Sqrt[(X - Y).(X - Y)]]
```
Out[54]=
12.0416

The distance between the points $\{2, 7, -6\}$ and $\{3, -5, 6\}$ in three dimensions is measured by:

In[55]:=
```
X = {2,7,-6}; Y = {3,-5,4};
N[Sqrt[(X - Y).(X - Y)]]
```
Out[55]=
15.6525

■ **B.4) The push of one vector in the direction of another, and the formula $X \bullet Y = ||X|| \, ||Y|| \cos[b]$, where b is the angle between X and Y**

Here are two vectors in two dimensions shown with their tails at $\{0, 0\}$ in true scale:

In[56]:=
```
X = {1.3,1.7}; Y = {1,0.5};
Show[Arrow[X,Tail->{0,0}],
Graphics[Text["X",X/2]],
Arrow[Y,Tail->{0,0},Red],
Graphics[Text["Y",Y/2]],
AspectRatio->Automatic,
Axes->True,AxesLabel->{"x","y"}];
```

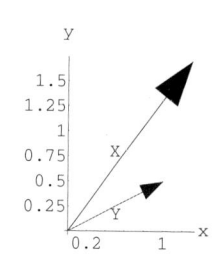

The question addressed here is how to measure the push of X in the direction of Y.

Lesson 3.01 Vectors Point the Way

The answer is that the push of X in the direction of Y is calculated by the clean formula:
$$X \text{pushalong} Y = \left(\frac{X \bullet Y}{Y \bullet Y}\right) Y.$$

Take a look:

In[57]:=
```
XpushalongY = ((X.Y)/(Y.Y)) Y;
Show[Arrow[X,Tail->{0,0}],
Graphics[Text["X",X/2]],
Arrow[XpushalongY ,Tail->{0,0},Red],
Graphics[Text["X push along Y",
(XpushalongY)/2]],AspectRatio->Automatic,
Axes->True,AxesLabel->{"x","y"}];
```

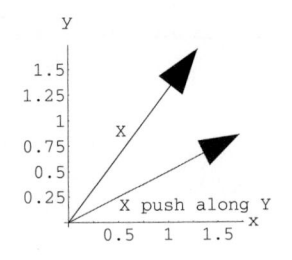

Here are two new vectors X and Y:

In[58]:=
```
X = {-8,6}; Y = {-4,0.6};
Show[Arrow[X,Tail->{0,0}],
Graphics[Text["X",X/2]],
Arrow[Y,Tail->{0,0},Red],
Graphics[Text["Y",Y/2]],
AspectRatio->Automatic,
Axes->True,AxesLabel->{"x","y"}];
```

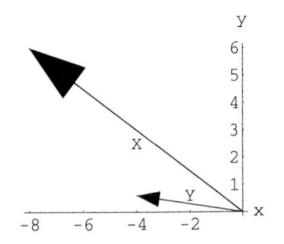

Here are X and $X \text{pushalong} Y = ((X \bullet Y)/(Y \bullet Y)) Y$:

In[59]:=
```
XpushalongY = ((X.Y)/(Y.Y)) Y;
Show[Arrow[X,Tail->{0,0}],
Graphics[Text["X",X/2]],
Arrow[XpushalongY ,Tail->{0,0},Red],
Graphics[Text["X push along Y",
(XpushalongY)/2]],AspectRatio->Automatic,
Axes->True,AxesLabel->{"x","y"}];
```

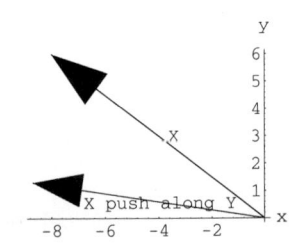

Try it for other X's and Y's until the plots make sense to you.

B.4.a) According to what was done above, you calculate the push of a vector X in the direction of another vector Y by calculating
$$X \text{pushalong} Y = \left(\frac{X \bullet Y}{Y \bullet Y}\right) Y.$$

> Explain where this formula comes from, and explain what the push of X in the direction of Y means.

Answer: Go with cleared vectors X and Y. Put
$$f[t] = ||X - tY|| = \sqrt{(X - tY) \bullet (X - tY)}.$$

The function $f[t]$ measures the distance between the tip of X and the tip of the vector tY when both vectors have their tails at $\{0, 0\}$.

In[60]:=
```
Clear[f,t,x,y,X,Y]
X = {x[1],x[2]}; Y = {y[1],y[2]};
f[t_] = Sqrt[(X - t Y).(X - t Y)]
```

Out[60]=
$$\text{Sqrt}[(x[1] - t\ y[1])^2 + (x[2] - t\ y[2])^2]$$

$f[t]$ is as small as it can be when $f'[t] = 0$:

In[61]:=
```
Solve[f'[t] == 0,t]
```

Out[61]=
$$\{\{t \to \frac{x[1]\ y[1] + x[2]\ y[2]}{y[1]^2 + y[2]^2}\}\}$$

This is the same as:

In[62]:=
```
bestt = ((X.Y)/(Y.Y))
```

Out[62]=
$$\frac{x[1]\ y[1] + x[2]\ y[2]}{y[1]^2 + y[2]^2}$$

This is the t that makes $f[t] = ||X - tY||$ as small as it can be. But you already know that $X \text{pushalong} Y = ((X \bullet Y)/(Y \bullet Y))Y$.

The upshot: The push of X in the direction of Y is that multiple of Y whose tip is closest to the tip of X when you put the tails of both vectors at $\{0, 0\}$.

This works in three dimensions as well.

B.4.b) Take a look at

→ X and $X\text{pushalong}Y$ with their tails at $\{0, 0\}$ and

→ $(X - X\text{pushalong}Y)$ with its tail at the tip of $X\text{pushalong}Y$

for two sample vectors X and Y in true scale:

Lesson 3.01 Vectors Point the Way

```
In[63]:=
X = {1,2}; Y = {4,1};
XpushalongY = ((X.Y)/(Y.Y)) Y;
Show[Arrow[X,Tail->{0,0}],Graphics[Text["X",X/2]],
 Arrow[XpushalongY ,Tail->{0,0},Red],
 Graphics[Text["X push along Y",(XpushalongY)/2]],
 Arrow[X - XpushalongY,Tail->XpushalongY,Red],
 Graphics[Line[{-0.2 Y, 0.8 Y}]],
 AspectRatio->Automatic,
 Axes->True,AxesLabel->{"x","y"}];
```

The line is the plot of the tips of relevant multiples tY of Y.

> Describe what you see and explain why you see it.

Answer: You see a right triangle. And this is what you'll see for any two vectors X and Y unless X and Y are parallel. Reason:

→ The push of X in the direction of Y is the multiple of Y whose tip is closest to the tip of X when you put the tails of both vectors at $\{0,0\}$.

The upshot: The tip of Xpushalong$Y$ is the closest point on the line to the tip of X. And the shortest distance between the line and the tip of X is the perpendicular distance.

B.4.c) Take another look at

→ X and Xpushalong$Y$ with their tails at $\{0,0\}$ and

→ $(X - X\text{pushalong}Y)$ with its tail at the tip of Xpushalong$Y$

for two sample vectors X and Y in true scale:

```
In[64]:=
X = {2.8,1.9}; Y = {1.2,0.2};
XpushalongY = ((X.Y)/(Y.Y)) Y;
plot = Show[Arrow[X,Tail->{0,0}],
 Graphics[Text["X",X/2]],
 Arrow[XpushalongY ,Tail->{0,0},Red],
 Graphics[Text["X push along Y",
 (XpushalongY)/2]],Arrow[X - XpushalongY,
 Tail->XpushalongY,Red],
 AspectRatio->Automatic,
 Axes->True,AxesLabel->{"x","y"}];
```

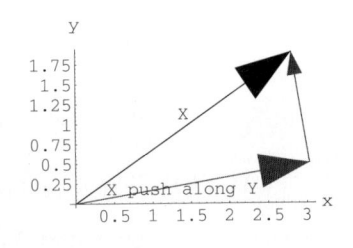

> Use the plot to help explain the formula
> $$(X \bullet Y) = ||X||\,||Y||\cos[b]$$
> where b is the angle between X and Y, $||X|| = \sqrt{X \bullet X}$ and $||Y|| = \sqrt{Y \bullet Y}$.

Answer: The goal is to explain the formula $X \bullet Y = ||X||\,||Y|| \cos[b]$, where b is the angle between X and Y when their tails are at the origin. Take another look. Remembering that
$$\cos[b] = \frac{\text{adjacent}}{\text{hypotenuse}},$$
read off
$$\cos[b] = \frac{t\,||Y||}{||X||}$$
where
$$t = \frac{X \bullet Y}{Y \bullet Y}$$
because
$$X \text{ pushalong } Y = \left(\frac{X \bullet Y}{Y \bullet Y}\right) Y.$$
Now you know why
$$||X|| \cos[b] = t\,||Y||.$$
Next, multiply both sides by $||Y||$ to get
$$||X||\,||Y|| \cos[b] = t\,||Y||^2.$$
This tells you that to explain why
$$X \bullet Y = ||X||\,||Y|| \cos[b],$$
you gotta explain why
$$t\,||Y||^2 = X \bullet Y.$$
This is easy because
$$t = (X \bullet Y) / (Y \bullet Y);$$
so
$$t\,||Y||^2 = t\,Y \bullet Y = \frac{X \bullet Y}{Y \bullet Y}(Y \bullet Y) = X \bullet Y.$$
That's it. This formula also works in three dimensions.

■ B.5) $X \bullet Y = 0$ means X is perpendicular to Y

B.5.a.i) Look at $X = \{1, 4/3, 7/2\}$ and $Y = \{-1/3, -5, 2\}$ with their tails at $\{0, 0, 0\}$:

In[65]:=
```
X = {1,4/3,7/2}; Y = {-1/3,-5,2};
threedims = ThreeAxes[1,.2];
Show[Arrow[X,Tail->{0,0,0},Red],
Arrow[Y,Tail->{0,0,0},Red],
threedims,PlotRange->All,
ViewPoint->CMView,Boxed->False];
```

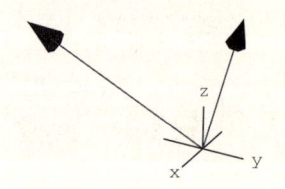

These vectors look like they might be perpendicular.

How can you tell for sure?

Answer: Just look at $X \bullet Y$:

In[66]:=
```
X.Y
```
Out[66]=
0

Now you know that $X \bullet Y = 0$; so you know for sure that X is indeed perpendicular to Y.

B.5.a.ii) Explain the statement: If $X \bullet Y = 0$, then X is perpendicular to Y.

Answer: Well, if $X \bullet Y = 0$, then since

$$(X \bullet Y) = \|X\| \, \|Y\| \cos[b],$$

you know that

$$\cos[b] = 0 = \cos\left[\frac{\pi}{2}\right] = \cos\left[\frac{-\pi}{2}\right].$$

This tells you that the angle between X and Y is a right angle.

B.5.b) How do you know that $X = \{6, 2\}$ is perpendicular to $Y = \{3, -9\}$?

Answer: Check to see that $X \bullet Y = 0$:

In[67]:=
```
X = {6,2}; Y = {3,-9};
X.Y
```

Out[67]=
0

Yep; $X = \{6, 2\}$ is perpendicular to $Y = \{3, -9\}$.

Tutorials

■ T.1) Velocity and acceleration

At time t, an object is at the location $P[t] = \{x[t], y[t]\}$.

In[1]:=
```
Clear[t,x,y,P]; P[t_] = {x[t],y[t]}
```

Out[1]=
{x[t], y[t]}

The velocity of the object at time t is given by:

In[2]:=
```
Clear[vel]; vel[t_] = D[P[t],t]
```

Out[2]=
{x'[t], y'[t]}

The velocity is a vector quantity. When you put its tail at $\{x[t], y[t]\}$, the velocity vector $\{x'[t], y'[t]\} = P'[t] = D[P[t], t]$ is tangent to the curve traced out by the motion of $P[t]$. This velocity vector points in the instantaneous direction that the object is going.

The speed of the object is the length of the velocity vector:

In[3]:=
```
Clear[speed]; speed[t_] = Sqrt[vel[t].vel[t]]
```

Out[3]=
$$\text{Sqrt}[x'[t]^2 + y'[t]^2]$$

The acceleration of the object is given by:

In[4]:=
```
Clear[accel]; accel[t_] = D[vel[t],t]
```

Out[4]=
{x''[t], y''[t]}

T.1.a) What does the acceleration vector measure?

Answer: Because the velocity vector is the derivative of the position vector, the velocity vector measures the rate of change of the position. Because the acceleration vector is the derivative of the velocity vector, the acceleration vector measures the rate of change of the velocity.

T.1.b) At time t, an object is at the point $P[t] = \{3\cos[t], 3\sin[t]\}$.

> Plot the motion of the object in true scale for $0 \leq t \leq 2\pi$ and include several velocity and acceleration vectors. Discuss what the plot reveals.

Answer:

In[5]:=
```
Clear[t,P];
P[t_] = {3 Cos[t],3 Sin[t]};
curveplot =
ParametricPlot[P[t],{t,0, 2 Pi},
PlotStyle->Thickness[0.01],
AspectRatio->Automatic,
AxesLabel->{"x","y"}];
```

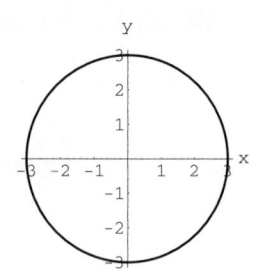

Circular motion—no big surprise. Now set up the velocity and acceleration vectors:

In[6]:=
```
Clear[vel,accel]; vel[t_] = D[P[t],t];
accel[t_] = D[vel[t],t];
Clear[velvector]
velvector[t_] := Arrow[vel[t],Tail-> P[t],Blue];
Clear[accelvector]
accelvector[t_] := Arrow[accel[t],Tail->P[t],Red];
Show[curveplot,velvector[Pi/4],accelvector[Pi/4],
AspectRatio->Automatic];
```

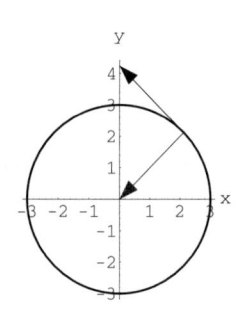

The velocity vector indicates the direction of the motion, and the acceleration vector indicates that the force on the object (force = mass × acceleration) is directed toward the origin, perpendicular to the velocity vector. Check out some more velocity and acceleration vectors:

In[7]:=
```
Show[curveplot,Table[velvector[t],
{t,1.2,7.2,.5}],Table[accelvector[t],
{t,1.2,7.2,.5}],AspectRatio->Automatic];
```

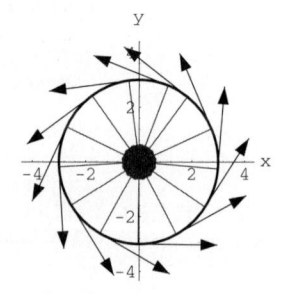

The acceleration vectors at $P[t]$ are all perpendicular to the velocity vectors at $P[t]$. This means the force on the object neither speeds up nor slows down the object. To confirm this, check out the speed:

In[8]:=
```
Clear[speed]
speed[t_] = Sqrt[Expand[vel[t].vel[t],Trig->True]]
```
Out[8]=
3

The object moves with a constant speed of three units of length per unit of time. Another way to see this is to look at the plot and note that all the velocity vectors have the same length.

T.1.c.i) At time t, an object is at the point $P[t] = \{7\cos[t], 4\sin[t]\}$.

> Plot the motion of the object in true scale for $0 \leq t \leq 2\pi$ and include several velocity and acceleration vectors. Discuss what the plot reveals.

Answer:

In[9]:=
```
Clear[t,P]
P[t_] = {7 Cos[t], 4 Sin[t] };
curveplot = ParametricPlot[P[t],{t,0,2 Pi},
PlotStyle->Thickness[0.01],
AspectRatio->Automatic];
```

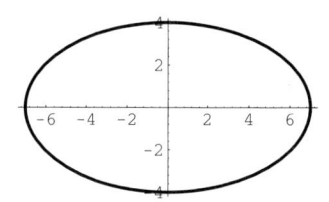

The object is moving on an ellipse. To see which way the object is moving, look at the velocity and acceleration vectors:

In[10]:=
```
Clear[vel,accel]; vel[t_] = D[P[t],t];
accel[t_] = D[vel[t],t];
Clear[velvector]
velvector[t_] := Arrow[vel[t],Tail-> P[t],Blue];
Clear[accelvector]
accelvector[t_] := Arrow[accel[t],Tail->P[t],Red];
Show[curveplot,Table[velvector[t],
{t,1.2,7.2,.5}],Table[accelvector[t],
{t,1.2,7.2,.5}],AspectRatio->Automatic];
```

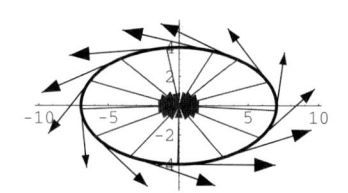

The velocity vectors have different lengths, and the acceleration vectors are not perpendicular to the velocity vectors. This signals that the speed is not constant.

Looking again, you can see that the acceleration vectors are working to speed up the object in the first and third quadrants, and the acceleration vectors are slowing the object down in the second and fourth quadrants.

Confirm with a plot of the speed:

Lesson 3.01 Vectors Point the Way

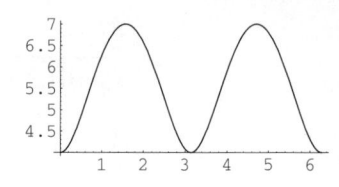

In[11]:=
```
Plot[Sqrt[vel[t].vel[t]],{t,0,2 Pi}];
```

Going faster, then slower, then faster, and then slower.

T.1.c.ii) Folks like to call the push of the acceleration vector in the direction of the velocity vector "the tangential component of the acceleration."

> Go back to the plot in part T.1.c.i) above and show some velocity vectors and the tangential components of the acceleration vectors. Discuss what your plot reveals.

Answer: Remember that the push of a vector X in the direction of another vector Y is calculated by $((X \bullet Y)/(Y \bullet Y))Y$.

Here come the tangential components of the acceleration vectors plotted with the velocity vectors on the curve:

In[12]:=
```
Clear[tancompaccel,tanaccelvector]
tancompaccel[t_] = ((accel[t].vel[t])/(vel[t].vel[t])) vel[t];
tanaccelvector[t_] := Arrow[tancompaccel[t],Tail->P[t],Red];
```

In[13]:=
```
Show[curveplot,Table[velvector[t],
{t,1.2,7.2,.5}],Table[tanaccelvector[t],
{t,1.2,7.2,.5}],AspectRatio->Automatic];
```

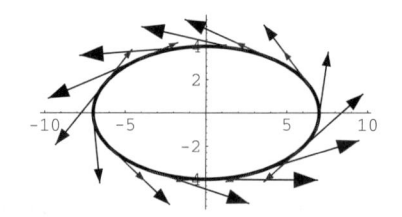

When the tangential components of the acceleration vectors are pushing in the same direction as the velocity vectors, the object is gaining speed. When the tangential components of the acceleration vectors are pushing in the direction opposite to the velocity vectors, the object is losing speed. Look at the brakes go on as the object goes into the sharp turns, and look at how the object speeds up as it comes out of the turns.

T.1.c.iii) Folks like to call the push of the acceleration vector in the direction perpendicular to the velocity vector by the name "the normal component of the acceleration."

> Go back to the plot in part T.1.c.i) above and show some acceleration vectors split into normal and tangential components. Discuss what your plot reveals.

Answer: You already have the tangential components of the acceleration vectors live in part T.1.c.ii) above. The perpendicular component is just

$$\text{accel}[t] - \text{tancompaccel}[t] :$$

In[14]:=
```
Clear[perpcompaccel,perpaccelvector]
perpcompaccel[t_] = accel[t] - tancompaccel[t];
perpaccelvector[t_] := Arrow[perpcompaccel[t],Tail->P[t],Red];
```

In[15]:=
```
Show[curveplot,Table[tanaccelvector[t],
  {t,1.2,7.2,0.5}],Table[perpaccelvector[t],
  {t,1.2,7.2,0.5}],AspectRatio->Automatic];
```

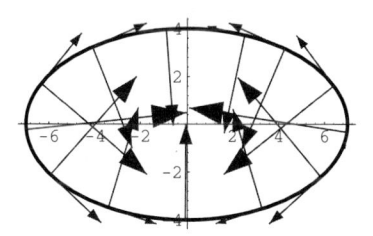

The tangential components govern the speed of the object, and the perpendicular components measure the tug on the object as it goes on its elliptical path. Note that the tug is greater in the sharp turns than it is on the flatter parts of the curve. Just as you expect.

■ T.2) Using the normal vector to bounce light beams off two-dimensional curves

When you have a parametric formula $P[t] = \{x[t], y[t]\}$ for a curve in two dimensions, then you can calculate a tangent vector at $P[t]$ by calculating

$$\tan[t] = \{x'[t], y'[t]\}.$$

You can also calculate a vector perpendicular to the curve at $P[t]$ by calculating

$$\text{normal}[t] = \{y'[t], -x[t]\}.$$

Fancy folks like to call the perpendicular vector by the name "normal vector." Here is a sample curve shown with a selection of tangents and normals:

In[16]:=
```
Clear[t,x,y,P]; Clear[tan,normal]; Clear[tanvector]; Clear[normalvector]
x[t_] = t + Sin[2 t]; y[t_] = 3 Cos[t];
P[t_] = {x[t],y[t]}; a = 0; b = 5;
curveplot = ParametricPlot[P[t],{t,a,b},
  PlotStyle->Thickness[0.01],DisplayFunction->Identity];
tan[t_] = {x'[t],y'[t]}; normal[t_] = {y'[t],-x'[t]};
tanvector[t_] := Arrow[tan[t],Tail-> P[t],Blue];
normalvector[t_] := Arrow[normal[t],Tail->P[t],Red];
```

Lesson 3.01 Vectors Point the Way

In[17]:=
```
Show[curveplot,Table[{tanvector[t],
  normalvector[t]},{t,a,b,(b - a)/6}],
  AspectRatio->Automatic,
  AxesLabel->{"x","y"},
  DisplayFunction->$DisplayFunction];
```

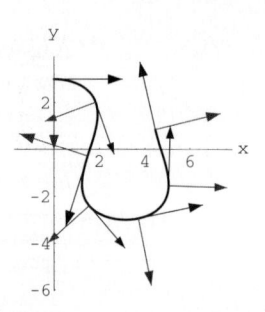

T.2.a) How did you know in advance that the normal vectors $\{y'[t], -x'[t]\}$ are guaranteed to be perpendicular to the tangent vectors $\{x'[t], y'[t]\}$?

Answer: The perpendicularity test says: Two vectors X and Y are perpendicular if $X \bullet Y = 0$.

Taking $X = \{x'[t], y'[t]\}$ and $Y = \{y'[t], -x'[t]\}$, you don't need *Mathematica* to see that

$$X \bullet Y = x'[t]\, y'[t] + y'[t]\, (-x'[t]) = 0.$$

T.2.b) Here is a curve plotted in nonparametric form:

In[18]:=
```
Clear[f,x]
f[x_] = 5 E^(-((x - 3))^2);
a = 0; b = 6;
curveplot = Plot[f[x],{x,a,b},
  PlotStyle->Thickness[0.01],
  AxesLabel->{"x","y"}];
```

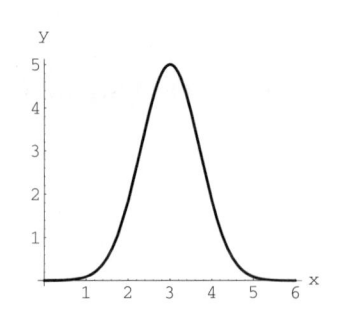

How do you stick some tangent and normal vectors onto this curve?

Answer: Every function specified as $y = f[x]$ can be thought of as a shorthand version of the parametric version, $P[x] = \{x, y\} = \{x, f[x]\}$, or, after changing the independent variable, $P[t] = \{t, f[t]\}$. That makes this job easy: Replot in parametric form, scale the tangent and normal vector, and run:

In[19]:=
```
Clear[t,x,y,P]; Clear[tan,normal]; Clear[tanvector]; Clear[normalvector]
x[t_] = t; y[t_] = f[t]; P[t_] = {x[t],y[t]};
curveplot = ParametricPlot[P[t],{t,a,b},PlotStyle->Thickness[0.01],
  DisplayFunction->Identity]; tan[t_] = {x'[t],y'[t]};
normal[t_] = {y'[t],-x'[t]}; scalefactor = 0.5;
tanvector[t_] := Arrow[scalefactor tan[t],Tail->P[t],Blue];
normalvector[t_] := Arrow[scalefactor normal[t],Tail->P[t],Red];
```

```
In[20]:=
  Show[curveplot,Table[{tanvector[t],
    normalvector[t]]},{t,a,b,(b - a)/6}],
  AspectRatio->Automatic,AxesLabel->{"x","y"},
  DisplayFunction->$DisplayFunction];
```

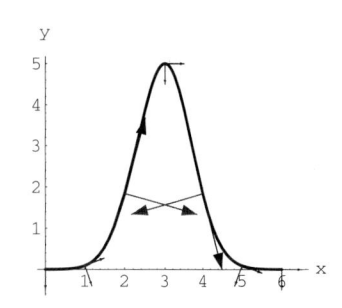

There ya go.

T.2.c) Here is the curve

$$f[x] = 1 - \cos[x] \qquad \text{for} \quad -\frac{\pi}{2} \leq x \leq \frac{\pi}{2}$$

shown with a light beam emanating from $\{2,3\}$ and hitting the curve at $\{0.5, f[0.5]\}$:

```
In[21]:=
  Clear[t,f,x,y,P]
  f[x_] = 1 - Cos[x]; x[t_] = t;
  y[t_] = f[t]; P[t_] = {x[t],y[t]}; a = -Pi/2; b = Pi/2;
  curveplot = ParametricPlot[P[t],{t,a,b},PlotStyle->Thickness[0.01],
    AxesLabel->{"x","y"},DisplayFunction->Identity];
  source = {2,3}; hit = {0.5,f[0.5]};
  label = Graphics[Text["source",source]];
  sourceplot = Graphics[{Red,PointSize[0.05],Point[source]}];
  incominglight = Arrow[hit - source,Tail->source,Red];
```

```
In[22]:=
  Show[curveplot,incominglight,label,
    sourceplot,PlotRange->All,
    DisplayFunction->$DisplayFunction];
```

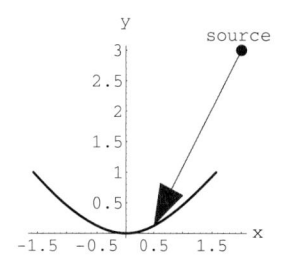

Plot where the light beam goes after it bounces off the curve.

Answer: The physical principle behind bouncing light is that the angle of incidence is the same as the angle of reflection.

To get an idea of how to make use of this physical principle, include the normal at the hit and run the vector from the hit to the source.

Lesson 3.01 Vectors Point the Way

In[23]:=
```
Clear[normal,normalvector,t]
X = source - hit; reverseincominglight = Arrow[X,Tail->hit,Red];
extralabels = {Graphics[Text["X",hit + (source - hit)/2]],
 Graphics[Text["normal[0.5]",hit + normal[0.5]/2]]};
normal[t_] = {y'[t],-x'[t]};
normalvector[t_] := Arrow[normal[t],Tail->hit,Blue];
```

In[24]:=
```
Show[curveplot,reverseincominglight,
 normalvector[0.5],extralabels,
 sourceplot,label,PlotRange->All,
 DisplayFunction->$DisplayFunction];
```

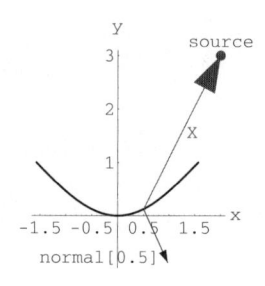

Agree that R is a vector specifying the direction of the reflected light and note that the angle R makes with normal[0.5] is the same as the angle X makes with normal. So you want

$$X \bullet \text{normal}[0.5] = ||X|| \, ||\text{normal}[0.5]|| \cos[b]$$

and

$$R \bullet \text{normal}[0.5] = ||R|| \, ||\text{normal}[0.5]|| \cos[b]$$

where b is the angle between X and normal[0.5].

When you make $R \bullet R = ||R||^2 = ||X||^2 = X \bullet X$, you get two equations to solve for the two slots of R:

In[25]:=
```
Clear[R,r1,r2]; R = {r1,r2};
equation1 = R.normal[0.5] == X.normal[0.5];
equation2 = R.R == X.X;
solutions = Solve[{equation1,equation2}]
```
Out[25]=
```
{{r1 -> -3.18283, r2 -> 0.632514}, {r1 -> 1.5, r2 -> 2.87758}}
```

One solution for R is X itself:

In[26]:=
```
X
```
Out[26]=
```
{1.5, 2.87758}
```

Discard it and go with:

In[27]:=
```
R = {r1,r2}/.solutions[[1]]
```

Out[27]=
{-3.18283, 0.632514}

Try it out:

In[28]:=
```
outgoinglight = Arrow[R,Tail->hit,Red];
Show[outgoinglight,
  curveplot,reverseincominglight,
  normalvector[0.5],extralabels,
  sourceplot,label,PlotRange->All,
  DisplayFunction->$DisplayFunction];
```

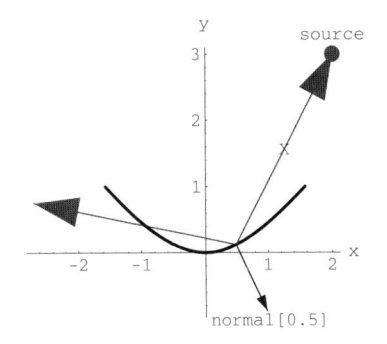

Beautiful. Here's the path of the light:

In[29]:=
```
Show[curveplot,incominglight,
  outgoinglight,label,
  sourceplot,PlotRange->All,
  DisplayFunction->$DisplayFunction];
```

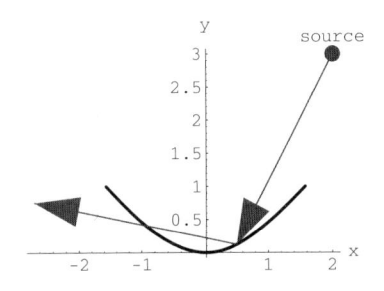

Lookin' just fine, thank you.

■ T.3) Lines

To specify a line, you need a point to run the line through and you need a vector to specify the direction of the line.

T.3.a.i) Come up with a parametric formula for the line that runs through $\{-2, -1\}$ and runs parallel to the vector $\{3, 1\}$.

Show the line, the vector, and the point in a single plot.

Give the equation of the line in the nonparametric form $y = f[x]$.

Answer: Here comes the parametric formula:

Lesson 3.01 Vectors Point the Way

In[30]:=
```
Clear[line,t]
point = {-2,-1}; parallelvector = {3,1};
line[t_] = point + t parallelvector
```
Out[30]=
```
{-2 + 3 t, -1 + t}
```

A parametric formula for the line that runs through $\{-2, -1\}$ and runs parallel to the vector $\{3, 1\}$ is

$$\{x[t], y[t]\} = \{-2, -1\} + t\{3, 1\}.$$

Note how the parametric formula displays the given point and the parallel vector.

Here comes the plot:

In[31]:=
```
directionvector = Arrow[parallelvector,
 Tail->point,Red];
pointplot = Graphics[{PointSize[0.04],
 Red,Point[point]}];
lineplot = ParametricPlot[line[t],{t,-2,3},
 PlotStyle->Thickness[0.01],
 AxesLabel->{"x","y"},DisplayFunction->Identity];
Show[lineplot,pointplot,directionvector,
 DisplayFunction->$DisplayFunction];
```

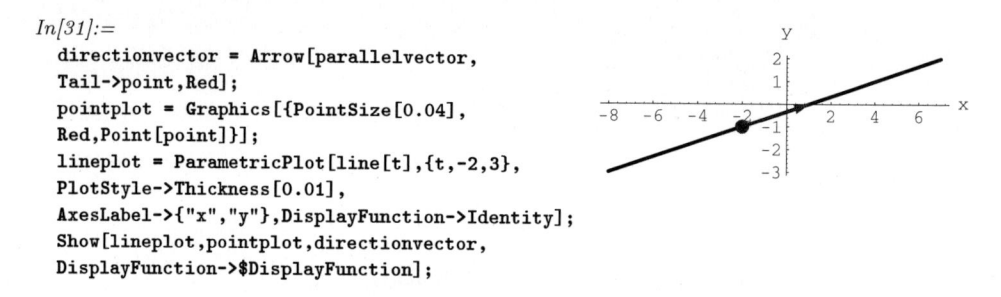

Here's how the plot looks when you stick the tail of the parallel vector at $\{0, 0\}$:

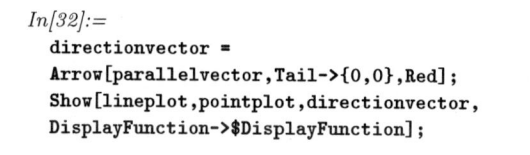

In[32]:=
```
directionvector =
 Arrow[parallelvector,Tail->{0,0},Red];
Show[lineplot,pointplot,directionvector,
 DisplayFunction->$DisplayFunction];
```

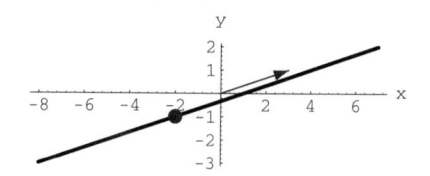

Parallel.

To come up with the formula of the line in the nonparametric form $y = f[x]$, look at:

In[33]:=
```
Clear[x,y]; equation = {x,y} == line[t]
```
Out[33]=
```
{x, y} == {-2 + 3 t, -1 + t}
```

Eliminate t by writing

$$t = \frac{x+2}{3} \quad \text{and} \quad t = y+1$$

and getting

$$\frac{x+2}{3} = y+1:$$

In[34]:=
```
squasht = Eliminate[equation,t]
```
Out[34]=
```
3 y == -1 + x
```

Now solve for y:

In[35]:=
```
Solve[squasht,y]
```
Out[35]=
$$\{\{y \to \frac{-(1 - x)}{3}\}\}$$

The nonparametric formula of this line is $y = (x-1)/3$.

T.3.a.ii) Come up with a parametric formula for the line that runs through $\{2, -3, -5\}$ and runs parallel to the vector $\{-1.5, 2.5, 3.3\}$. Show the line, the vector, and the point in a single plot.

Answer: You do it the same way you handle the 2D case. Here comes the parametric formula:

In[36]:=
```
point = {2,-3,-5};
parallelvector = {-1.5,2.5,3.3};
Clear[line,t]; line[t_] = point + t parallelvector
```
Out[36]=
```
{2 - 1.5 t, -3 + 2.5 t, -5 + 3.3 t}
```

A parametric formula for the line that runs through $\{2, -3, -5\}$ and runs parallel to the vector $\{-1.5, 2.5, 3.3\}$ is

$$\{x[t], y[t], z[t]\} = \{2, -3, -5\} + t\{-1.5, 2.5, 3.3\}$$

Note how the parametric formula conveniently displays the given point and the parallel vector.

Here comes the plot:

In[37]:=
```
directionvector = Arrow[parallelvector,
  Tail->point,Red];
pointplot = Graphics3D[{PointSize[0.04],
  Red,Point[point]}];
lineplot = ParametricPlot3D[line[t],{t,-2,3},
  DisplayFunction->Identity];
threedims = ThreeAxes[3,0.2];
Show[threedims,lineplot,
  pointplot,directionvector,
  ViewPoint->CMView,Boxed->False,
  DisplayFunction->$DisplayFunction];
```

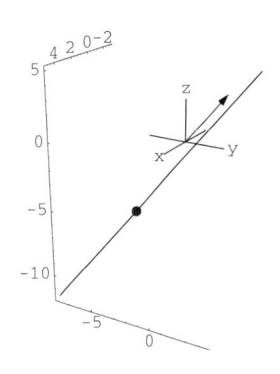

Here's how the plot looks when you stick the tail of the parallel vector at $\{0,0,0\}$:

```
In[38]:=
  directionvector =
    Arrow[parallelvector,Tail->{0,0,0},Red];
  Show[threedims,lineplot,
    pointplot,directionvector,
    ViewPoint->CMView,Boxed->False,
    DisplayFunction->$DisplayFunction];
```

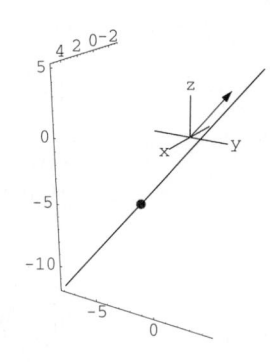

T.3.b.i) Here's a look at the two points $\{-2,3\}$ and $\{3,0\}$ in two dimensions:

```
In[39]:=
  point1 = {-2,3}; point2 = {3,0};
  pointplot =
    Graphics[{{PointSize[0.04],Red,Point[point1]},
      {PointSize[0.04],Red,Point[point2]}}];
  Show[pointplot,Axes->True,AxesLabel->{"x","y"}];
```

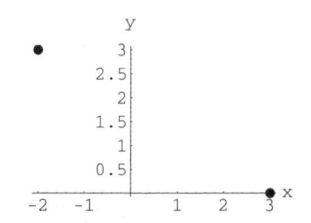

> Come up with a parametric formula for the line that runs through these two points. Show the line and the two points in a single plot.

Answer: This line is parallel to the vector that runs from $\{-2,3\}$ to $\{3,0\}$, which is given by

$$\{3,0\} - \{-2,3\} = \{5,-3\}.$$

```
In[40]:=
  parallelvector = point2 - point1
Out[40]=
  {5, -3}
```

Take a look:

```
In[41]:=
  Show[pointplot,
    Arrow[parallelvector,
      Tail->point1,Red],
    Axes->True,AxesLabel->{"x","y"}];
```

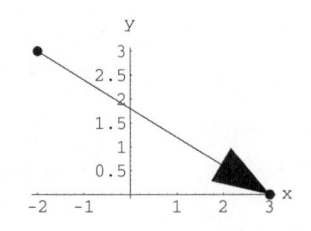

Here's a parametric formula:

In[42]:=
```
Clear[line,t]; line[t_] = point1 + t parallelvector
```

Out[42]=
```
{-2 + 5 t, 3 - 3 t}
```

You can see one of the points—namely, $\{-2, 3\}$—prominently displayed in this parametric formula. Here comes the plot:

In[43]:=
```
lineplot =
ParametricPlot[line[t],{t,-1,2},
PlotStyle->Thickness[0.01],
AxesLabel->{"x","y"},
DisplayFunction->Identity];
Show[lineplot,pointplot,
DisplayFunction->$DisplayFunction];
```

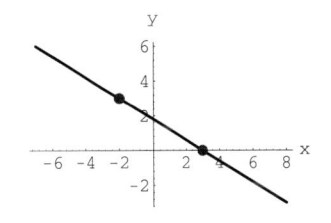

There you go.

T.3.b.ii) Here's a look at three points in three dimensions:

In[44]:=
```
point1 = {-1.51,2.32,-1.81};
point2 = {2.69,-1.14,4.43};
point3 = {0.59,0.59,1.31};
pointplot =
{Graphics3D[{PointSize[0.04],Red,Point[point1]}],
Graphics3D[{PointSize[0.04],Red,Point[point2]}],
Graphics3D[{PointSize[0.04],Red,Point[point3]}]};
threedims = ThreeAxes[1,.3];
Show[threedims,pointplot,ViewPoint->CMView,
PlotRange->All,Boxed->False];
```

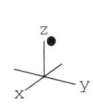

These three points certainly look like they're lined up in a straight line.

> How can you tell for sure?

Answer: Put a line between two of them and see whether the third point is on this line:

In[45]:=
```
Clear[line,t]; line[t_] = point1 + t (point2 - point1);
Solve[line[t] == point3,t]
```

Out[45]=
```
{{t -> 0.5}}
```

Compare:

In[46]:=
```
{point3, line[0.5]}
```

Out[46]=
```
{{0.59, 0.59, 1.31}, {0.59, 0.59, 1.31}}
```

Yep, the line goes through all three points. Another way of seeing this is to make a vector running from point1 to point2 and another vector running from point1 to point3:

In[47]:=
```
vector12 = point2 - point1
```

Out[47]=
```
{4.2, -3.46, 6.24}
```

In[48]:=
```
vector13 = point3 - point1
```

Out[48]=
```
{2.1, -1.73, 3.12}
```

In[49]:=
```
Solve[vector12 == t vector13,t]
```

Out[49]=
```
{{t -> 2.}}
```

This tells you that the vector running from point1 to point2 is parallel to the vector running from point1 to point3. This is enough to confirm that the three points all reside on a single line.

T.3.c) Look at:

In[50]:=
```
X = {2,10}; Y = {11,3};
setup = Show[Arrow[X,Tail->{0,0},Red],
Graphics[Text["X",X/2]],
Arrow[Y,Tail->{0,0},Red],
Graphics[Text["Y",Y/2]],
Axes->True,AxesLabel->{"x","y"}];
```

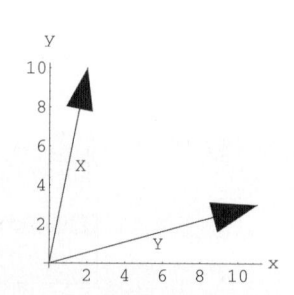

> Use vectors to find the midpoint of the line segment that runs from the tip of X to the tip of Y.

Answer: Remember that if you hook the tail of $X - Y$ to the tip of Y, then $X - Y$ runs from the tip of Y to the tip of X.

Why? Because $X = Y + (X - Y)$.

In[51]:=
```
Show[setup,Arrow[(X - Y),Tail->Y,Red],
 Graphics[Text["X-Y",Y + (X - Y)/2]]];
```

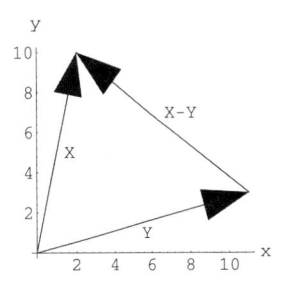

To get to the midpoint of the line segment connecting the tip of Y and the tip of X, you can travel on Y to the tip of Y and then travel along $X - Y$, but you should only go halfway. This means:

In[52]:=
```
midpoint = Y + (1/2)(X - Y)
```

Out[52]=
$$\{\frac{13}{2}, \frac{13}{2}\}$$

Check with a plot:

In[53]:=
```
midpointplot = Graphics[
 {Blue,PointSize[0.04],Point[midpoint]}];
lineplot = Graphics[Line[{X,Y}]];
Show[setup,lineplot,midpointplot];
```

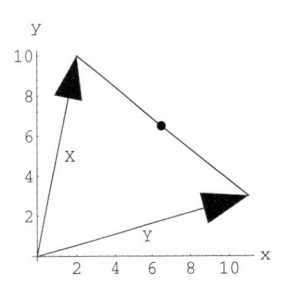

Copacetic.

■ T.4) Pursuits

This problem appears only in the electronic version.

■ T.5) Spying along the tangent

At time t, you are at the point $P[t]$ specified through the polar parameterization:

In[54]:=
```
Clear[P,x,y,r,t]
r[t_] = 3.12 - 0.65 Cos[t] + 0.32 Cos[2 t] - 0.83 Cos[3 t] + 1.92 Sin[t] -
 2.68 Sin[2 t] + 1.79 Sin[3 t];
P[t_] = r[t]{Cos[t],Sin[t]};
```

Take a look:

In[55]:=
```
path =
ParametricPlot[P[t],{t,0, 2 Pi},
PlotStyle->{{Thickness[0.01],Blue}},
AxesLabel->{"x","y"}];
```

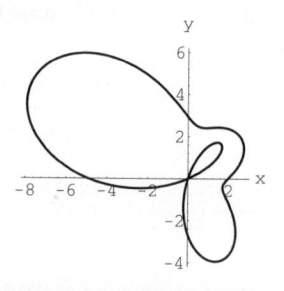

> Add your line of sight to the plot when you look forward in the instantaneous direction you are going when $t = 1$. Then describe, in terms of a clean formula, all the points you can see at the instant $t = 1$.

Answer: You are looking in the direction of the tangent vector $P'[1]$ with its tail at $P[1]$:

In[56]:=
```
point = Graphics[{Red,PointSize[0.03],Point[P[1]]}];
Show[path,point,Arrow[P'[1],Tail->P[1],Red]];
```

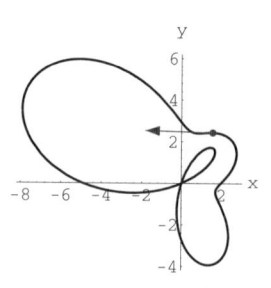

The points you can see are described by the clean formula

In[57]:=
```
Clear[lineofsight,s]
lineofsight[s_] = N[P[1] + s P'[1]]
```

Out[57]=
{1.56074 - 3.36648 s, 2.43071 + 0.103365 s}

where you take $s \geq 0$ because you are looking forward along the tangent vector. Here's part of your line of sight:

In[58]:=
```
sightplot =
ParametricPlot[lineofsight[s],{s,0,5},
DisplayFunction->Identity];
Show[path,point,sightplot,
Arrow[P'[1],Tail->P[1],Red],
DisplayFunction->$DisplayFunction];
```

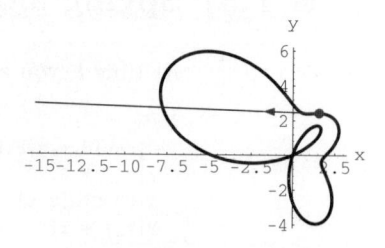

If you had eyes in the very back of your head, you would also be able to see points on this line:

In[59]:=
```
bothways =
  ParametricPlot[lineofsight[s],{s,-2,4},
  DisplayFunction->Identity];
Show[path,point,bothways,
  DisplayFunction->$DisplayFunction];
```

The line you see is what lots of the good old folks like to call a tangent line at $P[1]$.

Give It a Try

Experience with the starred (*) problems will be especially beneficial for understanding later lessons.

■ G.1) Vector and line fundamentals*

This problem appears only in the electronic version.

■ G.2) Measurements*

G.2.a) Measure the distance between $\{2, 7\}$ and $\{3, -5\}$ in two dimensions.

Measure the distance between $\{2, 7, -8\}$ and $\{3, -5, 9\}$ in three dimensions.

G.2.b.i) Here is a triangle sitting happily in three dimensions:

In[1]:=
```
point1 = {0,2,-4};
point2 = {-6/5, 41/10,-5/2};
point3 = {1,0,5};
triangle = Graphics3D[Polygon[{point1,point2,point3}]];
threedims = ThreeAxes[1,.2];
Show[threedims,triangle,
  ViewPoint->CMView,PlotRange->All,Boxed->False];
```

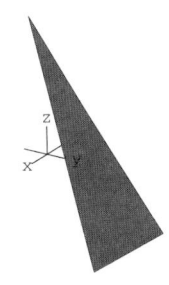

That triangle sure looks like a right triangle.

> Use a dot product to confirm or dispel this observation.

G.2.b.ii) Mrs. Stephens is one of those sly math teachers who uses *Mathematica* in her office but doesn't allow the students to use computers or calculators for class work. In fact, she doesn't even let her students know that she even has a computer. In preparation for yet another captivating lecture, she needs to generate a right triangle in three dimensions with one vertex at $\{0,0,0\}$. She locks her office and gets out her computer and types:

In[2]:=
```
X = {1,2,3}; Y = {0,1,1};
point1 = {0,0,0}; point2 = X;
point3 = ((X.Y)/(Y.Y)) Y;
triangle = Graphics3D[Polygon[{point1,point2,point3}]];
threedims = ThreeAxes[2,0.2];
Show[threedims,triangle,ViewPoint->CMView,
PlotRange->All,Boxed->False];
```

She knows in advance that, unless X and Y are parallel vectors, the triangle with vertices (corners) at $\{0,0,0\}$, the tip of $(X \bullet Y)/(Y \bullet Y)Y$ (with tail at $\{0,0,0\}$) and the tip of X (with tail at $\{0,0,0\}$) will always give her a right triangle.

> Why does this always work?

G.2.c) Here is a plot of a line and a point:

In[3]:=
```
Clear[L,t]
L[t_] = {-1.3, 5.6,-1.8} + t {1.2, 7.3,-3.6};
point = {2.4,0.7,-0.2};
lineplot = ParametricPlot3D[L[t],
{t,-1,1/2},DisplayFunction->Identity];
threedims = ThreeAxes[2.5,0.2];
Show[threedims,lineplot,
Graphics3D[{Red,PointSize[0.02],Point[point]}],
ViewPoint->CMView,PlotRange->All,Boxed->False,
DisplayFunction->$DisplayFunction];
```

> Measure the shortest distance between the line and the point.

■ G.3) With or against?*

G.3.a.i) Here are two vectors X and Y:

In[4]:=
```
X = {-0.5,2}; Y = {-2,1};
Show[Arrow[X,Tail->{0,0},Red],
Arrow[Y,Tail->{0,0},Red],
Graphics[{{Text["X",X/2]},
{Text["Y",Y/2]}}],Axes->Automatic,
AspectRatio->Automatic];
```

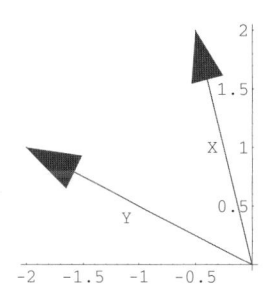

> How does the picture reveal with no calculation that $X \bullet Y > 0$?
>
> Confirm what you say by calculating $X \bullet Y$.
>
> Add to the plot the vector that measures the push of X in the direction of Y.
>
> Would you say that the push of X in the direction of Y is with Y or against Y?

G.3.a.ii) Here are two new vectors X and Y:

In[5]:=
```
X = {2,2}; Y = {-3,1};
Show[Arrow[X,Tail->{0,0},Red],
Arrow[Y,Tail->{0,0},Red],
Graphics[{{Text["X",X/2]},
{Text["Y",Y/2]}}],Axes->Automatic,
AspectRatio->Automatic];
```

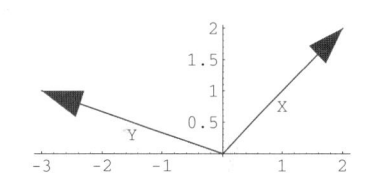

> How does the picture reveal with no calculation that $X \bullet Y < 0$?
>
> Confirm what you say by calculating $X \bullet Y$.
>
> Add to the plot the vector that measures the push of X in the direction of Y.
>
> Would you say that the push of X in the direction of Y is with Y or against Y?

G.3.a.iii) Given any two vectors X and Y, how does the sign of the dot product $X \bullet Y$ tell you whether the push of X in the direction of Y is with Y or against Y?

G.3.b) At time t with $0 \leq t \leq 2\pi$, Luke Skywalker is at the point
$$\{x[t], y[t]\} = \{\cos[t], 3\sin[t]\}.$$

Darth Vader has activated a force field that puts a force (= push or pull) equal to $\{x, -y\}$ on any object at $\{x, y\}$.

> At time t:
>
> → Is the force pushing with or against Skywalker's movement if
> $$\{x[t], -y[t]\} \bullet \{x'[t], y'[t]\} < 0?$$
>
> → Is the force pushing with or against Skywalker's movement if
> $$\{x[t], -y[t]\} \bullet \{x'[t], y'[t]\} > 0?$$
>
> For approximately which t's is Vader's force with young Skywalker?

G.3.c) Here is a pair of vectors.

In[6]:=
```
X = 3 {Cos[Pi/3],Sin[Pi/3]};
Y = {Cos[Pi/6],Sin[Pi/6]};
Show[Arrow[X,Tail->{0,0},Red],
Arrow[Y,Tail->{0,0},Red],
Graphics[{{Text["X",X/2]},
{Text["Y",Y/2]}}],Axes->Automatic,
AspectRatio->Automatic];
```

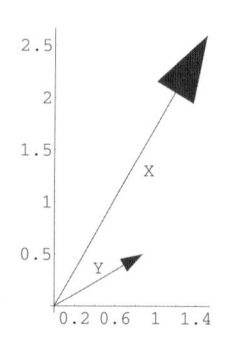

> Come up with vectors U and V with:
>
> → U parallel to Y
>
> → V perpendicular to Y
>
> → $X = U + V$.
>
> Once you have your vectors U and V, add them to the plot above by plotting U with its tail at $\{0, 0\}$ and V with its tail at the tip of U. Describe what you see.

■ G.4) Velocity and acceleration★

G.4.a) At time t with $0 \leq t \leq 6$, an object is at the position $P[t] = \{\cos[t], \cos[2\,t]\,\sin[t]\}$.

> Plot the curve and some of its velocity vectors and acceleration vectors at a selection of points.
>
> Use your plot to analyze the direction and the speed of the object. Then plot its speed as a function of t.

G.4.b) Ballistic projectiles (like a cannonball fired from a cannon) fired from the origin with muzzle velocity v_0 ft/sec and angle b with the horizontal are at the position

$$P[t] = \{v_0 \cos[b]\, t, v_0 \sin[b]\, t - 16\, t^2\}$$

t seconds after firing.

> Take $v_0 = 180$ ft/sec and $b = 3\pi/8$ and plot the trajectory, some of its velocity vectors, and some of its acceleration vectors at a selection of points.
>
> Explain anything of interest you note.

G.4.c) For $P[t] = \{2\sin[t], 6\sin[t/2]^2, 3\cos[t]\}$, plot in true scale the motion of the object for $1 \le t \le 6$, including several velocity vectors.

> Then, on a separate plot, show the motion of the object for $1 \le t \le 6$, including several tangential components of acceleration.
>
> On a third plot, show the motion of the object for $1 \le t \le 6$, including several tangential and normal components of acceleration.
>
> Discuss what each plot reveals.

G.4.d.i) An object is at $P[t] = \{9\sin[t], 3\cos[t]\}$ at time t.

> Plot in true scale the motion of the object for $0 \le t \le 3\pi/2$, including several velocity vectors.
>
> Then, on a separate plot, show in true scale the motion of the object for $0 \le t \le 3\pi/2$, including several tangential and normal components of acceleration.
>
> Discuss what each plot reveals.
>
> Finally, show the plots together and discuss relations between them.

G.4.d.ii) For $0 \le t \le 3\pi/2$, the object in part G.4.d.i) is constrained to move on an ellipse. Suppose, at the instant $t = 3\pi/2$, the object is released from all constraints and allowed to move of its own free will independent of any accelerations due to forces.

> Plot the path the object takes.

G.4.d.iii) The speed of the object in part G.4.d.i) at time t is defined to be the length of the velocity vector at time t; in other words speed$[t] = \sqrt{\text{velocity}[t] \bullet \text{velocity}[t]}$. Put

$$\text{unittan}[t] = \frac{\tan[t]}{\sqrt{\tan[t] \bullet \tan[t]}}$$

where $\tan[t] = \{x'[t], y'[t]\}$

> Calculate $D[\text{speed}[t], t]$ and compare it with acceleration$[t] \bullet$ unittan$[t]$. Try to explain why you think that the result is natural or weird.

G.4.d.iv)
> If you know speed$[0]$, then how do you give a formula for speed$[t]$ in terms of acceleration$[t] \bullet$ unittan$[t]$?

G.4.d.v)
> Can you get any information about speed$[t]$ if you know the tangential component of acceleration but lose all your information about the normal component of acceleration?

■ G.5) The coordinate axes and coordinate planes in three dimensions★

This problem appears only in the electronic version.

■ G.6) Serious plotting: Parametric planets

This problem appears only in the electronic version.

■ G.7) Lines★

G.7.a) Here is the line segment running from $\{-3, 2\}$ to $\{3, -4\}$.

In[7]:=
```
point1 = {-3, 2}; point2 = { 3,-4};
segment = Show[Graphics[{PointSize[0.02],Blue,
Point[point1],Point[point2]}],
Graphics[Line[{point1,point2}]],
Axes->Automatic,AxesOrigin->{0,0},
AxesLabel->{"x","y"}];
```

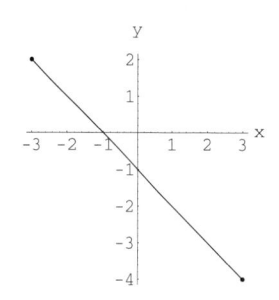

A parametric formula of the line through these two points is:

In[8]:=
```
Clear[L,t]; L[t_] = point1 + t(point2 - point1)
```
Out[8]=
```
{-3 + 6 t, 2 - 6 t}
```

> What value of t makes $L[t]$ land on point1?
>
> What value of t makes $L[t]$ land on point2?
>
> What value of t makes $L[t]$ land on the point halfway between point1 and point2 on the indicated segment?
>
> What values of t make $L[t]$ land on the indicated segment?
>
> Illustrate with plots similar to these:

In[9]:=
```
Show[segment,Arrow[L[0.57],
Tail->{0,0},Red]];
```

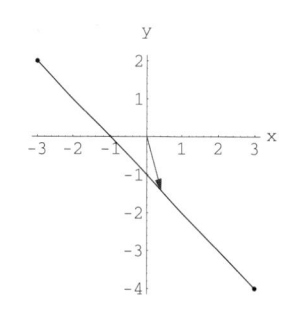

In[10]:=
```
Show[segment,Arrow[L[1.46],
Tail->{0,0},Red]];
```

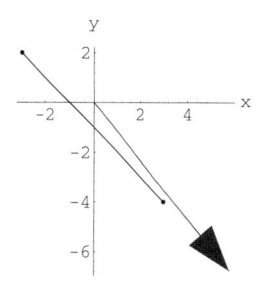

Lesson 3.01 Vectors Point the Way

G.7.b.i) Write down a parametric formula for the line that passes through the points $\{2, 0\}$ and $\{3, 4\}$.

Give a vector parallel to this line.

Give a vector perpendicular to this line.

G.7.b.ii) Write a parametric formula for the line that has xy-equation
$$(y+2) = 1.52\,(x-1).$$
Give a vector parallel to this line.

Give a vector perpendicular to this line.

Give a point on this line.

G.7.b.iii) Are the lines with parametric formulas
$$L_1[t] = \{2, 3\} + t\{-3, 5\}$$
and
$$L_2[t] = \{2, 3\} + t\{6, -10\}$$
the same line or different lines? Are the lines with vector equations
$$L_1[t] = \{2, 3\} + t\{-3, 5\}$$
and
$$L_2[t] = \{-4, 13\} + t\{-3, 5\}$$
the same line or different lines?

G.7.c.i) Give a parametric formula for the line passing through the point $\{1, 4, 5\}$ and moving away in the direction of the vector $\{2, 1, 1\}$.

G.7.c.ii) Give a parametric formula for the line in three dimensions through the tips of the vectors $X = \{2, 6, 2\}$ and $Y = \{4, 2, 1\}$ when their tails are at the origin.

G.7.d) Parametric equations for the line through $\{1, 2, 3\}$ parallel to the vector $\{5, 6, -4\}$ are:
$$x = 1 + 5t, \quad y = 2 + 6t, \quad \text{and} \quad z = 3 - 4t.$$
A friend taking the old-fashioned course tells you that the xyz-equations for this line are
$$\frac{x-1}{5} = \frac{y-2}{6} = \frac{z-3}{-4}.$$
Your friend is right.

> Why?

■ G.8) Lasers

Calculus&*Mathematica* thanks Professor Todd Will of Davidson College for suggesting this problem.

G.8.a.i) A Spartan missile and a Trojan missle are both flying at the same constant altitude. At time t, the Spartan missile is at the point:

In[11]:=
```
Clear[spartan,t]
spartan[t_] = {16.1 - 7 t + t^2, 13 t - 2 t^2}
```

Out[11]=
$$\{16.1 - 7t + t^2, \; 13t - 2t^2\}$$

At the same time t, the Trojan missile is at the point:

In[12]:=
```
Clear[trojan]
trojan[t_] = {26 - 13t + 2 t^2, 23 - 5t + t^2}
```

Out[12]=
$$\{26 - 13t + 2t^2, \; 23 - 5t + t^2\}$$

Here is a plot of the paths of the two missiles:

In[13]:=
```
spartanpath = ParametricPlot[spartan[t],{t,0,6},
  PlotStyle->{{Blue,Thickness[0.01]}},
  DisplayFunction->Identity];
trojanpath = ParametricPlot[trojan[t],{t,0,6},
  PlotStyle->{{Red,Thickness[0.01]}},
  DisplayFunction->Identity];
Show[spartanpath,trojanpath,
  DisplayFunction->$DisplayFunction];
```

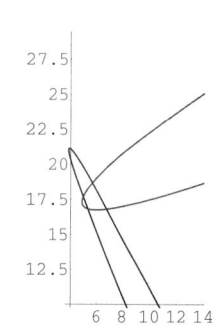

> Their paths cross, but do they crash?

G.8.a.ii) Continue with the set-up in part G.8.a.i) immediately above, but with additional information: Each of the missiles has a laser in its nose that can shoot straight ahead and zap instantaneously.

> Can either missile ever zap the other? If so, who can zap whom and when?

G.8.b) At time $t \geq 0$, your Luke Skywalker laser rocket scooter is at the position:

```
In[14]:=
    Clear[P,t]
    P[t_] = {6 - t + Sin[t], 10 - 2 t, 2 + 0.5 Sin[2 t]};
```

Here is its path for $0 \leq t \leq 5$:

```
In[15]:=
    path = ParametricPlot3D[Evaluate[P[t]],{t,0,5},
    DisplayFunction->Identity]; h = 6;
    xzplane = Graphics3D[Polygon[
    {{-h,0,0},{-h,0,h},{h,0,h},{h,0,0}}]];
    threedims = ThreeAxes[6,.4];
    setup = Show[threedims,path,xzplane,
    ViewPoint->CMView,Boxed->False,
    Axes->None,PlotRange->All,
    DisplayFunction->$DisplayFunction];
```

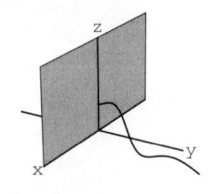

A laser beam emanates from the nose cone of your scooter and shoots out in in a straight line tangent to the path like this:

```
In[16]:=
    samplebeam = Arrow[4 P'[3],Tail->P[3],Red];
    Show[setup,samplebeam];
```

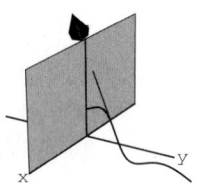

Note that the beam pierces the xz-plane.

> Imagine that the xz-plane is made of cardboard, and plot the curve burned into the xz-plane by your scooter's laser during the time interval $0 \leq t \leq 5$.

■ G.9) Parabolic reflectors, spherical reflectors, and elliptical reflectors

G.9.a.i) Here is part of the parabola $f[x] = x^2/8$:

```
In[17]:=
    Clear[f,x]; f[x_] = (x^2)/8;
    {x[t_],y[t_]} = {t,f[t]}; parabola =
    ParametricPlot[{x[t],y[t]},{t,-4,4},
    PlotStyle->{{Blue,Thickness[0.01]}},
    AxesLabel->{"x","y"}];
```

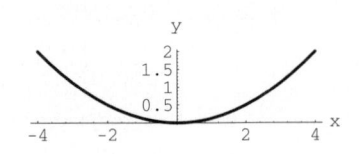

Three vertical light rays, emanating from $\{x[1], 6\}, \{x[2], 6\}$ and $\{x[3], 6\}$ hit the parabola at the points $\{x[1], y[1]\}, \{x[2], y[2]\}$ and $\{x[3], y[3]\}$ respectively:

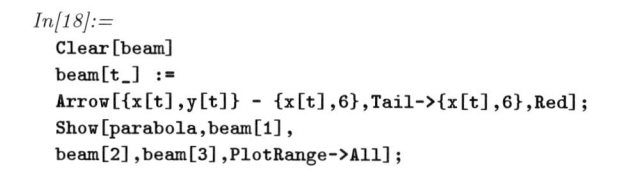

In[18]:=
```
Clear[beam]
beam[t_] :=
 Arrow[{x[t],y[t]} - {x[t],6},Tail->{x[t],6},Red];
Show[parabola,beam[1],
 beam[2],beam[3],PlotRange->All];
```

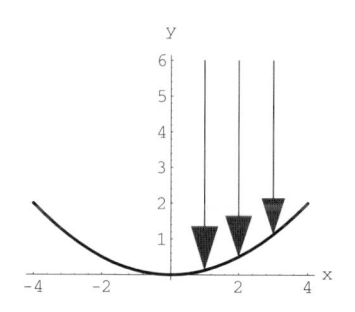

> On one plot, show the paths the three beams take after they have bounced off the parabola. Make sure you show enough of the reflected beams to see where each beam crosses the y-axis. Describe what you see, and mention anything interesting.

G.9.a.ii) You are given a positive number p, but you are not told what the specific value of p is. A vertical light beam comes from high above the plot of $f[x] = x^2/(4p)$ and bounces off the curve at a given point point $\{a, f[a]\}$. You are not told what the specific value of a is.

> Calculate, in terms of p, the point at which the reflected beam crosses the y-axis.

G.9.a.iii) Most traditional reference books say that the focus of the parabola $f[x] = x^2/(4p)$ is located at the point $\{0, p\}$ on the vertical-axis.

> Why do they say this? Why do folks use parabolas to build television satellite dish antennas to sell to rednecks in the boonies who want to watch wrestling and roller derby?

G.9.a.iv) The great Greek scientist, Archimedes (287–212 B.C.), was the first scientist to understand what parabolic mirrors can do. In fact, Archimedes once used parabolic mirrors to concentrate sunlight on the sails of attacking Roman ships, thereby burning them up before they could attack.

> How do you think Archimedes went about this?

Lesson 3.01 Vectors Point the Way

G.9.b) Here is part of the circle of radius 4 centered at $\{0,4\}$:

```
In[19]:=
  Clear[x,y,t]
  {x[t_],y[t_]} = {0,4} + 4 {Cos[t],Sin[t]};
  starter = 3 Pi/2 - Pi/3; stopper = 3 Pi/2 + Pi/3;
  circulardish = ParametricPlot[
  {x[t],y[t]},{t,starter,stopper},
  PlotStyle->{{Blue,Thickness[0.01]}},
  AspectRatio->Automatic,AxesLabel->{"x","y"}];
```

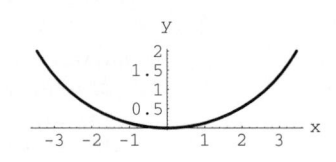

Three vertical light rays, emanating from $\{x[5], 6\}$, $\{x[5.3], 6\}$ and $\{x[5.6], 6\}$ hit the circular dish at the points $\{x[5], y[5]\}$, $\{x[5.3], y[5.3]\}$, and $\{x[5.6], y[5.6]\}$, respectively:

```
In[20]:=
  Clear[beam]; beam[t_] :=
  Arrow[{x[t],y[t]} - {x[t],6},Tail->{x[t],6},Red];
  Show[circulardish,beam[5],
  beam[5.3],beam[5.6],
  AspectRatio->Automatic,PlotRange->All];
```

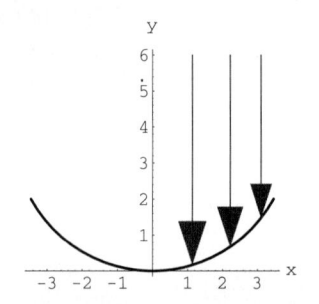

On one plot, show the path each of the three beams takes after it has bounced off the parabola. Make sure you show enough of the reflected beams to see where each beam crosses the y-axis.

Why don't spherical dish antennas work very well?

G.9.c.i) Here is the lower half of the ellipse

$$\left(\frac{x}{3}\right)^2 + \left(\frac{y}{2}\right)^2 = 1$$

shown with its two focuses at $\{\sqrt{3^2 - 2^2}, 0\}$ and $\{-\sqrt{3^2 - 2^2}, 0\}$:

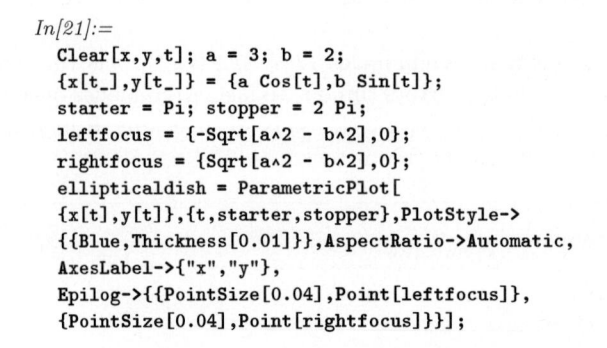

```
In[21]:=
  Clear[x,y,t]; a = 3; b = 2;
  {x[t_],y[t_]} = {a Cos[t],b Sin[t]};
  starter = Pi; stopper = 2 Pi;
  leftfocus = {-Sqrt[a^2 - b^2],0};
  rightfocus = {Sqrt[a^2 - b^2],0};
  ellipticaldish = ParametricPlot[
  {x[t],y[t]},{t,starter,stopper},PlotStyle->
  {{Blue,Thickness[0.01]}},AspectRatio->Automatic,
  AxesLabel->{"x","y"},
  Epilog->{{PointSize[0.04],Point[leftfocus]},
  {PointSize[0.04],Point[rightfocus]}}];
```

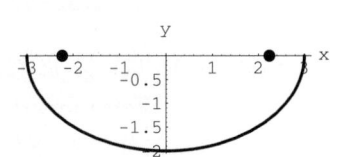

Three light rays emanating the left focus hit the elliptical dish at the points $\{x[4], y[4]\}, \{x[5], y[5]\}$ and $\{x[6], y[6]\}$:

In[22]:=
```
Clear[beam]
beam[t_] := Arrow[{x[t],y[t]} - leftfocus,
  Tail->leftfocus,Red];
Show[ellipticaldish,beam[4],beam[5],beam[6],
  AspectRatio->Automatic,PlotRange->All];
```

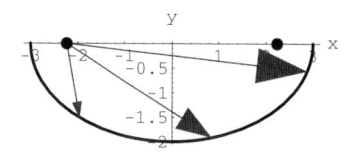

> On one plot, show the path each of the three beams takes after it has bounced off the parabola. Make sure you show enough of the reflected beams to see where each beam crosses the x-axis. Describe what you see.

G.9.c.ii) The ceiling of the rotunda in the United States Capitol building in Washington D.C. is in the form of an elliptical dish. Tourists are often surprised that they can sometimes hear very clearly what strangers well across the room are saying.

> Use what you did above to explain this spooky phenom.

G.9.c.iii) Lewis Carroll (actual name Charles Dodgson, 1832–1898) was one far-out fellow. In addition to writing the famous tale of "Alice in Wonderland," he was an accomplished mathematician and logician. As a prank, he invented an elliptical pool table with a mark at one focus and a hole at the other focus and challenged expert pool players to sink more balls in the holes than he.

> What advantage did he have over those who didn't know about reflecting properties of ellipses?

■ G.10) Pursuits by a robotic cowhand

Calculus&Mathematica thanks cattleman Thomas O. Smith of Homer, Illinois for some help with this problem.

The scene is the C&M Electronic Ranch, where all the cattle are prize winners and where electronic robots do all the work. One night, the prize bull breaks out of the pen, meanders around the ranch grounds, and then heads for the gate to the highway on the path plotted below:

In[23]:=
```
Clear[bull,t]
bull[t_] = 4 {1 + 0.1 t + Cos[0.3 t] ,6 - 0.05 t + Sin[0.4 t]};
bullroute = ParametricPlot[bull[t],{t,0,60},
  PlotStyle->{{Thickness[0.01],Red}},
```

Lesson 3.01 Vectors Point the Way

```
          AxesLabel->{"x","y"},DisplayFunction->Identity,
          Epilog->Text["Bull route",{25,30}]];
        pointplot = {Graphics[{PointSize[0.06],Point[bull[0]]}],
          Graphics[{PointSize[0.06],Point[bull[60]]}],
          Graphics[{PointSize[0.05],Point[{2,3}]}]};
        labels = {Graphics[Text["Bull at t = 0",bull[0],{0,-4}]],
          Graphics[Text["Gate to highway",bull[60],{0,3}]],
          Graphics[Text["Robot at t = 0",{2,3},{0,-2}]]};
```

In[24]:=
```
        setup = Show[bullroute,pointplot,
          labels,PlotRange->All,
          AxesOrigin->{0,0},AspectRatio->Automatic,
          DisplayFunction->$DisplayFunction];
```

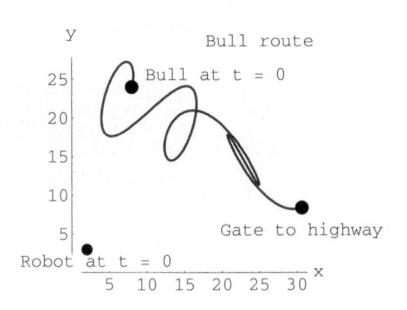

Here x and y are measured in yards and t is measured in seconds

At time $t = 0$, the robot is at position $\{2,3\}$ on the plot above. The robot is programmed to move so that if the robot's position at time t is $\{x[t], y[t]\}$, then the robot's velocity vector at that time is

$$\{x'[t], y'[t]\} = r\,(\text{bull}[t] - \{x[t], y[t]\})$$

where r is a positive number yet to be determined. This is good because:

→ The robot is always moving toward the bull.

→ The robot slows down when it gets near the bull so that the robot neither smashes into the bull nor stampedes the bull.

G.10.a.i) Given that the robot can lasso the bull anytime the robot gets within 4.5 yards of the bull, your job is to come up with a specific positive number r, as small as practical, so that the robot will be successful lassoing the bull before the bull gets to the gate to the highway.

G.10.a.ii) After you have settled on a good number r in part G.10.a.i), make a nice plot of the actual lassoing of the bull by the robot.

■ G.11) Stealth technology

This problem appears only in the electronic version.

LESSON 3.02

Perpendicularity

Basics

■ **B.1) The cross product $X \times Y$ of two 3D vectors is perpendicular to both X and Y**

Take two vectors X and Y from the same dimension.

Testing for perpendicularity by checking whether

$$X \times Y = 0$$

is quick and easy.

In three dimensions another product comes to the front. This product is also related to perpendicularity.

To calculate the cross product

$$X \times Y \quad \text{for } X = \{1, 2, 3\} \text{ and } Y = \{-2, 0, 7\},$$

you make a matrix with $\{i, j, k\}$ in the top (horizontal) row, with X in the middle (horizontal) row, and with Y in the bottom row:

In[1]:=
```
X = {1,2,3}; Y = {-2,0,7};
Clear[i,j,k]
MatrixForm[{{i,j,k},X,Y}]
```
Out[1]=
```
 i  j  k
 1  2  3
-2  0  7
```

Next, take the determinant of this matrix:

In[2]:=
```
Det[{{i,j,k},X,Y}]
```

Out[2]=
14 i - 13 j + 4 k

To arrive at the cross product $X \times Y$, you replace i by $\{1,0,0\}$, j by $\{0,1,0\}$, and k by $\{0,0,1\}$:

In[3]:=
```
XcrossY = Det[{{i,j,k},X,Y}]/.
  {i->{1,0,0},j->{0,1,0},k->{0,0,1}}
```

Out[3]=
{14, 13, 4}

The dot product is a number. The cross product is a vector. Here are are two new 3D vectors and their plots:

In[4]:=
```
X = {1,-2,1}; Y = {1,3,-1};
newtail= {1,-1,2}; XandY =
Show[threedims,Arrow[X,Tail->newtail],
Arrow[Y,Tail->newtail],
Graphics3D[Text["X",newtail + X/2]],
Graphics3D[Text["Y",newtail + Y/2]],
ViewPoint->CMView,PlotRange->All,
BoxRatios->Automatic];
```

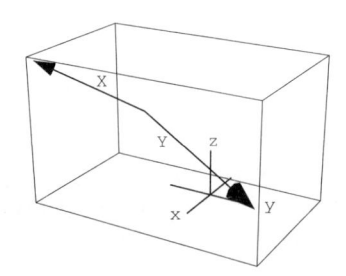

Now throw in the cross product $X \times Y$:

In[5]:=
```
Clear[i,j,k]
XcrossY = Det[{{i,j,k},X,Y}]/.
  {i->{1,0,0},j->{0,1,0},k->{0,0,1}};
Show[XandY,
Arrow[XcrossY,Tail->newtail,Red],
ViewPoint->CMView,BoxRatios->Automatic,
PlotRange->All];
```

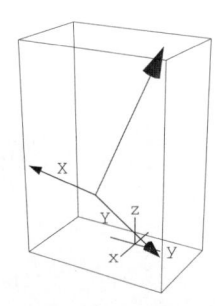

The cross product $X \times Y$ appears to be perpendicular to both X and Y. Rerun this with new vectors X and Y and new tails.

B.1.a) For the record, explain why it is that when you take any two three-dimensional vectors X and Y, then their cross product $X \times Y$ is perpendicular to both X and Y.

Answer: Clear the vectors and test with the dot product:

In[6]:=
```
Clear[X,x,Y,y]; Clear[i,j,k]
X ={x[1],x[2],x[3]}; Y ={y[1],y[2],y[3]};
XcrossY = Det[{{i,j,k},X,Y}]/.
  {i->{1,0,0},j->{0,1,0},k->{0,0,1}};
Expand[X.XcrossY]
```

Out[6]=
0

Ah-ha! No matter what X is, the vector $X \times Y$ is perpendicular to X.

In[7]:=
```
Expand[Y.XcrossY]
```

Out[7]=
0

No matter what X is, $X \times Y$ is perpendicular to Y. That's all there is to it. The joy of automated algebra!

■ B.2) Planes in 3D

Just as a line is determined by a point and one direction vector, a plane in three dimensions is determined by a point and TWO direction vectors.

Here's how you plot part of the plane determined by the following point and two direction vectors:

In[8]:=
```
Clear[planeplotter,u,v]
point = {1,0,1}; vector1 = {-1,1,0}; vector2 = {0.3,1,-1};
planeplotter[u_,v_] = point + u vector1 + v vector2;
{ulow,uhigh} = {-2,2}; {vlow,vhigh} = {-3,3};
plane = ParametricPlot3D[Evaluate[planeplotter[u,v]],
  {u,ulow,uhigh},{v,vlow,vhigh},DisplayFunction->Identity];
```

In[9]:=
```
threedims = ThreeAxes[6,0.2];
planeplot = Show[threedims,plane,
Axes->Automatic,PlotRange->All,
ViewPoint->CMView,BoxRatios->Automatic,
DisplayFunction->$DisplayFunction];
```

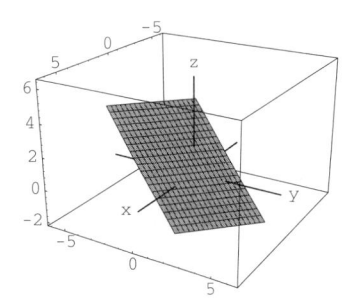

Look at the plane together with the point and the two vectors that determine this particular plane:

Lesson 3.02 Perpendicularity

In[10]:=
```
scalefactor = 2;
Show[planeplot,
Arrow[scalefactor vector1,Tail->point,Blue],
Arrow[scalefactor vector2,Tail->point,Blue],
Graphics3D[{PointSize[0.04],Point[point]}],
BoxRatios->Automatic];
```

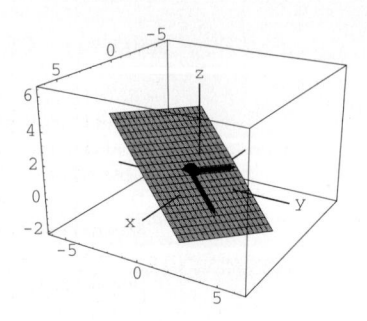

There you are. The plane is what you get by sticking the tails of vector1 and vector2 at the point and nailing a flat piece of plywood onto the braces made from those vectors. The function

$$\text{planeplotter}[u, v] = \text{point} + u \,\text{vector1} + v \,\text{vector2}$$

plots the plane by moving to the given point and then adding on multiples of the two determining vectors. Try it for a different point and a different pair of vectors:

In[11]:=
```
Clear[planeplotter,u,v]
point = {0,-1,0}; vector1 = {0,0,1}; vector2 = {1,0,1};
planeplotter[u_,v_] = point + u vector1 + v vector2;
{ulow,uhigh} = {-3,3}; {vlow,vhigh} = {-3,3};
plane = ParametricPlot3D[Evaluate[planeplotter[u,v]],
{u,ulow,uhigh},{v,vlow,vhigh},PlotPoints->{2,2},DisplayFunction->Identity];
```

In[12]:=
```
scalefactor = 2;
Show[plane,threedims,
Arrow[scalefactor vector1,Tail->point,Blue],
Arrow[scalefactor vector2,Tail->point,Blue],
Graphics3D[{PointSize[0.04],Point[point]}],
Axes->Automatic,PlotRange->All,
ViewPoint->CMView,Boxed->False,
BoxRatios->Automatic,
DisplayFunction->$DisplayFunction];
```

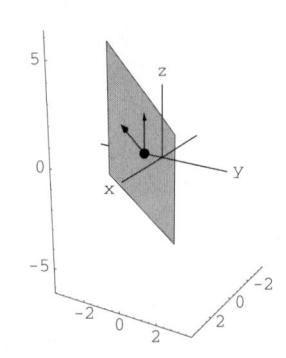

B.2.a) Plot an elliptical piece of the plane plotted above.

Answer: That's not too hard; you just modify the plotting function as follows:

In[13]:=
```
point = {0,-1,0}; vector1 = {0,0,1}; vector2 = {1,0,1};
Clear[planeplotter,r,t]
planeplotter[r_,t_] = point + 4 r Cos[t] vector1 + 6 r Sin[t] vector2;
{rlow,rhigh} = {0,1}; {tlow,thigh} = {0,2 Pi}; threedims = ThreeAxes[5,0.2];
plane = ParametricPlot3D[Evaluate[planeplotter[r,t]],{r,rlow,rhigh},{t,tlow,thigh},
PlotPoints->{2,Automatic},DisplayFunction->Identity];
```

In[14]:=
```
scalefactor = 3;
elliptical = Show[plane,threedims,
  Arrow[scalefactor vector1,Tail->point,Blue],
  Arrow[scalefactor vector2,Tail->point,Blue],
  Graphics3D[{PointSize[0.04],Point[point]}],
  Axes->Automatic,PlotRange->All,
  ViewPoint->CMView,Boxed->False,
  BoxRatios->Automatic,
  DisplayFunction->$DisplayFunction];
```

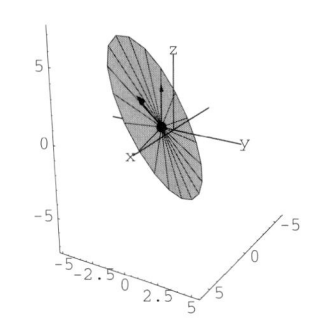

No sweat.

B.2.b) A normal vector for a given plane is a vector perpendicular to that plane.

> How do you come up with a normal (perpendicular) vector for a plane determined by a point and two direction vectors?

Answer: To get a normal vector for a plane, just take the cross product of two nonparallel direction vectors. Try it; you'll like it:

In[15]:=
```
Clear[x,y,z,r,t]; Clear[planeplotter,r,t]; Clear[i,j,k];
point = {0,0,-2}; vector1 = {-1,2,1}; vector2 = {-2,-1,-2};
normal = Det[{{i,j,k},vector1,vector2}]/.{i->{1,0,0},j->{0,1,0},k->{0,0,1}};
planeplotter[r_,t_] = point + 3 r Cos[t] vector1 + 5 r Sin[t] vector2;
{rlow,rhigh} = {0,1}; {tlow,thigh} = {0,2 Pi};
plane = ParametricPlot3D[Evaluate[planeplotter[r,t]],
  {r,rlow,rhigh},{t,tlow,thigh},PlotPoints->{2,Automatic},DisplayFunction->Identity];
```

In[16]:=
```
scalefactor = 3; Show[plane,
  Arrow[scalefactor vector1,Tail->point,Blue],
  Arrow[scalefactor vector2,Tail->point,Blue],
  Arrow[scalefactor normal,Tail->point,Red],
  Graphics3D[{PointSize[0.04],Point[point]}],
  Axes->Automatic,PlotRange->All,ViewPoint->
  CMView,Boxed->False,BoxRatios->Automatic,
  DisplayFunction->$DisplayFunction];
```

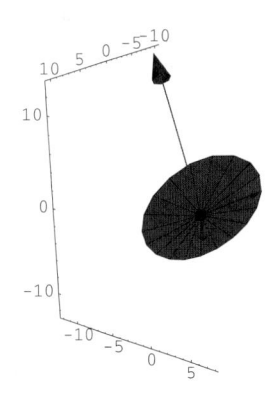

Lookin' fairly good.

B.2.c)
> How do you get the xyz-equation for a plane specified by a given point and two direction vectors?

Answer: You get it from the given point and the normal vector. To see what's involved, go with the plane determined by the following point and direction vectors:

Lesson 3.02 Perpendicularity

In[17]:=
```
Clear[x,y,z]; point = {1,2,3}; vector1 = {2,4,-1.2}; Clear[i,j,k]
vector2 = {-1,1.5,0};
normal = Det[{{i,j,k},vector1,vector2}]/.{i->{1,0,0},j->{0,1,0},k->{0,0,1}}
```
Out[17]=
{1.8, 1.2, 7.}

Saying that $\{x, y, z\}$ is on this plane is the same as saying that when you stick the tail of the normal at the given point $\{1, 2, 3\}$, then the normal is perpendicular to the vector with tail at $\{1, 2, 3\}$ and tip at $\{x, y, z\}$. So saying that $\{x, y, z\}$ is on this plane is the same as saying that

$$(\{x, y, z\} - \text{point}) \bullet \text{normal} = 0:$$

In[18]:=
```
xyzeqn = ({x,y,z} - point). normal == 0
```
Out[18]=
1.8 (-1 + x) + 1.2 (-2 + y) + 7. (-3 + z) == 0

Notice how the given point $\{1, 2, 3\}$ and the normal vector $\{1.8, 1.2, 7\}$ are conveniently displayed in the plane equation. You can solve this equation for z:

In[19]:=
```
zsolved = Solve[xyzeqn,z]
```
Out[19]=
{{z -> -0.142857 (-25.2 + 1.8 x + 1.2 y)}}

And you can use this to plot the part of the plane over any rectangle in the xy-plane you like.

In[20]:=
```
Clear[f]
f[x_,y_] = z/.zsolved[[1]]
```
Out[20]=
-0.142857 (-25.2 + 1.8 x + 1.2 y)

Here comes the plot of the part of this plane over the rectangle $-4 \leq x \leq 4$ and $0 \leq y \leq 6$ in the xy-plane:

In[21]:=
```
plane = Plot3D[f[x,y],{x,-4,4},{y,0,6},
  PlotPoints->{2,2},DisplayFunction->Identity];
Show[plane,Arrow[vector1,Tail->point,Blue],
  Arrow[vector2,Tail->point,Blue],
  Arrow[normal,Tail->point,Red],
  Graphics3D[{PointSize[0.04],Point[point]}],
  Axes->Automatic,PlotRange->All,AxesLabel->
  {"x","y","z"},ViewPoint->CMView,Boxed->False,
  BoxRatios->Automatic,
  DisplayFunction->$DisplayFunction];
```

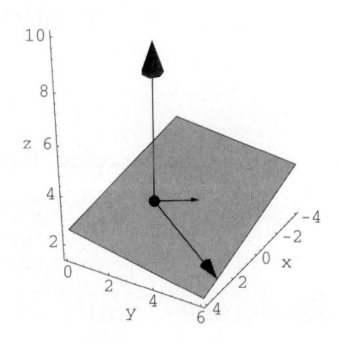

Bingo.

B.2.d) A normal vector and a point also determine a plane. To see what the plane is, put the tail of the normal vector at the point. The plane is what you get by putting a flat plywood sheet at the tail of the normal vector and aligning the plywood sheet to be perpendicular to the normal vector. In fact, if the normal vector and the point are related to the xyz equation for the plane by:

In[22]:=
```
Clear[R,S,T,a,b,c,x,y,z]
normal = {R,S,T}; point = {a,b,c};
xyzequation = (({x,y,z} - {a,b,c}).normal == 0)
```

Out[22]=
```
R (-a + x) + S (-b + y) + T (-c + z) == 0
```

> Explain where this comes from, and use it to plot the plane passing through the point $\{4, 2, 1\}$ with normal vector $\{2, -1, 1.5\}$.

Answer: Take another look:

In[23]:=
```
normal = {R,S,T}; point = {a,b,c};
xyzequation = (({x,y,z} - point).normal == 0)
```

Out[23]=
```
R (-a + x) + S (-b + y) + T (-c + z) == 0
```

This comes from the fact that saying that:

→ $\{x, y, z\}$ is on the plane passing through the given point $\{a, b, c\}$ with normal vector $\{R, S, T\}$ is the same as saying that:

→ the vector running from the given point $\{a, b, c\}$ to $\{x, y, z\}$ is perpendicular to the normal vector $\{R, S, T\}$.

This is the same as saying:

In[24]:=
```
xyzequation = (({x,y,z} - point).normal == 0)
```

Out[24]=
```
R (-a + x) + S (-b + y) + T (-c + z) == 0
```

To plot the plane passing through the point $\{4, 2, 1\}$ with normal vector $\{2, -0.8, 1.5\}$, go with:

In[25]:=
```
normal = {2, -0.8, 1.5}; point = {4,2,1};
xyzequation = (({x,y,z} - point).normal == 0)
```

Out[25]=
```
2 (-4 + x) - 0.8 (-2 + y) + 1.5 (-1 + z) == 0
```

Here comes the plot:

Lesson 3.02 Perpendicularity

In[26]:=
```
zsolved = Solve[xyzequation,z]
```
Out[26]=
```
{{z -> -1.33333 (-3.95 + x - 0.4 y)}}
```

In[27]:=
```
Clear[f]; f[x_,y_] = z/.zsolved[[1]]
```
Out[27]=
```
-1.33333 (-3.95 + x - 0.4 y)
```

In[28]:=
```
plane = Plot3D[f[x,y],{x,-1,7},{y,-4,6},
  DisplayFunction->Identity]; scalefactor = 4;
Show[plane,Arrow[scalefactor normal,
  Tail->point,Red],Graphics3D[{PointSize[0.04],
  Point[point]}],Axes->Automatic,
  AxesLabel->{"x","y","z"},PlotRange->All,
  ViewPoint->CMView,Boxed->False,
  BoxRatios->Automatic,DisplayFunction->$DisplayFunction];
```

The flat plywood sheet is perpendicular to the normal vector at the point:

In[29]:=
```
point
```
Out[29]=
```
{4, 2, 1}
```

And that's all there is to it.

■ B.3) Normal vectors for curved surfaces in 3D

Here is a curved surface in three dimensions given in parametric form:

In[30]:=
```
Clear[x,y,z,u,v]; x[u_,v_] = u + v/3; y[u_,v_] = v - u/4;
z[u_,v_] = (10 - u^2 - v^2)/3; {ulow,uhigh} = {-2,2}; {vlow,vhigh} = {-2,2};
surf = ParametricPlot3D[Evaluate[{x[u,v],y[u,v],z[u,v]}],
  {u,ulow,uhigh},{v,vlow,vhigh},DisplayFunction->Identity];
```

In[31]:=
```
surface = Insert[surf,EdgeForm[],{1,1,1}];
threedims = ThreeAxes[2,.2];
surfaceplot = Show[threedims,surface,
  Axes->Automatic,PlotRange->All,
  ViewPoint->CMView,BoxRatios->Automatic,
  DisplayFunction->$DisplayFunction];
```

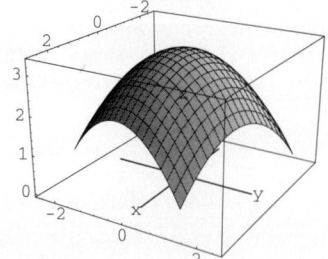

Basics (B.3)

B.3.a) Take a point on the surface and say how to build a vector that is perpendicular to the surface at that point.

Show off your work with a sample plot.

Answer: Any point on the surface can be described by $\{x[u,v], y[u,v], z[u,v]\}$ for specific choices of u and v. Here's the point you get by setting $u = 0.2$ and $v = 1.3$:

In[32]:=
```
uu = 0.2; vv = 1.3;
Clear[surfacepoint]
surfacepoint[u_,v_] = {x[u, v], y[u, v], z[u, v]};
surfacepoint[uu,vv]
```

Out[32]=
{0.633333, 1.25, 2.75667}

To build the perpendicular vector at this point, look at the point and two curves on the surface that meet at this point:

In[33]:=
```
Clear[ucurve,vcurve]
ucurve[u_] = surfacepoint[u,vv];
vcurve[v_] = surfacepoint[uu,v];
ucurveplot = ParametricPlot3D[ucurve[u],
  {u,ulow,uhigh},DisplayFunction->Identity];
vcurveplot = ParametricPlot3D[vcurve[v],
  {v,vlow,vhigh},DisplayFunction->Identity];
pointplot = Graphics3D[{PointSize[0.04],
  Point[surfacepoint[uu,vv]]}];
firstlook = Show[surfaceplot,ucurveplot,
  vcurveplot,pointplot];
```

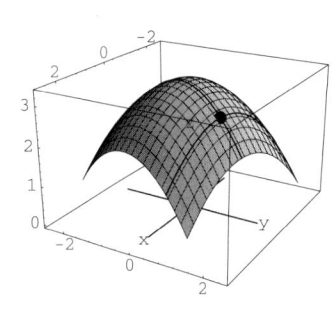

Now throw in the tangent vectors to each curve at the point in question:

In[34]:=
```
utanvector = D[ucurve[u],u]/.u->uu;
vtanvector = D[vcurve[v],v]/.v->vv;
tangents = {Arrow[utanvector,
  Tail->surfacepoint[uu,vv],Blue],
  Arrow[vtanvector,Tail->surfacepoint[uu,vv],Blue]};
secondlook = Show[firstlook,tangents];
```

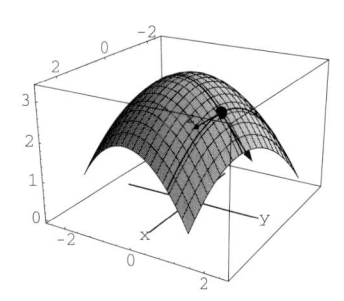

To build a vector perpendicular to the surface at the point in question, all you gotta do is take the cross product of the two tangent vectors:

Lesson 3.02 Perpendicularity

In[35]:=
```
Clear[i,j,k]
normal = Det[{{i,j,k},utanvector,vtanvector}]/.
  {i->{1,0,0},j->{0,1,0},k->{0,0,1}};
proudlook = Show[secondlook,Arrow[normal,
  Tail->surfacepoint[uu,vv],Red],BoxRatios->Automatic];
```

There it is! Proud and perpendicular. Most folks don't bother to show the curves and the tangents:

In[36]:=
```
cleanplot = Show[surfaceplot,
  Arrow[normal,Tail->surfacepoint[uu,vv],Red]];
```

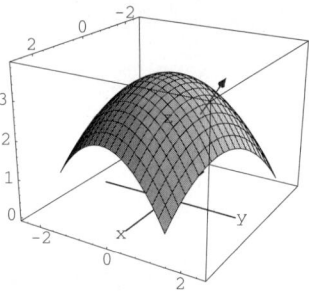

Math folks like to call this the normal vector at the point at the tail of the vector. Play with this by rerunning it with your own choices of *uu* and *vv*.

B.3.b) | Of what practical use is this normal vector?

Answer: You can use it to see what happens when you bounce light off a surface. You'll have a chance to give this a try later.

B.3.c.i) | Say, that presentation of what you do to get the normal vector was pretty windy.

Can you give some concise code that efficiently produces the normal vector at a desired point without all the wind?

Answer: Be happy to. To get the concise code that will work on any surface given in parametric form, just review what went on in part B.3.a) and try it out on a plot of the surface

$$z = f[x,y] = -x\,y:$$

Basics (B.3) 61

In[37]:=
```
Clear[f,x,y,z,u,v];
f[x_,y_] = - x y; x[u_,v_] = u Cos[v];
y[u_,v_] = u Sin[v]; z[u_,v_] = f[x[u,v],y[u,v]];
{ulow,uhigh} = {0,1.5}; {vlow,vhigh} = {0,2 Pi};
surface = ParametricPlot3D[Evaluate[{x[u,v],y[u,v],z[u,v]}],
{u,ulow,uhigh},{v,vlow,vhigh},DisplayFunction->Identity];
```

In[38]:=
```
threedims = ThreeAxes[2,.2];
surfaceplot = Show[threedims,surface,
Axes->Automatic,PlotRange->All,
ViewPoint->CMView,
DisplayFunction->$DisplayFunction];
```

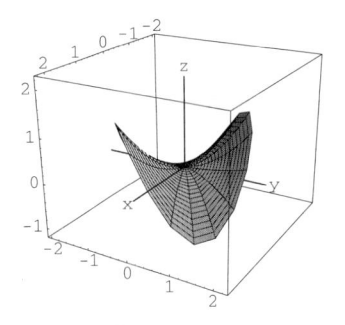

Here's a quick way of throwing in the normal vector at the following point:

In[39]:=
```
Clear[surfacepoint]; {uu,vv} = {0.5,15 Pi/16};
surfacepoint[u_,v_] = {x[u,v],y[u,v],z[u,v]}; surfacepoint[uu,vv]
```

Out[39]=
$$\{0.5 \, \text{Cos}[\frac{15 \, \text{Pi}}{16}], \, 0.5 \, \text{Sin}[\frac{15 \, \text{Pi}}{16}], \, -0.25 \, \text{Cos}[\frac{15 \, \text{Pi}}{16}] \, \text{Sin}[\frac{15 \, \text{Pi}}{16}]\}$$

Here comes the normal vector at this point:

In[40]:=
```
Clear[ucurve,vcurve]; Clear[i,j,k]
ucurve[u_] = surfacepoint[u,vv];
vcurve[v_] = surfacepoint[uu,v];
utanvector = D[ucurve[u],u]/.u->uu;
vtanvector = D[vcurve[v],v]/.v->vv;
normal = Det[{{i,j,k},utanvector,vtanvector}]/.
{i->{1,0,0},j->{0,1,0},k->{0,0,1}};
scalefactor = 2; onenormal = Show[surfaceplot,
Arrow[scalefactor normal,
Tail->surfacepoint[uu,vv],Red]];
```

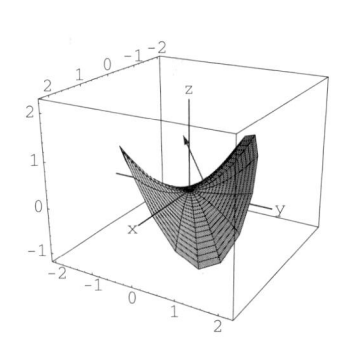

B.3.c.ii) How do you handle the situation in which the surface is not given parametrically?

Answer: Rather easily. Change it to a parametrically represented surface:

Here is the surface
$$z = f[x,y] \qquad \text{for } f[x,y] = 2 \, e^{-(x-y)^2}.$$

Lesson 3.02 Perpendicularity

```
In[41]:=
    Clear[x,y,f];
    f[x_,y_] = 2 E^(-(x - y)^2);
    surfaceplot =
    Plot3D[f[x,y],{x,0,2},{y,-1,3},
    Axes->Automatic,PlotRange->All,
    BoxRatios->Automatic,
    AxesLabel->{"x","y","z"},
    ViewPoint->CMView];
```

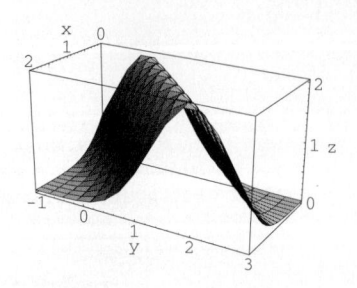

Here's how you plant the normal vector at the point $\{1.5, 1, f[1.5, 0]\}$ that sits on the surface directly above the point $\{1.5, 1, 0\}$ in the xy-plane.

```
In[42]:=
    {xx,yy} = {1.5,1};
    Clear[surfacepoint]
    surfacepoint[x_,y_] = {x,y,f[x,y]};
    surfacepoint[xx,yy]

Out[42]=
    {1.5, 1, 1.5576}
```

Here comes the normal vector at this point:

```
In[43]:=
    Clear[xcurve,ycurve]; Clear[i,j,k]
    xcurve[x_] = surfacepoint[x,yy];
    ycurve[y_] = surfacepoint[xx,y];
    xtanvector = D[xcurve[x],x]/.x->xx;
    ytanvector = D[ycurve[y],y]/.y->yy;
    normal = Det[{{i,j,k},xtanvector,ytanvector}]/.
    {i->{1,0,0},j->{0,1,0},k->{0,0,1}};
    scalefactor = 1; Show[surfaceplot,
    Arrow[scalefactor normal,
    Tail->surfacepoint[xx,yy],Red],
    BoxRatios->Automatic,Boxed->False];
```

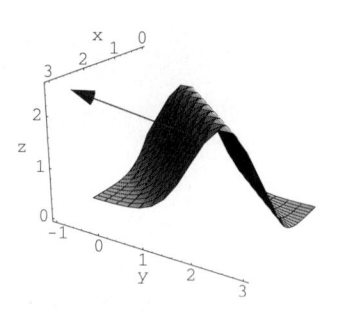

Rerun with different points on the surface until you're comfortable.

Tutorials

■ T.1) True scale plots

T.1.a) Look at this:

```
In[1]:=
    X = {1,2}; Y = {-8,4}; X.Y

Out[1]=
    0
```

This tells you that the two vectors are perpendicular. Now look at this:

In[2]:=
```
Show[Arrow[X,Tail->{0,0},Blue],
Arrow[Y,Tail->{0,0},Blue],Axes->True,
AxesLabel->{"x","y"},AspectRatio->1];
```

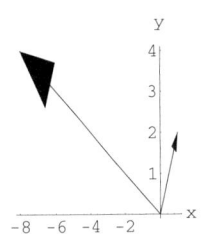

> Although you knew in advance that these vectors are perpendicular, they didn't plot out perpendicularly. What gives?

Answer: The two vectors weren't plotted in true scale. To guarantee a true scale plot in 2D graphics, use the plotting option AspectRatio->Automatic. This way you get the same scale on both axes. Try it out:

In[3]:=
```
Show[Arrow[X,Tail->{0,0},Blue],
Arrow[Y,Tail->{0,0},Blue],Axes->True,
AxesLabel->{"x","y"},AspectRatio->Automatic];
```

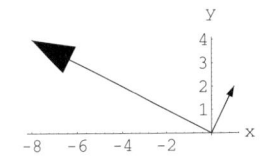

Much better.

In 2D plots, if you don't use AspectRatio->Automatic, then actual perpendicularity can be obscured. In 3D plots, if you don't use BoxRatios->Automatic, then actual perpendicularity can be obscured.

Moral: When you are studying anything related to perpendicularity, you need to plot in true scale.

■ T.2) Flatness and plotting

This problem appears only in the electronic version.

■ T.3) Unit vectors and perpendicularity: Plotting curves on planes

A unit vector is just a vector whose length is 1. Any vector X can be converted to a unit vector unless all its slots are zero. To wit:

In[4]:=
```
X = {-4,2}; unitX = X/Sqrt[X.X]
```

Lesson 3.02 Perpendicularity

Out[4]=

$\{\frac{-2}{\text{Sqrt}[5]}, \frac{1}{\text{Sqrt}[5]}\}$

Check:

In[5]:=
```
unitX.unitX
```
Out[5]=
1

and

In[6]:=
```
X = {-4,2,8}; unitX = X/Sqrt[X.X]
```
Out[6]=

$\{\frac{-2}{\text{Sqrt}[21]}, \frac{1}{\text{Sqrt}[21]}, \frac{4}{\text{Sqrt}[21]}\}$

Check:

In[7]:=
```
unitX.unitX
```
Out[7]=
1

Unit vectors are the handiest thing since zip-lock bags. Why? Wait and see.

T.3.a) Here is a true scale plot of $y = \sin[x]$ on the xy-plane:

In[8]:=
```
Clear[x]
ParametricPlot[{x,Sin[x]},{x,-Pi,3 Pi},
AspectRatio->Automatic,AxesLabel->{"x","y"}];
```

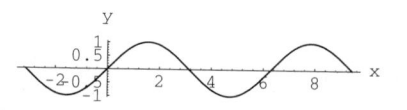

Plot a duplicate copy of this curve in true scale on the plane through $\{1, 1, -2\}$ determined by the direction vectors $\{0.3, 1, -0.2\}$ and $\{0, 0.4, -1\}$.

Answer:

In[9]:=
```
point = {1,1,2}; vector1 = {0.3,1,-0.2};
vector2 = {0, 0.4, -1};
Clear[i,j,k]
normal = Det[{{i,j,k},vector1,vector2}]/.
{i->{1,0,0},j->{0,1,0},k->{0,0,1}};
```
Out[9]=
{-0.92, 0.3, 0.12}

When you think about it for a minute, you see that the same plane is also determined by the given point and the vectors unitvector1 and unitnewvector2 calculated as follows:

In[10]:=
```
unitvector1 = vector1/Sqrt[vector1.vector1]
```
Out[10]=
{0.282216, 0.940721, -0.188144}

In[11]:=
```
Clear[i,j,k]
newvector2 = Det[{{i,j,k},vector1,normal}]/.{i->{1,0,0},j->{0,1,0},k->{0,0,1}};
unitnewvector2 = newvector2/Sqrt[newvector2.newvector2]
```
Out[11]=
{0.173656, 0.142783, 0.974401}

The great thing about unitvector1 and unitnewvector2 is that they are unit vectors and they are perpendicular:

In[12]:=
```
Chop[unitvector1.unitnewvector2]
```
Out[12]=
0

As perpendicular unit vectors, they define axes for plotting on the plane:

In[13]:=
```
Clear[x,y,z,s,t]
{x[s_,t_],y[s_,t_],z[s_,t_]} = point + s unitvector1 + t unitnewvector2;
{slow,shigh} = {-Pi,3 Pi}; {tlow,thigh} = {-3,3};
plane = ParametricPlot3D[Evaluate[{x[s,t],y[s,t],z[s,t]}],
  {s,slow,shigh},{t,tlow,thigh},PlotPoints->{2,2},DisplayFunction->Identity];
```

In[14]:=
```
planeplot = Show[plane,
Arrow[unitvector1,Tail->point],
Arrow[unitnewvector2,Tail->point],
Axes->Automatic,PlotRange->All,
ViewPoint->CMView,BoxRatios->Automatic,
DisplayFunction->$DisplayFunction];
```

Here is a true scale duplicate copy of the curve $y = \sin[x]$ for $-\pi \leq x \leq 3\pi$ plotted on this plane.

In[15]:=
```
sineplot = ParametricPlot3D[
  point+ s unitvector1 + Sin[s] unitnewvector2,
  {s,-Pi,3 Pi},DisplayFunction->Identity];
Show[planeplot,sineplot,
DisplayFunction->$DisplayFunction];
```

Keen.

Lesson 3.02 Perpendicularity

T.3.b) Here is a true scale plot of the circle $x^2 + y^2 = 4$ on the xy-plane:

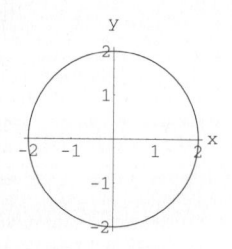

In[16]:=
```
Clear[t]
ParametricPlot[2 {Cos[t],Sin[t]},{t,0,2 Pi},
AxesLabel->{"x","y"}];
```

> Plot a duplicate copy of this circle on the plane through $\{1, 2, -1\}$ determined by the direction vectors $\{1, 0, 2\}$ and $\{-2, 1, 0\}$. Center the circle at $\{1, 2, -1\}$.

Answer:

In[17]:=
```
point = {1,2,-1}; vector1 = {1,0,2}; vector2 = {-2,1,0};
normal = Cross[vector1,vector2]
```

Out[17]=
$\{-2, -4, 1\}$

Notice the shorthand command for calculating the cross product. Feel free to use it.

When you think about it for a minute, you see that the same plane is also determined by the given point and the vectors unitvector1 and unitnewvector2 calculated as follows:

In[18]:=
```
unitvector1 = vector1/Sqrt[vector1.vector1]
```

Out[18]=
$\{\frac{1}{\sqrt{5}}, 0, \frac{2}{\sqrt{5}}\}$

In[19]:=
```
newvector2 = Cross[vector1,normal];
unitnewvector2 = newvector2/Sqrt[newvector2.newvector2]
```

Out[19]=
$\{\frac{8}{\sqrt{105}}, -\sqrt{\frac{5}{21}}, \frac{-4}{\sqrt{105}}\}$

The great thing about unitvector1 and unitnewvector2 is that they are unit vectors and they are perpendicular:

In[20]:=
```
Chop[unitvector1.unitnewvector2]
```

Out[20]=
0

As perpendicular unit vectors, they define axes for plotting on the plane, and allow you to plot the circle where you want it.

In[21]:=
```
Clear[circleplotter,t]; Clear[planeplotter,r,s];
circleplotter[t_] = point + 2Cos[t] unitvector1 + 2 Sin[t] unitnewvector2;
{tlow,thigh} = {0,2 Pi}; circle = ParametricPlot3D[Evaluate[circleplotter[t]],
  {t,tlow,thigh},DisplayFunction->Identity];
planeplotter[r_,s_] = point + r unitvector1 + s unitnewvector2;
{rlow,rhigh} = {-3,3}; {slow,shigh} = {-3,3};
plane = ParametricPlot3D[Evaluate[planeplotter[r,s]],
  {r,rlow,rhigh},{s,slow,shigh},PlotPoints->{2,2},DisplayFunction->Identity];
```

In[22]:=
```
Show[plane,circle,Arrow[unitvector1,Tail->point,Red],
Arrow[unitnewvector2,Tail->point,Red],
Axes->Automatic,PlotRange->All,
ViewPoint->CMView,BoxRatios->Automatic,
DisplayFunction->$DisplayFunction];
```

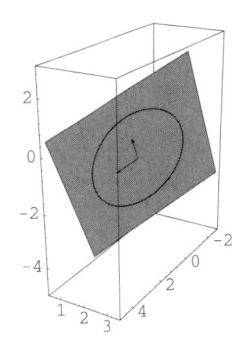

You can make the vectors touch the circle:

In[23]:=
```
Show[plane,circle,
Arrow[2 unitvector1,Tail->point,Red],
Arrow[2 unitnewvector2,Tail->point,Red],
Axes->Automatic,PlotRange->All,
ViewPoint->CMView,BoxRatios->Automatic,
DisplayFunction->$DisplayFunction];
```

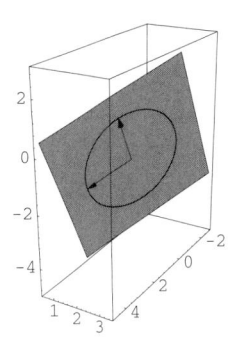

The reason this worked: The vectors unitvector1 and unitnewvector2 are perpendicular unit vectors. If you had used the original vectors vector1 and vector2, you would not have gotten a circle.

■ T.4) Unit vectors and perpendicularity: Main unit normals, binormals, tubes, horns, and corrugations

Unit vectors are just vectors whose lengths are 1. Unit vectors are useful because they convey the idea of direction only. You can take any vector and make it into a unit vector by dividing it by its length:

In[24]:=
```
X = {2,1};
unitX = X/Sqrt[X.X]
```

Lesson 3.02 Perpendicularity

Out[24]=
$\{\frac{2}{\text{Sqrt}[5]}, \frac{1}{\text{Sqrt}[5]}\}$

In[25]:=
```
Show[Arrow[X,Tail->{0,0},Blue],
 Arrow[unitX,Tail->{0,0},Red],Axes->Automatic];
```

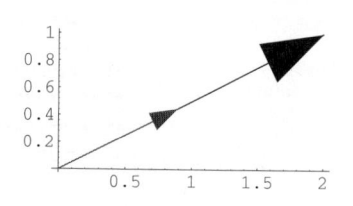

The unit vector points in the same direction as X, but its length is 1:

In[26]:=
```
Sqrt[unitX.unitX]
```

Out[26]=
1

T.4.a.i) Here is a curve in three dimensions with some of its unit tangents plotted in true scale:

In[27]:=
```
Clear[P,x,y,z,t]; Clear[unittan]
x[t_] = 3 Cos[t]; y[t_] = 3 Sin[t];
z[t_] = 4 Cos[t]; P[t_] = {x[t],y[t],z[t]};
curve = ParametricPlot3D[Evaluate[P[t]],
 {t,0,2 Pi},PlotPoints->30,DisplayFunction->Identity];
unittan[t_] = D[P[t],t]/Sqrt[D[P[t],t].D[P[t],t]];
unittanvectors = Table[Arrow[unittan[t],
 Tail->P[t],Blue],{t,0, 2 Pi, Pi/4}];
curveandtans = Show[curve,unittanvectors,
 ViewPoint->CMView,PlotRange->All,
 BoxRatios->Automatic,AxesLabel->{"x","y","z"},
 DisplayFunction->$DisplayFunction];
```

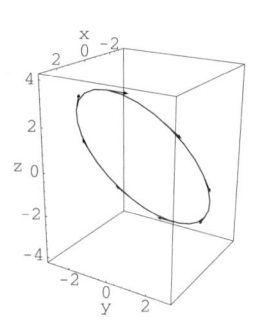

Add to the plot the vectors $D[\text{unittan}[t], t]$ with tails at the same places as the tails of the unit tangents. Describe what you see.

Answer:

In[28]:=
```
Clear[newvector]
newvector[t_] = D[unittan[t],t];
newvectors = Table[Arrow[newvector[t],Tail->P[t],Red],
{t,0, 2 Pi, Pi/4}]; Show[curveandtans,newvectors];
```

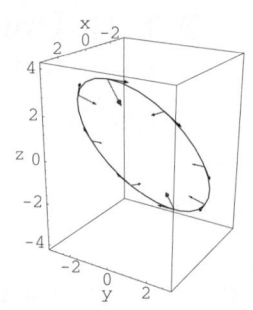

They look perpendicular to the curve! Check this to 20 accurate digits:

In[29]:=
```
Table[N[unittan[t].newvector[t],20],{t,0, 2 Pi, Pi/4}]
```

Out[29]=
{0, 0., 0, 0., 0, 0., 0, 0., 0}

They **are** perpendicular to the curve.

T.4.a.ii) | Is this outcome peculiar to this curve or does it happen for all other curves too?

Answer: Try it for a general curve in three dimensions:

In[30]:=
```
Clear[x,y,z,t,P,unittan,newvector]
P[t_] = {x[t],y[t],z[t]}; unittan[t_] = D[P[t],t]/Sqrt[D[P[t],t].D[P[t],t]];
newvector[t_] = D[unittan[t],t]; Simplify[unittan[t].newvector[t]]
```

Out[30]=
0

No matter what curve you have, the vector $D[\text{unittan}[t], t]$ is perpendicular to the curve at $P[t]$!

Most folks make this into a unit vector and call the result the main unit normal vector for the curve.

T.4.a.iii) | Does this work in two dimensions as well?

Answer: Try it and see:

In[31]:=
```
Clear[P,t]; Clear[mainnormal]; Clear[unittan];
P[t_] = {t/2 - 4 Sin[t],3 + 3 Cos[t]};
curve = ParametricPlot[Evaluate[P[t]],{t,0,10},
  PlotStyle->Thickness[0.01],DisplayFunction->Identity];
unittan[t_] = P'[t]/Sqrt[P'[t].P'[t]];
unittanvectors = Table[Arrow[unittan[t],Tail->P[t],Blue],{t,1,9}];
mainnormal[t_] = Simplify[D[unittan[t],t]];
mainnormalvectors = Table[Arrow[mainnormal[t],Tail->P[t],Red],{t,1,9}];
```

In[32]:=
```
Show[curve,unittanvectors,
  mainnormalvectors,
  AspectRatio->Automatic,
  DisplayFunction->$DisplayFunction];
```

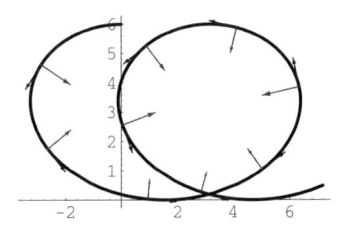

Works like a charm in two dimensions. When you're hot, you're hot.

T.4.b) Given a curve $P[t] = \{x[t], y[t], z[t]\}$:

→ The unit tangent

$$\text{unittan}[t] = \frac{D[P[t], t]}{\sqrt{P'[t] \bullet P'[t]}}$$

is tangent to the curve at $P[t]$.

→ The main normal

$$\text{mainnorm}[t] = D[\text{unittan}[t], t]$$

is perpendicular to the curve at $P[t]$.

→ The main unit normal

$$\text{mainunitnorm}[t] = \frac{D[\text{unittan}[t], t]}{\sqrt{D[\text{unittan}[t], t] \bullet D[\text{unittan}[t], t]}}$$

is also perpendicular to the curve at $P[t]$.

You can construct another normal vector by putting

$$\text{binormal}[t] = \text{unittan}[t] \times \text{mainunitnorm}[t].$$

This second normal vector always turns out to be a unit vector, for reasons you will learn when you give it a try.

Take a look:

The N[]'s are included in the specifications for the main unit normal and the binormal to make the plotting instructions below run as fast as possible.

In[33]:=
```
Clear[P,x,y,z,t,unittan,mainunitnormal,binormal]
x[t_] = t^2; y[t_] = 1 - 2 t; z[t_] = t; P[t_] = {x[t],y[t],z[t]};
curve = ParametricPlot3D[Evaluate[P[t]],{t,0,2},DisplayFunction->Identity];
unittan[t_] = P'[t]/Sqrt[P'[t].P'[t]];
unittanvectors = Table[Arrow[unittan[t],Tail->P[t],Blue],{t,0,2,0.5}];
mainunitnormal[t_] = N[D[unittan[t],t]/Sqrt[Expand[D[unittan[t],t].D[unittan[t],t]]]];
mainnormalvectors = Table[Arrow[mainunitnormal[t],Tail->P[t],Red],{t,0,2,0.5}];
binormal[t_] = N[Cross[unittan[t],mainunitnormal[t]]];
binormalvectors = Table[Arrow[binormal[t],Tail->P[t],Red],{t,0,2,0.5}];
```

In[34]:=
```
everything = Show[curve,unittanvectors,
mainnormalvectors,binormalvectors,
ViewPoint->CMView,PlotRange->All,
BoxRatios->Automatic,AxesLabel->{"x","y","z"},
DisplayFunction->$DisplayFunction];
```

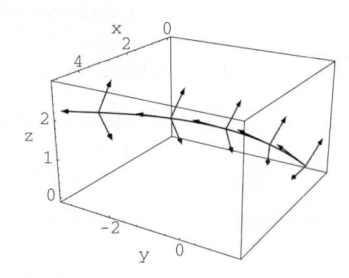

Neato. Some folks call these moving frames.

What's this stuff good for?

Answer: You can put in a couple of circles that

→ are centered on the curve, and

→ lie in planes that cut the curve perpendicularly:

In[35]:=
```
radius = 1;
circle1 = ParametricPlot3D[N[P[1.0] +
  radius Cos[s] mainunitnormal[1.0] +
  radius Sin[s] binormal[1.0]],
  {s,0,2 Pi},DisplayFunction->Identity];
circle2 = ParametricPlot3D[N[P[1.5] +
  radius Cos[s] mainunitnormal[1.5] +
  radius Sin[s] binormal[1.5]],
  {s,0,2 Pi},DisplayFunction->Identity];
Show[everything,circle1,circle2];
```

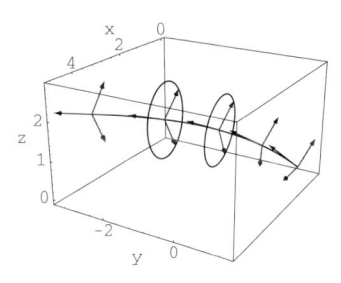

But why settle for just a couple of circles when you can go for the whole tube composed of all such circles?

In[36]:=
```
ParametricPlot3D[Evaluate[P[t] +
  radius Cos[s] mainunitnormal[t] +
  radius Sin[s] binormal[t]],
  {t,0,2},{s,0,2 Pi},
  ViewPoint->CMView,
  BoxRatios->Automatic,
  AxesLabel->{"x","y","z"}];
```

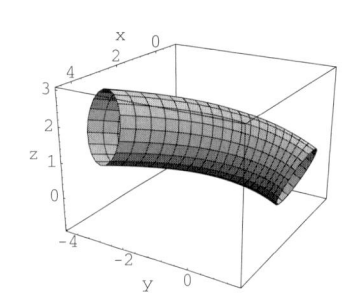

Or a horn:

In[37]:=
```
Clear[radius]
radius[t_] = (t/2)^2 + 0.1;
ParametricPlot3D[Evaluate[P[t] +
  radius[t] Cos[s] mainunitnormal[t] +
  radius[t] Sin[s] binormal[t]],{t,0,2},{s,0,2 Pi},
  ViewPoint->CMView,BoxRatios->Automatic,
  AxesLabel->{"x","y","z"}];
```

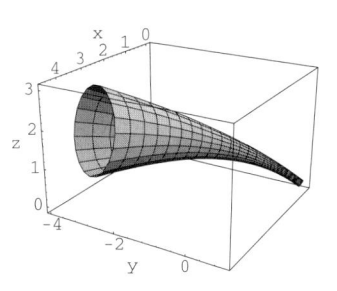

Or a corrugation:

In[38]:=
```
ParametricPlot3D[P[t] + s mainunitnormal[t] +
  Cos[3 s] binormal[t],{t,0,2},{s,-Pi,Pi},
  PlotPoints->{Automatic,25},ViewPoint->CMView,
  BoxRatios->Automatic,AxesLabel->{"x","y","z"}];
```

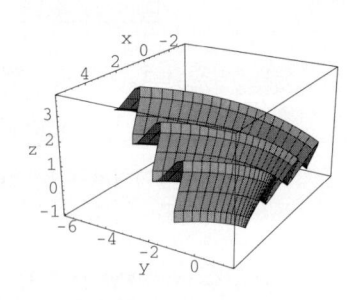

Every slice of this surface cut by any plane perpendicular to the original curve is the same cosine wave.

Give It a Try

Experience with the starred (★) problems will be useful for understanding developments later in the course.

■ G.1) Plane fundamentals★

G.1.a) Here are a point and a vector with the tail of the vector stuck at the given point:

In[1]:=
```
point = {0,-2,1}; vector = {2,3,4};
threedims = ThreeAxes[2,0.2];
pointandvector = Show[Arrow[vector,Tail->point,Red],
  threedims,Graphics3D[{PointSize[0.04],Point[point]}],
  ViewPoint->CMView,Axes->Automatic,
  AxesLabel->{"x","y","z"},PlotRange->All];
```

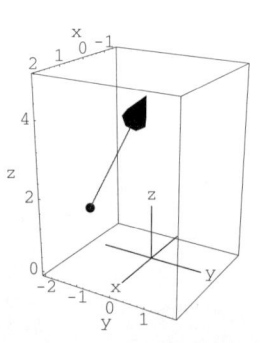

> Throw in a plot of a plane perpendicular to the given vector and passing through the given point into the plot above.
>
> Write down the xyz equations for this plane.

G.1.b.i) Here are three points:

In[2]:=
```
point1 = {-2, 1, 4}; point2 = {3, 1, 8};
point3 = {-9,- 5, 0};
pointandvector = Show[Graphics3D[
{Red,PointSize[0.04],Point[point1]}],
Graphics3D[{Red,PointSize[0.04],Point[point2]}],
Graphics3D[{Red,PointSize[0.04],Point[point3]}],
threedims,Axes->Automatic,PlotRange->All,
ViewPoint->CMView,AxesLabel->{"x","y","z"}];
```

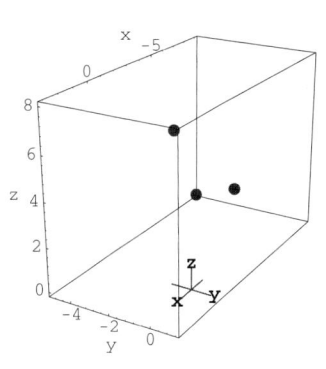

> Pick one of the points and add the two vectors running from this point to each of the other two points.
>
> Use what you see to plot the plane on which these three points all reside.
>
> Confirm your work by showing the three points on the plane plot.

G.1.b.ii) Here are two lines in 3D:

In[3]:=
```
Clear[line1,line2,s,t]
line1[t_] = {1,2,1} + t {-1,-1,1};
line2[s_] ={-1,0,1} + s {1,1,0};
line1plot = ParametricPlot3D[line1[t],{t,-3,3},PlotPoints->2,
DisplayFunction->Identity];
line2plot = ParametricPlot3D[line2[s],{s,-4,5},PlotPoints->2,
DisplayFunction->Identity];
```

In[4]:=
```
Show[line1plot,line2plot,
AxesLabel->{"x","y","z"},
ViewPoint->CMView,Axes->Automatic,
PlotRange->All,
DisplayFunction->$DisplayFunction];
```

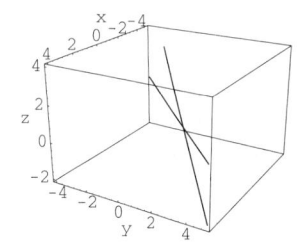

> At what point do these lines intersect?
>
> Plot the plane on which both of these lines reside.
>
> Confirm your work by showing the lines and the plane in the same plot.

G.1.c.i) What do you get when you intersect the planes with xyz-equations

$$x + y - z = 1 \quad \text{and} \quad -x + 2y + 3z = 6?$$

Give a parametric formula that describes the points on this intersection and confirm your work with a plot.

G.1.c.ii) How can you tell without plotting that the planes

$$x + y - z = 1 \quad \text{and} \quad x + 4y + 5z = 6$$

cut each other at right angles?

G.1.d) When you take the aggregate (collection, set) of all lines through the point $\{3, -4, 7\}$ that are also perpendicular to a given vector $\{a, b, c\}$, what do you get?

Give an equation that describes the points $\{x, y, z\}$ on this aggregate of lines.

■ G.2) Plotting on planes*

This problem appears only in the electronic version.

■ G.3) Serious 3D plots*

G.3.a) Here's a 3D curve:

```
In[5]:=
  Clear[P,t]
  P[t_] = {-t,t,t (8 - t)/2};
  arch = ParametricPlot3D[P[t],{t,0,8},
  Axes->Automatic,AxesLabel->{"x","y","z"},
  PlotRange->All,Boxed->False,
  ViewPoint->CMView,BoxRatios->Automatic];
```

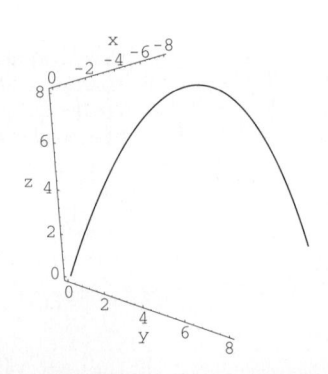

You are asked to plot the tube whose skin consists of all circles of radius 0.5 that

- → are centered on this curve and
- → lie in planes that cut this curve perpendicularly.

G.3.b) Here's the same 3D curve:

In[6]:=
```
Clear[P,t]
P[t_] = {-t,t,t (8 - t)/2};
arch = ParametricPlot3D[P[t],{t,0,8},
Axes->Automatic,AxesLabel->{"x","y","z"},
PlotRange->All,Boxed->False,
ViewPoint->CMView,BoxRatios->Automatic];
```

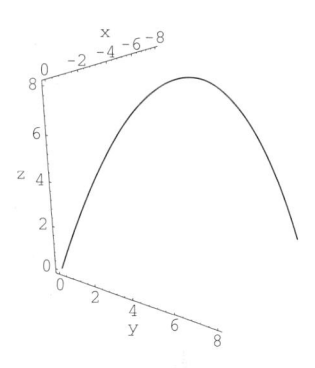

You are asked to plot a ribbon two units wide whose center curve coincides with the curve plotted above. Corrugate the ribbon if you like.

G.3.c) Here's a different 3D curve:

In[7]:=
```
Clear[curveplotter,t]
curveplotter[t_] = {-t,t,Sin[t]};
snake = ParametricPlot3D[
curveplotter[t],{t,0, 8},
Axes->Automatic,AxesLabel->{"x","y","z"},
PlotRange->All,Boxed->False,
ViewPoint->CMView,BoxRatios->Automatic];
```

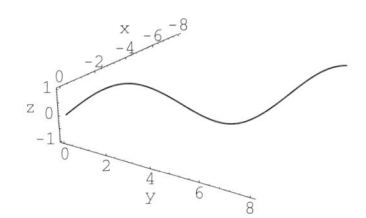

This time you're asked to add to this plot four additional curves with the property that if you slice the original curve by any plane perpendicular to the original curve, then:

- → The four additional curves will show up on this plane as dots at the corners of a square one unit on a side and
- → the original curve will show up as a dot in the center of the square.

G.3.d) Now it's time to take the shackles off.

Turn yourself loose and do something artistic and creative with a curve of your choice. Show off.

■ G.4) Experiments with linearizations*

This problem appears only in the electronic version.

■ G.5) Badger borings

Calculus&Mathematica thanks Professor Rod Smart of the University of Wisconsin for suggesting this problem. Professor Smart mentions that a company in Madison, Wisconsin actually called him to ask how to solve problems of this type.

G.5.a.i) You are the chief engineer at the Badger Steel Plate Company in Madison, Wisonsin. In comes an order for 750 square steel plates, each measuring 12 inches wide and 12 inches long. Go to the drawing board:

```
In[8]:=
    plate = Graphics[{SteelBlue,Thickness[0.01],
    Line[{{-6,-6},{-6,6},{6,6},{6,-6},{-6,-6}}]}];
    plateplot = Show[plate,Axes->True,AxesOrigin->{0,0},
    AspectRatio->Automatic,AxesLabel->{"x","y"}];
```

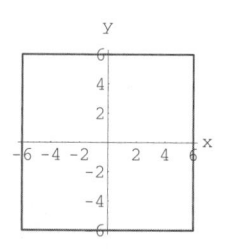

But here's the kicker: The plates are to have everything inside the ellipse

$$\left(\frac{x}{4}\right)^2 + \left(\frac{y}{2}\right)^2 = 1$$

cut out. Take a look:

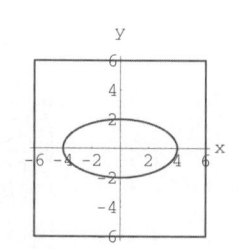

```
In[9]:=
    Clear[x,y,t]
    x[t_] = 4 Cos[t]; y[t_] = 2 Sin[t];
    hole = ParametricPlot[{x[t],y[t]},{t,0,2 Pi},
    PlotStyle->{{SteelBlue,Thickness[0.01]}},
    DisplayFunction->Identity];
    Show[plateplot,hole,
    DisplayFunction->$DisplayFunction];
```

You have a new robotic router that takes instructions from *Mathematica* and whose cutting center can be programmed to follow any curve you tell it to follow.

> If you are going to use a bit 1 inch in diameter, then what curve should you program in as the path of the center of the router to cut out the ellipse?
>
> After you have found the correct curve, add its plot to the plot above.

G.5.a.ii) Actually, the bit size of 1 inch in diameter used above was arbitrarily chosen by reaching into the drawer and pulling out a bit. You could always get by with a smaller bit. Why?

But you cannot use bits that are too large. Why?

Try to estimate the diameter of the largest bit that you could use to do the job.

■ G.6) Using the product rule to break acceleration vectors into normal and tangential components

G.6.a.i) At time t, an object is given to be at a point $P[t] = \{x[t], y[t]\}$. Its velocity at time t is just $\text{velocity}[t] = D[P[t], t]$. The unit tangent vector at $P[t]$ is
$$\text{unittan}[t] = \frac{\text{velocity}[t]}{\sqrt{\text{velocity}[t] \bullet \text{velocity}[t]}}.$$

Explain why
$$\text{velocity}[t] = \text{speed}[t]\, \text{unittan}[t]$$
where
$$\text{speed}[t] = \sqrt{\text{velocity}[t] \bullet \text{velocity}[t]}.$$

G.6.a.ii) The acceleration vector at $P[t]$ is given by
$$\text{accel}[t] = D[\text{velocity}[t], t].$$

Use the fact that
$$\text{velocity}[t] = \text{speed}[t]\, \text{unittan}[t]$$
and use the product rule for taking derivatives to say why
$$\text{accel}[t] = D[\text{speed}[t], t]\, \text{unittan}[t] + \text{speed}[t]\, \text{mainunitnormal}[t].$$

G.6.a.iii) When you write
$$\text{accel}[t] = D[\text{speed}[t], t]\, \text{unittan}[t] + \text{speed}[t]\, \text{mainunitnormal}[t],$$
then why is it fairly transparent that $D[\text{speed}[t], t]\, \text{unittan}[t]$ is the tangential component of the acceleration and
$$\text{speed}[t]\, \text{mainunitnormal}[t]$$
is the normal component of the acceleration?

G.6.a.iv) Explain why the following true scale plot came out the way it did and discuss the information the plot exhibits.

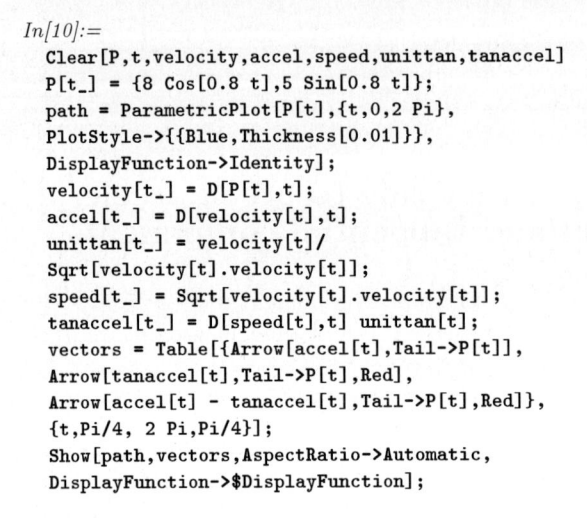

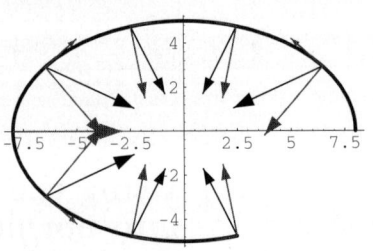

```
In[10]:=
  Clear[P,t,velocity,accel,speed,unittan,tanaccel]
  P[t_] = {8 Cos[0.8 t],5 Sin[0.8 t]};
  path = ParametricPlot[P[t],{t,0,2 Pi},
    PlotStyle->{{Blue,Thickness[0.01]}},
    DisplayFunction->Identity];
  velocity[t_] = D[P[t],t];
  accel[t_] = D[velocity[t],t];
  unittan[t_] = velocity[t]/
    Sqrt[velocity[t].velocity[t]];
  speed[t_] = Sqrt[velocity[t].velocity[t]];
  tanaccel[t_] = D[speed[t],t] unittan[t];
  vectors = Table[{Arrow[accel[t],Tail->P[t]],
    Arrow[tanaccel[t],Tail->P[t],Red],
    Arrow[accel[t] - tanaccel[t],Tail->P[t],Red]},
    {t,Pi/4, 2 Pi,Pi/4}];
  Show[path,vectors,AspectRatio->Automatic,
    DisplayFunction->$DisplayFunction];
```

■ G.7) Using the normal vector to bounce light beams off surfaces

G.7.a) Here's a piece of the surface
$$z = 4 - \left(\frac{x}{2}\right)^2 + y^2 :$$

```
In[11]:=
  Clear[f,x,y,z,r,t,surfaceplotter]
  f[x_,y_] = 4 - (x/2)^2 + (y/2)^2;
  x[r_,t_] = r Cos[t]; y[r_,t_] = r Sin[t];
  z[r_,t_] = f[x[r,t],y[r,t]];
  surfaceplotter[r_,t_] = {x[r,t],y[r,t],z[r,t]};
  surface = ParametricPlot3D[surfaceplotter[r,t],
    {r,0,2},{t,0, 2 Pi},ViewPoint->CMView,
    PlotRange->All,Boxed->False,
    AxesLabel->{"x","y","z"}];
```

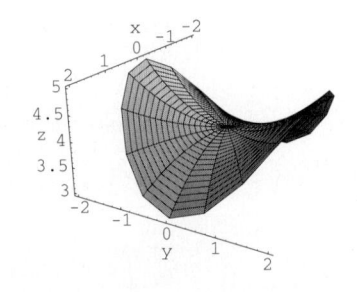

A light beam emanates from $\{-1, 1, 6\}$ and strikes the surface at:

```
In[12]:=
  hit = surfaceplotter[1,1.2]
```

Out[12]=
  {0.362358, 0.932039, 4.18435}

Here's a look:

```
In[13]:=
    lightsource = {-1,1,6};
    Show[surface,
    Arrow[hit - lightsource,Tail->lightsource,Red],
    PlotRange->All];
```

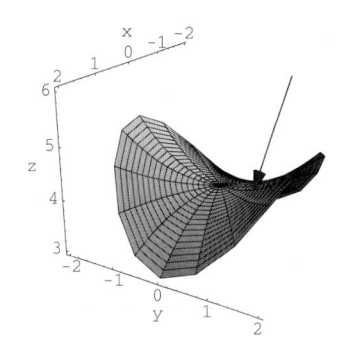

Your job is to plot the vector on which the reflected light moves.

■ G.8) Kissing circles and curvature

A curve in two dimensions is plotted by specifying two functions $x[t]$ and $y[t]$ and plotting $P[t] = \{x[t], y[t]\}$ for t running through a desired interval. A curve in three dimensions is plotted by specifying three functions $x[t], y[t]$ and $z[t]$ and plotting $P[t] = \{x[t], y[t], z[t]\}$ for t running through a desired interval. In both situations, you get the unit tangent vector at $P[t]$ through the formula:

$$\text{unittan}[t] = \frac{D[P[t], t]}{\sqrt{D[P[t], t] \bullet D[P[t], t]}}.$$

In both situations, the main normal at $P[t]$ is given by

$$\text{mainnormal}[t] = D[\text{unittan}[t], t].$$

The vector mainnormal[t] is a measurement of how fast the unit tangent is turning as t progresses. You can get an even better measurement of how the curve turns by calculating the instantaneous rate of change of the unit tangent as a function of arclength s measured on the curve from a specified reference point usually at one end of the curve. This measurement is intrinsic to the shape of the curve and does not depend on the specific parameterization of the curve you are using. The chain rule tells you that at $P[t]$, this measurement is given by

$$\begin{aligned}\text{turn}[t] &= D[\text{unittan}[t], s] \\ &= D[\text{unittan}[t], t]D[t, s] \\ &= \frac{D[\text{unittan}[t], t]}{\sqrt{D[P[t], t] \bullet D[P[t], t]}}\end{aligned}$$

because

$$D[t, s] = \frac{1}{\sqrt{x'[t]^2 + y'[t]^2}} = \frac{1}{\sqrt{D[P[t], t] \bullet D[P[t], t]}}.$$

80 **Lesson 3.02 Perpendicularity**

Now make note of the fact that

→ turn[t] is a vector that points in the same direction as mainnormal[t].

Why? Because it's a positive multiple of

$$D[\text{unittan}[t], t] = \text{mainunitnormal}[t].$$

This tells you that turn[t] is perpendicular to the curve at P[t]. The length of turn[t] measures the roundness of the curve at P[t].

G.8.a.i) Here, in true scale, is the curve $P[t] = \{3\cos[t], 2\sin[t]\}$ for $0 \leq t \leq \pi$:

In[14]:=
```
Clear[P,t]; P[t_] = {3 Cos[t],2 Sin[t]};
curve = ParametricPlot[P[t],{t,0,Pi},
PlotStyle->{{Blue,Thickness[0.01]}},
AspectRatio->Automatic];
```

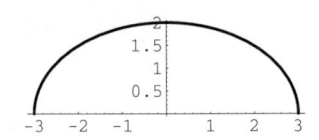

Here's the same curve together with a selection of vectors turn[t] with tails at P[t]:

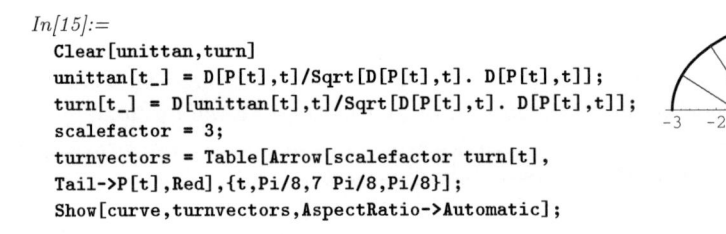

In[15]:=
```
Clear[unittan,turn]
unittan[t_] = D[P[t],t]/Sqrt[D[P[t],t] . D[P[t],t]];
turn[t_] = D[unittan[t],t]/Sqrt[D[P[t],t] . D[P[t],t]];
scalefactor = 3;
turnvectors = Table[Arrow[scalefactor turn[t],
Tail->P[t],Red],{t,Pi/8,7 Pi/8,Pi/8}];
Show[curve,turnvectors,AspectRatio->Automatic];
```

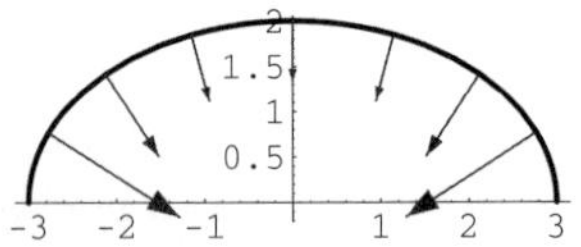

The turn vectors are shown three times longer than their actual lengths.

> Describe how the lengths of the turn vectors are related to flatness or roundness of the curve.
>
> Then plot the length $\|\text{turn}[t]\| = \sqrt{\text{turn}[t] \bullet \text{turn}[t]}$ for $0 \leq t \leq \pi$.
>
> Discuss the relations between the two plots.
>
> How does $\|\text{turn}[t]\|$ plot out over intervals where the curve is fairly flat?
>
> On what side of the curve does turn[t], with its tail at P[t], point?

G.8.a.ii) Here is the same curve shown in true scale with the circle of radius

$$\frac{1}{\|\text{turn}[a]\|} = \frac{1}{\sqrt{\text{turn}[a] \bullet \text{turn}[a]}}$$

centered at

$$P[a] + \frac{\text{turn}[a]}{\text{turn}[a] \bullet \text{turn}[a]} \quad \text{for } a = \frac{\pi}{4}:$$

In[16]:=
```
Clear[a,center,radius,circle,point]
center[a_] = P[a] + turn[a]/(turn[a].turn[a]);
radius[a_] = 1/Sqrt[turn[a].turn[a]];
circle[a_] =
Graphics[{Red,Circle[center[a],radius[a]]}];
point[a_] =
Graphics[{Red,PointSize[0.04],Point[P[a]]}];
a = Pi/4;
Show[curve,circle[a],point[a],
PlotRange->All,AspectRatio->Automatic];
```

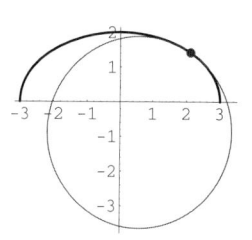

Here's what happens when you go with $a = 5\pi/6$:

In[17]:=
```
a = 5 Pi/6;
Show[curve,circle[a],point[a],
PlotRange->All,AspectRatio->Automatic];
```

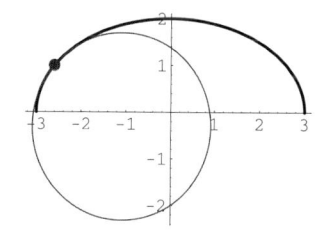

> Experiment with what happens when you go with other choices of a with $0 \le a \le \pi$, paying special attention to what you get when you go with $a = 0$, $\pi/2$, and π.
>
> Make a movie if you like.
>
> After you are done experimenting, say what you think is happening and try to explain why it is happening.

G.8.a.iii) Fancy folks call the circles you were studying in the last part by the name "osculating circle" at $P[a]$. You might prefer to call them "kissing circles" or "smooching circles" at $P[a]$. Most everyone calls

$$\|\text{turn}[a]\| = \sqrt{\text{turn}[a] \bullet \text{turn}[a]} = \frac{1}{\text{radius}}$$

(where radius stands for the radius of the kissing circle at $P[a]$) by the name "curvature" at $P[a]$.

> Explain what curvature tries to measure.
>
> Does a high curvature number at $P[a]$ mean that the curve is very flat or very rounded at $P[a]$?
>
> What does a low curvature number mean?

Lesson 3.02 Perpendicularity

G.8.a.iv) Here's a true scale parametric plot of part of the curve $y = 12/x$:

```
In[18]:=
  Clear[P,t]; P[t_] = {t,12 /t};
  curve = ParametricPlot[P[t],{t,1,10},
  PlotStyle->{{Blue,Thickness[0.01]}},
  AspectRatio->Automatic];
```

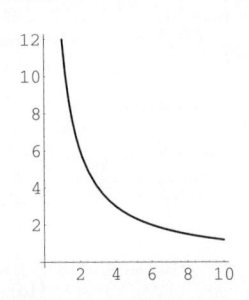

> Estimate the point on this curve at which the curvature is biggest and plot the kissing circle at that point.
>
> Then, on a separate plot, show the curve, the kissing circle at the point of biggest curvature, and a few other kissing circles. A good eye-ball estimate is OK.

G.8.b.i) Here's a 3D curve:

```
In[19]:=
  Clear[P,t]; P[t_] = {Cos[t],1.5 Sin[t],t/2};
  curve = ParametricPlot3D[P[t],{t,0,Pi},
  BoxRatios->Automatic,ViewPoint->CMView];
```

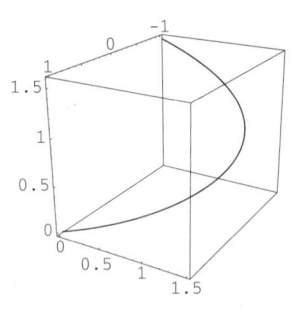

Here's the same 3D curve together with a selection of vectors unittan[t] and turn[t] with tails at $P[t]$:

```
In[20]:=
  Clear[unittan,turn]
  unittan[t_] = D[P[t],t]/
  Sqrt[ D[P[t],t]. D[P[t],t]];
  turn[t_] = D[unittan[t],t]/
  Sqrt[ D[P[t],t]. D[P[t],t]];
  unittanvectors = Table[Arrow[unittan[t],
  Tail->P[t],Blue],{t,Pi/8,7 Pi/8,Pi/8}];
  turnvectors = Table[Arrow[turn[t],Tail->P[t],Red],
  {t,Pi/8,7 Pi/8,Pi/8}];
  Show[curve,unittanvectors,turnvectors,
  ViewPoint->CMView,BoxRatios->Automatic];
```

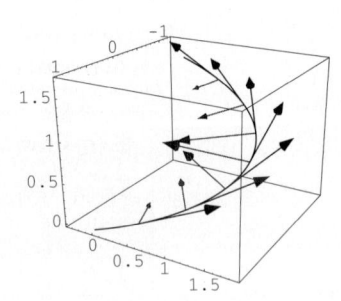

> Describe how you believe the lengths and directions of the turn vectors are related to the shape of the curve.

G.8.b.ii) You can do kissing circles in three dimensions:

Given a curve $P[t]$ and a value $t = a$, you center the circle at

$$\text{center} = P[a] + \frac{\text{turn}[a]}{\text{turn}[a] \bullet \text{turn}[a]}$$

and put

$$\text{radius} = \frac{1}{||\text{turn}[a]||} = \frac{1}{\sqrt{\text{turn}[a] \bullet \text{turn}[a]}}$$

and now you gotta choose two perpendicular unit vectors to determine the plane in which the circle resides. You want the circle to be tangent to the plane; so one good unit vector to go with is

$$\text{vector1}[a] = \text{unittan}[a].$$

The other vector you want is:

$$\text{vector2}[a] = \frac{\text{turn}[a]}{\sqrt{\text{turn}[a] \bullet \text{turn}[a]}}.$$

Try it out on the same curve as plotted above:

In[21]:=
```
Clear[P,t]; Clear[unittan,turn]
P[t_] = {Cos[t],1.5 Sin[t],t/2};
curve = ParametricPlot3D[P[t],{t,0,Pi},
  DisplayFunction->Identity];
unittan[t_] = D[P[t],t]/Sqrt[D[P[t],t]. D[P[t],t]];
turn[t_] = D[unittan[t],t]/Sqrt[D[P[t],t]. D[P[t],t]];
Clear[t,a,center,radius,point,vector1,vector2,kisser]
point[a_] = Graphics3D[{Red,PointSize[0.03],Point[P[a]]}];
center[a_] = P[a] + turn[a]/(turn[a].turn[a]);
radius[a_] = 1/Sqrt[turn[a].turn[a]];
vector1[a_] = unittan[a];
vector2[a_] = turn[a]/Sqrt[turn[a].turn[a]];
kisser[a_] := ParametricPlot3D[center[a] +
  radius[a](Cos[t] vector1[a] + Sin[t] vector2[a]),
  {t,0,2 Pi},DisplayFunction->Identity]; a = Pi/4;
Show[curve,point[a],kisser[a],
  ViewPoint->CMView,BoxRatios->Automatic,
  DisplayFunction->$DisplayFunction];
```

Another:

In[22]:=
```
a = 2 Pi/3;
Show[curve,point[a],kisser[a],
ViewPoint->CMView,BoxRatios->Automatic,
DisplayFunction->$DisplayFunction];
```

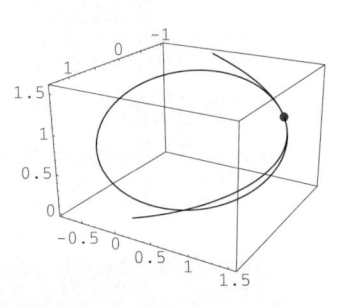

Estimate the point on this curve at which the curvature is biggest and plot the kissing circle at that point.

■ G.9) Measurements with the cross product★

This problem appears only in the electronic version.

■ G.10) Thumbs up or thumbs down?

G.10.a) This is a chance for you to learn how the cross product $X \times Y$ is oriented with respect to X and Y. By this point, you know that $X \times Y$ is perpendicular to both X and Y. But you can say a little more after you do some experimentation. Here are several samples of plots of 3D vectors X, Y and the cross product $X \times Y$:

In[23]:=
```
X = {1,0,0}; Y = {0,1,0}; XcrossY = Cross[X,Y];
Show[Arrow[X,Tail->{0,0,0},Blue],
Graphics3D[Text["X",X/2+{0,0,0.1}]],
Arrow[Y,Tail->{0,0,0},Blue],
Graphics3D[Text["Y",Y/2+{0,0,0.1}]],
Arrow[XcrossY,Tail->{0,0,0},Red],
Graphics3D[Text["XcrossY",(XcrossY)/2]],
PlotRange->All,BoxRatios->Automatic,
ViewPoint->CMView,Axes->None,Boxed->False];
```

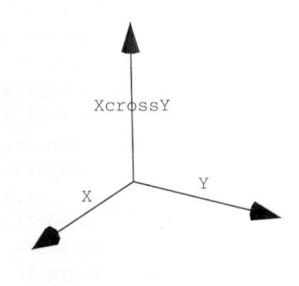

If you grab onto the shaft of $X \times Y$ with your right hand to use your fingers to push X onto Y through the smaller of the two possible angles, does your thumb point with $X \times Y$ or against $X \times Y$?

In[24]:=
```
X = {1,0.4,0}; Y = {0,1,1.2}; XcrossY = Cross[X,Y];
Show[Arrow[X,Tail->{0,0,0},Blue],
Graphics3D[Text["X",X/2+{0,0,0.1}]],
Arrow[Y,Tail->{0,0,0},Blue],
Graphics3D[Text["Y",Y/2+{0,0,0.1}]],
Arrow[XcrossY,Tail->{0,0,0},Red],
Graphics3D[Text["XcrossY",(XcrossY)/2]],
PlotRange->All,BoxRatios->Automatic,
ViewPoint->CMView,Axes->None,Boxed->False];
```

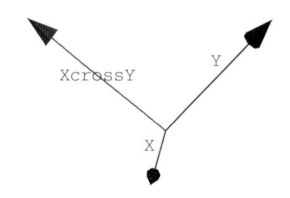

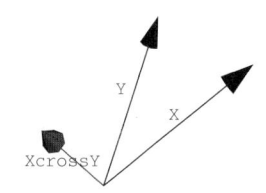

> If you grab onto the shaft of $X \times Y$ with your right hand to use your fingers to push X onto Y through the smaller of the two possible angles, does your thumb point with $X \times Y$ or against $X \times Y$?

In[25]:=
```
X = {-1,1,1}; Y = {-1,0,1.3}; XcrossY = Cross[X,Y];
Show[Arrow[X,Tail->{0,0,0},Blue],
Graphics3D[Text["X",X/2+{0,0,0.1}]],
Arrow[Y,Tail->{0,0,0},Blue],
Graphics3D[Text["Y",Y/2+{0,0,0.1}]],
Arrow[XcrossY,Tail->{0,0,0},Red],
Graphics3D[Text["XcrossY",(XcrossY)/2]],
PlotRange->All,BoxRatios->Automatic,
ViewPoint->CMView,Axes->None,Boxed->False];
```

> If you grab onto the shaft of $X \times Y$ with your right hand as to use your fingers to push X onto Y through the smaller of the two possible angles, does your thumb point with $X \times Y$ or against $X \times Y$?

In[26]:=
```
X = {1,1,0}; Y = {1,-0.5,0.3}; XcrossY = Cross[X,Y];
Show[Arrow[X,Tail->{0,0,0},Blue],
Graphics3D[Text["X",X/2]],
Arrow[Y,Tail->{0,0,0},Blue],
Graphics3D[Text["Y",Y/2]],
Arrow[XcrossY,Tail->{0,0,0},Red],
Graphics3D[Text["XcrossY",(XcrossY)/2]],
ViewPoint->CMView,Axes->None,Boxed->False];
```

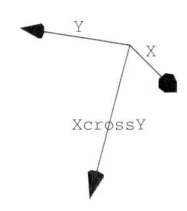

> If you grab onto the shaft of $X \times Y$ with your right hand as to use your fingers to push X onto Y through the smaller of the two possible angles, does your thumb point with $X \times Y$ or against $X \times Y$? Do more experiments like these and then describe how $X \times Y$ is oriented with respect to X and Y.

G.10.b) Given two three-dimensional vectors X and Y, then how do you express $Y \times X$ in terms of $X \times Y$?

Try some examples before you jump to a conclusion.

Use your ideas from part G.10.a) to say why this is natural.

LESSON 3.03

The Gradient

Basics

■ **B.1)** The gradient:
$$\text{gradf}[x,y] = \{D[f[x,y],x], D[f[x,y],y]\}$$
and the chain rule
$$D[f[x[t],y[t]],t] = \text{gradf}[x[t],y[t]] \cdot \{x'[t], y'[t]\}$$

Here is the chain rule for a function of one variable:

In[1]:=
```
Clear[f,x,t]; D[f[x[t]],t]
```
Out[1]=
```
f'[x[t]] x'[t]
```

This tells you that
$$\frac{df[x[t]]}{dt} = f'[x[t]]\, x'[t],$$

just as you have known for some time. Here is the chain rule for a function of two variables:

In[2]:=
```
Clear[f,x,t]; D[f[x[t],y[t]],t]
```
Out[2]=
```
        (0,1)                  (1,0)
y'[t] f     [x[t], y[t]] + x'[t] f     [x[t], y[t]]
```

87

Lesson 3.03 The Gradient

The notation
$$f^{(1,0)}[x,y] \quad \text{stands for} \quad D[f[x,y],x]$$
and
$$f^{(0,1)}[x,y] \quad \text{stands for} \quad D[f[x,y],y].$$

Some folks like to use the notation
$$\frac{\partial f[x,y]}{\partial x} = f^{(1,0)}[x,y] = D[f[x,y],x]$$

$$\frac{\partial f[x,y]}{\partial y} = f^{(0,1)}[x,y] = D[f[x,y],y].$$

So this output gives you three ways of writing the same thing:
$$\frac{df[x[t],y[t]]}{dt} = D[f[x[t],y[t]],x]\,x'[t] + D[f[x[t],y[t]],y]\,y'[t]$$

$$= \left(\frac{\partial f[x[t],y[t]]}{\partial x}\right) x'[t] + \left(\frac{\partial f[x[t],y[t]]}{\partial y}\right) y'[t]$$

$$= f^{(1,0)}[x[t],y[t]]\,x'[t] + f^{(0,1)}[x[t],y[t]]\,y'[t].$$

This is the chain rule for functions of two variables.

B.1.a.i) How do you calculate the gradient of a function $f[x,y]$?

Answer: The gradient of $f[x,y]$ is a 2D vector given by:

In[3]:=
```
Clear[f,gradf,x,y]
gradf[x_,y_] = {D[f[x,y],x], D[f[x,y],y]}
```
Out[3]=
$$\{f^{(1,0)}[x,y],\ f^{(0,1)}[x,y]\}$$

In other words, the gradient of a function $f[x,y]$ is given by
$$\text{gradf}[x,y] = \{D[f[x,y],x], D[f[x,y],y]\}$$

$$= \left\{\frac{\partial f[x,y]}{\partial x}, \frac{\partial f[x,y]}{\partial y}\right\}$$

$$= \{f^{(1,0)}[x,y], f^{(0,1)}[x,y]\}.$$

The gradient is important enough to carry the whole course from this point until its conclusion.

B.1.a.ii) How do you calculate the gradient of a function $f[x,y,z]$?

Answer: The gradient of $f[x, y, z]$ is a 3D vector given by:

In[4]:=
```
Clear[f,gradf,x,y,z]
gradf[x_,y_,z_] = {D[f[x,y,z],x], D[f[x,y,z],y],D[f[x,y,z],z]}
```

Out[4]=
$\{f^{(1,0,0)}[x, y, z], f^{(0,1,0)}[x, y, z], f^{(0,0,1)}[x, y, z]\}$

In other words, the gradient of a function $f[x, y, z]$ is given by

$$\text{gradf}[x, y, z] = \{D[f[x, y, z], x], D[f[x, y, z], y], D[f[x, y, z], z]\}$$
$$= \left\{ \frac{\partial f[x, y, z]}{\partial x}, \frac{\partial f[x, y, z]}{\partial y}, \frac{\partial f[x, y, z]}{\partial z} \right\}$$
$$= \{f^{(1,0,0)}[x, y, z], f^{(0,1,0)}[x, y, z], f^{(0,0,1)}[x, y, z]\}.$$

Lots of times you'll see the notation
$$\nabla f[x, y] = \text{gradf}[x, y]$$
or
$$\nabla f[x, y, z] = \text{gradf}[x, y, z].$$

B.1.b) Here is *Mathematica*'s calculation of the derivative of $f[x[t], y[t]]$ with respect to t:

In[5]:=
```
Clear[f,x,y,t]; D[f[x[t],y[t]],t]
```

Out[5]=
$y'[t] \, f^{(0,1)}[x[t], y[t]] + x'[t] \, f^{(1,0)}[x[t], y[t]]$

Most folks call this formula the chain rule. You can think of this as a dot product
$$\text{gradf}[x[t], y[t]] \bullet \{x'[t], y'[t]\} :$$

In[6]:=
```
Clear[gradf]; gradf[x_,y_] = {D[f[x,y],x],D[f[x,y],y]};
gradf[x[t],y[t]].{x'[t],y'[t]}
```

Out[6]=
$y'[t] \, f^{(0,1)}[x[t], y[t]] + x'[t] \, f^{(1,0)}[x[t], y[t]]$

Does the chain rule formula
$$D[f[x[t], y[t], z[t]], t] = \text{gradf}[x[t], y[t], z[t]] \bullet \{x'[t], y'[t], z'[t]\}$$
work for functions of three variables?

Answer: Try it and see. Here is $\text{gradf}[x, y, z]$:

In[7]:=
```
Clear[f,x,y,z,t]
gradf[x_,y_,z_] = {D[f[x,y,z],x],D[f[x,y,z],y],D[f[x,y,z],z]}
```

Lesson 3.03 The Gradient

Out[7]=
$\{f^{(1,0,0)}[x, y, z], f^{(0,1,0)}[x, y, z], f^{(0,0,1)}[x, y, z]\}$

Here is $D[f[x[t], y[t], z[t]], t]$:

In[8]:=
```
D[f[x[t],y[t],z[t]],t]
```

Out[8]=
$z'[t] f^{(0,0,1)}[x[t], y[t], z[t]] +$
$y'[t] f^{(0,1,0)}[x[t], y[t], z[t]] +$
$x'[t] f^{(1,0,0)}[x[t], y[t], z[t]]$

Here is $\text{gradf}[x[t], y[t], z[t]] \bullet \{x'[t], y'[t], z'[t]\}$:

In[9]:=
```
gradf[x[t],y[t],z[t]].{x'[t],y'[t],z'[t]}
```

Out[9]=
$z'[t] f^{(0,0,1)}[x[t], y[t], z[t]] +$
$y'[t] f^{(0,1,0)}[x[t], y[t], z[t]] +$
$x'[t] f^{(1,0,0)}[x[t], y[t], z[t]]$

Check whether $D[f[x[t], y[t], z[t]], t] = \text{gradf}[x[t], y[t], z[t]] \bullet \{x'[t], y'[t], z'[t]\}$:

In[10]:=
```
D[f[x[t],y[t],z[t]],t] == gradf[x[t],y[t],z[t]].{x'[t],y'[t],z'[t]}
```

Out[10]=
True

You bet your sweet ear that the formula

$$D[f[x[t], y[t], z[t]], t] = \text{gradf}[x[t], y[t], z[t]] \bullet \{x'[t], y'[t], z'[t]\}$$

works for functions of three variables. In fact, analogous versions of this formula work for functions of any number of variables.

B.1.c.i) Here is *Mathematica*'s calculation of $D[f[\sin[3t], e^{2t}], t]$ for a cleared function $f[x, y]$:

In[11]:=
```
Clear[f,x,y,t]
x[t_] = Sin[3 t]; y[t_] = E^(2 t);
D[f[x[t],y[t]],t]
```

Out[11]=
$2 E^{2t} f^{(0,1)}[\sin[3 t], E^{2t}] + 3 \cos[3 t] f^{(1,0)}[\sin[3 t], E^{2t}]$

> Use the fact that $D[f[x[t], y[t]], t] = \text{gradf}[x[t], y[t]] \bullet \{x'[t], y'[t]\}$ to explain the *Mathematica* output.

Answer:

In[12]:=
```
Clear[gradf]; gradf[x_,y_] = {D[f[x,y],x],D[f[x,y],y]}
```

Out[12]=
$\{f^{(1,0)}[x, y], f^{(0,1)}[x, y]\}$

In[13]:=
```
Clear[tan]; tan[t] = D[{x[t],y[t]},t]
```

Out[13]=
$\{3 \cos[3 t], 2 E^{2t}\}$

The chain rule says $D[f[x[t], y[t]], t] = \text{gradf}[x[t], y[t]] \bullet \tan[t]$:

In[14]:=
```
gradf[x[t],y[t]].tan[t]
```

Out[14]=
$2 E^{2t} f^{(0,1)}[\sin[3 t], E^{2t}] + 3 \cos[3 t] f^{(1,0)}[\sin[3 t], E^{2t}]$

Check:

In[15]:=
```
D[f[x[t],y[t]],t]
```

Out[15]=
$2 E^{2t} f^{(0,1)}[\sin[3 t], E^{2t}] + 3 \cos[3 t] f^{(1,0)}[\sin[3 t], E^{2t}]$

Yes ma'am.

B.1.c.ii) Here is *Mathematica*'s calculation of $D[f[x[t], y[t]], t]$ for $f[x, y] = \sin[x^2 y^3]$ and cleared functions $x[t]$ and $y[t]$:

In[16]:=
```
Clear[f,x,y,t]; f[x_,y_] = Sin[x^2 y^3];
D[f[x[t],y[t]],t]
```

Out[16]=
$\cos[x[t]^2 y[t]^3] (2 x[t] y[t]^3 x'[t] + 3 x[t]^2 y[t]^2 y'[t])$

> Use the fact that $D[f[x[t], y[t]], t] = \text{gradf}[x[t], y[t]] \bullet \{x'[t], y'[t]\}$ to explain the *Mathematica* output.

Answer:

In[17]:=
```
Clear[gradf]; gradf[x_,y_] = {D[f[x,y],x],D[f[x,y],y]}
```

Out[17]=
$\{2 x y^3 \cos[x^2 y^3], 3 x^2 y^2 \cos[x^2 y^3]\}$

Lesson 3.03 The Gradient

In[18]:=
```
Clear[tan]; tan[t] = D[{x[t],y[t]},t]
```
Out[18]=
{x'[t], y'[t]}

The chain rule says $D[f[x[t], y[t]], t] = \text{gradf}[x[t], y[t]] \bullet \tan[t]$:

In[19]:=
```
gradf[x[t],y[t]].tan[t]
```
Out[19]=
$2 \cos[x[t]^2 \, y[t]^3] \, x[t] \, y[t]^3 \, x'[t] + 3 \cos[x[t]^2 \, y[t]^3] \, x[t]^2 \, y[t]^2 \, y'[t]$

Check:

In[20]:=
```
D[f[x[t],y[t]],t]
```
Out[20]=
$\cos[x[t]^2 \, y[t]^3] \, (2 \, x[t] \, y[t]^3 \, x'[t] + 3 \, x[t]^2 \, y[t]^2 \, y'[t])$

You can do these by hand with no sweat.

■ B.2) Level curves, level surfaces, and the gradient as normal vector

Anytime you are examining a function, you can learn a lot by plotting some of its level curves.

In[21]:=
```
Clear[f,x,y]; f[x_,y_] = 5 - (x^2 + x y + y^2)
```
Out[21]=
$5 - x^2 - x \, y - y^2$

Here is a plot of some level curves

$$f[x, y] = c$$

for various c's as selected by *Mathematica*. You are looking down at the surface $z = f[x, y]$ from the positive z-direction.

In[22]:=
```
a = -1; b = 1;
levelcurves =
ContourPlot[Evaluate[f[x,y]],{x,a,b},{y,a,b},
ContourSmoothing->Automatic,
AxesLabel->{"x","y"}];
```

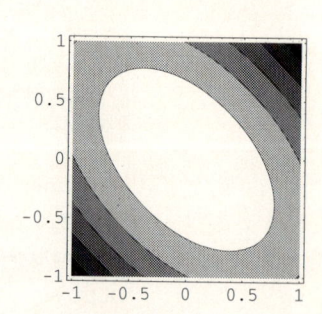

The lighter shading indicates larger values of $f[x, y]$.

Here is the 3D plot of the surface $z = f[x, y]$:

In[23]:=
```
surfaceplot =
ParametricPlot3D[{x,y,f[x,y]},
{x,a,b},{y,a,b},ViewPoint->CMView,
AxesLabel->{"x","y","f[x,y]"}];
```

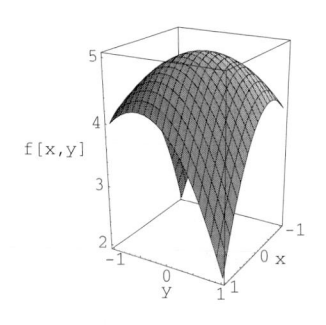

The level curves are plots of the shapes of various trips on the surface that keep the height $f[x, y]$ constant for the whole trip.

Here is a plot of the shape of the trip that keeps $f[x, y] = 4.5$:

In[24]:=
```
c = 4.5;
Show[levelcurves,Contours->{c},
ContourShading->False];
```

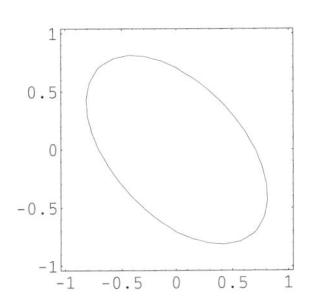

Here is the actual plot of the same trip that lies on the intersection of the surface and the plane $z = 4.5$:

In[25]:=
```
c = 4.5;
plane =
ParametricPlot3D[{x,y,c},
{x,a,b},{y,a,b},PlotPoints->{2,2},
DisplayFunction->Identity];
Show[surfaceplot,plane,
ViewPoint->CMView,
DisplayFunction->$DisplayFunction];
```

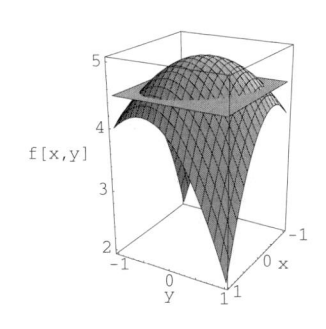

B.2.a) Here's a new function $f[x, y] = 2x^2 + xy + y^2$ together with its gradient:

In[26]:=
```
Clear[x,y,f,gradf]; f[x_,y_] = 2 x^2 + x y + y^2;
gradf[x_,y_] = {D[f[x,y],x],D[f[x,y],y]}
```

Out[26]=
{4 x + y, x + 2 y}

Here is a part of the level curve $f[x,y] = f[1.5,2]$ shown with gradf[1.5, 2] plotted with its tail at $\{1.5, 2\}$:

In[27]:=
```
levelcurve =
ContourPlot[f[x,y],{x,0,3},{y,1,4},
Contours->{f[1.5,2]},ContourShading->False,
DisplayFunction->Identity];
point = {1.5,2};
gradient = Arrow[gradf[1.5,2],Tail->point,Red];
Show[levelcurve,gradient,PlotRange->All,
AspectRatio->Automatic,AxesLabel->{"x","y"},
DisplayFunction->$DisplayFunction];
```

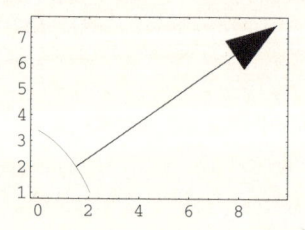

That gradient vector is perpendicular to the level curve.

> Is this just an accident?

Answer: This is no accident.

Here's why: If you parameterize the level curve

$$f[x,y] = c$$

with a parameterization $\{x[t], y[t]\}$ and then you put

$$g[t] = f[x[t], y[t]],$$

then you get $g[t] = c$ no matter what t is. Consequently,

$$g'[t] = 0$$

no matter what t is. But the chain rule says

$$g'[t] = \text{gradf}[x[t], y[t]] \bullet \{x'[t], y'[t]\}.$$

So,

$$\text{gradf}[x[t], y[t]] \bullet \{x'[t], y'[t]\} = 0$$

no matter what t is. This means that the gradient, gradf$[x[t], y[t]]$, is perpendicular to the tangent vector $\{x'[t], y'[t]\}$ at $\{x[t], y[t]\}$. As a result, no matter what $\{x, y\}$ you go to on a level curve $f[x, y] = c$, the gradient, gradf$[x, y]$, is perpendicular to the level curve at $\{x, y\}$.

Read that again! Check it out again with a new function:

Here is a true scale plot of the level curve

$$f[x, y] = x^2 + 4y^2 = 36$$

and some of the gradients, gradf[x, y], with tails at some points {x, y} on the level curve. The level curve $f[x, y] = x^2 + 4y^2 = 36$ is the ellipse
$$\left(\frac{x}{6}\right)^2 + \left(\frac{y}{3}\right)^2 = 1.$$

In[28]:=
```
Clear[x,y,f,gradf]; f[x_,y_] = x^2 + 4 y^2;
gradf[x_,y_] = {D[f[x,y],x],D[f[x,y],y]};
Clear[t]; {x[t_],y[t_]} = {6 Cos[t],3 Sin[t]};
levelcurve = ParametricPlot[{x[t],y[t]},{t,0,2 Pi},
  PlotStyle->{{Blue,Thickness[0.01]}},DisplayFunction->Identity];
scalefactor = 0.4; gradients = Table[Arrow[scalefactor gradf[x[t],y[t]],
  Tail->{x[t],y[t]},Red],{t,0,2 Pi,2 Pi/12}];
```

In[29]:=
```
Show[levelcurve,gradients,
  AspectRatio->Automatic,
  AxesLabel->{"x","y"},PlotRange->All,
  DisplayFunction->$DisplayFunction];
```

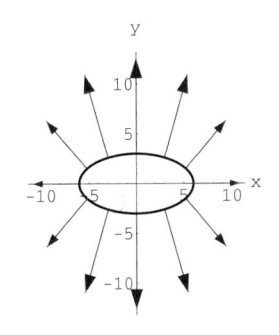

Perpendicular as all get-out. It will work anytime and anyplace.

B.2.b) Does this work in three dimensions?

Answer: Check it out. Here is part of the level surface $f[x, y, z] = z - xy = 1$.

In[30]:=
```
Clear[f,gradf,x,y,z]; f[x_,y_,z_] = z - x y;
gradf[x_,y_,z_] = {D[f[x,y,z],x],D[f[x,y,z],y],D[f[x,y,z],z]};
Solve[f[x,y,z] == 1,z]
```

Out[30]=
```
{{z -> 1 + x y}}
```

In[31]:=
```
levelsurface = ParametricPlot3D[
  {x,y,1 + x y},{x,-1, 1},{y,-1,1},
  ViewPoint->CMView,
  AxesLabel->{"x","y","z"}];
```

You can get one of the many curves on this surface by taking the surface plotting function $\{x, y, 1 + x\, y\}$ and setting $x = \cos[t]/2$ and $y = \sin[t]$.

Here come a lot of gradients with tails on this curve:

In[32]:=
```
Clear[x,y,z,t]
{x[t_],y[t_],z[t_]} =
{x,y,1 + x y}/.{x->Cos[t]/2,y->Sin[t]};
gradients = Table[Arrow[gradf[x[t],y[t],z[t]],
Tail->{x[t],y[t],z[t]},Red],{t,0,2 Pi,2 Pi/12}];
Show[levelsurface,gradients];
```

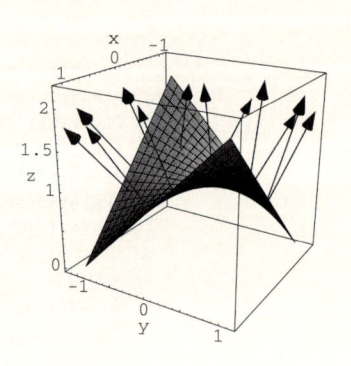

The gradients are just as normal as Beaver, Wally, June, and Ward.

■ **B.3) The gradient points in the direction of greatest initial increase.**
The negative gradient points in the direction of greatest initial decrease.

B.3.a) Here is a function and its gradient:

In[33]:=
```
Clear[x,y,f,gradf]; f[x_,y_] = -5 x^2 + x^4 y/4 + 2 y + 10;
gradf[x_,y_] = {D[f[x,y],x],D[f[x,y],y]}
```

Out[33]=
$$\{-10\, x + x^3\, y,\ 2 + \frac{x^4}{4}\}$$

You are sitting at:

In[34]:=
```
{a,b} = {1.5,1.8};
```

At this point the function's value is:

In[35]:=
```
Apply[f,{a,b}]
```

Out[35]=
4.62812

The instruction $\text{Apply}[f, \{a, b\}]$ accomplishes the same thing as the instruction $f[a, b]$:

In[36]:=
```
f[a,b]
```
Out[36]=
```
4.62812
```

Here is a plot of part of the level curve $f[x,y] = f[a,b]$:

In[37]:=
```
levelcurve = ContourPlot[f[x,y],
  {x,a - 2,a + 2},{y,b - 2,b + 2},
  Contours->{f[a,b]},
  ContourShading->False,
  ContourSmoothing->True,
  DisplayFunction->Identity];
labels = {Graphics[{PointSize[0.07],
  Point[{a,b}]}],Graphics[
  Text["{a,b}",{a,b},{-1.5,0}]]};
Show[levelcurve,labels,
  DisplayFunction->$DisplayFunction];
```

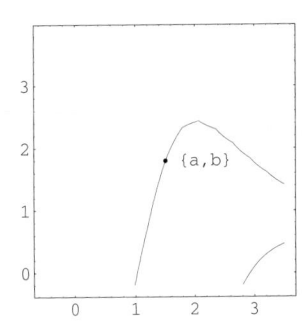

You can leave the $\{a,b\}$ in the direction of any point on the circle of radius 1 centered at $\{a,b\}$.

In[38]:=
```
circle =
Graphics[{Red,Thickness[0.01],Circle[{a,b},1]}];
Show[levelcurve,labels,circle,
  DisplayFunction->$DisplayFunction];
```

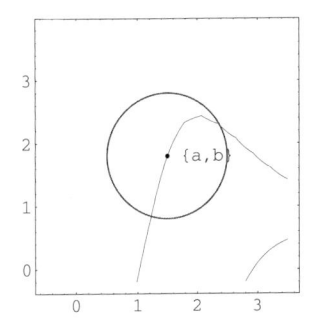

> Which direction should you go to get the greatest initial increase of the function?
>
> Which direction should you go to get the greatest initial decrease of the function?

Answer: Calculus&*Mathematica* thanks 1991 C&M student Tony Pulokas of the University of Illinois for suggesting this way of looking at the problem.

Two unit vectors come to mind. They are unit tangent and unit normal vectors to the level curve at $\{a,b\}$. One unit normal is the unit vector in the direction of the gradient.

In[39]:=
```
unitgradnormal = gradf[a,b]/Sqrt[gradf[a,b].gradf[a,b]]
```

Out[39]=
{-0.93911, 0.343617}

You can get a unit tangent vector by switching the components and hitting one of them with a minus one.

In[40]:=
```
unittan = {unitgradnormal[[2]],-unitgradnormal[[1]]}
```

Out[40]=
{0.343617, 0.93911}

In[41]:=
```
setup = Show[levelcurve,labels,circle,
Arrow[unitgradnormal,Tail->{a,b},Blue],
Arrow[unittan,Tail->{a,b},Blue],
DisplayFunction->$DisplayFunction];
```

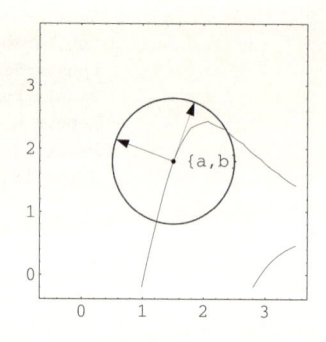

You can leave the point in the direction of any vector with tail at $\{a,b\}$ and tip on the circle of radius 1 centered at $\{a,b\}$. Here are some possibilities:

In[42]:=
```
Table[Show[setup,Arrow[Cos[t] unitgradnormal + Sin[t] unittan,
Tail->{a,b}],DisplayFunction->$DisplayFunction,
AspectRatio->Automatic],{t,0,2 Pi - Pi/4,Pi/4}];
```

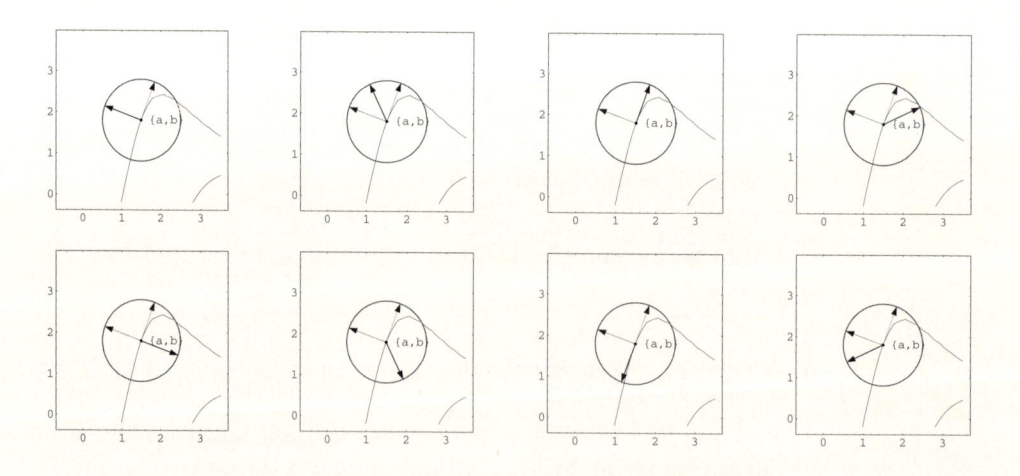

Above you saw the tips of the vectors $\cos[t]$ unitnormal $+ \sin[t]$ unittan with tails at $\{a,b\}$ sweep out the circle as t went from 0 to 2π. The question is how to set t so that the vector $\cos[t]$ unitnormal $+ \sin[t]$ unittan with tail at $\{a,b\}$ points in the direction of greatest initial change of $f[x,y]$ as you leave $\{a,b\}$.

Think for a second and you will probably agree that going in the direction of unittan is a bad idea because it snuggles against the level curve and $f[x, y]$ doesn't change at all on this level curve. In fact, any direction that has a nonzero tangential component is a poor choice, because the tangential component won't do much for changing the function.

To get the greatest initial change in $f[x, y]$, you should leave $\{a, b\}$ in the direction of $\cos[t]$ unitnorm $+ \sin[t]$ unittan with t set so that the unittan term is zeroed out. This means you set $t = 0$ or π, and this leaves you the choice of directions: unitnormal or $-$unitnormal. Because the unitnormal points in the same direction as the gradient, to get the greatest initial change in $f[x, y]$ you should leave $\{a, b\}$ in the direction of either $\text{gradf}[a, b]$ or $-\text{gradf}[x, y]$. Let's see which one:

In[43]:=
```
Clear[s]
{Apply[f,{a,b} + s gradf[a,b]],
Apply[f,{a,b}],
Apply[f,{a,b} + s (-gradf[a,b])]}/.s->0.1
```
Out[43]=
{12.4803, 4.62812, -3.60418}

In[44]:=
```
{Apply[f,{a,b} + s gradf[a,b]],
Apply[f,{a,b}],
Apply[f,{a,b} + s (-gradf[a,b])]}/.s->0.01
```
Out[44]=
{5.529, 4.62812, 3.72469}

In[45]:=
```
{Apply[f,{a,b} + s gradf[a,b]],
Apply[f,{a,b}],
Apply[f,{a,b} + s (-gradf[a,b])]}/.s->0.001
```
Out[45]=
{4.71843, 4.62812, 4.53779}

In[46]:=
```
{Apply[f,{a,b} + s gradf[a,b]],
Apply[f,{a,b}],
Apply[f,{a,b} + s (-gradf[a,b])]}/.s->0.0001
```
Out[46]=
{4.63716, 4.62812, 4.61909}

The gradient points in the direction of greatest initial increase. The negative gradient points in the direction of the greatest initial decrease. This will happen for any function at any point at which the gradient is not $\{0, 0\}$.

■ B.4) Using linearizations to help to explain the chain rule

This problem appears only in the electronic version.

Tutorials

■ T.1) The total differential

If you are given a function $f[x]$ of one variable, then you can write the differential

$$df = f'[x]\,dx.$$

This is really just a suggestive notation, and it is handy for hand calculations:

→ You divide both sides of $df = f'[x]\,dx$ by dt to get

$$\frac{df}{dt} = f'[x]\frac{dx}{dt}.$$

Interpret this as

$$\frac{df[x[t]]}{dt} = f'[x[t]]\frac{dx[t]}{dt} = f'[x[t]]\,x'[t].$$

This is the chain rule for functions of one variable.

T.1.a.i) How do you carry off this stunt for a function $f[x, y]$?

Answer: Very easily. Just write

$$df = D[f[x, y], x]\,dx + D[f[x, y], y]\,dy.$$

Divide both sides by dt to get

$$\frac{df}{dt} = D[f[x, y], x]\frac{dx}{dt} + D[f[x, y], y]\frac{dy}{dt}.$$

This gives you the chain rule:

In[1]:=
   `Clear[f,x,y,t]; D[f[x[t],y[t]],t]`

Out[1]=
   $y'[t]\,f^{(0,1)}[x[t],y[t]] + x'[t]\,f^{(1,0)}[x[t],y[t]]$

Remember $D[f[x, y], x] = f^{(1,0)}[x, y]$ and $D[f[x, y], y] = f^{(0,1)}[x, y]$. It pays to notice once again that

$$\frac{df}{dt} = \frac{D[f[x,y],x]\,dx}{dt} + \frac{D[f[x,y],y]\,dy}{dt}$$

is nothing more than

$$\text{gradf}[x, y] \bullet \{x'[t], y'[t]\},$$

which is the chain rule.

T.1.a.ii) Use the total differential
$$df = D[f[x,y],x]\,dx + D[f[x,y],y]\,dy$$
to help do a hand calculation of $D[\sin[x[t]^2\,y[t]],t]$.

Answer by hand calculation: Take $f[x,y] = \sin[x^2\,y]$. So
$$df = \left(\cos[x^2\,y]\,2\,x\,y\right)\,dx + \left(\cos[x^2\,y]\,x^2\right)\,dy.$$
Divide both sides by dt to get:
$$\frac{df}{dt} = \left(\cos[x^2\,y]\,2\,x\,y\right)\frac{dx}{dt} + \left(\cos[x^2\,y]\,x^2\right)\frac{dy}{dt}.$$
Interpret the result as
$$\frac{df[x[t],y[t]]}{dt} = \left(\cos[x[t]^2\,y[t]]\,2\,x[t]\,y[t]\right)\frac{dx[t]}{dt} + \left(\cos[x[t]^2\,y[t]]\,x[t]^2\right)\frac{dy[t]}{dt}.$$
This is in agreement with:

In[2]:=
```
Clear[x,y,t]; Expand[D[Sin[x[t]^2 y[t]],t]]
```
Out[2]=
$$2\,\text{Cos}[x[t]^2\,y[t]]\,x[t]\,y[t]\,x'[t] + \text{Cos}[x[t]^2\,y[t]]\,x[t]^2\,y'[t]$$

This is a nice technique that's especially handy when you are working with a pencil on the back of an old envelope.

T.1.b) Use the total differential to give a hand derivation of the product rule
$$D[x[t]\,y[t]\,z[t],t] = x'[t]\,y[t]\,z[t] + x[t]\,y'[t]\,z[t] + x[t]\,y[t]\,z'[t].$$

Answer by hand: Put $f[x,y,z] = x\,y\,z$. This gives
$$df = y\,z\,dx + x\,z\,dy + x\,y\,dz.$$
Divide both sides by dt to get
$$\frac{df}{dt} = \frac{y\,z\,dx}{dt} + \frac{x\,z\,dy}{dt} + \frac{x\,y\,dz}{t}.$$
Interpret this as
$$\frac{d\,(x[t]y[t]z[t])}{dt} = y[t]\,z[t]\,x'[t] + x[t]\,z[t]\,y'[t] + x[t]\,y[t]\,z'[t]$$
$$= x'[t]\,y[t]\,z[t] + x[t]\,y'[t]\,z[t] + x[t]\,y[t]\,z'[t].$$

Check:

In[3]:=
```
Clear[x,y,z,t]; D[x[t] y[t] z[t],t]
```
Out[3]=
```
y[t] z[t] x'[t] + x[t] z[t] y'[t] + x[t] y[t] z'[t]
```

Yep.

T.1.c) Use the total differential to help give a formula for the derivative with respect to t of
$$g[t] = \int_{\cos[t]}^{\sin[t]} h[s]\, ds$$
where $h[s]$ is an unspecified function.

Answer: Put $f[x, y] = \int_y^x h[s]\, ds$. The total differential is
$$df = D[f[x,y], x]\, dx + D[f[x,y], y]\, dy = h[x]\, dx - h[y]\, dy.$$
Divide through by dt to get
$$\frac{df}{dt} = \frac{h[x]\, dx}{dt} - \frac{h[y]\, dy}{dt}.$$
Now put $x = \sin[t]$ and $y = \cos[t]$ to get
$$D\left[\int_{\cos[t]}^{\sin[t]} h[s]\, ds, t\right] = h[\sin[t]] \cos[t] - h[\cos[t]] (-\sin[t])$$
$$= h[\sin[t]] \cos[t] + h[\cos[t]] \sin[t].$$

Would you have gotten it right without using the total differential? Probably not.

■ T.2) What's the chain rule good for?

T.2.a) Is the chain rule formula for functions of more than one variable useful for calculating derivatives of specific functions?

Answer: Not really. In specific situations, the chain rule for more than one variable is not really needed because if you make the substitutions first and then differentiate, all you need is the chain rule for functions of one variable.

Case in point: If you want to differentiate $\cos[x[t]\, y[t]^3]$ with respect to t, you can use the chain rule for a function of one variable and the product rule (from the early part of the course) to get

$$\frac{d\left(\cos[x[t]\,y[t]^3]\right)}{dt} = -\sin[x[t]\,y[t]^3]\,\frac{d\left(x[t]\,y[t]^3\right)}{dt}$$
$$= -\sin[x[t]\,y[t]^3]\left(x[t]\,3\,y[t]^2\,y'[t] + x'[t]\,y[t]^3\right).$$

You could also use the two-variable chain rule from this lesson. But the point is that you don't really need the two-variable chain rule to do this problem.

T.2.b) | If the chain rule formula for functions of more than one variable is not particularly useful for calculating derivatives of specific functions, then what the heck is it good for?

Answer: Instead of being a great calculational tool, the chain rule for functions of more than one variable is a great theoretical tool. Reason: The chain rule for functions of more than one variable helps you unlock the magic of the gradient. Cases in point:

→ The chain rule gave you the basis for explaining why the gradient is perpendicular to level curves and surfaces.

→ Once you knew this, it was not a big jump to see why the gradient points in the direction of greatest initial increase.

Should you forget about the chain rule? Certainly not.

■ T.3) The gradient and maximization and minimization: The FindMinimum instruction

When you plot a surface, your eyes look for crests and dips. Tops of crests are the tops of hills. Bottoms of dips are the deepest points in depressions that could collect rain water. Fancy folks call the tops of crests by the name "local maxima." The same folks call the bottoms of the dips by the name "local minima." Most folks, fancy and down to earth, are interested in the maximum, which is located at the top of the very highest crest. The same folks are also interested in the minimum, which is located at the bottom of the deepest dip.

T.3.a.i) | How do you know that if $\{x_0, y_0, f[x_0, y_0]\}$ sits at the top of a crest on the surface $z = f[x, y]$, then $\text{gradf}[x_0, y_0] = \{0, 0\}$?

Answer: If you are at $\{x_0, y_0\}$ and you leave $\{x_0, y_0\}$ in the direction of $\text{gradf}[x_0, y_0]$, then $f[x, y]$ initially goes **up** unless

$$\text{gradf}[x_0, y_0] = \{0, 0\}.$$

Lesson 3.03 The Gradient

The upshot: If gradf$[x_0, y_0]$ is not $\{0,0\}$, then $\{x_0, y_0, f[x_0, y_0]\}$ cannot sit at the top of a crest on the surface $z = f[x, y]$.

In other words, if $\{x_0, y_0, f[x_0, y_0]\}$ sits at the top of a crest on the surface $z = f[x, y]$, then gradf$[x_0, y_0] = \{0, 0\}$.

T.3.a.ii) How do you know that if $\{x_0, y_0, f[x_0, y_0]\}$ sits at the bottom of a dip on the surface $z = f[x, y]$, then gradf$[x_0, y_0] = \{0, 0\}$?

Answer: If you are at $\{x_0, y_0\}$ and you leave $\{x_0, y_0\}$ in the direction of $-$gradf$[x_0, y_0]$, then $f[x, y]$ initially goes **down** unless

gradf$[x_0, y_0] = \{0, 0\}$.

The upshot: If gradf$[x_0, y_0]$ is not $\{0, 0\}$, then $\{x_0, y_0, f[x_0, y_0]\}$ cannot sit at the bottom of a dip on the surface $z = f[x, y]$.

In other words, if $\{x_0, y_0, f[x_0, y_0]\}$ sits at the bottom of a dip on the surface $z = f[x, y]$, then gradf$[x_0, y_0] = \{0, 0\}$.

T.3.b.i) Find the maximizers and minimizers (if any) of

$$f[x, y] = 2.1\, x^2 + 5.9\, y^2 - 7.2\, x + 3.3\, y + 8.7.$$

Answer: Look at the formula

$$f[x, y] = 2.1\, x^2 + 5.9\, y^2 - 7.2\, x + 3.3\, y + 8.7.$$

For large $|x|$ and $|y|$, the dominant terms $2.1\, x^2 + 5.9\, y^2$ make $f[x, y]$ really huge. This means $f[x, y]$ has no maximum value. It also means that there is no way for $f[x, y]$ to ever get near $-\infty$.

So $f[x, y]$ has a minimum (at the bottom of the deepest dip), and the minimizer must be a point at which gradf$[x, y] = 0$:

In[4]:=
```
Clear[x,y,f,gradf]
f[x_,y_] = 2.1 x^2 + 5.9 y^2 - 7.2 x + 3.3 y + 8.7;
gradf[x_,y_] = {D[f[x,y],x],D[f[x,y],y]};
Solve[gradf[x,y] == {0,0},{x,y}]
```
Out[4]=
{{x -> 1.71429, y -> -0.279661}}

The minimizer is $\{1.71429, -0.279661\}$ and the minimum value is

$f[1.71429, -0.279661]$:

In[5]:=
```
Apply[f,{1.71429,-0.279661}]
```

Out[5]=
2.06713

Take a look:

In[6]:=
```
deepestdip = {1.71429,-0.279661, 2.06713};
flagpole = Graphics3D[{Red,Thickness[0.01],
Line[{deepestdip,{1.7,-0.3,10}}]}] ;
surface = Plot3D[f[x,y],{x,1.7 -1,1.7 + 1},
{y,-0.3 - 1,-0.3 + 1},DisplayFunction->Identity];
Show[surface,flagpole,AxesLabel->{"x","y","z"},
ViewPoint->CMView,DisplayFunction->$DisplayFunction];
```

The flagpole is planted at the bottom of the deepest dip.

T.3.b.ii) Find the maximizers and minimizers (if any) of
$$f[x,y] = 0.2\,x^4 + 0.1\,y^4 - 8.2\,x\,y - 18.4\,x.$$

Answer: Look at the formula $f[x,y] = 0.2\,x^4 + 0.1\,y^4 + 8.2\,x\,y - 18.4\,x$. For large $|x|$ and $|y|$, the dominant terms $0.2\,x^4 + 0.1\,y^4$ make $f[x,y]$ really huge. This means $f[x,y]$ has no maximum. It also means that there is no way for $f[x,y]$ to ever get near $-\infty$; so $f[x,y]$ has a minimum and the minimizer must be a point at which $\text{gradf}[x,y] = 0$:

In[7]:=
```
Clear[x,y,f,gradf]
f[x_,y_] = 0.2 x^4 + 0.1 y^4 + 8.2 x y - 18.4 x;
gradf[x_,y_] = {D[f[x,y],x],D[f[x,y],y]};
Solve[gradf[x,y] == {0,0},{x,y}]
```

Out[7]=
```
{{x -> 4.07802, y -> -4.37255},
 {x -> -2.53906 + 3.06993 I, y -> -3.16285 - 2.96988 I},
 {x -> -2.53906 - 3.06993 I, y -> -3.16285 + 2.96988 I},
 {x -> -0.653039 - 3.63228 I, y -> -0.250639 - 4.22198 I},
 {x -> -0.653039 + 3.63228 I, y -> -0.250639 + 4.22198 I},
 {x -> -0.564143, y -> 2.26142},
 {x -> 2.66448 + 1.68863 I, y -> 2.62211 - 3.03901 I},
 {x -> 2.66448 - 1.68863 I, y -> 2.62211 + 3.03901 I},
 {x -> -2.45864, y -> 3.69388}}
```

Tossing out the complex candidates gives three candidates for the minimizer:

In[8]:=
```
candidate1 = {4.07802,-4.37255};
candidate2 = {-0.564143,2.26142};
candidate3 = {-2.45864,3.69388};
```

Compare:

In[9]:=
```
{Apply[f,candidate1],Apply[f,candidate2],Apply[f,candidate3]}
```

Out[9]=
{-129.385, 2.55454, -3.30667}

It's no contest. The minimizer is:

In[10]:=
candidate1

Out[10]=
{4.07802, -4.37255}

The minimum value is:

In[11]:=
Apply[f,candidate1]

Out[11]=
-129.385

Not much to it.

T.3.c.i) Find the maximizers and minimizers (if any) of
$$f[x, y] = 1.2\, e^{x^2} + 2.9\, e^{y^2} - 8.2\, x\, y^2.$$
Show off your results with a plot.

Answer: Look at the formula $f[x,y] = 1.2\, e^{x^2} + 2.9\, e^{y^2} - 8.2\, x\, y^2$. For large $|x|$ and $|y|$, the dominant terms $1.2\, e^{x^2} + 2.9\, e^{y^2}$ make $f[x,y]$ hugely positive. This means $f[x,y]$ has no maximum. It also means that there is no way for $f[x,y]$ to ever get near $-\infty$; so $f[x,y]$ has a minimum value and the minimizer must be a point at which $\text{gradf}[x,y] = 0$:

In[12]:=
```
Clear[f,gradf,x,y]
f[x_,y_] = 1.2 E^(x^2) + 2.9 E^(y^2) - 8.2 x y^2;
gradf[x_,y_] = {D[f[x,y],x],D[f[x,y],y]}
```

Out[12]=
$$\{2.4\, E^{x^2} x - 8.2\, y^2,\ 5.8\, E^{y^2} y - 16.4\, x\, y\}$$

In[13]:=
Solve[gradf[x,y] == {0,0},{x,y}]

Solve::ifun:
  Warning: Inverse functions are being used by Solve, so
    some solutions may not be found.

Solve::tdep:
  The equations appear to involve transcendental
    functions of the variables in an essentially
    non-algebraic way.

Out[13]=
$$\text{Solve}[\{2.4\, E^{x^2} x - 8.2\, y^2,\ 5.8\, E^{y^2} y - 16.4\, x\, y\} == \{0, 0\}, \{x, y\}]$$

Dung. *Mathematica* had to bail out. But this is no reason for you to give up. Take another look at the formula:

In[14]:=
```
f[x,y]
```

Out[14]=
$$1.2 E^{x^2} + 2.9 E^{y^2} - 8.2 x y^2$$

For the smaller values of $|x|$ and $|y|$, the term $8.2\, x\, y^2$ has a lot of influence. Also, to drive $f[x,y]$ down, you'll want $x > 0$. Look at plot of some level curves of $f[x,y]$ for the smaller values of $|x|$ and $|y|$ with $x > 0$:

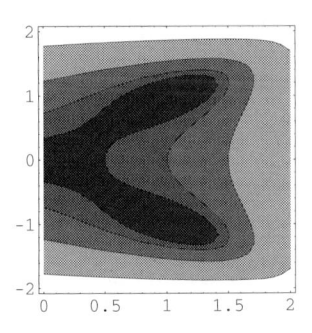

In[15]:=
```
ContourPlot[f[x,y],{x,0,2},{y,-2,2}];
```

The lighter the shading, the higher the function.

The minimizer(s) lurk inside the dark zone. Look again:

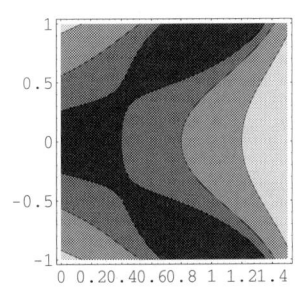

In[16]:=
```
ContourPlot[f[x,y],{x,0,1.5},{y,-1,1}];
```

This is evidence pointing to two minimizers. Look at the formula:

In[17]:=
```
f[x,y]
```

Out[17]=
$$1.2 E^{x^2} + 2.9 E^{y^2} - 8.2 x y^2$$

And look at:

In[18]:=
```
f[x,-y]
```

Out[18]=

$$1.2\, E^{x^2} + 2.9\, E^{y^2} - 8.2\, x\, y^2$$

Aha! $f[x,y] = f[x,-y]$. This tells you that if you locate one minimizer at $\{x^*, y^*\}$, then $\{x^*, -y^*\}$ is also a minimizer. Look high:

In[19]:=
```
ContourPlot[f[x,y],{x,0.8,1.3},{y,0.7,1.2}];
```

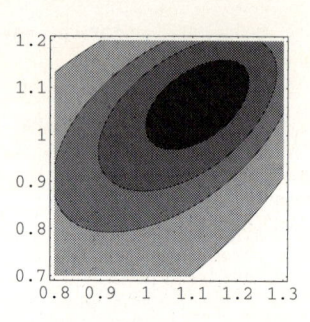

You've got one of the minimizers trapped! It's somewhere in the dark zone near $\{1.05, 1.1\}$.

Now move in for the kill with the *Mathematica* instruction FindMinimum starting with an initial guess of $\{1.05, 1.1\}$:

In[20]:=
```
minimizer = FindMinimum[f[x,y],{x,1.05},{y,1.1}]
```

Out[20]=
```
{2.80399, {x -> 1.12131, y -> 1.07421}}
```

Mathematica is telling you that it has located a bottom of a dip at the point $\{1.12131, 1.07421, 2.80399\}$. Test the gradient at $\{1.12131, 1.07421\}$:

In[21]:=
```
Apply[gradf,{1.12131, 1.07421}]
```

Out[21]=
```
{-0.0000252787, 0.000119753}
```

Great; it's darn close to $\{0,0\}$. Now you can say with considerable authority that your estimate is that the minimizers are

$$\{1.12131, 1.07421\} \quad \text{and} \quad \{1.12131, -1.07421\}$$

and the minimum value of $f[x,y]$ is:

In[22]:=
```
Apply[f,{1.12131, 1.07421}]
```

Out[22]=
2.80399

In[23]:=
```
Apply[f,{1.12131, -1.07421}]
```

Out[23]=
2.80399

Notice that this is the same as the first slot of the output of the FindMinimum instruction.

Take a look at the fruits of your labor:

In[24]:=
```
h = 0.3;
deepestdip1 = {1.12131, 1.07421, 2.80399};
deepestdip2 = {1.12131, -1.07421, 2.80399};
flagpoles = {Graphics3D[{Red,Thickness[0.01],
  Line[{deepestdip1,{1.1,1.07,10}}]}],
  Graphics3D[{Red,Thickness[0.01],
  Line[{deepestdip2,{1.1,-1.07,10}}]}]};
surface = Plot3D[f[x,y],{x,1.1 - h,1.1 + 0.8 h},
  {y,-1.07 - h,1.07 + h},PlotPoints->{20,15},
  DisplayFunction->Identity];
Show[surface,flagpoles,AxesLabel->{"x","y","z"},
  ViewPoint->CMView,DisplayFunction->$DisplayFunction];
```

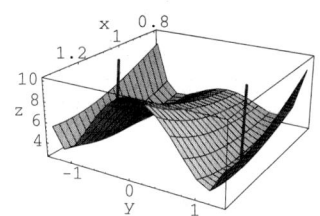

The two flagpoles are planted at the bottoms of the two dips, which are equally deep. Here's a look from a different viewpoint:

In[25]:=
```
Show[surface,flagpoles,
AxesLabel->{"x","y","z"},
ViewPoint->{10,0,1},
DisplayFunction->$DisplayFunction];
```

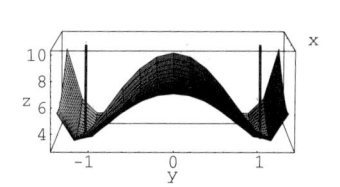

T.3.c.ii) Roughly speaking, how did the FindMinimum instruction find the minimizer in part T.3.c.i)?

Answer: Here're the function and its gradient.

In[26]:=
```
Clear[f,gradf,x,y]
f[x_,y_] = 1.2 E^(x^2) + 2.9 E^(y^2) - 8.2 x y;
gradf[x_,y_] = {D[f[x,y],x],D[f[x,y],y]}
```

Out[26]=
$\{2.4 E^{x^2} x - 8.2 y, -8.2 x + 5.8 E^{y^2} y\}$

Lesson 3.03 The Gradient

Starting with an initial guess at $\{1.05, 1.1\}$, the FindMinimum instruction found a minimizer at:

In[27]:=
```
minimizer = {1.12131, 1.07421}
```

Out[27]=
```
{1.12131, 1.07421}
```

To get to this point, *Mathematica* tried to move from $\{1.05, 1.11\}$ on a path whose tangent vectors always point with the negative gradient of $f[x, y]$. This is a good strategy because when you go in the direction of the negative gradient, you drive $f[x, y]$ down.

Fancy folks call this by the name "steepest descent."

T.3.c.iii) Can you use FindMinimum to help find maximizers?

Answer: Sure. The maximizers of $f[x, y]$ are the minimizers of $-f[x, y]$. You can use FindMinimum$[-f[x, y], \{x, a\}, \{y, b\}]$ to start your search at $\{a, b\}$ for maximizers of $f[x, y]$. This time *Mathematica* will try to start at $\{a, b\}$ and go on a path whose tangent vectors always point with the positive gradient of $f[x, y]$.

T.3.c.iv) Is there a trick to using FindMinimum?

Answer: Sometimes using FindMinimum is as tricky as one of our former presidents who had the nickname "Tricky Dick."

To see what can happen, go with

$$f[x, y] = 4 e^{-(x^2+y^2)} + 8 e^{-((x-2.5)^2 + 2(y-2)^2)}.$$

This function is never negative or zero, but as $|x|$ and $|y|$ get big, $f[x, y]$ gets really close to 0. Consequently, $f[x, y]$ has no minimum value but has at least one maximizer.

Look at a plot of some level curves:

In[28]:=
```
Clear[f,gradf,x,y]
f[x_,y_] = 4 E^(-(x^2 + y^2)) +
   8 E^(-((x-2.5)^2 + 2 (y-2)^2));
ContourPlot[f[x,y],{x,-3,3},{y,-3,3}];
```

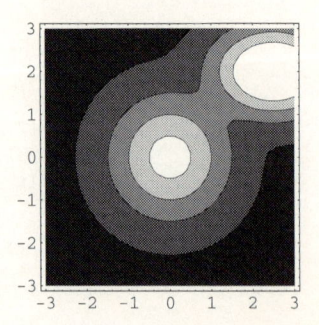

The lighter the shading, the higher $f[x, y]$ is. Crests have been trapped near $\{0, 0\}$ and $\{2, 2\}$.

Search for the first maximizer by starting at $\{2, 2\}$:

In[29]:=
    `FindMinimum[-f[x,y],{x,2},{y,2}]`

Out[29]=
    {-8.00014, {x -> 2.49996, y -> 1.99998}}

In[30]:=
    `candidate1 = {2.49996,1.99998}`

Out[30]=
    {2.49996, 1.99998}

Test it:

In[31]:=
    `Clear[gradf]; gradf[x_,y_] = {D[f[x,y],x],D[f[x,y],y]};`
    `Apply[gradf,candidate1]`

Out[31]=
    {-0.0000673367, 0.0000741272}

Fairly good. Search for a maximizer by starting at $\{0, 0\}$:

In[32]:=
    `FindMinimum[-f[x,y],{x,0},{y,0}]`

Out[32]=
    {-4.00001, {x -> 3.23856 10^{-6}, y -> 5.18174 10^{-6}}}

In[33]:=
    `candidate2 = {3.23856 10^(-6),5.18174 10^(-6)}`

Out[33]=
    {3.23856 10^{-6}, 5.18174 10^{-6}}

Test it:

In[34]:=
    `Apply[gradf,candidate2]`

Out[34]=
    {-3.21154 10^{-9}, -5.54416 10^{-9}}

Good. Two maximizers? Take a look at a plot:

Lesson 3.03 The Gradient

In[35]:=
```
surface = ParametricPlot3D[
  {x,y,f[x,y]},{x,-2,4},{y,-2,4},
  ViewPoint->CMView,PlotRange->All,
  AxesLabel->{"x","y","z"}];
```

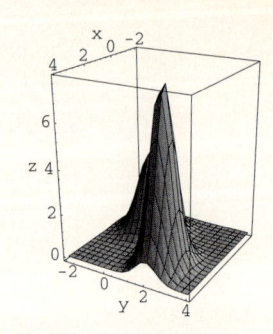

Take a look from a different viewpoint:

In[36]:=
```
Show[surface,ViewPoint->{8,-8,0}];
```

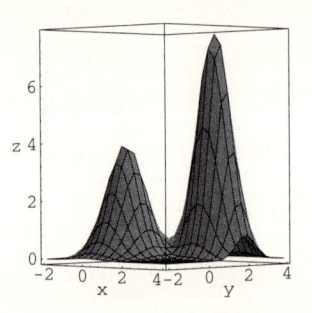

There are two crests. One exhibits the true maximum value; the other is just a silly pretender sitting at the top of its own short crest. The true maximizer sits on the top of the highest crest, master of all.

Fancy dudes call the pretender a "relative maximizer" or a "local maximizer." The same fancy dudes call the true maximizers or minimizers "global maximizers" or "global minimizers."

Compare:

In[37]:=
```
{Apply[f,candidate1],Apply[f,candidate2]}
```
Out[37]=
  {8.00014, 4.00001}

The maximizer is:

In[38]:=
```
candidate1
```
Out[38]=
  {2.49996, 1.99998}

The maximum value is 8.0014.

The trick to using FindMinimum$[g[x,y],\{x,a\},\{y,b\}]$ is to pick a starting point $\{a,b\}$ close enough to the true minimizer of $g[x,y]$. Sometimes this isn't as easy as you might wish.

T.3.c.v) Can you use FindMinimum for functions of more than two variables?

Answer: Yep.

■ T.4) Eye-balling a function for max-min

T.4.a) Does
$$f[x, y] = \frac{x^4}{10} - 3xy + \frac{y^4}{7} - 8x + 9y$$
have a maximizer or a minimizer?

Answer: Look at the formula
$$f[x, y] = \frac{x^4}{10} - 3xy + \frac{y^4}{7} - 8x + 9y.$$

When $|x|$ and $|y|$ are large then the dominant terms $x^4/10 + y^4/7$ make $f[x,y]$ really huge, so the global scale plot of $f[x,y]$ looks like a cup. This means $f[x,y]$ has no tallest crest but does have a deepest dip.

As a result, $f[x,y]$ has no maximum value but does have a minimum value.

T.4.b) Does
$$f[x, y] = e^{(-x^2 - y^2)} \left(x^6 + 7y^2 + 4 \right)$$
have a maximizer or a minimizer?

Answer: Look at the formula $f[x,y] = e^{(-x^2-y^2)} \left(x^6 + 7y^2 + 4 \right)$. When $|x|$ and $|y|$ are large, then $f[x,y]$ is incredibly close to 0. Also, $f[x,y] > 0$ for all x's and y's.

As a result, the surface $z = f[x,y]$ has a biggest crest, but no dip below 0. Consequently, $f[x,y]$ has a maximizer but no minimizer.

T.4.c) Does
$$f[x, y] = \frac{x^3 + 7y^2}{1 + x^4 + y^6}$$
have a maximizer or a minimizer?

Answer: Look at the formula $f[x,y] = (x^3 + 7y^2)/(1 + x^4 + y^6)$. When $|x|$ and $|y|$ are large, then the dominant terms in the denominator make $f[x,y]$ incredibly close to 0. Also $f[x,0] > 0$ for $x > 0$ and $f[x,0] < 0$ for $x < 0$. Now you know that $f[x,y]$ has positive and negative values and $f[x,y]$ is arbitrarily close to 0 as $|x|$ and $|y|$ become large.

Consequently, the surface $z = f[x,y]$ has a biggest crest and a deepest dip. This means $f[x,y]$ has both a maximizer and a minimizer.

■ T.5) Data fit

This problem appears only in the electronic version.

■ T.6) Lagrange's method for constrained maximization and minimization

This problem appears only in the electronic version.

Give It a Try

Experience with the starred (⋆) problems will be useful for understanding developments later in the course.

■ G.1) The gradient points in the direction of greatest initial increase⋆

G.1.a) Go with

$$f[x,y,z] = \log[3x^2 + y^2 + z^2].$$

> In which direction should you leave the point $\{2.1, -3.7, 4.3\}$ to get the greatest possible initial increase of $f[x,y,z]$?

G.1.b.i) Set

$$f[x,y] = 2\sin[0.3\,x\,y]$$

and look at a plot of $f[x[t], y[t]]$ for

$$\{x[t], y[t]\} = \{0.1, 2\} + t\,\text{gradf}[0.1, 2] \quad \text{and} \quad 0 \le t \le 3:$$

```
In[1]:=
    Clear[f,gradf,x,y,t]
    f[x_,y_] = 2 Sin[0.3 x y];
    gradf[x_,y_] = {D[f[x,y],x],D[f[x,y],y]};
    {x[t_],y[t_]} = {0.1,2} + t gradf[0.1,2];
    Plot[f[x[t],y[t]],{t,0,3},
    PlotStyle->{{Blue,Thickness[0.02]}},
    PlotRange->All,AxesLabel->{"t",""}];
```

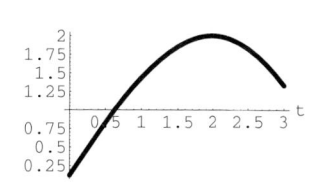

> The knowledge of what fact about gradients gives you the ability to predict, before seeing the plot, that the curve must go **up** before it can go down?
>
> Experiment with other functions $f[x, y]$.
>
> Explain why the plot will go up initially no matter what function $f[x, y]$ you go with, as long as gradf$[0.1, 2]$ is not $\{0, 0\}$.

G.1.b.ii) Set

$$f[x, y, z] = 0.6 \cos[1.8\, x\, y\, z^2]$$

and look at a plot of $f[x[t], y[t], z[t]]$ for

$$\{x[t], y[t], z[t]\} = \{0.3, -0.8, 1.2\} - t\,\text{gradf}[0.3, -0.8, 1.2] \qquad \text{and} \qquad 0 \le t \le 1:$$

```
In[2]:=
    Clear[f,gradf,x,y,z,t]
    f[x_,y_,z_] = 0.6 Cos[1.8 x y z^2];
    gradf[x_,y_,z_] = {D[f[x,y,z],x],D[f[x,y,z],y],D[f[x,y,z],z]};
    {x[t_], y[t_], z[t_]} = {0.3,-0.8,1.2} - t gradf[0.3,-0.8,1.2];
```

```
In[3]:=
    Plot[f[x[t],y[t],z[t]],{t,0,1},
    PlotStyle->{{Blue,Thickness[0.02]}},
    AxesLabel->{"t",""}];
```

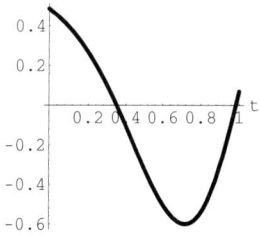

> The knowledge of what fact about gradients gives you the ability to predict, before seeing the plot, that the curve must go **down** before it can go up?
>
> Experiment with other functions $f[x, y, z]$.
>
> Explain why the plot will go down initially no matter what function $f[x, y, z]$ you go with as long as gradf$[0.3, -0.8, 1.2]$ is not $\{0, 0, 0\}$.

G.1.c.i) Here is the curve

$$\{x[t], y[t]\} = \left\{\cos[t] + 1, \frac{3}{2}\sin[t]\right\} \quad \text{for} \quad 0 < t < 2\pi,$$

together with tangent vectors at the points $\{x[\pi/2], y[\pi/2]\}$ and $\{x[3\pi/2], y[3\pi/2]\}$, all in true scale:

In[4]:=
```
Clear[x,y,t]; Clear[tanvect]
{x[t_],y[t_]} = {Cos[t] + 1, 3 Sin[t]/2};
curveplot = ParametricPlot[{x[t],y[t]},{t,0,2 Pi},
PlotStyle->{{Blue,Thickness[0.01]}},
AxesLabel->{"x","y"},DisplayFunction->Identity];
tanvect[t_] := Arrow[{x'[t],y'[t]},Tail->{x[t],y[t]},Red];
setup = Show[curveplot,tanvect[Pi/2],tanvect[3 Pi/2],
AspectRatio->Automatic, DisplayFunction->$DisplayFunction];
```

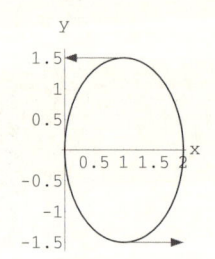

Now go with

$$f[x, y] = 4 - (x + 1)^2 - y^2$$

and add to the plot the gradient vectors gradf$[x[\pi/2], y[\pi/2]]$ with tail at $\{x[\pi/2], y[\pi/2]\}$ and gradf$[x[3\pi/2], y[3\pi/2]]$ with tail at $\{x[3\pi/2], y[3\pi/2]\}$.

Remembering that

$$D[f[x[t], y[t]], t] = \text{gradf}[x[t], y[t]] \bullet \{x'[t], y'[t]\},$$

use the information displayed in your plot to explain how the plot tells you that the derivative $D[f[x[t], y[t]], t]$ is positive at the higher point and negative at the lower point.

G.1.c.ii) Now add to the plot above the tangent vector and the gradient vector at the point $\{x[\pi], y[\pi]\}$.

Again, remembering that

$$D[f[x[t], y[t]], t] = \text{gradf}[x[t], y[t]] \bullet \{x'[t], y'[t]\},$$

use what you see to explain why the derivative $D[f[x[t], y[t]], t]$ is equal to zero at $t = \pi$.

Confirm what you say with a calculation of the derivative at that point.

G.1.d.i) You can describe any line through the point $\{1, 1\}$ by selecting a unit vector $\{a, b\}$ and putting:

$$\{x[t], y[t]\} = \{1, 1\} + t\{a, b\}.$$

> Go with
> $$f[x, y] = x^2 y$$
> and say how you should set a and b so that the derivative of $f[x[t], y[t]]$ with respect to t is as big as possible when $t = 0$.

G.1.d.ii) Here is a true scale plot of the level curve
$$x^2 y = 1$$
together with a line through $\{1, 1\}$:

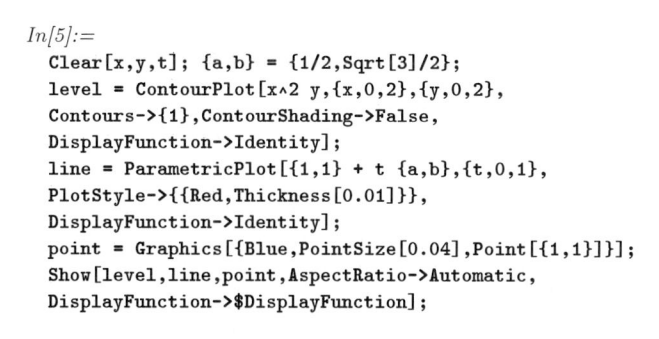

In[5]:=
```
Clear[x,y,t]; {a,b} = {1/2,Sqrt[3]/2};
level = ContourPlot[x^2 y,{x,0,2},{y,0,2},
Contours->{1},ContourShading->False,
DisplayFunction->Identity];
line = ParametricPlot[{1,1} + t {a,b},{t,0,1},
PlotStyle->{{Red,Thickness[0.01]}},
DisplayFunction->Identity];
point = Graphics[{Blue,PointSize[0.04],Point[{1,1}]}];
Show[level,line,point,AspectRatio->Automatic,
DisplayFunction->$DisplayFunction];
```

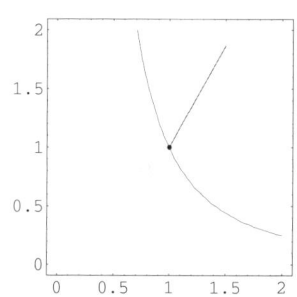

> Copy, paste, and edit the code above so that it plots the curve and the line you found in the last part.
>
> Describe what you see and explain why you see it.

■ G.2) The gradient is perpendicular to the level curves and surfaces★

G.2.a)
> Plot part of the level curve
> $$f[x, y] = y - \sin[x] = 0$$
> and throw in a few normal vectors.

G.2.b) The sphere of radius 2 centered at $\{1, 2, 0\}$ is the level surface of the function
$$f[x, y, z] = (x - 1)^2 + (y - 2)^2 + z^2 = 4.$$

It's not easy to use $f[x, y, z]$ to plot this sphere, but plotting it parametrically is not bad if you use:

In[6]:=
```
Clear[s,t,sphereplotter]
sphereplotter[s_,t_] = {1, 2, 0} + 2 {Sin[s] Cos[t], Sin[s] Sin[t], Cos[s]};
```

Check:

In[7]:=
```
Clear[f,x,y,z]
f[x_,y_,z_] = (x - 1)^2 + ( y - 2)^2 + z^2;
Expand[Apply[f,sphereplotter[s,t]],Trig->True]
```

Out[7]=
4

Good.

Here comes the plot of the top cap of the sphere with a selection of gradient vectors

$$\nabla f[x,y,z] = \text{gradf}[x,y,z]$$

at some selected points on the cap:

In[8]:=
```
Clear[gradf,gradvect]; gradf[x_,y_,z_] =
{D[f[x,y,z],x],D[f[x,y,z],y],D[f[x,y,z],z]};
topcap = ParametricPlot3D[Evaluate[
sphereplotter[s,t]],{s,0,Pi/2},{t,0,2 Pi},
DisplayFunction->Identity];
gradvect[s_,t_] := Arrow[Apply[gradf,
sphereplotter[s,t]],Tail->sphereplotter[s,t],Blue];
Show[topcap,Table[gradvect[s,t],{t,0,7 Pi/4,Pi/4},
{s,0,Pi/2,Pi/4}],ViewPoint->CMView,
AxesLabel->{"x","y","z"},Boxed->False,
BoxRatios->Automatic,PlotRange->All,
DisplayFunction->$DisplayFunction];
```

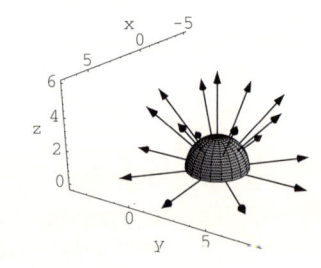

> Describe the relationship between the gradient vectors and the surface and explain why you could have predicted the outcome in advance.

G.2.c.i) > You are looking for the highest and lowest possible z coordinates of points $\{x, y, z\}$ on a the level surface
>
> $$f[x, y, z] = 1$$
>
> for a function $f[x, y, z]$.
>
> Explain:
>
> At the point $\{x_1, y_1, z_1\}$ on the level surface with the highest z coordinate or the lowest z coordinate, the first two slots of $\text{gradf}[x_1, y_1, z_1]$ must be 0.

G.2.c.ii) The points $\{x, y, z\}$ on the sphere $x^2 + y^2 + z^2 = 1$ with the highest and lowest z coordinates are $\{0, 0, 1\}$ and $\{0, 0, -1\}$.

Here's how you use the information in part G.2.c.i) to have *Mathematica* spit these points out:

In[9]:=
```
Clear[f,x,y,z]; f[x_,y_,z_] = x^2 + y^2 + z^2;
gradf[x_,y_,z_] = {D[f[x,y,z],x],D[f[x,y,z],y],D[f[x,y,z],z]};
eqn1 = (f[x,y,z] == 1); eqn2 = (D[f[x,y,z],x] == 0);
eqn3 = (D[f[x,y,z],y] == 0); Solve[{eqn1,eqn2,eqn3}]
```
Out[9]=
{{z -> 1, y -> 0, x -> 0}, {z -> -1, y -> 0, x -> 0}}

The level curve $f[x, y] = 3x^2 - xy + y^2 = 1$ is an ellipse in 2D:

In[10]:=
```
Clear[f,x,y]; f[x_,y_] = 3 x^2 - x y + y^2;
level = ContourPlot[f[x,y],{x,-1,1},{y,-1.3,1.3},
Contours->{1},ContourSmoothing->True,
ContourShading->False];
```

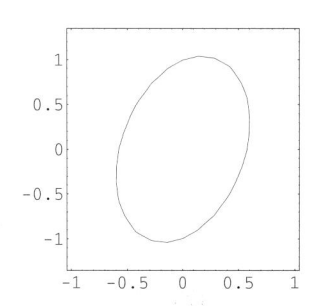

> Adapt the idea immediately above to nail down the points $\{x, y\}$ on this ellipse with the highest and lowest y-coordinates.

■ G.3) The heat seeker*

G.3.a.i) When you go with

$$f[x, y] = 4 - (x - 3)^2 + (y - 5)^2,$$

you can spot the maximizer at $\{3, 5\}$.

In[11]:=
```
Clear[f,gradf,x,y];
f[x_,y_] = 4 - (x - 3)^2 + (y - 5)^2;
gradf[x_,y_] = {D[f[x,y],x],D[f[x,y],y]};
N[Solve[gradf[x,y] == {0,0},{x,y}]]
```
Out[11]=
{{x -> 3., y -> 5.}}

Look at this plot of gradf$[x, y]$ at various points on the circle of radius 1 centered at the maximizer at $\{3, 5\}$:

Lesson 3.03 The Gradient

```
In[12]:=
    maximizer = {3,5};
    maxpoint = {Graphics[{PointSize[0.05],
    Point[maximizer]}],Graphics[Text["maximizer",
    maximizer,{0,-2}]]}; radius = 1; Clear[x,y,t]
    {x[t_],y[t_]} = maximizer + radius {Cos[t],Sin[t]};
    scalefactor = 0.25; gradients = Table[
    Arrow[scalefactor(gradf[x[t],y[t]]),
    Tail->{x[t],y[t]},Red],{t,0,12,0.5}];
    Show[maxpoint,gradients,Axes->True];
```

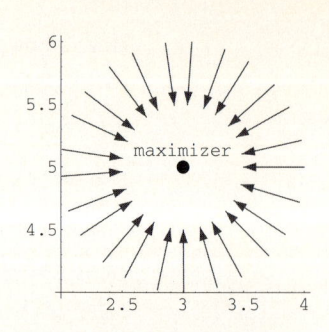

> Why do you think that those scaled gradient vectors are pointing the way they are?

G.3.a.ii) When you go with
$$f[x,y] = (x-1)^2 + (y-2)^2,$$
you can spot the minimizer at $\{1,2\}$.

```
In[13]:=
    Clear[f,gradf,x,y]; f[x_,y_] = (x - 1)^2 + (y - 2)^2;
    gradf[x_,y_] = {D[f[x,y],x],D[f[x,y],y]};
    N[Solve[gradf[x,y] == {0,0},{x,y}]]
Out[13]=
    {{x -> 1., y -> 2.}}
```

Look at this plot of $-\text{gradf}[x,y]$ at various points on the circle of radius 0.5 centered at the minimizer at $\{1,2\}$:

```
In[14]:=
    minimizer = {1,2};
    minpoint = {Graphics[{PointSize[0.04],
    Point[minimizer]}],Graphics[Text["minimizer",
    minimizer,{0,-2}]]}; radius = 0.5; Clear[x,y,t]
    {x[t_],y[t_]} = minimizer + radius {Cos[t],Sin[t]};
    scalefactor = 0.25; negativegradients = Table[
    Arrow[scalefactor (-gradf[x[t],y[t]]),
    Tail->{x[t],y[t]},Red],{t,0,12,0.5}];
    Show[minpoint,negativegradients,Axes->True];
```

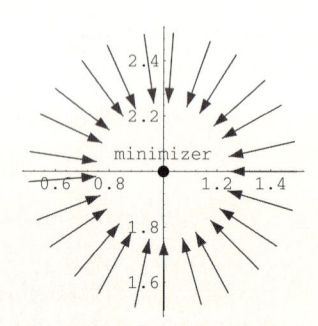

> Why do you think that those scaled negative gradient vectors are pointing the way they are?

G.3.b.i) The Calculus&Mathematica Missile Co. is working on some primitive heat-seeking devices, and you are chief engineer of the TAD (Target Acquistion Division). The current problem under study is to program a device to go to the hottest point in a temperature distribution.

For instance, if
$$\text{temp}[x, y] = \frac{100}{1 + (x - 2.5)^2 + 2(y - 3.5)^2}$$
measures the temperature at a point $\{x, y\}$, then the hottest point is $\{2.5, 3.5\}$ because the denominator is smallest at this point. You can use the gradient to try to make a heat-seeking device that starts at $\{0, 0\}$ and tries to seek out the hottest spot $\{2.5, 3.5\}$.

Here's a look:

In[15]:=
```
Clear[temp,gradtemp,x,y]
temp[x_,y_] = 100/(1 + (x - 2.5)^2 + 2 (y - 3.5)^2);
gradtemp[x_,y_] = {D[temp[x,y],x],D[temp[x,y],y]};
hottestpoint = {2.5,3.5}; start = {0,0};
hotpt= Graphics[{{PointSize[0.04],Red,Point[
hottestpoint]},Text["hottest",hottestpoint,{0,4}]}];
startpt= Graphics[{{PointSize[0.04],Blue,
Point[start]},Text["start",start,{-1.5,-1}]}];
setup = Show[hotpt,startpt,PlotRange->All,
Axes->True,AxesOrigin->{0,0},AspectRatio->1];
```

The heat seeker can't tell the exact location of the hot spot, but it can sense the gradient of the temp$[x, y]$ merely by noting the hottest direction at a point $\{x, y\}$.

> If the heat seeker is at a point $\{x, y\}$, why should you program the heat seeker so that it leaves $\{x, y\}$ in the direction of gradtemp$[x, y]$?

G.3.b.ii) Given:

→ The heat seeker can update its direction every instant.

→ The heat seeker is programmed so that it leaves $\{x, y\}$ in the direction of gradtemp$[x, y]$.

> Explain why the following plot displays a good approximation of the heat seeker's actual path when the seeker starts at $\{0, 0\}$:

In[16]:=
```
Clear[Derivative,x,y,t]
equationx = (x'[t] == gradtemp[x[t],y[t]][[1]]);
equationy = (y'[t] == gradtemp[x[t],y[t]][[2]]);
starterx = (x[0] == 0); startery = (y[0] == 0); endtime = 10;
approxsolutions = NDSolve[{equationx,equationy,starterx,
startery},{x[t],y[t]},{t,0,endtime}];
Clear[seeker]; seeker[t_] = {x[t]/.approxsolutions[[1]],
y[t]/.approxsolutions[[1]]}; seekerplot = ParametricPlot[seeker[t],
{t,0,endtime},PlotStyle->{{Thickness[0.02]}},DisplayFunction->Identity];
```

Lesson 3.03 The Gradient

$In[17]:=$
```
Show[setup,seekerplot,
 PlotRange->All,
 DisplayFunction->$DisplayFunction];
```

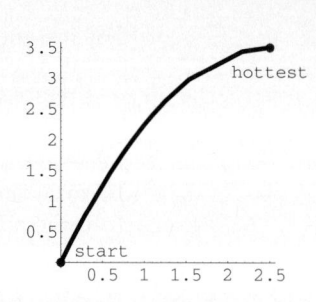

G.3.b.iii) Given:

→ The heat seeker can update its direction every instant.

→ The heat seeker is programmed so that it leaves $\{x, y\}$ in the direction of gradtemp$[x, y]$.

> Give a plot of the heat seeker's path when the seeker starts at $\{0, 2\}$.

G.3.c.i) The people over at the assembly division tell you that the heat seeker can't be built to update its direction at every instant. Instead, it will update its direction many times, but it will move on straight line segments between direction updates. Your group at TAD reacts to this information by programming the heat seeker as follows:

→ If the heat seeker is at $\{x_{k-1}, y_{k-1}\}$, then the heat seeker moves to a new point

$$\{x_k, y_k\} = \{x_{k-1}, y_{k-1}\} + \text{jump gradtemp}[x_{k-1}, y_{k-1}]$$

where the jump is a positive number selected by trial and error.

> For appropriately small jump numbers, why is this a good update?

G.3.c.ii) Now go from theory to practice. Start at $\{0, 0\}$ and program the heat seeker with jump $= 0.1$ and 20 updates:

$In[18]:=$
```
start = {0,0}; hotpt = Graphics[{{PointSize[0.04],Red,
Point[hottestpoint]},Text["hottest",hottestpoint,{0,4}]}];
startpt=Graphics[{{PointSize[0.04],Blue,
Point[start]},Text["start",start,{-1.5,-1}]}];
jump = 0.1; updates = 20; Clear[next,point,k]
next[{x_,y_}] = {x,y} + jump gradtemp[x,y];
point[0] = start; point[k_] :=
point[k] = N[next[point[k-1]]];
path = Graphics[{Thickness[0.01],
Line[Table[point[k],{k,0,updates}]]}];
Show[hotpt,startpt,path,Axes->Automatic,
AxesLabel->{"x","y"},AspectRatio->Automatic];
```

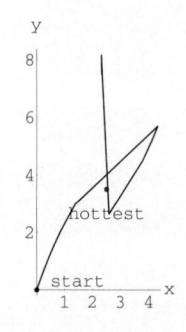

Good start, but the heat seeker lost its cool. It nearly ran right over the hottest spot without even stopping to say hello. Reprogram it with a smaller jump and more updates:

In[19]:=
```
jump = 0.04; updates = 40; Clear[next,point,k]
next[{x_,y_}] = {x,y} + jump gradtemp[x,y];
point[0] = start;
point[k_] := point[k] = N[next[point[k-1]]];
path = Graphics[{Thickness[0.01],
  Line[Table[point[k],{k,0,updates}]]}];
Show[hotpt,startpt,path,Axes->Automatic,
  AxesLabel->{"x","y"},AspectRatio->Automatic];
```

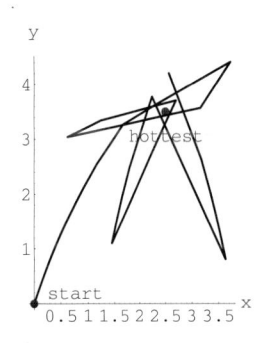

This time the heat seeker got close, but it blew its cool just as it was about to accomplish its desires.

> Use trial and error to program in a jump size and an update number that send the heat seeker steadily to the hot spot so it can ignite its warhead and blow that hot spot to smithereens.

G.3.c.iii)
> One way to increase the efficiency of the heat seeker is to use one of the larger jump sizes at first and run it until it goes bats. Then use the last good point generated as a new starting point with a new, reduced jump size and run again.
>
> Try this out, starting at $\{0,0\}$ on the same function as above and incorporating any additional ideas that come to you.

G.3.d)
> Use jumps and updates, as above, to guide the heat seeker starting at $\{0,0\}$ to the hottest point for:
>
> *In[20]:=*
> ```
> Clear[temp,gradtemp,x,y]
> temp[x_,y_] = 100/(1 + ( 2 x - y)^2 + (x - 1)^2);
> ```

■ G.4) Doing 'em by hand★

The ability to do these problems is a good sign of gradient literacy. You should be able to do them by hand over coffee, writing with a pencil on a napkin or the back of an envelope.

Lesson 3.03 The Gradient

G.4.a) Lots of folks like to use the notations
$$\nabla f[x, y] = \text{grad} f[x, y]$$
and
$$\nabla f[x, y, z] = \text{grad} f[x, y, z].$$

> Calculate the gradient $\nabla f[x, y]$ by hand for
> $$f[x, y] = e^{2x-3y}.$$
> Calculate the gradient $\nabla f[x, y, z]$ by hand for
> $$f[x, y, z] = \cos\left[\frac{x^2 y}{z}\right].$$

G.4.b.i) Lots of folks like to use the notations
$$f^{(1,0)}[x, y] = \frac{\partial f[x, y]}{\partial x} = D[f[x, y], x]$$
and
$$f^{(0,1)}[x, y] = \frac{\partial f[x, y]}{\partial y} = D[f[x, y], y].$$

> Calculate
> $$\frac{\partial f[x, y]}{\partial x} = f^{(1,0)}[x, y]$$
> by hand for $f[x, y] = x^2 + y^2 + 1$.

G.4.b.ii)
> Calculate
> $$\frac{\partial f[x, y]}{\partial y} = f^{(0,1)}[x, y]$$
> by hand for $f[x, y] = e^{x-4y}$.

G.4.b.iii)
> Calculate
> $$\frac{\partial f[x, y, z]}{\partial z} = f^{(0,0,1)}[x, y, z]$$
> by hand for $f[x, y, z] = \sin[x^2 y^3 z^4]$.

G.4.c) Here is
$$\frac{\partial f[x,y]}{\partial x} = f^{(1,0)}[x,y]$$
for $f[x,y] = e^{\sin[xy-x]}$:

In[21]:=
```
Clear[f,x,y]; f[x_,y_] = E^Sin[x y - x]; D[f[x,y],x]
```
Out[21]=
$$-\left(\frac{(1 - y)\,\text{Cos}[x - x\,y]}{E^{\text{Sin}[x - x\,y]}}\right)$$

And here is
$$\frac{\partial f[1.8, 0.7]}{\partial x} = f^{(1,0)}[1.8, 0.7]$$
for the same function:

In[22]:=
```
D[f[x,y],x]/.{x->1.8,y->0.7}
```
Out[22]=
-0.153877

> What does this calculation tell you about what happens to $f[x,y]$ when you hold y at 0.7 but make x just a teensy-weensy bit bigger than 1.8?

G.4.d) Use the total differential of $f[x,y,z] = x\,y\,z$ to give a hand derivation of the formula
$$D[x[t]\,y[t]\,z[t], t] = x'[t]\,y[t]\,z[t] + x[t]\,y'[t]\,z[t] + x[t]\,y[t]\,z'[t].$$

G.4.e) Explain why the following notation gives you nothing more than the total differential of f. Think about the relationship between the gradient and the total differential.
$$\nabla f[x,y] \bullet \{dx, dy\} = \left\{\frac{\partial f[x,y]}{\partial x}, \frac{\partial f[x,y]}{\partial y}\right\} \bullet \{dx, dy\}.$$

G.4.f) Comment on the analogy among the chain rule formulas:
→ $D[f[x[t]], t] = f'[x[t]]\,x'[t]$
→ $D[f[x[t], y[t]], t] = \text{gradf}[x[t], y[t]] \bullet \{x'[t], y'[t]\}$
→ $D[f[x[t], y[t], z[t]], t] = \text{gradf}[x[t], y[t], z[t]] \bullet \{x'[t], y'[t], z'[t]\}.$

G.4.g) The fundamental formula of calculus and the chain rule for functions of one variable give you the clean formula $\int_a^b g'[x[t]]\, x'[t]\, dt = g[x[b]] - g[x[a]]$.

> Use the fundamental formula of calculus and the chain rule for functions of two variables to give a clean formula for $\int_a^b \text{gradf}[x[t], y[t]] \cdot \{x'[t], y'[t]\}\, dt$.

■ G.5) The highest crests and the deepest dips

G.5.a) Way back in 1958, a fellow named Beale came up with a function whose plot looks like an alpine valley. Here is a fairly careful rendition of the plot of his function:

In[23]:=
```
Clear[f,x,y]
f[x_,y_] = (1.500 - x (1 - y^2))^2 +
  (2.250 - x (1 - y^3))^2 + (2.625 - x (1 - y^4))^2;
Plot3D[f[x,y],{x,-3,6},{y,-2,2},
  PlotPoints->{20,20},PlotRange->{0,20},ClipFill->None,
  ViewPoint->CMView,AxesLabel->{"x","y","z"}];
```

Look at:

In[24]:=
```
Clear[gradf]
gradf[x_,y_] = {D[f[x,y],x],D[f[x,y],y]};
candidates = Chop[NSolve[gradf[x,y] == {0,0}]]
```

Out[24]=
```
{{x -> 0, y -> -1.37528},
 {x -> 0.0962699 - 0.0274921 I, y -> -1.26456 - 1.76695 I},
 {x -> 0.0962699 + 0.0274921 I, y -> -1.26456 + 1.76695 I},
 {x -> 1.97765 + 0.37012 I, y -> -0.518203 - 0.826085 I},
 {x -> 1.97765 - 0.37012 I, y -> -0.518203 + 0.826085 I},
 {x -> 2.24351, y -> -0.44166}, {x -> 0, y -> -0.240932 - 1.30684 I},
 {x -> 0, y -> -0.240932 + 1.30684 I}, {x -> 2.125, y -> 0},
 {x -> 5.31639, y -> 0.84052}, {x -> 0, y -> 1.}}
```

> Use what you see above to locate the deepest point in this valley.

G.5.b) Lots of folks like to call the function

$$f[x, y] = 10\left(y - x^2\right)^2 + \left(1 - x^2\right)^2$$

"Rosenbrock's banana," because its level curves look like bananas and because it was first studied by someone named Rosenbrock. Take a look:

In[25]:=
```
Clear[f,x,y,r,t]
f[x_,y_] = 10(y - x^2)^2 + (1 - x^2)^2;
chiquita = ContourPlot[f[x,y],{x,-2,2},{y,-4,6},
PlotRange->{0,50}];
```

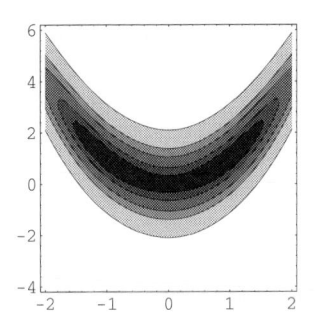

Here's a 3D plot of the banana function:

In[26]:=
```
Plot3D[f[x,y],{x,-2,2},{y,-4,6},
ViewPoint->CMView,AxesLabel->{"x","y","z"}];
```

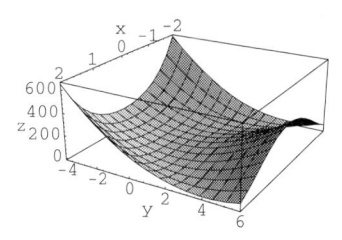

Note the parabolic valley running through the surface.

> Use gradf$[x,y]$ to find the lowest points (= deepest dips) in this valley. After you've got the locations, plant flagpoles at these locations so everyone can see them.

G.5.c) The function studied in this problem was adapted from the book *Process Optimization with Applications in Metallurgy and Chemical Engineering,* by W. H. Ray and J. Szekely, Wiley-Interscience, New York, 1973.

Engineers at the C&M Engineering and Foundry Company are analyzing the profit per hour from a solid-state reaction carried on at a high temperature t. In their work, they come across a function like this:

In[27]:=
```
Clear[f,x,t]
f[x_,t_] = (10 (1 - 0.7 x) x t^3 E^(0.01 t) -
 E^(0.02 t) - 5 x(1 - x) - 0.1 t^4)/10^18;
```

You are working in the math division of C&M, and in comes some e-mail from the engineers asking for the x and t that make this function as large as possible. The engineers mention that x is a variable the engineers call "void fraction" and t is the temperature at which the reaction is to run.

This function takes you aback because there is no hope of setting its gradient equal to 0 and then solving for x and t. You call down to the engineers and ask what reasonable ranges on the temperature t and x are. They reply that the reaction can run at temperatures between 2200 and 2500 and x must be between 0.3 and 1.0. You thank them, and then you immediately look at:

In[28]:=
```
ContourPlot[f[x,t],{x,0.3,1.0},{t,2200,2500}];
```

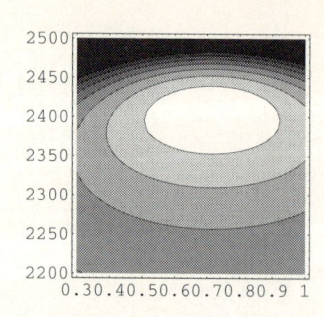

You take one look at this and say, "Yippee!" because now you know how to use *Mathematica* to come up with the optimal x and t.

> Do it.

G.5.d.i)
> Analyze the formula
> $$f[x, y] = \frac{\sin[2\,x] + \sin[3\,y]}{0.5 + x^2 + y^2}$$
> and say how you know that the surface $z = f[x, y]$ definitely has a deepest dip and a highest crest.

G.5.d.ii)
> Use contour plots, surface plots, and the FindMinimum instruction to help you come up with your own best estimates of the exact locations of the highest crest and the deepest dip on the surface you studied in part G.5.d.i) immediately above.
>
> Once you've located them, plot the surface with flagpoles planted at the top of the highest crest and at the bottom of the deepest dip.

■ G.6) Closest points, gradients, and Lagrange's method

This problem appears only in the electronic version.

■ G.7) The Cobb-Douglas manufacturing model for industrial engineering

Manufacturing costs are usually split into capital (CEO stock options, golden parachutes, limos, private jets, other overhead, equipment, etc.) and labor (salaries and sweat). Manufacturers can manipulate these expenses in many ways. They can automate heavily, or they can cut capital costs by increasing the labor force.

Economists Cobb and Douglas became famous because they noticed that if a manufacturing process uses x units of capital and y units of labor, then the output function of a manufacturing process is usually approximated by

$$f[x,y] = A\, x^{k/m}\, y^{(m-k)/m}$$

for some fixed constants A, k, and m that depend on the goods being manufactured.

In[29]:=
```
Clear[f,x,y,A,k,m,t]; f[x_,y_] = A x^(k/m) y^((m - k)/m)
```

Out[29]=
```
   k/m  (-k + m)/m
A x    y
```

This makes some sense because if you look at:

In[30]:=
```
{f[t x,t y], t f[x,y]}
```

Out[30]=
```
        k/m      (-k + m)/m        k/m  (-k + m)/m
{A (t x)    (t y)          , A t x    y           }
```

Then a little mental algebra revolving around the fact that

$$t^{k/m}\, t^{(-k+m)/m} = t^{k/m + (-k+m)/m} = t^{m/m} = t$$

tells you that

$$f[t\,x, t\,y] = t\, f[x,y],$$

and this means that the output is directly proportional to the inputs of capital and labor.

In this problem, you are a big shot in The C&M Manufacturing Co. and are preparing a production run of a certain product. The bean counters over in the finance office say that for a production run, capital costs will be $129.15 for each unit x of capital and $95.74 dollars for each unit y of labor. So the overall cost in dollars of a production run is

$$129.15\, x + 95.74\, y.$$

The company has budgeted exactly $50,000 on this production run. This gives you the budget line

$$129.15\, x + 95.74\, y = 50000.$$

The industrial engineers have figured out that that x units of labor and y units of capital result in a production run of

$$f[x,y] = 149.2\, x^{3/4}\, y^{1/4}$$

units.

In[31]:=
```
Clear[f,gradf,x,y,t]; f[x_,y_] = 149.2 x^(3/4) y^(1/4)
```
Out[31]=
$$149.2 \, x^{3/4} \, y^{1/4}$$

In[32]:=
```
gradf[x_,y_] = {D[f[x,y],x],D[f[x,y],y]}
```
Out[32]=
$$\left\{ \frac{111.9 \, y^{1/4}}{x^{1/4}}, \frac{37.3 \, x^{3/4}}{y^{3/4}} \right\}$$

Go for a plot of the budget line $129.15\,x + 95.74\,y = 50000$. To get a useful plot of the line, pick off the two extreme possibilities:

In[33]:=
```
nolabor = Solve[(129.15 x + 95.74 y == 50000)/.x->0]
alllabor = Solve[(129.15 x + 95.74 y == 50000)/.y->0]
```
Out[33]=
```
{{y -> 522.248}}
{{x -> 387.147}}
```

In[34]:=
```
point1 = {0,522.25}
point2 = {387.15,0}
```
Out[34]=
```
{0, 522.25}
{387.15, 0}
```

In[35]:=
```
{x[t_],y[t_]} = point1 + t (point2 - point1)
```
Out[35]=
```
{387.15 t, 522.25 - 522.25 t}
```

As t runs from 0 to 1, the points $\{x[t], y[t]\}$ sweep out the budget line from one extreme to the other.

The extreme at the left uses only labor and the extreme at the right uses only capital.

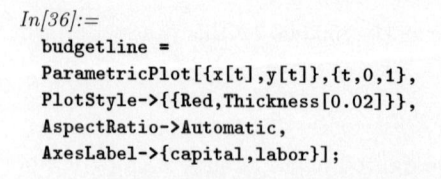

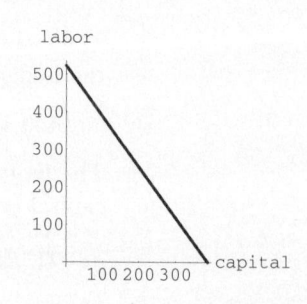

In[36]:=
```
budgetline =
ParametricPlot[{x[t],y[t]},{t,0,1},
PlotStyle->{{Red,Thickness[0.02]}},
AspectRatio->Automatic,
AxesLabel->{capital,labor}];
```

The points $\{x[t], y[t]\}$ on this line represent all possible expenditures {capital,labor} that conform to the budget of 50,000 dollars.

> Find the t^* between 0 and 1 such that the production $f[x[t], y[t]]$ is maximized at $t = t^*$, and report the optimal values
>
> $$\{x[t^*], y[t^*]\} = \{\text{capital units, labor units}\}$$
>
> that give the most production for the given budget.

G.7.a.i) Here are some gradient vectors $\text{gradf}[x[t], y[t]]$ with their tails at a choice of points $\{x[t], y[t]\}$ on the line plotted above; the plot is in true scale:

```
In[37]:=
  grads = Table[Arrow[gradf[x[t],y[t]],
    Tail->{x[t],y[t]},Blue],
    {t,0.1,0.95,0.85/8}];
  Show[budgetline,grads,
  AspectRatio->Automatic,AxesLabel->None];
```

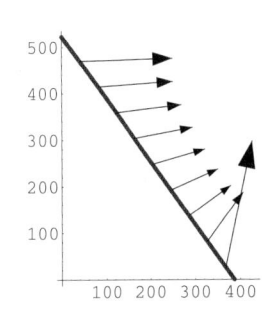

> Take the maximizing t^* you got in the last part and add the plot of
>
> $$2\,\text{gradf}[x[t^*], y[t^*]]$$
>
> with tail at
>
> $$\{x[t^*], y[t^*]\}$$
>
> to the plot immediately above.
>
> (The scale factor 2 is tacked on so that you can easily distinguish this gradient vector from the others.)
>
> Describe what you see and then discuss the information conveyed by the plots of the other gradient vectors above.

G.7.a.ii)
> Use the chain rule formula
>
> $$D[f[x[t], y[t]], t] = \text{gradf}[x[t], y[t]] \bullet \{x'[t], y'[t]\}$$
>
> to explain why your plot of $2\,\text{gradf}[x[t^*], y[t^*]]$ with tail at $\{x[t^*], y[t^*]\}$ turned out the way it did.

■ G.8) Data fit in two variables: Plucking a guitar string

This problem appears only in the electronic version.

■ G.9) Linearizations and total differentials*

This problem appears only in the electronic version.

■ G.10) Keeping track of constituent costs

This problem appears only in the electronic version.

■ G.11) The great pretender

This problem appears only in the electronic version.

LESSON 3.04

2D Vector Fields and Their Trajectories

Bascis

■ B.1) Vector fields and their trajectories

B.1.a) A vector field is a function that spits out vectors. You make a 2D vector field by taking two regular functions, $m[x,y]$ and $n[x,y]$ and throwing them into the two slots:

In[1]:=
```
Clear[Field,m,n,x,y];
m[x_,y_] = 0.5 (y - 1); n[x_,y_] = 0.3 (x - 0.5);
Field[x_,y_] = {m[x,y],n[x,y]}
```

Out[1]=
{0.5 (-1 + y), 0.3 (-0.5 + x)}

You plot a vector field by plotting the vector, $Field[x,y]$, with its tail at $\{x,y\}$ for a selection of points $\{x,y\}$:

In[2]:=
```
vectorfieldplot =
Table[Arrow[Field[x,y],Tail->{x,y},Blue],
{x,-3,3,0.5},{y,-3,3,0.5}];
Show[vectorfieldplot,Axes->True,AxesLabel->{"x","y"}];
```

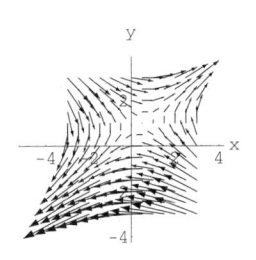

A mad rush to the lower left and the upper right.

133

B.1.a.i) What's a good way of interpreting this plot?

Answer: Look at what went into the plot. The plot shows field vectors Field$[x,y]$ plotted with their tails at $\{x,y\}$ for a selection of points $\{x,y\}$. The selection of points was:

In[3]:=
```
pointplot =
Table[Graphics[{Red,PointSize[0.02],Point[{x,y}]}],
{x,-3,3,0.5},{y,-3,3,0.5}];
Show[pointplot,Axes->True,AxesLabel->{"x","y"}];
```

The plot shows field vectors, Field$[x,y]$, plotted with their tails at the points $\{x,y\}$ shown above. Take a look:

In[4]:=
```
Show[pointplot,vectorfieldplot,Axes->True,
AxesLabel->{"x","y"}];
```

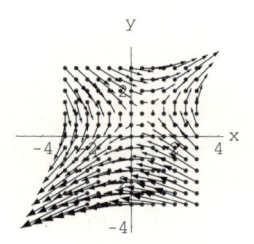

Now you're ready to interpret the plot. Think of the whole xy-plane as fluid flowing with currents and eddies.

→ The vectors represent the flow of the fluid at its tails.

→ The length of each vector indicates the speed of the fluid flow at its tail.

→ The direction of each vector indicates the direction of the flow.

In the plot above, some of the fluid is flowing off to the lower left and some of it is flowing off to the upper right.

The upshot: The field vector Field$[x,y]$ with tail at $\{x,y\}$ represents the speed and the direction with which a cork at $\{x,y\}$ moves away from $\{x,y\}$ as it is caught by the flow.

B.1.a.ii) Continue to go with the same vector field, Field$[x,y]$, as used in part B.1.a.i) above and look at this plot:

In[5]:=
```
Clear[Derivative,x,y,t]; Clear[trajectory]
{a,b} = {2,-1}; starterpoint = {a,b};
equationx = (x'[t] == m[x[t],y[t]]);
equationy = (y'[t] == n[x[t],y[t]]);
starterx = (x[0] == a); startery = (y[0] == b);
endtime = 6; approxsolutions = NDSolve[{equationx,
equationy,starterx,startery},{x[t],y[t]},{t,0,endtime}];
trajectory[t_] = {x[t]/.approxsolutions[[1]],
y[t]/.approxsolutions[[1]]}; trajectoryplot =
ParametricPlot[trajectory[t],{t,0,endtime},
PlotStyle->{{Red,Thickness[0.008]}},
DisplayFunction->Identity]; starterplot =
Graphics[{Red,PointSize[0.06],Point[starterpoint]}];
Show[vectorfieldplot,starterplot,trajectoryplot,
PlotRange->All,DisplayFunction->$DisplayFunction];
```

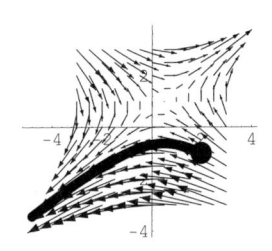

Some folks like to call the path you see a trajectory of the vector field. Other folks call this path a streamline of the vector field.

What's going on here?

Answer: Run it again with a different starting point.

In[6]:=
```
{a,b} = {3,-0.5}; starterpoint = {a,b};
Clear[Derivative,x,y,t]
equationx = (x'[t] == m[x[t],y[t]]);
equationy = (y'[t] == n[x[t],y[t]]);
starterx = (x[0] == a); startery = (y[0] == b);
endtime = 6; approxsolutions = NDSolve[
{equationx,equationy,starterx,startery},
{x[t],y[t]},{t,0,endtime}]; Clear[newtrajectory];
newtrajectory[t_] = {x[t]/.approxsolutions[[1]],
y[t]/.approxsolutions[[1]]};
newtrajectoryplot = ParametricPlot[newtrajectory[t],
{t,0,endtime},PlotStyle->{{Red,Thickness[0.008]}},
DisplayFunction->Identity]; newstarterplot =
Graphics[{Red,PointSize[0.06],Point[starterpoint]}];
Show[vectorfieldplot,newstarterplot,newtrajectoryplot,
PlotRange->All,DisplayFunction->$DisplayFunction];
```

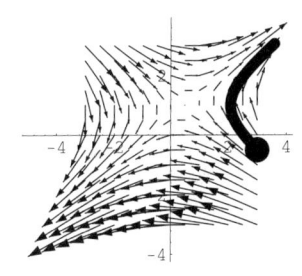

See both plots together:

In[7]:=
```
Show[vectorfieldplot,starterplot,newstarterplot,
trajectoryplot,newtrajectoryplot,PlotRange->All,
DisplayFunction->$DisplayFunction];
```

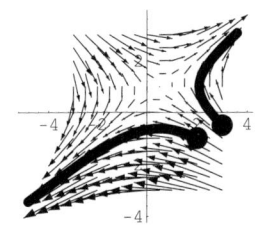

Now get down to brass tacks.

These plots depict the path of a cork dropped in the flow defined by

$$\text{Field}[x, y] = \{m[x, y], n[x, y]\}$$

at the starter point.

Reason: The specifications in the differential equation

$$\text{starterpoint} = \{a, b\},$$
$$\text{equationx} = (x'[t] = m[x[t], y[t]]),$$
$$\text{equationy} = (y'[t] = n[x[t], y[t]]),$$
$$\text{starterx} = (x[0] = a)$$

and

$$\text{startery} = (y[0] = b)$$

tell you that the path will start at $\{a, b\}$ and that at each point $\{x[t], y[t]\}$ on the path, the field vector

$$\text{Field}[x[t], y[t]] = \{m[x[t], y[t]], n[x[t], y[t]]\}$$

with tail at $\{x[t], y[t]\}$ is tangent to the path.

In fact, at each point $\{x[t], y[t]\}$ on the path, the field vector

$$\text{Field}[x[t], y[t]] = \{m[x[t], y[t]], n[x[t], y[t]]\}$$

is the velocity vector of the cork as it floats on its merry way.

Take a look:

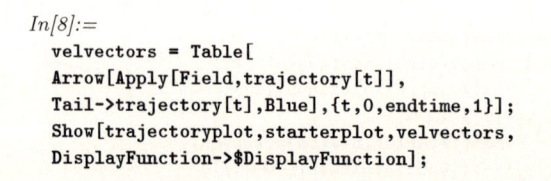

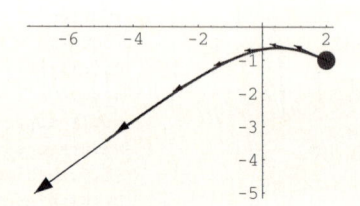

In[8]:=
```
velvectors = Table[
  Arrow[Apply[Field,trajectory[t]],
  Tail->trajectory[t],Blue],{t,0,endtime,1}];
Show[trajectoryplot,starterplot,velvectors,
  DisplayFunction->$DisplayFunction];
```

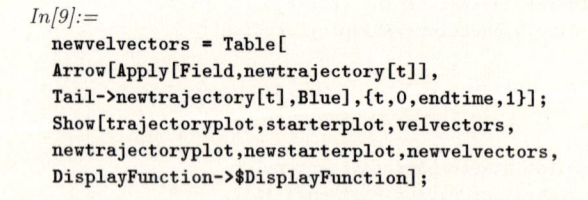

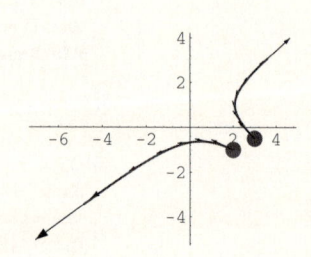

In[9]:=
```
newvelvectors = Table[
  Arrow[Apply[Field,newtrajectory[t]],
  Tail->newtrajectory[t],Blue],{t,0,endtime,1}];
Show[trajectoryplot,starterplot,velvectors,
  newtrajectoryplot,newstarterplot,newvelvectors,
  DisplayFunction->$DisplayFunction];
```

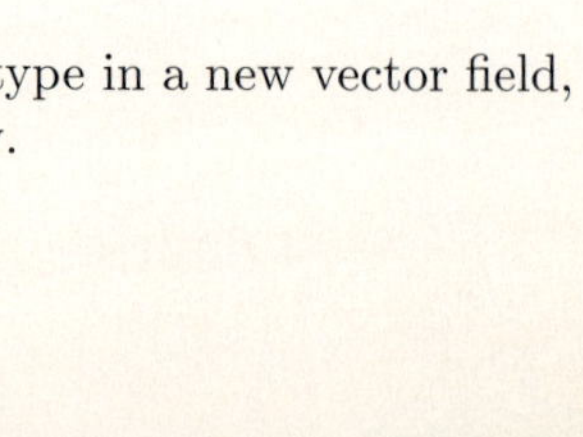

To really get a lot out of this, go back to the beginning, type in a new vector field, and redo everything above until you too go with the flow.

Bascis (B.2)

B.1.a.iii) How does the vector field

$$\text{Field}[x, y] = \{m[x, y], n[x, y]\}$$

govern the actual path of the cork dropped into the flow?

Answer: If the cork finds itself at a position $\{x, y\}$, then it has to go with the flow. The flow at $\{x, y\}$ is in the direction of:

In[10]:=
```
Field[x,y]
```

Out[10]=
```
{0.5 (-1 + y), 0.3 (-0.5 + x)}
```

To go with the flow, the cork must leave $\{x, y\}$ in the direction of the vector Field$[x, y]$. As the cork progresses, it corrects its direction instantaneously at each point to keep on the course that the flow determines.

As you saw above, the result is that if $\{x, y\}$ is any point on the actual path of the cork, then the vector Field$[x, y]$ with its tail at $\{x, y\}$ is tangent to the actual path of the cork.

■ B.2) Flow of vector fields along curves; flow of vector fields across curves: Visual inspection

B.2.a) Describe the net flow across the ellipse parameterized by

$$P[t] = \{6 \cos[t], 4 \sin[t]\} \qquad \text{with } 0 \leq t \leq 2\pi$$

of a fluid whose velocity is given by the vector field

$$\text{Field}[x, y] = \{x - 1, y\}.$$

Answer: Enter the vector field:

In[11]:=
```
Clear[Field,m,n,x,y]
m[x_,y_] = x - 1; n[x_,y_] = y;
Field[x_,y_] = {m[x,y],n[x,y]}
```

Out[11]=
```
{-1 + x, y}
```

Here's a look at the curve and the vector field:

Lesson 3.04 2D Vector Fields and Their Trajectories

In[12]:=
```
vectorfieldplot = Table[Arrow[
Field[x,y],Tail->{x,y},Blue],
{x,-6,6,1},{y,-4,4,2}]; Clear[t]
x[t_] = 6 Cos[t]; y[t_] = 4 Sin[t];
curveplot = ParametricPlot[{x[t],y[t]},{t,0,2 Pi},
PlotStyle->{{Red,Thickness[0.01]}},
DisplayFunction->Identity];
Show[vectorfieldplot,curveplot,Axes->True,
AspectRatio->Automatic,AxesLabel->{"x","y"},
DisplayFunction->$DisplayFunction];
```

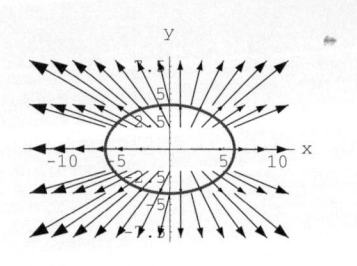

As you can see, this curve isn't a trajectory of the given field.

Your goal here is to analyze what the flow is doing at the points right on this curve. To do this, look at the field vectors at points right on the curve and ignore what's happening elsewhere:

In[13]:=
```
jump = Pi/8;
fieldvectors = Table[Arrow[
Field[x[t],y[t]],Tail->{x[t],y[t]},
Blue],{t,0,2 Pi - jump,jump}];
Show[curveplot,fieldvectors,AxesOrigin->{0,0},
AspectRatio->Automatic,AxesLabel->{"x","y"},
DisplayFunction->$DisplayFunction];
```

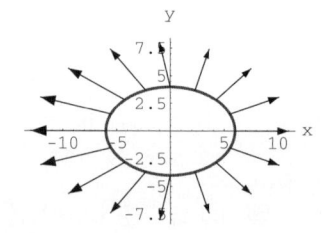

These vectors indicate the direction and the speed of the flow at points of the curve at the tails of the field vectors. No doubt about it. The net flow of this vector field across this curve is from inside to outside.

B.2.b) Describe the net flow across the ellipse parameterized by

$$P[t] = \{6 \cos[t], 4 \sin[t]\} \quad \text{with } 0 \leq t \leq 2\pi$$

of a fluid whose velocity is given by the vector field

$$\text{Field}[x, y] = \{y - 1, y + x\}.$$

Answer: Enter the vector field:

In[14]:=
```
Clear[Field,m,n,x,y]
m[x_,y_] = y - 1; n[x_,y_] = y + x;
Field[x_,y_] = {m[x,y],n[x,y]}
```

Out[14]=
```
{-1 + y, x + y}
```

Here's a look at the curve and the vector field:

Bascis (B.2)

In[15]:=
```
vectorfieldplot = Table[Arrow[Field[x,y],Tail->{x,y},Blue],
  {x,-6,6,1},{y,-4,4,2}]; Clear[t]
x[t_] = 6 Cos[t]; y[t_] = 4 Sin[t];
curveplot = ParametricPlot[{x[t],y[t]},{t,0,2 Pi},
  PlotStyle->{{Red,Thickness[0.01]}},
  DisplayFunction->Identity]; Show[vectorfieldplot,
  curveplot,Axes->True,AspectRatio->Automatic,
  AxesLabel->{"x","y"},DisplayFunction->$DisplayFunction];
```

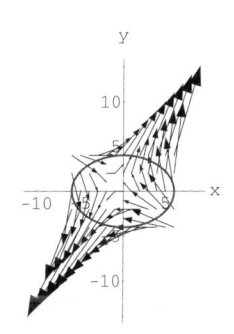

The curve and the vector field are two independent creatures. One has no influence on the other.

Your goal here is to analyze what the flow is doing at the points right on this curve.

To do this, look at the field vectors at points right on the curve, and ignore what's happening elsewhere:

In[16]:=
```
jump = Pi/8;
fieldvectors = Table[Arrow[Field[x[t],y[t]],
  Tail->{x[t],y[t]},Blue],{t,0,2 Pi - jump,jump}];
Show[curveplot,fieldvectors,AxesOrigin->{0,0},
  AspectRatio->Automatic,AxesLabel->{"x","y"},
  DisplayFunction->$DisplayFunction];
```

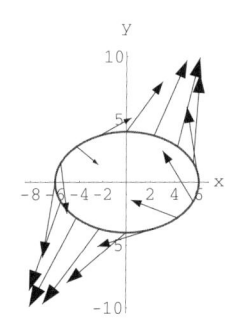

These vectors indicate the direction and the speed of the flow at points of the curve at the tails of the field vectors.

No doubt about it. The net flow of this vector field across this curve is from inside to outside.

B.2.c) Describe the net flow across and along the ellipse parameterized by

$$P[t] = \{5\cos[t], 3\sin[t]\} \quad \text{with } 0 \leq t \leq 2\pi$$

of a fluid whose velocity is given by the vector field

$$\text{Field}[x, y] = \{x - y, y/2\}.$$

Answer: Enter the vector field:

In[17]:=
```
Clear[Field,m,n,x,y]
m[x_,y_] = x - y; n[x_,y_] = y/2;
Field[x_,y_] = {m[x,y],n[x,y]}
```

Lesson 3.04 *2D Vector Fields and Their Trajectories*

Out[17]=

$\{x - y, \dfrac{y}{2}\}$

Here's a look at the curve and the vector field:

In[18]:=
```
Clear[t]; vectorfieldplot = Table[Arrow[Field[
  x,y],Tail->{x,y},Blue],{x,-6,6,1},{y,-4,4,2}];
x[t_] = 5 Cos[t]; y[t_] = 3 Sin[t];
curveplot = ParametricPlot[{x[t],y[t]},{t,0,2 Pi},
  PlotStyle->{{Red,Thickness[0.01]}},
  DisplayFunction->Identity]; Show[vectorfieldplot,
  curveplot,Axes->True,AspectRatio->Automatic,
  AxesLabel->{"x","y"},DisplayFunction->$DisplayFunction];
```

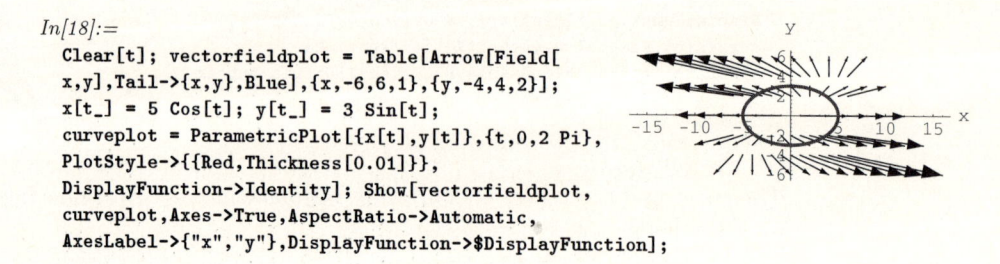

Your goal here is to analyze what the flow is doing at the points right on this curve. To do this, look at the field vectors at points right on the curve and ignore what's happening elsewhere:

In[19]:=
```
jump = Pi/8;
fieldvectors = Table[Arrow[Field[x[t],y[t]],
  Tail->{x[t],y[t]},Blue],{t,0,2 Pi - jump,jump}];
Show[curveplot,fieldvectors,AxesOrigin->{0,0},
  AspectRatio->Automatic,AxesLabel->{"x","y"},
  DisplayFunction->$DisplayFunction];
```

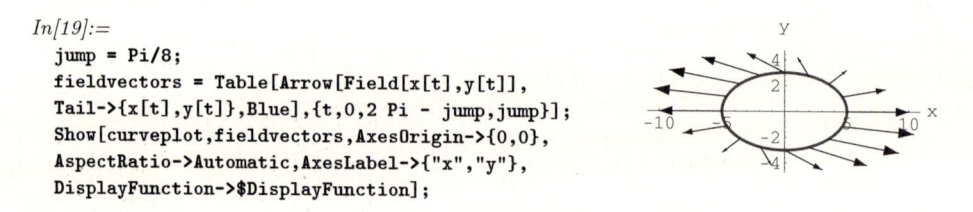

These vectors indicate the direction and the speed of the flow at points of the curve at the tails of the field vectors. Looking at the plot, you can see that:

→ The net flow of this vector field across this curve is from inside to outside.

→ The net flow of this vector field along this curve is counterclockwise.

■ B.3) Flow of vector fields along curves; flow of vector fields across curves

B.3.a.i) Try to describe the net flow across and along the ellipse parameterized by

$$P[t] = \{5 \cos[t], 3 \sin[t]\} \quad \text{with } 0 \leq t \leq 2\pi$$

of a fluid whose velocity is given by the vector field

$$\text{Field}[x, y] = \{x/4 + y - 1, 0.5\,(x + y^2)\}.$$

Answer: Enter the vector field:

In[20]:=
```
Clear[Field,m,n,x,y]
m[x_,y_] = x/4 + y - 1; n[x_,y_] = 0.5(x + y^2);
Field[x_,y_] = {m[x,y],n[x,y]}
```

Out[20]=
$$\{-1 + \frac{x}{4} + y, \ 0.5 \ (x + y^2)\}$$

Here's a look at the curve and the vector field:

In[21]:=
```
vectorfieldplot = Table[Arrow[Field[x,y],Tail->{x,y},Blue],
 {x,-6,6,1},{y,-4,4,2}]; Clear[t]
x[t_] = 5 Cos[t]; y[t_] = 3 Sin[t];
curveplot = ParametricPlot[{x[t],y[t]},{t,0,2 Pi},
 PlotStyle->{{Red,Thickness[0.01]}},
 DisplayFunction->Identity]; Show[vectorfieldplot,
 curveplot,Axes->True,AspectRatio->Automatic,
 AxesLabel->{"x","y"},DisplayFunction->$DisplayFunction];
```

Your goal here is to analyze what the flow is doing at the points right on this curve. To do this, look at the field vectors at points right on the curve and ignore what's happening elsewhere:

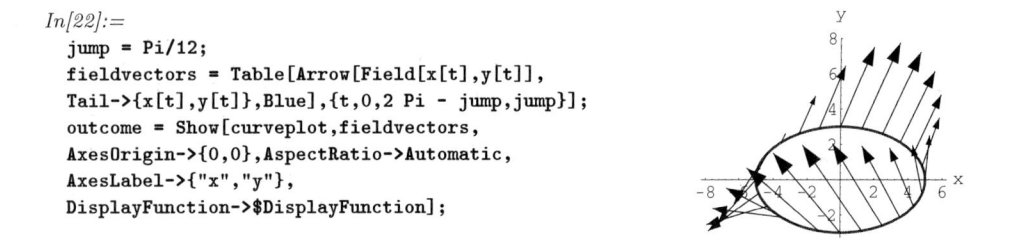

In[22]:=
```
jump = Pi/12;
fieldvectors = Table[Arrow[Field[x[t],y[t]],
 Tail->{x[t],y[t]},Blue],{t,0,2 Pi - jump,jump}];
outcome = Show[curveplot,fieldvectors,
 AxesOrigin->{0,0},AspectRatio->Automatic,
 AxesLabel->{"x","y"},
 DisplayFunction->$DisplayFunction];
```

It's hard to tell from this plot whether the net flow along the curve is clockwise or counterclockwise.

And it's hard to tell from this plot whether the net flow across the curve is from outside to inside or from inside to outside. You can spot some features:

→ The plot indicates clockwise flow along the curve at the top and the bottom and counterclockwise flow along the curve on the left and on the right.

→ The plot also indicates that the flow across the curve is from inside to outside at the top and on the left, but, on the bottom, the flow across the curve is from outside to inside.

B.3.a.ii) Here's the plot of

$$\text{Field}[x, y] = \left\{ \frac{x}{4} + y - 1, 0.5 \left(x + y^2 \right) \right\}$$

at points on the curve in part B.3.a.i) above:

In[23]:=
```
Show[outcome];
```

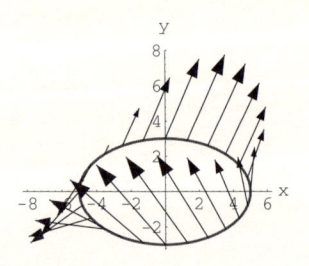

Look at the components of these field vectors in the direction of the tangent vectors to the curve:

In[24]:=
```
Clear[tangent,tancomponent]
tangent[t_] = {x'[t],y'[t]};
tancomponent[t_] = (((Field[x[t],y[t]].tangent[t])/
    (tangent[t].tangent[t])) tangent[t]);
actualflowalong = Table[Arrow[tancomponent[t],
    Tail->{x[t],y[t]},Blue],{t,0,2 Pi - jump,jump}];
flowalongplot = Show[curveplot,actualflowalong,
AxesOrigin->{0,0},AspectRatio->Automatic,
AxesLabel->{"x","y"},DisplayFunction->$DisplayFunction];
```

This is not a plot of the tangent vectors for the parameterization of the curve. This is a plot of the components of the field vectors in the direction of the tangent vectors for the curve.

> What does this plot of the tangential components of the field vectors tell you?

Answer: No doubt about it. The net flow of this vector field along this curve is clockwise.

B.3.a.iii)
> What do you do to make a visual estimation of the flow of the vector field in part B.3.a.i) above across the ellipse?

Answer: Plot the components of the field vectors in directions perpendicular to the curve. To do this, look at:

In[25]:=
```
Clear[tangent,normal]
tangent[t_] = {x'[t],y'[t]}; normal[t_] = {y'[t],- x'[t]};
normal[t].tangent[t]
```

Out[25]=
0

The upshot:
$$\text{normal}[t] = \{y'[t], -x'[t]\}$$
is perpendicular to the curve at the point $\{x[t], y[t]\}$.

Here comes the plot of the normal components of the field vectors on the curve:

In[26]:=
```
Clear[normalcomponent]
normalcomponent[t_] = (((Field[x[t],y[t]].normal[t])/
    (normal[t].normal[t])) normal[t]);
actualflowacross = Table[Arrow[normalcomponent[t],
    Tail->{x[t],y[t]},SteelBlue],{t,0,2 Pi - jump,jump}];
flowacrossplot = Show[curveplot,actualflowacross,
AxesOrigin->{0,0},AspectRatio->Automatic,
AxesLabel->{"x","y"},DisplayFunction->$DisplayFunction];
```

There is some flow from outside to inside, but there's a heck of a lot more flow from inside to outside.

No doubt about it; the net flow of this vector field across this curve is from inside to outside.

Tutorials

■ T.1) Flow across and flow along: Visual inspection

Here's a vector field:

In[1]:=
```
Clear[Field,m,n,x,y]
m[x_,y_] = 1; n[x_,y_] = y - Sin[x];
Field[x_,y_] = {m[x,y],n[x,y]}
```

Out[1]=
```
{1, y - Sin[x]}
```

Here's a curve:

In[2]:=
```
Clear[t]
{x[t_],y[t_]} = {0.8 Cos[t],0.5 Sin[t]};
curveplot = ParametricPlot[{x[t],y[t]},{t,0,2 Pi},
PlotStyle->{{Thickness[0.01],Red}},
AxesLabel->{"x","y"}];
```

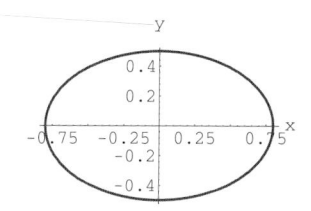

T.1.a.i) Give a visual analysis of the flow of this vector field across this curve.

Lesson 3.04 2D Vector Fields and Their Trajectories

Answer: First, plot the vector field on the curve:

In[3]:=
```
jump = Pi/12;
fieldvectors = Table[Arrow[Field[x[t],y[t]],
  Tail->{x[t],y[t]},Blue],{t,0,2 Pi - jump,jump}];
outcome = Show[curveplot,fieldvectors];
```

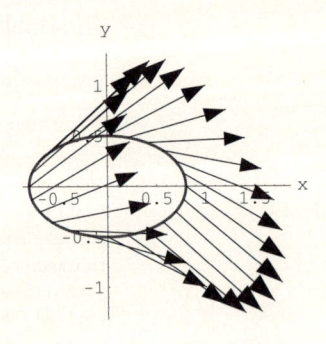

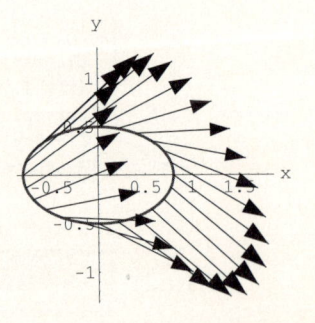

It looks like the net flow of this vector field across this curve is from inside to outside, but there is some flow from outside to inside on the left. To get a more accurate picture of the flow of this vector field across this curve, look at the normal components of the field vectors plotted above.

In[4]:=
```
Clear[normal]
normal[t_] = {y'[t],-x'[t]};
Clear[normalcomponent]
normalcomponent[t_] = (((Field[x[t],y[t]].normal[t])/
  (normal[t].normal[t])) normal[t]);
actualflowacross = Table[Arrow[normalcomponent[t],
  Tail->{x[t],y[t]}],{t,0,2 Pi - jump,jump}];
flowacrossplot = Show[curveplot,actualflowacross];
```

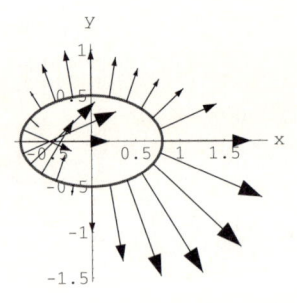

Now there's no doubt about it; the net flow of this vector field across this curve is from inside to outside.

T.1.a.ii) Give a visual analysis of the flow of this vector field along this curve.

Answer: First, plot the vector field on the curve:

In[5]:=
```
jump = Pi/12;
fieldvectors = Table[Arrow[Field[x[t],y[t]],
  Tail->{x[t],y[t]},Blue],{t,0,2 Pi - jump,jump}];
outcome = Show[curveplot,fieldvectors];
```

It looks like the net flow of this vector field along this curve is clockwise, but there is some counterclockwise flow along this curve at the bottom.

To get a more accurate picture of the flow of this vector field along this curve, look at the tangential components of the field vectors plotted above.

In[6]:=
```
Clear[tangent]
tangent[t_] = {x'[t],y'[t]};
Clear[tancomponent]
tancomponent[t_] = (((Field[x[t],y[t]].tangent[t])/
  (tangent[t].tangent[t])) tangent[t]);
actualflowalong = Table[Arrow[tancomponent[t],
  Tail->{x[t],y[t]}],{t,0,2 Pi - jump,jump}];
flowalongplot = Show[curveplot,actualflowalong];
```

This is not a plot of the tangent vectors for the parameterization of the curve. This is a plot of the components of the field vectors in the direction of the tangent vectors for the curve.

Now there's no doubt about it; clockwise flow along this curve overwhelms the counterclockwise flow. The net flow of this vector field along this curve is clockwise.

T.1.a.iii) This problem appears only in the electronic version.

■ T.2) Differential equations and their associated vector fields

Calculus&Mathematica is pleased to say that this problem was greatly influenced by the book *Differential Equations, A Dynamical Systems Approach*, Part 1 by J. H. Hubbard and B. H. West (Springer-Verlag, New York,1991). If you like this problem, then you'll want to experiment with the vector fields in the first chapter of that book.

T.2.a) When you have a differential equation like
$$y'[x] = 2x - y[x],$$
you can make the associated vector field:
$$\text{DEField}[x, y] = \{1, 2x - 7\}:$$

In[7]:=
```
Clear[DEField,x,y];
DEField[x_,y_] = {1, 2 x - y}; scalefactor = 0.2;
DEFieldplot = Table[Arrow[scalefactor DEField[x,y],
  Tail->{x,y},Blue],{x,-4,4,8/10},{y,-4,4,8/10}];
Show[DEFieldplot,Axes->Automatic,AxesLabel->{"x","y"}];
```

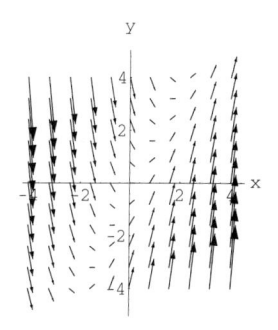

Lesson 3.04 2D Vector Fields and Their Trajectories

The flow of this field pulls all the trajectories down from the upper left, turns them around, and spits them out at the upper right. Try it:

In[8]:=
```
Clear[m,n]
{m[x_,y_],n[x_,y_]} = DEField[x,y]
```

Out[8]=
```
{1, 2 x - y}
```

In[9]:=
```
{a,b} = {-3.6,4}; starterpoint = {a,b};
Clear[Derivative,x,y,t]
equationx = (x'[t] == m[x[t],y[t]]);
equationy = (y'[t] == n[x[t],y[t]]);
starterx = (x[0] == a); startery = (y[0] == b);
endtime = 7; approxsolutions = NDSolve[
{equationx,equationy,starterx,startery},
{x[t],y[t]},{t,0,endtime}]; Clear[trajectory]
trajectory[t_] = {x[t]/.approxsolutions[[1]],
y[t]/.approxsolutions[[1]]};
trajectoryplot = ParametricPlot[trajectory[t],
{t,0,endtime},PlotStyle->{{Red,Thickness[0.01]}},
DisplayFunction->Identity]; starterplot =
Graphics[{Red,PointSize[0.06],Point[starterpoint]}];
trajectory = Show[DEFieldplot,starterplot,
trajectoryplot,PlotRange->All,Axes->Automatic,
AxesLabel->{"x","y"},DisplayFunction->$DisplayFunction];
```

What's the significance of the trajectories in the DEField?

Answer: The trajectory plotted above is nothing but a plot of a solution of the given differential equation

$$y'[x] = 2\,x - y[x] \quad \text{with } y[-3.6] = 4.$$

To see why, look at the ingredients of the trajectory plotter above:

In[10]:=
```
{a,b}
```

Out[10]=
```
{-3.6, 4}
```

In[11]:=
```
equationx = (x'[t] == m[x[t],y[t]])
```

Out[11]=
```
x'[t] == 1
```

In[12]:=
```
equationy = (y'[t] == n[x[t],y[t]])
```

Out[12]=
```
y'[t] == 2 x[t] - y[t]
```

In[13]:=
```
starterx = (x[0] == a)
```
Out[13]=
```
x[0] == -3.6
```

In[14]:=
```
startery = (y[0] == b)
```
Out[14]=
```
y[0] == 4
```

These equations tell you that this trajectory is the same as the plot of the solution of the differential equation $y'[x] = 2\,x - y[x]$ with $y[-3.6] = 4$.

Here's the plot of the solution of the same differential equation

$$y'[x] = 2\,x - y[x] \quad \text{with } y[-2] = 0$$

as a trajectory in this vector field:

In[15]:=
```
Clear[Derivative,x,y,t]; Clear[trajectory]
{a,b} = {-2,0}; starterpoint = {a,b};
equationx = (x'[t] == m[x[t],y[t]]);
equationy = (y'[t] == n[x[t],y[t]]);
starterx = (x[0] == a); startery = (y[0] == b);
begintime = -1; endtime = 7;
approxsolutions = NDSolve[{equationx,equationy,
  starterx,startery},{x[t],y[t]},{t,begintime,endtime}];
trajectory[t_] = {x[t]/.approxsolutions[[1]],
y[t]/.approxsolutions[[1]]}; trajectoryplot =
ParametricPlot[trajectory[t],{t,begintime,endtime},
PlotStyle->{{Red,Thickness[0.01]}},
DisplayFunction->Identity]; starterplot =
Graphics[{Red,PointSize[0.06],Point[starterpoint]}];
trajectory = Show[DEFieldplot,starterplot,
trajectoryplot,PlotRange->All,
DisplayFunction->$DisplayFunction];
```

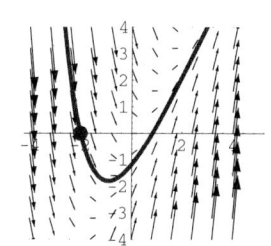

T.2.b) Here's another look at the vector field coming from the differential equation

$$y'[x] = 2\,x - y[x] :$$

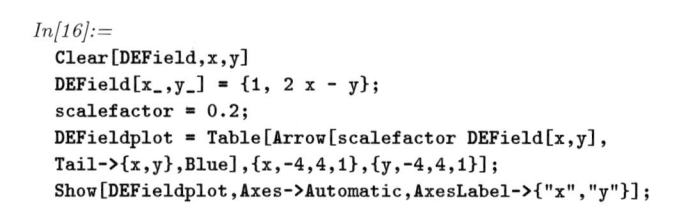

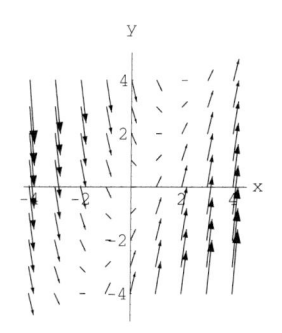

In[16]:=
```
Clear[DEField,x,y]
DEField[x_,y_] = {1, 2 x - y};
scalefactor = 0.2;
DEFieldplot = Table[Arrow[scalefactor DEField[x,y],
Tail->{x,y},Blue],{x,-4,4,1},{y,-4,4,1}];
Show[DEFieldplot,Axes->Automatic,AxesLabel->{"x","y"}];
```

> What does this plot tell you about the solutions of this differential equation?

Answer: Good question. It tells you a lot.

It tells you that no matter how you set a and b so that $\{a, b\}$ is within the plot range above, the solution of
$$y'[x] = 2x - y[x] \quad \text{with } y[a] = b$$
will go down on the left and then will turn itself around and go up on the right.

This plot tells you one heckuva lot more than does the formula coming from:

In[17]:=
```
Clear[x,y,a,b]
DSolve[{y'[x] == 2 x - y[x],y[a] == b},y[x],x]
```

Out[17]=
$$\{\{y[x] \to -2 + (2 - 2a + b) E^{a-x} + 2x\}\}$$

but this formula does tell you that any solution of
$$y'[x] = 2x - y[x] \quad \text{with } y[a] = b$$
does get sucked onto the line
$$y = 2x - 2$$
when x is large and positive. You can also see this fact in the DEField plot:

In[18]:=
```
sucker = Plot[2 x - 2,{x,-1,4},PlotStyle->
  {{Thickness[0.02],Red}},DisplayFunction->Identity];
Show[DEFieldplot,sucker,Axes->Automatic,
  AxesLabel->{"x","y"},
  DisplayFunction->$DisplayFunction];
```

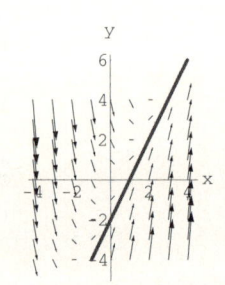

That's mathematics at work.

■ **T.3) Flow across a curve and the sign of the dot product**
$$\text{Field}[x[t], y[t]] \bullet \{y'[t], -x'[t]\}$$
Flow along a curve and the sign of the dot product
$$\text{Field}[x[t], y[t]] \bullet \{x'[t], y'[t]\}$$

Given a two-dimensional curve parameterized by $\{x[t], y[t]\}$ with $a \leq t \leq b$, you can get a tangent vector at $\{x[t], y[t]\}$ that points in the direction of the parameterization by setting

$$\text{tangent}[t] = \{x'[t], y'[t]\}.$$

You can also get a normal vector at $\{x[t], y[t]\}$ by setting

$$\text{normal}[t] = \{y'[t], -x'[t]\}.$$

Try them out:

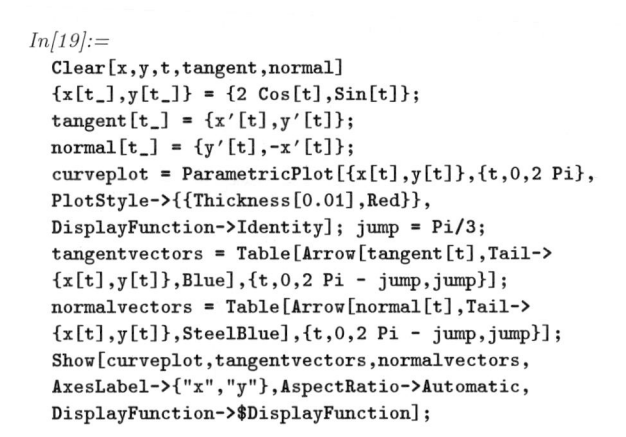

In[19]:=
```
Clear[x,y,t,tangent,normal]
{x[t_],y[t_]} = {2 Cos[t],Sin[t]};
tangent[t_] = {x'[t],y'[t]};
normal[t_] = {y'[t],-x'[t]};
curveplot = ParametricPlot[{x[t],y[t]},{t,0,2 Pi},
  PlotStyle->{{Thickness[0.01],Red}},
  DisplayFunction->Identity]; jump = Pi/3;
tangentvectors = Table[Arrow[tangent[t],Tail->
  {x[t],y[t]},Blue],{t,0,2 Pi - jump,jump}];
normalvectors = Table[Arrow[normal[t],Tail->
  {x[t],y[t]},SteelBlue],{t,0,2 Pi - jump,jump}];
Show[curveplot,tangentvectors,normalvectors,
  AxesLabel->{"x","y"},AspectRatio->Automatic,
  DisplayFunction->$DisplayFunction];
```

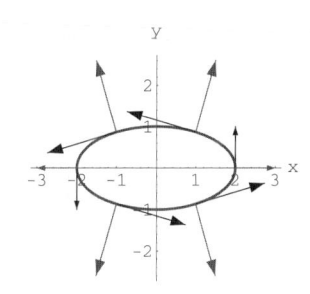

The tangent vectors, $\{x'[t], y'[t]\}$, tell you that the curve is parameterized in the counterclockwise way. The normal vectors, $\{y'[t], -x'[t]\}$, point to the outside of the curve.

Try another curve:

In[20]:=
```
Clear[x,y,t,tangent,normal]
{x[t_],y[t_]} = {Cos[t](1 - 0.5 Sin[3 t]),Sin[t]};
tangent[t_] = {x'[t],y'[t]};
normal[t_] = {y'[t],-x'[t]};
curveplot = ParametricPlot[{x[t],y[t]},{t,0,2 Pi},
  PlotStyle->{{Thickness[0.01],Red}},
  DisplayFunction->Identity]; jump = Pi/3;
tangentvectors = Table[Arrow[tangent[t],Tail->
  {x[t],y[t]},Blue],{t,0,2 Pi - jump,jump}];
normalvectors = Table[Arrow[normal[t],Tail->
  {x[t],y[t]},SteelBlue],{t,0,2 Pi - jump,jump}];
Show[curveplot,tangentvectors,normalvectors,
  AxesLabel->{"x","y"},AspectRatio->Automatic,
  DisplayFunction->$DisplayFunction];
```

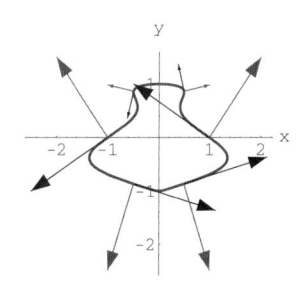

Again, the tangent vectors $\{x'[t], y'[t]\}$ tell you that the curve is parameterized in the counterclockwise way. The normal vectors $\{y'[t], -x'[t]\}$ point to the outside of the curve.

Try another curve:

Lesson 3.04 2D Vector Fields and Their Trajectories

In[21]:=
```
Clear[x,y,t,tangent,normal]
{x[t_],y[t_]} = {t (2 Pi - t), 2 Sin[t]};
tangent[t_] = {x'[t],y'[t]};
normal[t_] = {y'[t],-x'[t]};
curveplot = ParametricPlot[{x[t],y[t]},{t,0,2 Pi},
PlotStyle->{{Thickness[0.01],Red}},
DisplayFunction->Identity]; jump = Pi/3;
tangentvectors = Table[Arrow[tangent[t],Tail->
{x[t],y[t]},Blue],{t,0,2 Pi - jump,jump}];
normalvectors = Table[Arrow[normal[t],Tail->
{x[t],y[t]},SteelBlue],{t,0,2 Pi - jump,jump}];
Show[curveplot,tangentvectors,normalvectors,
AxesLabel->{"x","y"},DisplayFunction->$DisplayFunction];
```

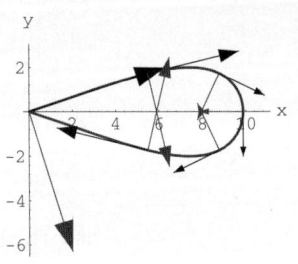

This time, the tangent vectors $\{x'[t], y'[t]\}$ tell you that the curve is parameterized in the clockwise way. This time, the normal vectors $\{y'[t], -x'[t]\}$ point to the inside of the curve.

T.3.a.i) Is it true that:

→ When a closed curve is parameterized by $\{x[t], y[t]\}$ in the counterclockwise way, then the normal vectors $\{y'[t], -x'[t]\}$ always point to the outside of the curve?

→ When a closed curve is parameterized by $\{x[t], y[t]\}$ in the clockwise way, then the normal vectors $\{y'[t], -x'[t]\}$ always point to the inside of the curve?

Answer: Yes. In fact, as you walk around the curve in the direction of the parameterization, the normal vector $\{y'[t], -x'[t]\}$ always points to your right.

T.3.a.ii) Given a curve parameterized by $\{x[t], y[t]\}$ with $a \leq t \leq b$, you have your choice of normal vectors at $\{x[t], y[t]\}$:

You can opt for

$$\{y'[t], -x'[t]\},$$

or you can go with

$$D[\text{unittan}[t], t] \qquad \text{where unittan}[t] = \frac{\{x'[t], y'[t]\}}{\sqrt{x'[t]^2 + y'[t]^2}}.$$

Discuss the merits of each.

Answer: For hand work, $\{y'[t], -x'[t]\}$ is more easily calculated than is $D[\text{unittan}[t], t]$. The choice of which to use depends on what you want to do.

→ If you want a normal that always points to the right as you advance on the curve in the direction of the parameterization, go with $\{y'[t], -x'[t]\}$.

→ If you want a normal that always points with the curvature, go with $D[\text{unittan}[t], t]$.

You probably know how handy $D[\text{unittan}[t], t]$ can be. If you want to see how handy $\{y'[t] - x'[t]\}$ can be, go on to the next problem.

T.3.b) Here's a vector field:

In[22]:=
```
Clear[Field,m,n,x,y]
m[x_,y_] = 1.4 x^2; n[x_,y_] = y - Sin[x];
Field[x_,y_] = {m[x,y],n[x,y]}
```

Out[22]=
$\{1.4\, x^2,\ y - \text{Sin}[x]\}$

Here's a curve:

In[23]:=
```
Clear[t]
{x[t_],y[t_]} = {Cos[t],0.5 Sin[t]};
curveplot = ParametricPlot[{x[t],y[t]},{t,0,2 Pi},
PlotStyle->{{Thickness[0.01],Red}},
AxesLabel->{"x","y"}];
```

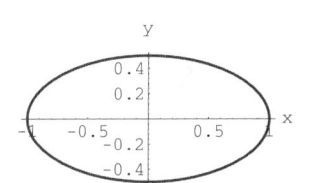

Look at this plot of some of the unit tangent vectors

$$\text{unittangent}[t] = \frac{\{x'[t], y'[t]\}}{\sqrt{x'[t]^2 + y'[t]^2}}$$

pointing in the direction of the parameterization:

In[24]:=
```
Clear[t,unittangent]
unittangent[t_] = {x'[t],y'[t]}/Sqrt[x'[t]^2 + y'[t]^2];
jump = Pi/8; unittangents = Table[Arrow[
unittangent[t],Tail->{x[t],y[t]},Red],
{t,0,2 Pi - jump,jump}];
Show[curveplot,unittangents];
```

This plot tells you that the curve is parameterized in the counterclockwise way. And this tells you that the normal vector $\{y'[t], -x'[t]\}$ at $\{x[t], y[t]\}$ points to the outside of the curve. Now look at this plot of

$\text{Field}[x[t], y[t]] \bullet \{y'[t], -x'[t]\}$:

In[25]:=
```
dotplot = Plot[Field[x[t],y[t]].{y'[t],-x'[t]},
{t,0,2 Pi},PlotStyle->{{Thickness[0.01],Blue}},
AxesLabel->{"t",""},AspectRatio->1/2];
```

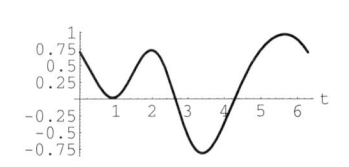

Lesson 3.04 2D Vector Fields and Their Trajectories

> Interpret this plot.

Answer: This is a plot of
$$\text{Field}[x[t], y[t]] \bullet \{y'[t], -x'[t]\}.$$
When
$$\text{Field}[x[t], y[t]] \bullet \{y'[t], -x'[t]\} > 0,$$
the flow of the vector field across the curve is in the direction of the outward normal vector $\{y'[t], -x'[t]\}$. When $\text{Field}[x[t], y[t]] \bullet \{y'[t], -x'[t]\} < 0$, the flow of the vector field across the curve is opposite to the direction of the outward normal vector $\{y'[t], -x'[t]\}$.

Since the normals $\{y'[t], -x'[t]\}$ point out from the curve, the upshot is this:

At a point $\{x[t], y[t]\}$ on the curve, the flow across the curve is:

→ from inside to outside when $\text{Field}[x[t], y[t]] \bullet \{y'[t], -x'[t]\} > 0$

→ from outside to inside when $\text{Field}[x[t], y[t]] \bullet \{y'[t], -x'[t]\} < 0$.

Armed with this insight, take another look at the plot of $\text{Field}[x[t], y[t]] \bullet \{y'[t], -x'[t]\}$.

In[26]:=
    Show[dotplot];

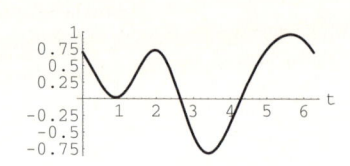

Eye-ball estimates tell you that the flow of this vector field across this curve is

→ from inside to outside at points $\{x[t], y[t]\}$ with $0 \le t < 2.5$ and $4.3 < t \le 2\pi$

→ from outside to inside at points $\{x[t], y[t]\}$ with $2.7 < t < 4.1$.

Using FindRoot, you can get more accurate estimates if you like.

Check it out:

In[27]:=
```
out1 = Table[Arrow[Field[x[t],y[t]],Tail->{x[t],y[t]},
  Blue],{t,0,2.5,(2.5 - 0)/6}];
out2 = Table[Arrow[Field[x[t],y[t]],Tail->{x[t],y[t]},
  Blue],{t,4.3,2 Pi,(2 Pi - 4.3)/6}];
in1 = Table[Arrow[Field[x[t],y[t]],Tail->{x[t],y[t]},
  Blue],{t,2.7,4.1,(4.1 - 2.7)/6}];
Show[curveplot,out1,out2];
```

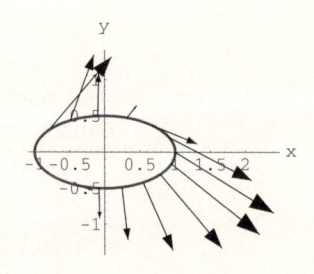

Outward flow across the curve at the plotted points on the curve. This is a good representative plot of the part of this vector field that flows from inside to outside across this curve. Now look at:

In[28]:=
```
Show[curveplot,in1];
```

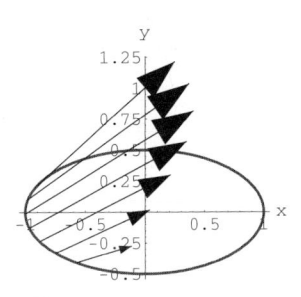

Inward flow across the curve at the plotted points on the curve. This is a good representative plot of the part of this vector field that flows from outside to inside across this curve.

And you set up both plots by doing nothing more than eye-balling a plot of
$$\text{Field}[x[t], y[t]] \bullet \{y'[t], -x'[t]\}.$$
Math works.

T.3.c) Alert C&M participants will note that this problem is nothing but a copy and paste job of part T.3.b) above.

Here's a vector field:

In[29]:=
```
Clear[Field,m,n,x,y]
m[x_,y_] = 3 x - y^2; n[x_,y_] = 2 y;
Field[x_,y_] = {m[x,y],n[x,y]}
```

Out[29]=
```
            2
{3 x - y , 2 y}
```

Here's a curve:

In[30]:=
```
Clear[t]
{x[t_],y[t_]} = {Cos[t](Cos[t]^2 + 1), Sin[t]};
curveplot = ParametricPlot[{x[t],y[t]},{t,0,2 Pi},
PlotStyle->{{Thickness[0.01],Red}},
AxesLabel->{"x","y"}];
```

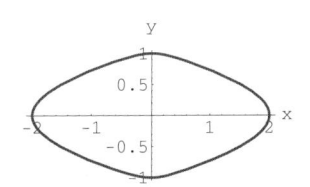

Look at this plot of some of the unit tangent vectors
$$\text{unittangent}[t] = \frac{\{x'[t], y'[t]\}}{\sqrt{x'[t]^2 + y'[t]^2}}$$
pointing in the direction of the parameterization:

In[31]:=
```
Clear[t,unittangent]
unittangent[t_] = {x'[t],y'[t]}/
Sqrt[x'[t]^2 + y'[t]^2]; jump = Pi/3;
unittangents = Table[Arrow[unittangent[t],Tail->
{x[t],y[t]},Blue],{t,0,2 Pi - jump,jump}];
Show[curveplot,unittangents];
```

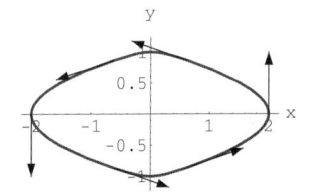

Lesson 3.04 2D Vector Fields and Their Trajectories

This plot tells you that the curve is parameterized in the counterclockwise way. Now look at this plot of

$$\text{Field}[x[t], y[t]] \bullet \{x'[t], y'[t]\}:$$

In[32]:=
```
dotplot = Plot[Field[x[t],y[t]].{x'[t],y'[t]},
  {t,0,2 Pi},PlotStyle->{{Thickness[0.01],Blue}},
  AxesLabel->{"t",""},AspectRatio->1/2];
```

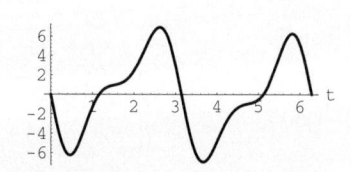

Interpret this plot.

Answer: This is a plot of

$$\text{Field}[x[t], y[t]] \bullet \{x'[t], y'[t]\}.$$

When

$$\text{Field}[x[t], y[t]] \bullet \{x'[t], y'[t]\} > 0,$$

the flow of the vector field along the curve is in the direction of the tangent vector $\{x'[t], y'[t]\}$. When $\text{Field}[x[t], y[t]] \bullet \{x'[t], y'[t]\} < 0$, the flow of the vector field along the curve is opposite the direction of the tangent vector $\{x'[t], y'[t]\}$.

Because the tangents $\{x'[t], y'[t]\}$ point in the counterclockwise direction, the upshot is: At a point $\{x[t], y[t]\}$ on the curve, the flow along the curve is:

→ counterclockwise when $\text{Field}[x[t], y[t]] \bullet \{x'[t], y'[t]\} > 0$

→ clockwise when $\text{Field}[x[t], y[t]] \bullet \{x'[t], y'[t]\} < 0$.

Armed with this insight, take another look at the plot of

$$\text{Field}[x[t], y[t]] \bullet \{x'[t], y'[t]\}:$$

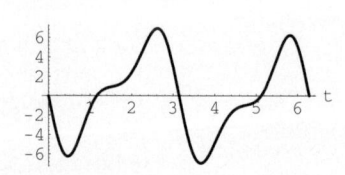

In[33]:=
```
Show[dotplot];
```

Eye-ball estimates tell you that the flow of this vector field along this curve is

→ clockwise at points $\{x[t], y[t]\}$ with $0 \leq t \leq 1$ and $3.1 \leq t \leq 5.1$

→ counterclockwise at points $\{x[t], y[t]\}$ with $1.2 \leq t \leq 3$ and $5.2 \leq t \leq 2\pi$.

Using FindRoot, you can get more precise estimates if you like. Check it out:

In[34]:=
```
clockwise1 = Table[Arrow[Field[x[t],y[t]],Tail->
  {x[t],y[t]},Blue],{t,0,1,(1 - 0)/6}];
clockwise2 = Table[Arrow[Field[x[t],y[t]],Tail->
  {x[t],y[t]},Blue],{t,3.1,5.1,(5.1 - 3.1)/6}];
counterclockwise1 = Table[Arrow[Field[x[t],y[t]],
  Tail->{x[t],y[t]},Blue],{t,1.2,3,(3 - 1.2)/6}];
counterclockwise2 = Table[Arrow[Field[x[t],y[t]],
  Tail->{x[t],y[t]},Blue],{t,5.2,2 Pi,(2 Pi - 5.2)/6}];
Show[curveplot,clockwise1,clockwise2];
```

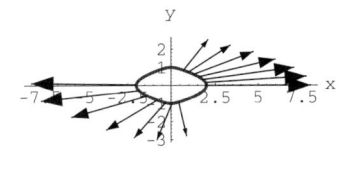

Clockwise flow along the curve at the plotted points on the curve. This is a good representative plot of the part of this vector field that flows in the clockwise direction along this curve. Now look at:

In[35]:=
```
Show[curveplot,counterclockwise1,
  counterclockwise2];
```

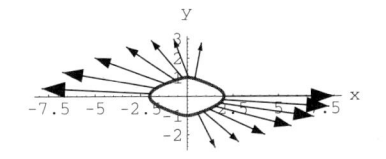

Counterclockwise flow along the curve at the plotted points on the curve. This is a good representative plot of the part of this vector field that flows in the counterclockwise direction along this curve. And you set up both plots by doing nothing more than eye-balling a plot of

$$\text{Field}[x[t], y[t]] \bullet \{x'[t], y'[t]\}.$$

Math works again.

■ T.4) The 2D electric field

In the electric interpretation, a vector field

$$\text{ElectricField}[x, y] = \{m[x, y], n[x, y]\}$$

is a vector such that

$$\text{ElectricField}[x, y] \bullet U$$

measures the instantaneous voltage drop when you leave $\{x, y\}$ in the direction of a unit vector U. This is the same as saying ElectricField$[x, y]$ represents the force the field places on a unit charge at the point $\{x, y\}$.

T.4.a) Here's a plot of the electric field resulting from a point charge of size 2 at $\{0.50, 1.50\}$:

In[36]:=
```
Clear[ElectricField,x,y]
q = 2; {a,b} = {0.50,1.50};
ElectricField[x_,y_] = q {x - a,y - b}/((x - a)^2 + (y - b)^2)
```

Lesson 3.04 2D Vector Fields and Their Trajectories

Out[36]=
$$\left\{\frac{2(-0.5+x)}{(-0.5+x)^2+(-1.5+y)^2}, \frac{2(-1.5+y)}{(-0.5+x)^2+(-1.5+y)^2}\right\}$$

Here's how this electric field looks:

In[37]:=
```
Show[Graphics[{Red,PointSize[0.08],Point[{a,b}]}],
Table[Arrow[ElectricField[x,y],Tail->{x,y},Blue],
{x,-1,3},{y,0,4}],Axes->Automatic];
```

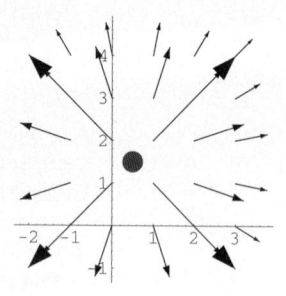

> How do you interpret this plot?

Answer: The electric field vector, ElectricField$[x,y]$, plotted as above, with tail at $\{x,y\}$ points in the direction of the biggest voltage drop, and the length of the vector represents the size of the voltage drop in that direction. Reason: If you leave $\{x,y\}$ in the direction of a unit vector U, then the instantaneous voltage drop is

$$\text{ElectricField}[x,y] \bullet U = ||\text{ElectricField}[x,y]||\,||U||\,\cos[\text{angle between}]$$
$$= ||\text{ElectricField}[x,y]||\,\cos[\text{angle between}]$$

because $||U|| = 1$. This tells you that the instantaneous voltage drop is as big as possible when

$$\text{ElectricField}[x,y] \bullet U = ||\text{ElectricField}[x,y]||\,\cos[\text{angle between}]$$

is as big as possible. This happens when $\cos[\text{angle between}] = 1$, which is the same as saying that

angle between $= 0$.

The upshot: To get the biggest instantaneous voltage drop, you take

$$U = \frac{\text{ElectricField}[x,y]}{||\text{ElectricField}[x,y]||}.$$

The direction of the biggest instantaneous voltage drop is the direction of ElectricField$[x,y]$.

The size of the biggest instantaneous voltage drop is

$$\frac{\text{ElectricField}[x,y] \bullet \text{ElectricField}[x,y]}{||\text{ElectricField}[x,y]||} = \frac{||\text{ElectricField}[x,y]||^2}{||\text{ElectricField}[x,y]||}$$
$$= ||\text{ElectricField}[x,y]||.$$

This means that ElectricField$[x, y]$ points in the direction of the biggest instantaneous voltage drop, and that the instantaneous voltage drop in this direction is $||\text{ElectricField}[x, y]||$.

> **T.4.b)** What do you get when you put a charge of size 1.5 at $\{-0.875, 0.375\}$ and a charge of size -1.5 at $\{0.875, 0.375\}$?

Answer: You get something pretty interesting. Start by adding the electric fields resulting from the two point charges:

In[38]:=
```
Clear[ElectricField1,ElectricFieldF2,
CombinedElectricField,x,y]
q1 = 1.5; {a1,b1} = {-0.875,0.375};
ElectricField1[x_,y_] = q1 {x - a1,y - b1}/
((x - a1)^2 + (y - b1)^2); q2 = -1.5;
{a2,b2} = {0.875,0.375}; ElectricField2[x_,y_] =
q2 {x - a2,y - b2}/((x - a2)^2 + (y - b2)^2);
CombinedElectricField[x_,y_] =
ElectricField1[x,y] + ElectricField2[x,y];
Show[Graphics[{Red,PointSize[0.08],Point[{a1,b1}]}],
Graphics[{Blue,PointSize[0.08],Point[{a2,b2}]}],
Table[Arrow[CombinedElectricField[x,y],Tail->{x,y}],
{x,-3,3},{y,-2,4}],Axes->Automatic];
```

That's a little hard to interpret.

To get a better idea, look at the unit electric field, which shows the direction of the greatest instantaneous voltage drop but does not show the size of the drop.

In[39]:=
```
Clear[unitfieldvector]
unitfieldvector[x_,y_] = CombinedElectricField[x,y]/
Sqrt[CombinedElectricField[x,y].CombinedElectricField[x,y]];
Show[Graphics[{Red,PointSize[0.08],Point[{a1,b1}]}],
Graphics[{Blue,PointSize[0.08],Point[{a2,b2}]}],
Table[Arrow[unitfieldvector[x,y],Tail->{x,y}],
{x,-3,3,0.5},{y,-2,4,0.5}],Axes->Automatic];
```

Hot plot. Well worth waiting for. The juice is pouring away from the positive charge toward the negative charge just like it does in the movies.

■ T.5) Troubleshooting plots of vector fields

This problem appears only in the electronic version.

Give It a Try

Experience with the starred (⋆) problems will be useful for understanding developments later in the course.

■ G.1) Looking for sinks (drains)⋆

G.1.a) Calculus&*Mathematica* thanks former C&M student Jennifer Lee Cassidy for suggesting this problem.

Here is a vector field and a plot:

In[1]:=
```
Clear[Field,x,y]
Field[x_,y_] = {-0.036 x E^(x^2) + 0.123 y,
0.123 x - 0.087 y E^(y^2)};
fieldplot = Table[Arrow[Field[x,y]/(1+Norm[Field[x,y]]),
Tail->{x,y},Blue],{x,-1.25,1.25,2.5/10},
{y,-1.25,1.25,2.5/10}];
Show[fieldplot,Axes->True,AxesLabel->{"x","y"}];
```

As you can see from the plot, if you drop a cork into this flow at any starting point other than $\{0,0\}$ within the plot, then it will float along to one of two special points and stop dead.

> Plot some trajectories in an effort to find the approximate locations of these two special points.

G.1.b) Here's an unscaled plot of a two-dimensional vector field:

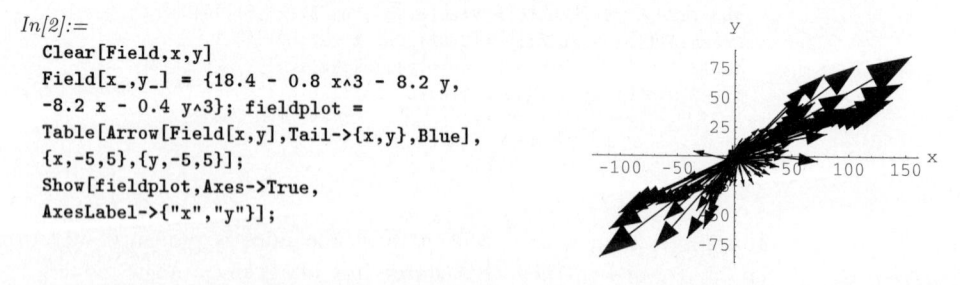

In[2]:=
```
Clear[Field,x,y]
Field[x_,y_] = {18.4 - 0.8 x^3 - 8.2 y,
-8.2 x - 0.4 y^3}; fieldplot =
Table[Arrow[Field[x,y],Tail->{x,y},Blue],
{x,-5,5},{y,-5,5}];
Show[fieldplot,Axes->True,
AxesLabel->{"x","y"}];
```

This plot sucks. You can't see what's happening.

> Give a scaled plot of this vector field at the same points to show it off to good advantage.

Pretend that this field represents fluid flow, and use your plot to determine whether there are any drains within the plot.

If you can't spot them visually, run some trajectories and see where they lead you.

■ G.2) Flow along and flow across*

Here's the vector field
$$\text{Field}[x, y] = \{x + \frac{y}{3}, x - \frac{y}{2}\}$$
plotted at selected points on the circle of radius 2 centered at $\{1, 0\}$:

In[3]:=
```
Clear[Field,m,n,x,y]; Clear[t]
m[x_,y_] = x + y/3; n[x_,y_] = x - y/2;
Field[x_,y_] = {m[x,y],n[x,y]};
{x[t_],y[t_]} = {1,0} + 2{Cos[t],Sin[t]};
curveplot = ParametricPlot[{x[t],y[t]},{t,0,2 Pi},
PlotStyle->{{Thickness[0.01],Red}},
DisplayFunction->Identity]; jump = Pi/12;
fieldoncurve = Table[Arrow[Field[x[t],y[t]],
Tail->{x[t],y[t]},Blue],{t,0,2 Pi - jump,jump}];
Show[curveplot,fieldoncurve,AxesOrigin->{0,0},
AspectRatio->Automatic,AxesLabel->{"x","y"},
DisplayFunction->$DisplayFunction];
```

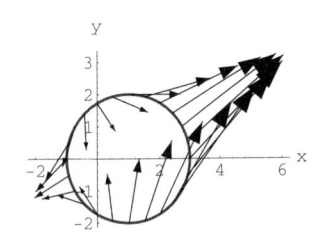

Here are the tangential components of the field vectors shown above:

In[4]:=
```
Clear[tangent,tancomponent]
tangent[t_] = {x'[t],y'[t]};
tancomponent[t_] = (((Field[x[t],y[t]].tangent[t])/
(tangent[t].tangent[t])) tangent[t]);
actualflowalong = Table[Arrow[tancomponent[t],
Tail->{x[t],y[t]},Blue],{t,0,2 Pi - jump,jump}];
flowalongplot = Show[curveplot,actualflowalong,
AxesOrigin->{0,0},AspectRatio->Automatic,
AxesLabel->{"x","y"},DisplayFunction->$DisplayFunction];
```

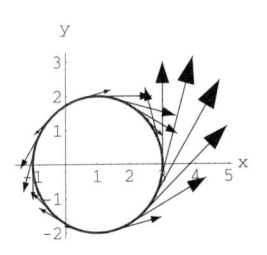

G.2.a.i) You make the call: Is the net flow of this vector field along this curve clockwise or counterclockwise?

G.2.a.ii) Go with the same vector field and the same curve as above, but this time look at normal components of the vector field at points on the curve:

Lesson 3.04 2D Vector Fields and Their Trajectories

In[5]:=
```
Clear[normal,normalcomponent]
normal[t_] = {y'[t],-x'[t]};
normalcomponent[t_] = (((Field[x[t],y[t]].normal[t])/
    (normal[t].normal[t])) normal[t]);
actualflowacross = Table[Arrow[normalcomponent[t],
    Tail->{x[t],y[t]},Blue],{t,0,2 Pi - jump,jump}];
flowacrossplot = Show[curveplot,actualflowacross,
    AspectRatio->Automatic,AxesLabel->{"x","y"},
    DisplayFunction->$DisplayFunction];
```

> You make the call: Is the net flow of this vector field across this curve from inside to outside or from outside to inside?

G.2.b) Here's a new vector field plotted at a selection of points on an ellipse:

In[6]:=
```
Clear[Field,m,n,x,y,t]
m[x_,y_] = (x - y)/3; n[x_,y_] = (x + y)/2;
Field[x_,y_] = {m[x,y],n[x,y]};
{x[t_],y[t_]} = {2,1} + {2 Cos[t],Sin[t]};
curveplot = ParametricPlot[{x[t],y[t]},{t,0,2 Pi},
    PlotStyle->{{Thickness[0.01],Red}},
    DisplayFunction->Identity]; jump = Pi/12;
fieldoncurve = Table[Arrow[Field[x[t],y[t]],
    Tail->{x[t],y[t]},Blue],{t,0,2 Pi - jump,jump}];
Show[curveplot,fieldoncurve,AxesOrigin->{0,0},
    AspectRatio->Automatic,AxesLabel->{"x","y"},
    DisplayFunction->$DisplayFunction];
```

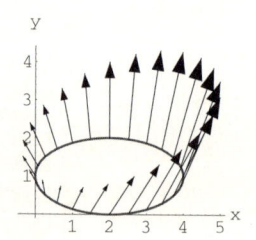

> Plot, as in part G.2.a), the tangential and normal components of the field vectors on this curve and then make the calls:
>
> → Is the net flow of this vector field along this curve clockwise or counterclockwise?
>
> → Is the net flow of this vector field across this curve from inside to outside or from outside to inside?

G.2.c) Here's a new vector field plotted at a selection of points on the circle of radius 1 centered at $\{0,0\}$:

In[7]:=
```
Clear[Field,m,n,x,y,t]
m[x_,y_] = x^3 - 3 x y^2; n[x_,y_] = -3 x^2 y + y^3;
Field[x_,y_] = {m[x,y],n[x,y]};
{x[t_],y[t_]} = {Cos[t], Sin[t]};
curveplot = ParametricPlot[{x[t],y[t]},{t,0,2 Pi},
PlotStyle->{{Thickness[0.01],Red}},
DisplayFunction->Identity]; jump = Pi/12;
fieldoncurve = Table[Arrow[Field[x[t],y[t]],
Tail->{x[t],y[t]},Blue],{t,0,2 Pi - jump,jump}];
Show[curveplot,fieldoncurve,Axes->True,
AspectRatio->Automatic,AxesLabel->{"x","y"},
DisplayFunction->$DisplayFunction];
```

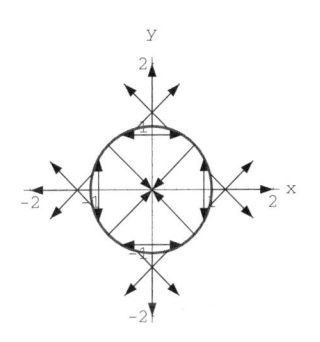

> Plot, as in part G.2.a), the tangential and normal components of these field vectors on this curve and then make the calls:
>
> → Is the net flow of this vector field along this curve clockwise, counterclockwise, or is the flow of this vector field along this curve 0 in the sense that the counterclockwise flow balances the clockwise flow?
>
> → Is the net flow of this vector field across this curve from inside to outside or from outside to inside, or is the flow of this vector field across this curve 0, in the sense that the inward flow balances the outward flow?

■ G.3) Normals, tangents, and dot plots*

G.3.a.i) | When a closed curve is parameterized by $\{x[t], y[t]\}$, then as you advance along the curve in the direction of the parameterization, which way do the tangent vectors $\{x'[t], y'[t]\}$ at $\{x[t], y[t]\}$ point—in the direction you are going or in the direction opposite to the direction you are going?

G.3.a.ii) | When a closed curve is parameterized by $\{x[t], y[t]\}$ in the counterclockwise way, then as you advance along the curve in the direction of the parameterization, which way do the normal vectors $\{y'[t], -x'[t]\}$ at $\{x[t], y[t]\}$ point—to the right to the outside of the curve or to the left to the inside of the curve?

G.3.a.iii) | When a closed curve is parameterized by $\{x[t], y[t]\}$ in the clockwise way, then as you advance along the curve in the direction of the parameterization, which way do the normal vectors $\{y'[t], -x'[t]\}$ at $\{x[t], y[t]\}$ point—to the right to the inside of the curve or to the left to the outside of the curve?

G.3.b.i) Here's a vector field:

In[8]:=
```
Clear[Field,m,n,x,y]; m[x_,y_] = 1; n[x_,y_] = Sin[x y];
Field[x_,y_] = {m[x,y],n[x,y]}
```

Out[8]=
{1, Sin[x y]}

Here's a curve:

In[9]:=
```
Clear[t]
{x[t_],y[t_]} = {Cos[t],Sin[t] + Sin[2 t]/3};
curveplot = ParametricPlot[{x[t],y[t]},{t,0,2 Pi},
PlotStyle->{{Thickness[0.01],Red}},
AxesLabel->{"x","y"}];
```

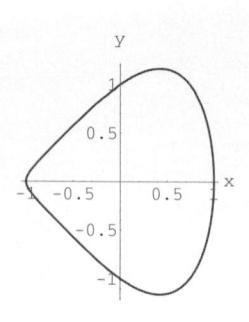

Look at this plot of some of the unit tangent vectors

$$\text{unittangent}[t] = \frac{\{x'[t], y'[t]\}}{\sqrt{x'[t]^2 + y'[t]^2}}$$

pointing in the direction of the parameterization:

In[10]:=
```
Clear[t,unittangent]
unittangent[t_] = {x'[t],y'[t]}/Sqrt[x'[t]^2 + y'[t]^2];
jump = Pi/8;
unittangents = Table[Arrow[unittangent[t],
Tail->{x[t],y[t]},Red],{t,0,2 Pi - jump,jump}];
Show[curveplot,unittangents];
```

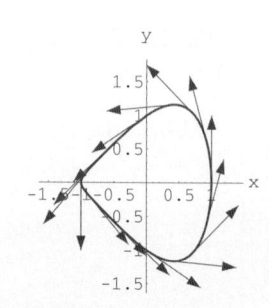

Now look at this plot of

$\text{Field}[x[t], y[t]] \bullet \text{unittangent}[t]$:

In[11]:=
```
dotplot = Plot[Field[x[t],y[t]].unittangent[t],
{t,0,2 Pi},PlotStyle->{{Thickness[0.01],Blue}},
AxesLabel->{"t",""},AspectRatio->1/2];
```

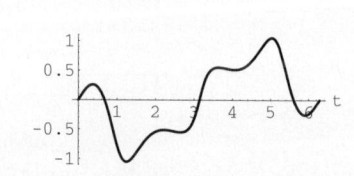

Interpret this plot.

G.3.b.ii) Go with the same vector field and curve as in part G.3.b.i) and look at this plot of
$$\text{Field}[x[t], y[t]] \bullet \{y'[t], -x'[t]\}:$$

In[12]:=
```
dotplot = Plot[Field[x[t],y[t]].{y'[t],-x'[t]},
  {t,0,2 Pi},PlotStyle->{{Thickness[0.01],Blue}},
  AxesLabel->{"t",""},AspectRatio->1/2];
```

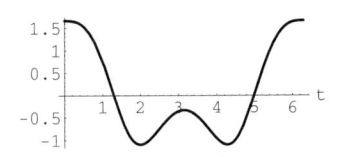

> Interpret this plot.
>
> Does your answer change if you replace
>
> $\{y'[t], -x'[t]\}$
>
> by the normalized vector,
>
> $$\dfrac{\{y'[t], -x'[t]\}}{\sqrt{x'[t]^2 + y'[t]^2}} \;?$$

■ **G.4) The most important vector field of them all: The gradient field★**

One way to come up with a vector field is to take a function $f[x, y]$ and put
$$\text{Field}[x, y] = \text{grad} f[x, y] = \nabla f[x, y].$$

Try it out:

In[13]:=
```
Clear[f,gradf,Field,x,y]
f[x_,y_] = 1 - (x^2 + 1.5 y^2)/2;
gradf[x_,y_] = {D[f[x,y],x],D[f[x,y],y]};
Field[x_,y_] = gradf[x,y]
```

Out[13]=
```
{-x, -1.5 y}
```

This is the gradient field of
$$f[x, y] = 1 - \dfrac{x^2 + 1.5\, y^2}{2}.$$

Note that $\{x, y\} = \{0, 0\}$ maximizes $f[x, y]$.

Here's a plot of the scaled gradient field of $f[x, y]$ shown with the maximizer at $\{0, 0\}$:

Lesson 3.04 2D Vector Fields and Their Trajectories

$In[14]:=$
```
maximizerplot = Graphics[{Red,
    PointSize[0.04],Point[{0,0}]}];
scalefactor = 0.2;
gradfieldplot = Table[Arrow[scalefactor gradf[x,y],
    Tail->{x,y},Blue],{x,-2,2,0.5},{y,-2,2,0.5}];
Show[maximizerplot,gradfieldplot,Axes->Automatic,
    AxesLabel->{"x","y"}];
```

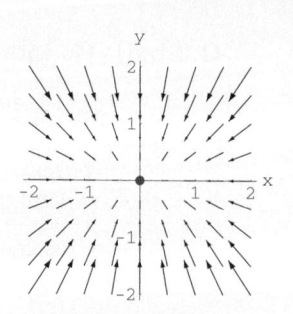

G.4.a.i) Why did this happen?

Where are the trajectories in this gradient field headed?

What do you learn about a function $f[x,y]$ by looking at a plot of its gradient field?

G.4.a.ii) This time go with the function

$$f[x,y] = \frac{2}{1+(y-2)^2+1.5(x-1)^2}.$$

Say why you are sure that $\{x,y\} = \{1,2\}$ maximizes $f[x,y]$.

Then plot the (scaled) gradient field of $f[x,y]$ near $\{1,2\}$.

Describe what you see and explain why you see it.

G.4.a.iii) You are given a certain function $f[x,y]$ and the information that $\{a,b,f[a,b]\}$ sits on the top of a crest of the surface given by $z = f[x,y]$. You center a small circle at $\{a,b\}$ and plot the gradient field of $f[x,y]$ on this circle.

Is the net flow of the gradient field of $f[x,y]$ across this circle from inside to outside or from outside to inside?

G.4.b.i) This time go with

$$f[x,y] = e^{(x^2+2y^2)/9}.$$

$In[15]:=$
```
Clear[f,gradf,Field,x,y]; f[x_,y_] = E^((x^2 + 2 y^2)/9);
gradf[x_,y_] = {D[f[x,y],x],D[f[x,y],y]}; Field[x_,y_] = gradf[x,y]
```
$Out[15]=$
$$\left\{\frac{2 E^{(x^2+2y^2)/9} x}{9}, \frac{4 E^{(x^2+2y^2)/9} y}{9}\right\}$$

This is the gradient field of $f[x,y] = e^{(x^2+2y^2)/9}$.

Note that $\{x,y\} = \{0,0\}$ minimizes $f[x,y]$. Here's a plot of the scaled gradient field of $f[x,y]$ shown with the minimizer at $\{0,0\}$:

In[16]:=
```
minimizerplot = Graphics[{Red,PointSize[0.04],
  Point[{0,0}]}]; scalefactor = 0.8;
gradfieldplot = Table[Arrow[scalefactor gradf[x,y],
  Tail->{x,y},Blue],{x,-1,1,0.25},{y,-1,1,0.25}];
Show[minimizerplot,gradfieldplot,Axes->Automatic,
  AxesLabel->{"x","y"}];
```

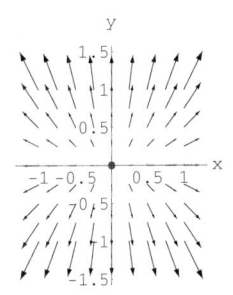

> Why did this happen?
>
> What are the trajectories in this gradient field trying to get away from?

G.4.b.ii) You are given a certain function $f[x,y]$ and the information that $\{a,b,f[a,b]\}$ sits on the bottom of a dip of the surface given by $z = f[x,y]$. You center a small circle at $\{a,b\}$ and plot the gradient field of $f[x,y]$ on this circle.

> Is the net flow of the gradient field of $f[x,y]$ across this circle from inside to outside or from outside to inside?

G.4.c) Here's a plot of the surface

$$z = f[x,y] = \frac{\sin[2x] + \sin[3y]}{0.5 + x^2 + y^2}$$

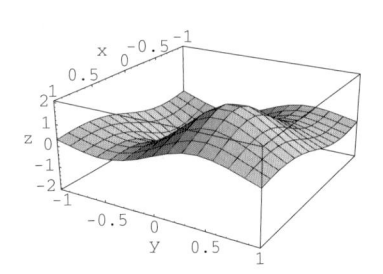

for $-1 \leq x \leq 1$ and $-1 \leq y \leq 1$:

In[17]:=
```
Clear[f,x,y]
f[x_,y_] = (Sin[2 x] + Sin[3 y])/
  (0.5 + x^2 + y^2);
Plot3D[f[x,y],{x,-1,1},{y,-1,1},
  AxesLabel->{"x","y","z"},ViewPoint->CMView];
```

Here's a plot of the (scaled) gradient field of the same function for the same x's and the same y's:

166 Lesson 3.04 2D Vector Fields and Their Trajectories

```
In[18]:=
    Clear[gradf]
    gradf[x_,y_] = {D[f[x,y],x],D[f[x,y],y]};
    scalefactor = 0.07;
    gradfieldplot = Table[Arrow[scalefactor gradf[x,y],
    Tail->{x,y},Blue],{x,-1,1,0.2},{y,-1,1,0.2}];
    Show[gradfieldplot,Axes->Automatic,AxesLabel->{"x","y"}];
```

> What information about the crests and dips on the surface does this plot of the gradient field of $f[x,y]$ convey?
>
> Use the information in this plot and the FindMinimum instruction to locate the position $\{a,b,f[a,b]\}$ of the top of the highest crest, and the position $\{c,d,f[c,d]\}$ of the bottom of the deepest dip on the part of the surface $z = f[x,y]$ plotted above.

G.4.d) Here's a contour plot of
$$f[x,y] = 1.2\,e^{x^2} + 2.9\,e^{y^2} - 8.2\,x\,y^2$$
for $0.8 \le x \le 1.3$ and $0.7 \le y \le 1.2$:

```
In[19]:=
    Clear[f,x,y]
    f[x_,y_] = 1.2 E^(x^2) + 2.9 E^(y^2) - 8.2 x y^2;
    ContourPlot[f[x,y],{x,0.8,1.3},{y,0.7,1.2}];
```

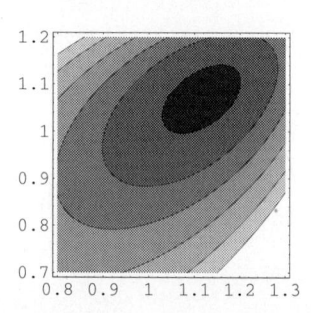

There's a dip bottoming out at $\{a,b,f[a,b]\}$ for some $\{a,b\}$ near $\{1,1\}$. Look at a plot of the (scaled) negative gradient in the vicinity of $\{1,1\}$.

```
In[20]:=
    Clear[negativegradient]
    negativegradf[x_,y_] = -{D[f[x,y],x],D[f[x,y],y]};
    scalefactor = 0.01;
    neggradfieldplot = Table[Arrow[
    scalefactor negativegradf[x,y],
    Tail->{x,y},Blue],{x,0.9,1.3,0.4/8},{y,0.8,1.2,0.4/8}];
    Show[neggradfieldplot,Axes->Automatic,
    AxesLabel->{"x","y"}];
```

Look at that flow.

Use what you see in the plot above, and then use the FindMinimum instruction to get a good estimate of the point $\{a, b, f[a, b]\}$ at the bottom of the dip.

Add a plot of $\{a, b\}$ to the plot of the negative gradient field above.

Describe what you see and say why you see it. How do trajectories in this negative gradient field look when they pass through the part of the negative gradient field plotted above?

■ G.5) Differential equations and their associated vector fields*

G.5.a) Here's the vector field coming from the differential equation
$$y'[x] = y[x]^2 - x :$$

In[21]:=
```
Clear[DEField,x,y]
DEField[x_,y_] = {1,y^2 - x}; scalefactor = 0.2;
DEFieldplot = Table[Arrow[scalefactor DEField[x,y],
  Tail->{x,y},Blue],{x,-2,6,8/10},{y,-2,2,4/10}];
Show[DEFieldplot,Axes->Automatic,
AxesLabel->{"x","y"}];
```

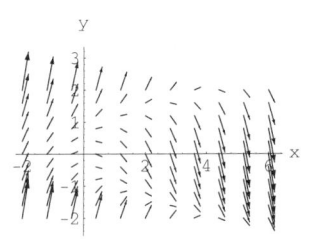

Look at the slopes of those field vectors.

Add plots of the solutions of
$$y'[x] = y[x]^2 - x \quad \text{with } y[-1.5] = -2$$
and
$$y'[x] = y[x]^2 - x \quad \text{with } y[0] = -1$$
as trajectories of the vector field to the DEField plot above, and discuss the relation between the plots of the solutions and the DEField.

Finally, examine the flow of this vector field, and use it to describe the behavior of solutions of the differential equation
$$y'[x] = y[x]^2 - x$$
that pass through the part of the xy-plane plotted above.

G.5.b.i) Here's the vector field coming from the differential equation
$$y'[x] = y[x]\left(\frac{1 - y[x]}{2}\right) :$$

Lesson 3.04 2D Vector Fields and Their Trajectories

In[22]:=
```
Clear[DEField,x,y]
DEField[x_,y_] = {1,y (1 - y/2)}; scalefactor = 0.3;
DEFieldplot = Table[Arrow[scalefactor DEField[x,y],
   Tail->{x,y},Blue],{x,0,8,8/10},{y,0,4,4/10}];
cutoff = Graphics[{Red,Thickness[0.01],
Line[{{-0.5,2},{8.5,2}}]}];
Show[cutoff,DEFieldplot,Axes->True,AxesLabel->
{"x","y"},PlotLabel->"Logistic field"];
```

Use what you see to describe the fundamental difference between the behavior of a solution of

$$y'[x] = y[x]\left(\frac{1-y[x]}{2}\right)$$

whose plot goes through a point $\{a,b\}$ with $b > 2$ and the behavior of a solution of

$$y'[x] = y[x]\left(\frac{1-y[x]}{2}\right)$$

whose plot goes through a point $\{a,b\}$ with $0 < b < 2$.

G.5.b.ii) Here's more of the vector field coming from the logistic differential equation

$$y'[x] = y[x]\left(\frac{1-y[x]}{2}\right):$$

In[23]:=
```
Clear[DEField,x,y]
DEField[x_,y_] = {1,y (1 - y/2)}; scalefactor = 0.3;
DEFieldplot = Table[Arrow[scalefactor DEField[x,y],
   Tail->{x,y},Blue],{x,0,10,1},{y,-2,4,4/10}];
cutoff = Graphics[{Red,Thickness[0.01],
Line[{{-0.5,2},{8.5,2}}]}];
Show[DEFieldplot,Axes->True,AxesLabel->{"x","y"}];
```

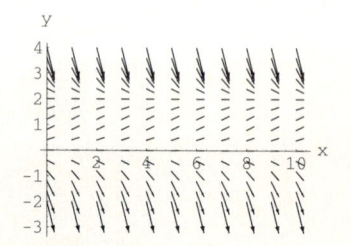

Now you see three distinct types of solutions of

$$y'[x] = y[x]\left(\frac{1-y[x]}{2}\right).$$

Describe the three distinct types of solutions in words.

G.5.c) Plot the vector field coming from the differential equation

$$y'[x] = x\,y[x]$$

for a healthy selection of $\{x, y\}$'s with $-4 \leq x \leq 4$ and $-3 \leq y \leq 3$. Apply an appropriate scale factor so you can see what's going on.

Look at the resulting plot of the vector field and describe how solutions with $y[-4] > 0$ are fundamentally different from solutions with $y[-4] < 0$.

■ G.6) Trajectories: Can they cross?

G.6.a) Calculus&*Mathematica* thanks Professor Stephanie Alexander of the University of Illinois for helpful conversations about these problems.

Can two trajectories of a vector field ever cross over each other like this:

Why or why not?

G.6.b) You have a differential equation

$$y'[x] = f[x, y[x]].$$

If you have a formula for $f[x, y]$ and you start with $y[0] = 1$, then you can use NDSolve to plot out a function $y[x]$ with

$$y[0] = 1$$

and

$$y'[x] = f[x, y[x]].$$

On the other hand, if you start with $y[0] = 2$, then you can use NDSolve to plot out a function $y[x]$ with $y[0] = 2$ and $y'[x] = f[x, y[x]]$.

The question here is: Can the plots of these two solutions ever cross over each other like this:

Why or why not?

■ G.7) Drifting along with a tumbleweed

G.7.a) Wind is swirling through the desert region

$$-4 \leq x \leq 4, 0 \leq y \leq 8.$$

At a point $\{x, y\}$ in this region, the velocity of the wind is

$$\text{wind}[x, y] = \left\{ 0.18 + 0.6 \, \sin[y], \, \frac{0.12 \, y}{1 + x^2} \right\}.$$

Here is a picture of the wind as you look down from a helicopter at this desert region:

In[24]:=
```
Clear[wind,x,y]
wind[x_,y_] = {0.18 + 0.6 Sin[y],0.12 y /(1 + x^2)};
windplot = Show[Table[Arrow[wind[x,y],
Tail->{x,y}],Blue],{x,-4,4},{y,0,8}],Axes->Automatic,
AxesLabel->{"x","y"}];
```

A tumbleweed drifting along was spotted at the point $\{-2, 2\}$.

> Add a plot of the approximate path of this drifting tumbleweed to the plot above.

■ G.8) Logistic harvesting revisited

Those who have been with Calculus&*Mathematica* from the beginning may recognize this problem.

Take the logistic equation

$$y'[t] = a \, y[t] \left(\frac{1 - y[t]}{b} \right)$$

with $0 < a$ and $0 < b$. And start with $0 < y[0] < b$.

As t advances from 0, you are guaranteed that $y[t]$ grows with some pep until $y[t]$ gets near b. Once $y[t]$ gets near b, then $y[t]$ settles into global scale with

$$\lim_{t \to \infty} y[t] = b.$$

Check it out with $a = 0.23$ and $b = 13$:

In[25]:=
```
a = 0.23; b = 13; Clear[DEField,t,y]
DEField[t_,y_] = {1, a y (1 - y/b)};
scalefactor = 2.5; DEFieldplot = Table[Arrow[
  scalefactor DEField[t,y],Tail->{t,y},Blue],
{t,0,52,52/12},{y,0,b,b/10}];
bline = Graphics[{Red,Line[{{0,b},{52,b}}]}];
Show[DEFieldplot,bline,Axes->Automatic,
AxesLabel->{"t","y"},AspectRatio->1/2];
```

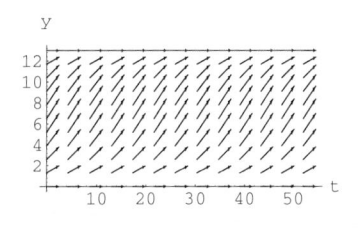

You can see that the trajectories start low on the left and are sucked onto the line $y = b$ plotted at the very top:

In[26]:=
```
endtime = 52; starter = 2;
Clear[t,y,Derivative,fakey]
solution = NDSolve[{y'[t] == a y[t] (1 - y[t]/b),
  y[0] == starter},y[t],{t,0,endtime}];
fakey[t_] = y[t]/.solution[[1]];
solutionplot = Plot[{fakey[t]},{t,0,endtime},
PlotStyle->{{Red,Thickness[0.01]}},
DisplayFunction->Identity];
Show[DEFieldplot,solutionplot,bline,Axes->Automatic,
AxesLabel->{"t","y"},AspectRatio->1/2,
DisplayFunction->$DisplayFunction];
```

Try it again with $y[0] = 5$:

In[27]:=
```
starter = 5;
Clear[t,y,Derivative,fakey]
solution = NDSolve[{y'[t] == a y[t] (1 - y[t]/b),
  y[0] == starter},y[t],{t,0,endtime}];
fakey[t_] = y[t]/.solution[[1]];
solutionplot = Plot[{fakey[t]},{t,0,endtime},
PlotStyle->{{Red,Thickness[0.01]}},
DisplayFunction->Identity];
Show[DEFieldplot,solutionplot,bline,Axes->Automatic,
AxesLabel->{"t","y"},AspectRatio->1/2,
DisplayFunction->$DisplayFunction];
```

You get a reasonable interpretation of the logistic differential equation

$$y'[t] = a\, y[t] \left(\frac{1 - y[t]}{b}\right) \qquad \text{with } 0 < y[0] < b$$

by imagining that $y[t]$ is number of catfish (in thousands) in a given lake on a catfish farm t weeks after the lake was stocked with $y[0]$ catfish. As time goes on, the catfish population increases until it reaches its steady-state population of b catfish. But the catfish farmer doesn't grow catfish as pets; the farmer is in business

Lesson 3.04 2D Vector Fields and Their Trajectories

to harvest catfish and sell them so that hungry people can fry them up and then wash them down with a couple of cold beers or iced teas.

G.8.a.i) Measure time t in weeks, assume that the farmer wants to harvest r fish per week, and explain why

$$y'[t] = a\,y[t]\left(\frac{1-y[t]}{b}\right) - r$$

lays the base for a reasonable model.

G.8.a.ii) Go with the model

$$y'[t] = a\,y[t]\left(\frac{1-y[t]}{b}\right) - r,$$

where t measures the number of weeks that harvesting goes on, $a = 0.23, b = 13$ (thousand fish), and the weekly harvest is $r = 0.5$ (thousand fish). Here is the associated DEField:

In[28]:=
```
Clear[DEField,t,y]
a = 0.23; b = 13; r = 0.5;
DEField[t_,y_] = {1, a y (1 - y/b) - r};
scalefactor = 2.5; DEFieldplot = Table[
Arrow[scalefactor DEField[t,y],
Tail->{t,y},Blue],{t,0,52,52/12},{y,0,b,b/10}];
bline = Graphics[{Red,Line[{{0,b},{52,b}}]}];
Show[DEFieldplot,bline,Axes->Automatic,
AxesLabel->{"t","y"},AspectRatio->1/2];
```

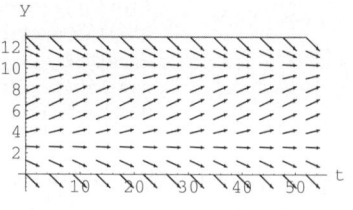

Examine the vector field and use what you see to estimate the smallest number of fish the lake must contain at the beginning ($t = 0$) to sustain the weekly harvest of $r = 0.5$ (thousand fish) for a full 52 consecutive weeks.

G.8.a.iii) Here's what happens with a harvest of $r = 1.5$ (thousand fish) per week from the same lake:

In[29]:=
```
r = 1.5;
Clear[DEField,t,y]
DEField[t_,y_] = {1, a y (1 - y/b) - r};
scalefactor = 2.5; DEFieldplot =
Table[Arrow[scalefactor DEField[t,y],
Tail->{t,y},Blue],{t,0,52,52/12},{y,0,b,b/10}];
bline= Graphics[{Red,Line[{{0,b},{52,b}}]}];
Show[DEFieldplot,bline,Axes->Automatic,
AxesLabel->{"t","y"},AspectRatio->1/2];
```

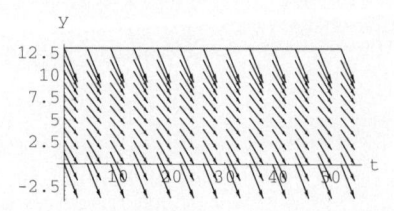

Harvesting at the level $r = 1.5$ (thousand fish) per week is a very bad idea because all the trajectories that start on the far left are going negative before the end of the 52-week period under study. This means that before the end of the 52-week period, all the fish will be gone and harvesting will have to stop no matter how many fish you start out with.

> Your job is to use DEField plots to estimate the largest possible weekly harvest rate r (in thousands) and a number
>
> $y[0] = $ starter
>
> so that, if the lake starts out containing "starter" fish, then the harvest can continue for the full 52-week period.
>
> Once you have your estimates for r and $y[0] = $ starter, show the resulting DEField together with the plot of the solution of
>
> $$y'[t] = a\, y[t] \left(\frac{1 - y[t]}{b}\right) - r \quad \text{with } y[0] = \text{starter}.$$

■ G.9) Water flow with spigots and drains

This problem appears only in the electronic version.

■ G.10) 2D Electrical fields

G.10.a) > Plot and describe the two-dimensional electric field resulting from equal charges of the same sign at $\{1.5, 0\}$ and $\{-1.5, 0\}$. Include a plot of the unit vector field.

G.10.b) A dipole can be approximated by two charges of the same magnitude but opposite sign separated by a small distance. Dipoles are especially important in atomic theory. Richard Feynman explained it this way:

"Although an atom or molecule remains neutral in an external electronic field, there is a tiny separation of its positive and negative charges, and it becomes a microscopic dipole."

Richard P. Feynman, Robert B. Leighton, Matthew Sands, *The Feynman Lectures on Physics*, Addison-Wesley 1964.

> Plot and describe the electric field resulting from a positive charge at $\{0.01, 0\}$ and the opposite charge of the same magnitude at $\{-0.01, 0\}$.
>
> Include a plot of the unit vector field.

LESSON 3.05

Flow Measurements by Integrals

Basics

■ **B.1)** Measuring flow across a curve with the integral

$$\int_{tlow}^{thigh} \text{Field}[x[t], y[t]] \cdot \{y'[t], -x'[t]\} \, dt$$

Here's a vector field:

In[1]:=
```
Clear[Field,m,n,x,y]
m[x_,y_] = 0.8 Sin[y]; n[x_,y_] = 0.4 x + 0.1 y;
Field[x_,y_] = {m[x,y],n[x,y]}
```

Out[1]=
{0.8 Sin[y], 0.4 x + 0.1 y}

Here's a curve:

In[2]:=
```
Clear[t]
{x[t_],y[t_]} = {2 Cos[t],Sin[t]};
tlow = 0; thigh = 2 Pi;
curveplot = ParametricPlot[{x[t],y[t]},{t,tlow,thigh},
PlotStyle->{{Thickness[0.01],Red}},
AxesLabel->{"x","y"}];
```

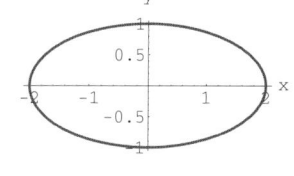

Pick a point on the curve, say $\{x[\pi/4], y[\pi/4]\}$, and look at the field vector at this point:

175

Lesson 3.05 Flow Measurements by Integrals

In[3]:=
```
tt = Pi/4; Clear[setup]
setup[t_] := Show[curveplot,
Arrow[Field[x[t],y[t]],Tail->{x[t],y[t]},Blue],
DisplayFunction->Identity];
Show[setup[tt],DisplayFunction->$DisplayFunction];
```

Now look at the curve with a little segment of fluid centered at the base of the field vector:

In[4]:=
```
Clear[fluidbit,center,height,width,h]
unittan[t_] = {x'[t],y'[t]}/Sqrt[x'[t]^2 + y'[t]^2];
center[t_] = {x[t],y[t]};
fluidbit[center_,t_,h_] :=
Graphics[{Thickness[0.015],Blue,
Line[{center - h unittan[t],center + h unittan[t]}]}];
h = 0.2; Show[setup[tt],fluidbit[center[tt],tt,h],
PlotRange->All,DisplayFunction->$DisplayFunction];
```

When you check out how the fluid segment moves in a short time, you'll see approximately this:

In[5]:=
```
flowtime = 0.4;
Show[setup[tt],
fluidbit[center[tt] + flowtime Field[x[tt],
y[tt]],tt,h],
DisplayFunction->$DisplayFunction];
```

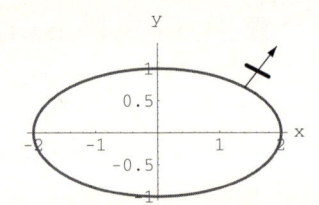

To get an even better idea, put a microscope on it:

In[6]:=
```
Show[setup[tt],fluidbit[center[tt],tt,h],PlotRange->
{{x[tt] - 4 h,x[tt] + 4 h},{y[tt] - 4 h,y[tt] + 4 h}},
DisplayFunction->$DisplayFunction];
Show[setup[tt],fluidbit[center[tt] + flowtime Field[x[tt],y[tt]],tt,h],
PlotRange->{{x[tt] - 4 h,x[tt] + 4 h},{y[tt] - 4 h,y[tt] + 4 h}},
DisplayFunction->$DisplayFunction];
```

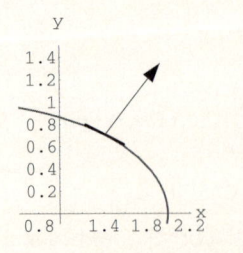

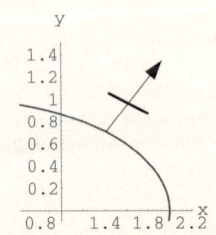

Now look at both together:

In[7]:=
```
Show[setup[tt],fluidbit[center[tt],tt,h],
fluidbit[center[tt] + flowtime Field[x[tt],y[tt]],tt,h],
PlotRange->{{x[tt] - 4 h,x[tt] + 4 h},
{y[tt] - 4 h,y[tt] + 4 h}},
DisplayFunction->$DisplayFunction];
```

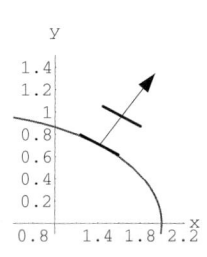

The approximate amount of fluid that flowed across the curve in this short time is measured by the area of the parallelogram defined by the two line segments shown above. Put in the vector

flowtime (Field$[x[tt], y[tt]]$ • unitnormal$[tt]$) unitnormal$[tt]$

for $tt = \pi/4$ and flowtime $= 0.4$ as above:

In[8]:=
```
Clear[unitnormal]
unitnormal[t_] := {y'[t],-x'[t]}/Sqrt[y'[t]^2 + x'[t]^2];
Show[setup[tt],fluidbit[center[tt],tt,h],
fluidbit[center[tt] + flowtime Field[x[tt],y[tt]],tt,h],
Arrow[flowtime(Field[x[tt],
y[tt]].unitnormal[tt]) unitnormal[tt],
Tail->{x[tt],y[tt]},Blue],PlotRange->
{{x[tt] - 4 h,x[tt] + 4 h},{y[tt] - 4 h,
y[tt] + 4 h}},DisplayFunction->$DisplayFunction];
```

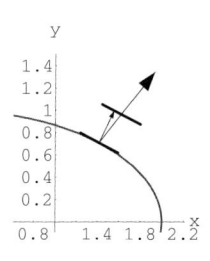

Bingo! The length of the new vector plotted above is

flowtime (unitnormal$[tt]$ • Field$[x[tt], y[tt]]$) .

Putting

$ds =$ length of the fluid segment,

you can say that in this flowtime, approximately

flowtime (unitnormal$[tt]$ • Field$[x[tt], y[tt]]$) ds

units of fluid flow across the segment of the curve in the above plot.

The flow goes with the normal $\{y'[tt], -x'[tt]\}$ at this point because

Field$[x[tt], y[tt]]$ • unitnormal$[tt] > 0$:

In[9]:=
```
N[Field[x[tt],y[tt]].unitnormal[tt]]
```
Out[9]=
0.801631

Lesson 3.05 Flow Measurements by Integrals

See what happens at another point on the curve:

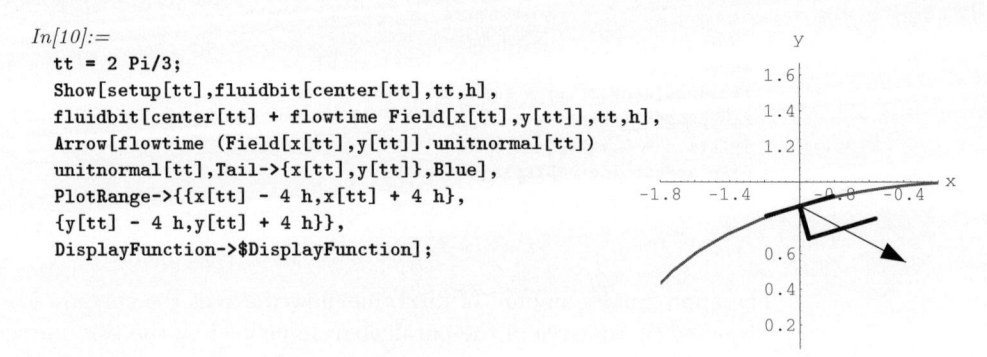

In[10]:=
```
tt = 2 Pi/3;
Show[setup[tt],fluidbit[center[tt],tt,h],
fluidbit[center[tt] + flowtime Field[x[tt],y[tt]],tt,h],
Arrow[flowtime (Field[x[tt],y[tt]].unitnormal[tt])
unitnormal[tt],Tail->{x[tt],y[tt]},Blue],
PlotRange->{{x[tt] - 4 h,x[tt] + 4 h},
{y[tt] - 4 h,y[tt] + 4 h}},
DisplayFunction->$DisplayFunction];
```

Again, putting ds = length of the fluid segment, you can say that in this flowtime, approximately

$$\text{flowtime}(\text{Field}[x[tt], y[tt]] \cdot \text{unittnormal}[tt]) \, ds$$

units of fluid flow across the curve in the above plot. The flow is opposite the direction of the normal $\{y'[tt], -x'[tt]\}$ at this point because

$$\text{Field}[x[tt], y[tt]] \bullet \text{unittnormal}[tt] < 0:$$

In[11]:=
```
N[Field[x[tt],y[tt]].unitnormal[tt]]
```

Out[11]=
-0.470122

B.1.a.i) Explain why this tells you that the integral

$$\int_{\text{tlow}}^{\text{thigh}} \text{Field}[x[t], y[t]] \bullet \{y'[t], -x'[t]\} \, dt :$$

In[12]:=
```
tlow = 0; thigh= 2 Pi;
NIntegrate[Field[x[t],y[t]].{y'[t],-x'[t]},{t,tlow,thigh}]
```

Out[12]=
0.628319

measures the net amount of fluid flowing across this curve per unit of flowtime.

Answer: Fixing a particular t, you know that in a short flowtime approximately

$$\text{flowtime}(\text{Field}[x[t], y[t]] \bullet \text{unittnormal}[t]) \, ds$$

units of fluid flow across the small segment of the curve of length ds centered at $\{x[t], y[t]\}$.

Because you integrate $\sqrt{x'[t]^2 + y'[t]^2}$ to measure the length on the curve, parameterized by $\{x[t], y[t]\}$, you know that $ds = \sqrt{x'[t]^2 + y'[t]^2}\, dt$. This tells you that, in a short flowtime, approximately

$$\text{flowtime}\,(\text{Field}[x[t], y[t]] \bullet \text{unitnormal}[t])\,\sqrt{x'[t]^2 + y'[t]^2}\, dt$$

units of fluid flow across the small segment of the curve of length ds centered at $\{x[t], y[t]\}$.

But

$$\text{unitnormal}[t] = \frac{\{y'[t], -x'[t]\}}{\sqrt{x'[t]^2 + y'[t]^2}},$$

so

$$(\text{Field}[x[t], y[t]] \bullet \text{unittnormal}[t])\,\sqrt{x'[t]^2 + y'[t]^2} = \text{Field}[x[t], y[t]] \bullet \{y'[t], -x'[t]\}.$$

This tells you that, in a short flowtime, approximately

$$\text{flowtime}(\text{Field}[x[t], y[t]] \bullet \{y'[t], -x'[t]\})\, dt$$

units of fluid flow across the small segment of the curve.

Now, cover the whole curve with non-overlapping little segments of length ds as above. Adding the individual measurements together, you get the approximate flow-across measurement

$$\text{flowtime Sum}[\text{Field}[x[t], y[t]] \bullet \{y'[t], -x'[t]\}\, dt, \{t, \text{tlow}, \text{thigh} - dt, dt\}].$$

As dt closes in on 0, these approximate measurements close in on the exact measurement

$$\text{flowtime} \int_{\text{tlow}}^{\text{thigh}} \text{Field}[x[t], y[t]] \bullet \{y'[t], -x'[t]\}\, dt$$

of the net flow across the curve in the given flowtime.

To arrive at the measurement of the net flow across the whole curve per time unit, divide by flowtime to learn that

$$\int_{\text{tlow}}^{\text{thigh}} \text{Field}[x[t], y[t]] \bullet \{y'[t], -x'[t]\}\, dt$$

measures the net flow over the curve per time unit.

If measurements are in gallons and seconds, the calculation:

In[13]:=
```
NIntegrate[Field[x[t],y[t]].{y'[t],-x'[t]},{t,0, 2 Pi}]
```

Out[13]=
0.628319

tells you that the net flow of this vector field across the curve above is 0.628319 gallons per second.

B.1.a.ii) Take another look at the integral

$$\int_{\text{tlow}}^{\text{thigh}} \text{Field}[x[t], y[t]] \bullet \{y'[t], -x'[t]\} \, dt$$

for the vector field and the curve specified above:

In[14]:=
```
NIntegrate[Field[x[t],y[t]].{y'[t],-x'[t]},{t,tlow,thigh}]
```

Out[14]=
0.628319

This tells you that the net flow of this vector field across this curve is about 0.63 fluid units per time unit.

> Why does this also tell you that the net flow is from inside to outside?

Answer: Look at this plot of

$$\text{Field}[x[t], y[t]] \bullet \{y'[t], -x'[t]\}$$

for tlow $\leq t \leq$ thigh:

In[15]:=
```
Plot[Field[x[t],y[t]].{y'[t],-x'[t]},
 {t,0,2 Pi},PlotStyle->
 {{Thickness[0.01],Blue}},
 AxesLabel->{"t","Field.normal"}];
```

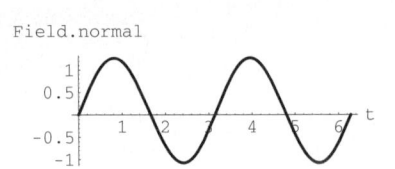

$$\int_{\text{tlow}}^{\text{thigh}} \text{Field}[x[t], y[t]] \bullet \{y'[t], -x'[t]\} \, dt = \text{area above} - \text{area below}.$$

The area between the curve and the t-axis can be interpreted as:

→ area above the t-axis = total flow of this vector field across this curve in the direction of the normals $\{y'[t], -x'[t]\}$.

→ area below the t-axis = total flow of this vector field across this curve opposite the direction of the normals $\{y'[t], -x'[t]\}$.

In the case above, $\int_{\text{tlow}}^{\text{thigh}} \text{Field}[x[t], y[t]] \bullet \{y'[t], -x'[t]\} \, dt > 0$.

In[16]:=
```
NIntegrate[Field[x[t],y[t]].{y'[t],-x'[t]},{t,tlow,thigh}]
```

Out[16]=
0.628319

This tells you that the flow in the direction of the normals is greater than the flow against the normals. Take a look at the normals:

In[17]:=
```
Show[curveplot,Table[Arrow[{y'[t],-x'[t]},Tail->
{x[t],y[t]},Red],{t,tlow,thigh,(thigh-tlow)/8}]];
```

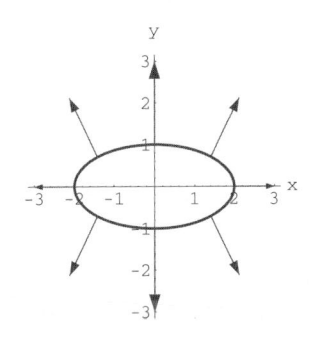

The normals point out away from the curve.

The upshot: The net flow of the given vector field across this curve is from inside to outside because
$$\int_{tlow}^{thigh} \text{Field}[x[t], y[t]] \bullet \{y'[t], -x'[t]\}\, dt > 0.$$

B.1.b.i) Here's a new vector field:

In[18]:=
```
Clear[Field,m,n,x,y]
m[x_,y_] = 2 x y; n[x_,y_] = x^2 - y^2; Field[x_,y_] = {m[x,y],n[x,y]}
```

Out[18]=
$\{2\, x\, y,\; x^2 - y^2\}$

Here's a new curve:

In[19]:=
```
Clear[t]
{x[t_],y[t_]} = {Cos[t]^3,Sin[t]^3};
tlow = 0; thigh = 2 Pi;
curveplot = ParametricPlot[{x[t],y[t]},{t,tlow,thigh},
PlotStyle->{{Thickness[0.01],Red}},
AxesLabel->{"x","y"}];
```

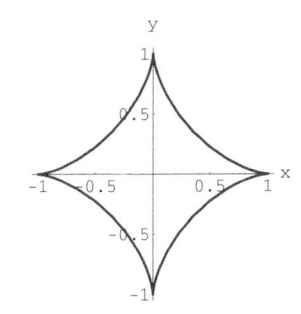

Use the integral
$$\int_{tlow}^{thigh} \text{Field}[x[t], y[t]] \bullet \{y'[t], -x'[t]\}\, dt$$
to analyze the net flow of this vector field across this curve.

Answer:

In[20]:=
```
Integrate[Field[x[t],y[t]].{y'[t],-x'[t]},{t,tlow,thigh}]
```

Out[20]=
0

This tells you that the inside-to-outside flow across this curve exactly balances the outside-to-inside flow across the curve. The net flow of this vector field across this curve is 0.

B.1.b.ii) Here's another vector field:

In[21]:=
```
Clear[Field,m,n,x,y]
m[x_,y_] = x + y; n[x_,y_] = x^2 + y^2; Field[x_,y_] = {m[x,y],n[x,y]}
```

Out[21]=
$\{x + y, x^2 + y^2\}$

Here's another new curve:

In[22]:=
```
Clear[t]
{x[t_],y[t_]} = {1,0} + {Sin[Pi t] + Cos[4 Pi t]/5,
    Cos[Pi t] + Sin[5 Pi t]/8};
tlow = 0; thigh = 2;
curveplot = ParametricPlot[{x[t],y[t]},
    {t,tlow,thigh},PlotStyle->{{Thickness[0.01],Red}},
    AxesLabel->{"x","y"}];
```

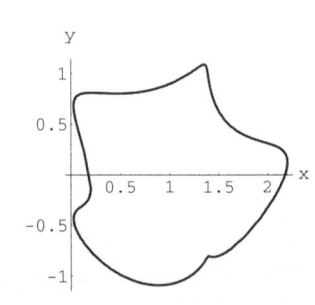

Use the integral

$$\int_{tlow}^{thigh} \text{Field}[x[t], y[t]] \bullet \{y'[t], -x'[t]\}\, dt$$

to analyze the net flow of this vector field across this curve.

Answer: The parameterization of the curve involves squirrelly functions, so try NIntegrate[ ]:

In[23]:=
```
NIntegrate[Field[x[t],y[t]].{y'[t],-x'[t]},{t,tlow,thigh},AccuracyGoal->2]
```

Out[23]=
-2.82743

This tells you

$$\int_{tlow}^{thigh} \text{Field}[x[t], y[t]] \bullet \{y'[t], -x'[t]\}\, dt < 0.$$

The net flow across this curve is in the opposite direction of the normals.

See which way the normals point:

```
In[24]:=
  scalefactor = 0.2;
  Show[curveplot,Table[
  Arrow[scalefactor {y'[t],-x'[t]},
  Tail->{x[t],y[t]},Red],
  {t,tlow,thigh,(thigh-tlow)/8}]];
```

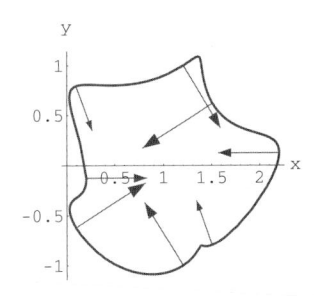

The normals point to the inside of the curve, and the net flow is opposite to the direction of these inward-pointing normals.

The result: The net flow of this vector field across this curve is from inside to outside.

Math happens again.

■ B.2) Measuring flow along a curve with the integral

$$\int_{tlow}^{thigh} \text{Field}[x[t], y[t]] \bullet \{x'[t], y'[t]\} \, dt$$

B.2.a) Given a vector field Field$[x, y]$ and a curve specified through parametric equations $\{x[t], y[t]\}$ with tlow $\leq t \leq$ thigh, you use the integral

$$\int_{tlow}^{thigh} \text{Field}[x[t], y[t]] \bullet \{y'[t], -x'[t]\} \, dt$$

to measure the flow of the field across the curve.

> What do you measure when you calculate
> $$\int_{tlow}^{thigh} \text{Field}[x[t], y[t]] \bullet \{x'[t], y'[t]\} \, dt?$$

Answer: $\{y'[t], -x'[t]\}$ is a normal vector, and when you calculate

$$\int_{tlow}^{thigh} \text{Field}[x[t], y[t]] \bullet \{y'[t], -x'[t]\} \, dt,$$

you measure net flow across the curve.

Analogously, $\{x'[t], y'[t]\}$ is a tangent vector, and when you calculate

$$\int_{tlow}^{thigh} \text{Field}[x[t], y[t]] \bullet \{x'[t], y'[t]\} \, dt,$$

you measure **net flow along the curve**.

B.2.b) Here's a vector field:

In[25]:=
```
Clear[Field,m,n,x,y,t]
m[x_,y_] = y Sin[x]; n[x_,y_] = -x Cos[y]; Field[x_,y_] = {m[x,y],n[x,y]}
```

Out[25]=
{y Sin[x], -(x Cos[y])}

And here is a curve C:

In[26]:=
```
tlow = 0; thigh = 1;
x[t_] = 6 t (1 - t);
y[t_] = 6 t^2 Cos[Pi t/2];
curveplot = ParametricPlot[{x[t],y[t]},{t,0,1},
PlotStyle->{{Red,Thickness[0.01]}},
AxesLabel->{"x","y"},PlotRange->All];
```

Use the integral

$$\int_{tlow}^{thigh} \text{Field}[x[t], y[t]] \bullet \{x'[t], y'[t]\} \, dt$$

to determine whether the net flow of Field$[x, y]$ along C is clockwise or counterclockwise.

Answer: Make the flow-along-the-curve measurement

$$\int_{tlow}^{thigh} \text{Field}[x[t], y[t]] \bullet \{x'[t], y'[t]\} \, dt :$$

In[27]:=
```
NIntegrate[Field[x[t],y[t]].{x'[t],y'[t]},{t,tlow,thigh}]
```

Out[27]=
-1.46585

Negative. This means that the flow of Field$[x, y]$ along C is against the direction of the tangent vectors $\{x'[t], y'[t]\}$ of this parameterization of C. Take a look at some of these tangent vectors:

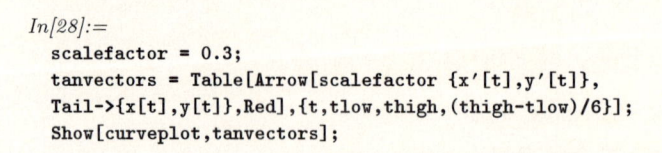

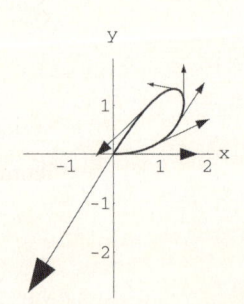

In[28]:=
```
scalefactor = 0.3;
tanvectors = Table[Arrow[scalefactor {x'[t],y'[t]},
Tail->{x[t],y[t]},Red],{t,tlow,thigh,(thigh-tlow)/6}];
Show[curveplot,tanvectors];
```

This curve is parameterized in the counterclockwise way. Because
$$\int_{tlow}^{thigh} \text{Field}[x[t], y[t]] \bullet \{x'[t], y'[t]\} \, dt$$
turned out negative, this tells you that the flow of this field along this curve is clockwise.

Confirm with a plot of some of the tangential components of the field vectors on the curve:

In[29]:=
```
Clear[tan,tancomp]
tan[t_] = {x'[t],y'[t]};
tancomp[t_] = ((Field[x[t],y[t]].tan[t])/
(tan[t].tan[t])) tan[t];
Show[curveplot,Table[Arrow[tancomp[t],Tail->
{x[t],y[t]},Blue],{t,tlow,thigh,
(thigh- tlow)/12}]];
```

Yessiree, Bob. Net clockwise flow, just as the measurement predicted.

■ B.3) Measurements by path integrals

$$\oint_C m[x, y] \, dx + n[x, y] \, dy$$

B.3.a.i) Here's some crazy looking notation:
$$\int_C m[x, y] \, dx + n[x, y] \, dy.$$
Folks call this a path integral.

Just what is a path integral?

Answer: A path integral needs the following ingredients:

→ It needs two functions $m[x, y]$ and $n[x, y]$.

→ It needs a curve C with a specified direction.

The resulting path integral is written as $\int_C m[x, y] \, dx + n[x, y] \, dy$.

The path integral
$$\int_C m[x, y] \, dx + n[x, y] \, dy$$

is calculated by evaluating the old-fashioned integral

$$\int_{tlow}^{thigh} \left(m[x[t], y[t]] x'[t] + n[x[t], y[t]] y'[t] \right) dt$$

where $\{x[t], y[t]\}$ with tlow $\leq t \leq$ thigh is any parameterization of C that gives the curve the same direction as the specified direction of C.

Look familiar? When you work with closed curves (like deformed circles with no loops), folks all across our planet have agreed to specify the counterclockwise direction. There is a nifty piece of notation to do this. For closed curves C without loops,

$$\oint_C m[x, y] \, dx + n[x, y] \, dy$$

means that the parameterization you use to evaluate the path integral is counterclockwise.

B.3.a.ii) Calculate

$$\oint_C m[x, y] \, dx + n[x, y] \, dy$$

for the case in which

$$m[x, y] = y - x,$$

$$n[x, y] = 2 x y,$$

and C is the ellipse

$$\left(\frac{x-1}{3} \right)^2 + \left(\frac{y}{2} \right)^2 = 1.$$

Answer: Enter the field and go with a counterclockwise parameterization of the ellipse:

```
In[30]:=
  Clear[x,y,m,n,t]
  m[x_,y_] = y - x; n[x_,y_] = 2 x y;
  {x[t_],y[t_]} = {1,0} + {3 Cos[t],2 Sin[t]};
  tlow = 0; thigh = 2 Pi;
  curveplot = ParametricPlot[{x[t],y[t]},{t,tlow,thigh},
  PlotStyle->{{Thickness[0.01],Red}},
  AxesLabel->{"x","y"}];
```

Check to be sure that the parameterization is counterclockwise:

In[31]:=
```
Show[curveplot,Table[Arrow[{x'[t],y'[t]},Tail->
{x[t],y[t]},Red],{t,tlow,thigh,(thigh-tlow)/6}]];
```

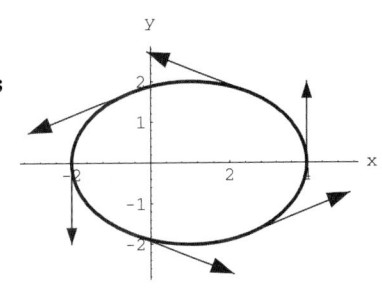

Good.

Here is the calculation of $\oint_C m[x,y]dx + n[x,y]dy$:

In[32]:=
```
NIntegrate[m[x[t],y[t]] x'[t] + n[x[t],y[t]] y'[t],{t,tlow,thigh}]
```

Out[32]=
-18.8496

Done.

B.3.b.i) Most folks say the path integral

$$\oint_C m[x,y]\,dx + n[x,y]\,dy$$

measures the flow of the vector field

$$\text{Field}[x,y] = \{m[x,y], n[x,y]\}$$

along a closed curve C.

They go on to say that if

$$\oint_C m[x,y]\,dx + n[x,y]\,dy > 0,$$

then the net flow of Field$[x,y] = \{m[x,y], n[x,y]\}$ along C is counterclockwise, but if

$$\oint_C m[x,y]\,dx + n[x,y]\,dy < 0,$$

then the net flow of Field$[x,y] = \{m[x,y], n[x,y]\}$ along C is clockwise.

Where do they get this idea?

Answer: Remember:

$$\oint_C m[x,y]\,dx + n[x,y]\,dy$$

demands a counterclockwise parameterization.

Lesson 3.05 Flow Measurements by Integrals

Also remember: When a closed curve C is parameterized in the counterclockwise way with a parameterization $\{x[t], y[t]\}$ with tlow $\leq t \leq$ thigh, then the tangent vectors $\{x'[t], y'[t]\}$ automatically point in the counterclockwise direction.

When you measure the flow of a vector field

$$\text{Field}[x, y] = \{m[x, y], n[x, y]\}$$

along such a curve, you calculate

$$\int_{\text{tlow}}^{\text{thigh}} \text{Field}[x[t], y[t]] \bullet \{x'[t], y'[t]\} dt$$

$$= \int_{\text{tlow}}^{\text{thigh}} \{m[x[t], y[t]], n[x[t], y[t]]\} \bullet \{x'[t], y'[t]\} \, dt$$

$$= \int_{\text{tlow}}^{\text{thigh}} m[x[t], y[t]] \, x'[t] + n[x[t], y[t]] \, y'[t] \, dt$$

$$= \oint_C m[x, y] \, dx + n[x, y] \, dy.$$

The result:

$$\oint_C m[x, y] \, dx + n[x, y] \, dy$$

measures the flow of the vector field

$$\text{Field}[x, y] = \{m[x, y], n[x, y]\}$$

around a closed curve C.

Consequently, if

$$\oint_C m[x, y] \, dx + n[x, y] \, dy > 0,$$

then the net flow of $\text{Field}[x, y] = \{m[x, y], n[x, y]\}$ around C is counterclockwise, but if

$$\oint_C m[x, y] \, dx + n[x, y] \, dy < 0,$$

then the net flow of $\text{Field}[x, y] = \{m[x, y], n[x, y]\}$ around C is clockwise.

B.3.b.ii) Calculate

$$\oint_C 5 x y \, dx + \left(x^3 + y^2\right) \, dy$$

where C is the circle

$$x^2 + (y - 1)^2 = 4$$

and interpret the result.

Answer:

$$\oint_C 5\,x\,y\,dx + (x^3 + y^2)\,dy$$

measures the net flow of the vector field

$$\text{Field}[x, y] = \{5\,x\,y, x^3 + y^2\}$$

along the circle.

Here are

→ a counterclockwise parameterization of C and
→ the calculation of $\oint_C 5\,x\,y\,dx + (x^3 + y^2)\,dy$:

In[33]:=
```
Clear[m,n,x,y,t]
{m[x_,y_],n[x_,y_]} = {5 x y,x^3 + y^2};
{x[t_],y[t_]} = {0,1} + 2 {Cos[t],Sin[t]};
tlow = 0; thigh = 2 Pi;
NIntegrate[m[x[t],y[t]] x'[t] + n[x[t],y[t]] y'[t],
{t,tlow,thigh}]
```

Out[33]=
37.6991

Big-time positive. This tells you that the net flow of the vector field

$$\text{Field}[x, y] = \{5\,x\,y, x^3 + y^2\}$$

along the circle is strongly counterclockwise.

Check it out with a plot if you don't believe it.

B.3.c.i) Most folks say the path integral

$$\oint_C -n[x,y]\,dx + m[x,y]\,dy$$

measures the flow of the vector field

$$\text{Field}[x, y] = \{m[x,y], n[x,y]\}$$

across a closed curve C. They go on to say that if

$$\oint_C -n[x,y]\,dx + m[x,y]\,dy > 0,$$

then the net flow of

$$\text{Field}[x, y] = \{m[x,y], n[x,y]\}$$

across C is from inside to outside, but if

$$\oint_C -n[x,y]\,dx + m[x,y]\,dy < 0,$$

then the net flow of the vector field

$$\text{Field}[x,y] = \{m[x,y], n[x,y]\}$$

across C is from outside to inside.

> **Where do they get this idea?**

Answer: Remember: $\oint_C -n[x,y]\,dx + m[x,y]\,dy$ demands a counterclockwise parameterization. Also remember: When a closed curve C is parameterized in the counterclockwise way with a parameterization

$$\{x[t], y[t]\} \qquad \text{with tlow} \le t \le \text{thigh},$$

then the normal vectors $\{y'[t], -x'[t]\}$ automatically point out away from the inside to the outside of the curve. When you measure the flow of a vector field

$$\text{Field}[x,y] = \{m[x,y], n[x,y]\}$$

across such a curve, you calculate

$$\int_{\text{tlow}}^{\text{thigh}} \text{Field}[x[t], y[t]] \bullet \{y'[t], -x'[t]\}\,dt$$

$$= \int_{\text{tlow}}^{\text{thigh}} \{m[x[t], y[t]], n[x[t], y[t]]\} \bullet \{y'[t], -x'[t]\}\,dt$$

$$= \int_{\text{tlow}}^{\text{thigh}} m[x[t], y[t]]y'[t] - n[x[t], y[t]]x'[t]\,dt$$

$$= \oint_C -n[x,y]\,dx + m[x,y]\,dy.$$

The result:

$$\oint_C -n[x,y]\,dx + m[x,y]\,dy$$

measures the flow of the vector field

$$\text{Field}[x,y] = \{m[x,y], n[x,y]\}$$

across C.

Consequently, if

$$\oint_C -n[x,y]\,dx + m[x,y]\,dy > 0,$$

then the net flow of $\text{Field}[x,y] = \{m[x,y], n[x,y]\}$ across C is from inside to outside, but if

$$\oint_C -n[x,y]\,dx + m[x,y]\,dy < 0,$$

then the net flow of $\text{Field}[x,y] = \{m[x,y], n[x,y]\}$ across C is from outside to inside.

B.3.c.ii) Calculate

$$\oint_C e^y \, dx - e^x \, dy$$

where C is the circle

$$x^2 + (y - 0.5)^2 = 0.7$$

and give two interpretations of the measurement.

Answer: Here are

→ a counterclockwise parameterization of C and

→ the calculation of $\oint_C e^y \, dx - e^x \, dy$:

In[34]:=
```
Clear[m,n,x,y,t]
{m[x_,y_],n[x_,y_]} = {E^y,-E^x};
{x[t_],y[t_]} = {0,0.5} + Sqrt[0.7] {Cos[t],Sin[t]};
tlow = 0; thigh = 2 Pi;
NIntegrate[m[x[t],y[t]] x'[t] + n[x[t],y[t]] y'[t],{t,tlow,thigh}]
```

Out[34]=
-6.3496

Negative.

→ Flow-along measurement interpretation:

$$\oint_C m[x,y] \, dx + n[x,y] \, dy$$

measures the net flow of $\text{Field}[x, y] = \{m[x, y], n[x, y]\}$ along C. The path integral calculated here was

$$\oint_C e^y \, dx - e^x \, dy < 0.$$

So the net flow of the vector field

$$\text{Field}[x, y] = \{e^y, -e^x\}$$

along the circle is clockwise.

→ Flow-across measurement interpretation:

$$\oint_C -n[x,y] \, dx + m[x,y] \, dy$$

measures the net flow of $\text{Field}[x, y] = \{m[x, y], n[x, y]\}$ across C. The path integral calculated here was

$$\oint_C e^y \, dx - e^x \, dy < 0.$$

So the net flow of the vector field

$$\text{Field}[x,y] = \{-e^x, -e^y\}$$

across the circle is is from outside to inside.

Note carefully that the two interpretations involve different vector fields.

■ B.4) Directed curves; path integrals

$$\int_C m[x,y]\,dx + n[x,y]\,dy,$$

path independence, and gradient fields

B.4.a) Lots of folks say that a parameterization gives a curve a direction. What do they mean by this?

Answer: The direction your parameterization goes specifies a direction for the curve.

Here's a curve $\{x[t], y[t]\}$ with a few scaled unit tangent vectors $\{x'[t], y'[t]\}$:

In[35]:=
```
Clear[x,y,t,direction]; tlow = 0.25; thigh = 1.25;
{x[t_],y[t_]} = {1 + 3 Sin[2 t]^2 Cos[t],t E^t};
curveplot1 = ParametricPlot[{x[t],y[t]},{t,tlow,thigh},
  PlotStyle->{{Red,Thickness[0.01]}},
  DisplayFunction->Identity,
  Epilog->{Text["start",{x[tlow],y[tlow]}],
  Text["end",{x[thigh],y[thigh]}]}];
scalefactor = 0.25; jump = (thigh - tlow)/6;
tanvectors = Table[Arrow[scalefactor {x'[t],y'[t]},
  Tail->{x[t],y[t]},Red],{t,tlow,thigh - jump,jump}];
direction1 = Show[curveplot1, tanvectors,
  AxesOrigin->{0,0},AxesLabel->{"x","y"},
  PlotRange->{{0,5},{0,5}},PlotLabel->"Direction 1",
  DisplayFunction->$DisplayFunction];
```

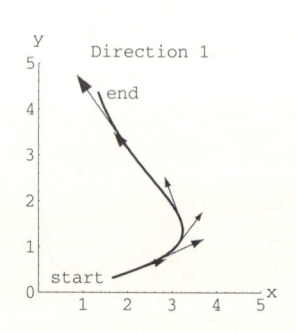

You can plot the same physical curve in the reverse direction by changing the parameterization:

In[36]:=
```
Clear[xx,yy,t]; a = tlow; b = thigh;
{xx[t_],yy[t_]} = {x[b - t(b - a)],y[b - t(b - a)]};
curveplot2 = ParametricPlot[{xx[t],yy[t]},{t,0,1},
PlotStyle->{{Red,Thickness[0.01]}},
DisplayFunction->Identity,
Epilog->{Text["start",{xx[0],yy[0]}],
Text["end",{xx[1],yy[1]}]}];
scalefactor = 0.25; jump = (1 - 0)/5;
tanvectors = Table[Arrow[scalefactor {xx'[t],yy'[t]},
Tail->{xx[t],yy[t]},Red],{t,0,1 - jump,jump}];
direction2 = Show[curveplot2, tanvectors,
AxesOrigin->{0,0},AxesLabel->{"x","y"},
PlotRange->{{0,5},{0,5}},PlotLabel->"Direction 2",
DisplayFunction->$DisplayFunction];
```

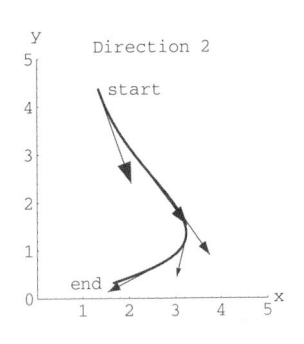

The curve is physically the same curve, but this new parameterization directs it to run from high to low.

If the curve C is closed (like a deformed circle) and has no loops, then there is no natural start or end. Your parametrization gives a start, an end, and a clockwise or counterclockwise direction. It's all up to you.

Here is a closed curve parameterized in the counterclockwise direction with the start point the same as the end point:

In[37]:=
```
Clear[x,y,t,direction]; tlow = 0; thigh = 2 Pi;
{x[t_],y[t_]} = {(1 + Sin[t]^2) Cos[t],
(0.5 + 2 Cos[t]^2) Sin[t]};
curveplot1 = ParametricPlot[{x[t],y[t]},{t,tlow,thigh},
PlotStyle->{{Red,Thickness[0.01]}},
DisplayFunction->Identity,
Epilog->{Text["start",{x[tlow],y[tlow]},{0,-2}],
Text["end",{x[thigh],y[thigh]},{0,2}]}];
scalefactor = 0.25; jump = (thigh - tlow)/8;
tanvectors = Table[Arrow[scalefactor {x'[t],y'[t]},
Tail->{x[t],y[t]},Red],{t,tlow,thigh - jump,jump}];
counterclockwise = Show[curveplot1, tanvectors,
AxesOrigin->{0,0},AxesLabel->{"x","y"},
PlotLabel->"Counterclockwise",PlotRange->All,
DisplayFunction->$DisplayFunction];
```

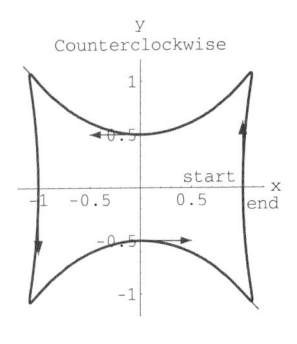

Here is the same curve with a clockwise parameterization with the start point the same as the end point:

Lesson 3.05 Flow Measurements by Integrals

In[38]:=
```
Clear[xx,yy,t]
{xx[t_],yy[t_]} = {x[thigh - t],y[thigh - t]};
curveplot2 = ParametricPlot[{xx[t],yy[t]},{t,tlow,thigh},
 PlotStyle->{{Red,Thickness[0.01]}},
 DisplayFunction->Identity,
 Epilog->{Text["start",{xx[tlow],yy[tlow]},{0,-2}],
 Text["end",{xx[thigh],yy[thigh]},{0,2}]}];
scalefactor = 0.25; jump = (thigh - tlow)/8;
tanvectors = Table[Arrow[scalefactor {xx'[t],yy'[t]},
 Tail->{xx[t],yy[t]},Red],{t,tlow,thigh - jump,jump}];
clockwise = Show[curveplot2, tanvectors,
 AxesOrigin->{0,0},AxesLabel->{"x","y"},
 PlotLabel->"Clockwise",PlotRange->All,
 DisplayFunction->$DisplayFunction];
```

For closed curves without loops, the clockwise and counterclockwise directions are the only choices you have.

B.4.b) Calculate
$$\int_C x\, e^{xy}\, dx + y\, e^{xy}\, dy$$
where C is the part of the parabola $y = x^2$ starting at $\{0,0\}$ and ending at $\{2,4\}$.

Interpret the meaning of the result.

Answer: Here's everything you need:

In[39]:=
```
Clear[m,n,x,y,t]
m[x_,y_] = x E^(x y); n[x_,y_] = y E^(x y);
tlow = 0; thigh = 2;
{x[t_],y[t_]} = {t,t^2};
curveplot = ParametricPlot[{x[t],y[t]},{t,tlow,thigh},
 PlotStyle->{{Red,Thickness[0.01]}},AxesLabel->{"x","y"},
 PlotLabel->"C and its direction",
 Epilog->{Text["start",{x[tlow],y[tlow]},{-2,-2}],
 Text["end",{x[thigh],y[thigh]},{2,2}]}];
```

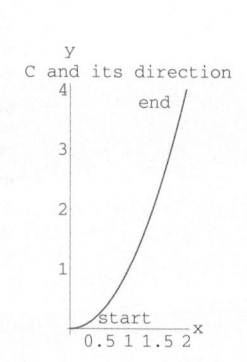

Yep; the curve C runs on the parabola $y = x^2$ it starts at $\{0,0\}$ and ends at $\{2,4\}$. Here comes the calculation of $\int_C x\, e^{xy}\, dx + y\, e^{xy}\, dy$:

In[40]:=
```
NIntegrate[m[x[t],y[t]] x'[t] + n[x[t],y[t]] y'[t],
 {t,tlow,thigh}]
```

Out[40]=
4312.75

Humongously positive.

→ Flow-along interpretation:

$$\int_C x\,e^{xy}\,dx + y\,e^{xy}\,dy$$

measures the net flow of the vector field

$$\text{Field}[x,y] = \{x\,e^{xy}, y\,e^{xy}\}$$

along C. The net flow of this vector field is strongly in the direction of the parameterization of the curve (from low to high).

→ Flow-across interpretation:

$$\int_C x\,e^{xy}\,dx + y\,e^{xy}\,dy$$

measures the net flow of the vector field

$$\text{Field}[x,y] = \{y\,e^{xy}, -x\,e^{xy}\}$$

across C. The net flow of this vector field across C is strongly in the direction of the normals $\{y'[t], -x'[t]\}$. These normals point to the right as you advance along the curve in the direction of the parameterization, so the net flow of the vector field $\text{Field}[x,y] = \{y\,e^{xy}, -x\,e^{xy}\}$ across C is from above C to below C.

B.4.c.i) Here is a cleared gradient field:

In[41]:=
```
Clear[f,x,y,m,n,gradf,Field]
gradf[x_,y_] = {D[f[x,y],x],D[f[x,y],y]};
{m[x_,y_],n[x_,y_]} = gradf[x,y];
Field[x_,y_] = {m[x,y],n[x,y]}
```

Out[41]=
$\{f^{(1,0)}[x, y], f^{(0,1)}[x, y]\}$

Here is *Mathematica*'s calculation of

$$\int_C m[x,y]\,dx + n[x,y]\,dy$$

for a cleared parameterization of a curve C that starts at $\{x[\text{tlow}], y[\text{tlow}]\}$ and ends at $\{x[\text{thigh}], y[\text{thigh}]\}$:

In[42]:=
```
Clear[t,tlow,thigh]
Integrate[m[x[t],y[t]]x'[t] + n[x[t],y[t]] y'[t],
{t,tlow,thigh}]
```

Out[42]=
f[x[thigh], y[thigh]] - f[x[tlow], y[tlow]]

> Explain where the answer comes from.

Answer: Put:

In[43]:=
```
Clear[g]; g[t_] = f[x[t],y[t]];
```

Compare:

In[44]:=
```
D[g[t],t] == m[x[t],y[t]] x'[t] + n[x[t],y[t]] y'[t]
```

Out[44]=
True

This tells you
$$g'[t] = m[x[t], y[t]]\, x'[t] + n[x[t], y[t]]\, y'[t].$$

The fundamental formula of calculus tells you

$$f[x[\text{thigh}], y[\text{thigh}]] - f[x[\text{tlow}], y[\text{tlow}]]$$
$$= g[\text{thigh}] - g[\text{tlow}]$$
$$= \int_{\text{tlow}}^{\text{thigh}} g'[t]\, dt$$
$$= \int_{\text{tlow}}^{\text{thigh}} m[x[t], y[t]]\, x'[t] + n[x[t], y[t]]\, y'[t]\, dt$$
$$= \int_C m[x, y]\, dx + n[x, y]\, dy.$$

So,

$$\int_C m[x, y]\, dx + n[x, y]\, dy = f[x[\text{thigh}], y[\text{thigh}]] - f[x[\text{tlow}], y[\text{tlow}]]$$

in the case that $\{m[x, y], n[x, y]\} = \text{gradf}[x, y]$.

The explanation is over.

B.4.c.ii) Now you know why you are guaranteed that

$$f[x_1, y_1] - f[x_0, y_0] = \int_C m[x, y]\, dx + n[x, y]\, dy$$

for any curve C starting at $\{x_0, y_0\}$ and ending at $\{x_1, y_1\}$ provided

$$\{m[x, y], n[x, y]\} = \text{gradf}[x, y]\ (= \nabla f[x, y])$$

for a function $f[x, y]$.

> What calculational advantage do you get from this?
>
> What theoretical advantage do you get from this?

Answer:

→ Your calculational advantage:

Here's a gradient field:

In[45]:=
```
Clear[f,x,y,gradf,m,n]
f[x_,y_] = Sin[Pi x y];
gradf[x_,y_] = {D[f[x,y],x],D[f[x,y],y]};
{m[x_,y_],n[x_,y_]} = gradf[x,y]
```

Out[45]=
{Pi y Cos[Pi x y], Pi x Cos[Pi x y]}

Because
$$\{m[x,y], n[x,y]\} = \text{gradf}[x,y],$$
you are guaranteed that if C is any curve running from $\{0,0\}$ to $\{1, 5/2\}$, then $\int_C m[x,y]\, dx + n[x,y]\, dy$ is given by:

In[46]:=
```
f[1,5/2] - f[0,0]
```

Out[46]=
1

Your calculational advantage is that you don't have to set up any parameterizations to calculate the path integral $\int_C m[x,y]\, dx + n[x,y]\, dy$.

→ Your theoretical advantage:

When you know that
$$\{m[x,y], n[x,y]\} = \text{gradf}[x,y],$$
then you know that $\int_C m[x,y]\, dx + n[x,y]\, dy$ **does not depend on the route of the path** C takes from its start to its end. In fact, the value of
$$\int_C m[x,y]\, dx + n[x,y]\, dy$$
depends **only** on the starting point and the ending point of C.

B.4.c.iii) If
$$\text{Field}[x,y] = \text{gradf}[x,y]$$
for a function $f[x,y]$, then how do you know that the flow of Field$[x,y]$ along any closed curve is 0?

Answer: Put
$$\text{Field}[x,y] = \{m[x,y], n[x,y]\} = \text{gradf}[x,y].$$

You are guaranteed that

$$f[x_1, y_1] - f[x_0, y_0] = \int_C m[x,y]\, dx + n[x,y]\, dy$$

for any curve C starting at $\{x_0, y_0\}$ and ending at $\{x_1, y_1\}$.

But for a closed curve C (like a deformed circle), you know that $\{x_0, y_0\} = \{x_1, y_1\}$. So

$$\oint_C m[x,y]\, dx + n[x,y]\, dy = f[x_1, y_1] - f[x_0, y_0] = f[x_0, y_0] - f[x_0, y_0] = 0$$

in the case that C is a closed curve. That's all there is to it.

Try it out for $\{m[x,y], n[x,y]\} = \text{gradf}[x,y]$ with

$$f[x,y] = x^4\, y^2$$

and with C the circle of radius 0.5 centered at $\{0,0\}$:

In[47]:=
```
Clear[f,x,y,m,n,gradf,Field,t]
f[x_,y_] = x^4 y^2;
gradf[x_,y_] = {D[f[x,y],x],D[f[x,y],y]};
{m[x_,y_],n[x_,y_]} = gradf[x,y];
Field[x_,y_] = {m[x,y],n[x,y]};
tlow = 0; thigh = 2 Pi;
{x[t_],y[t_]} = 0.5 {Cos[t],Sin[t]};
Integrate[m[x[t],y[t]]x'[t] + n[x[t],y[t]]y'[t],{t,tlow,thigh}]
```
Out[47]=
0

Just as theory predicted.

Tutorials

■ T.1) Backward and forward

T.1.a.i) Explain why if C_1 and C_2 are the same physical curve, but

→ the starting point of C_1 is the ending point of C_2 and

→ the ending point of C_1 is the starting point of C_2,

then for any $m[x,y]$ and $n[x,y]$ that come down the pike, you will always get

$$\int_{C_1} m[x,y]\, dx + n[x,y]\, dy = -\int_{C_2} m[x,y]\, dx + n[x,y]\, dy.$$

In other words, if you reverse the direction, then you reverse the sign of the path integral.

Answer: Both $\int_{C_1} m[x,y]\,dx + n[x,y]\,dy$ and $\int_{C_2} m[x,y]\,dx + n[x,y]\,dy$ measure the flow of the vector field

$$\text{Field}[x,y] = \{m[x,y], n[x,y]\}$$

along the same curve. But the interpretation is different in each case, because the unit tangent vectors on C_1 point in the direction exactly opposite of those on C_2. The opposite direction of the tangent vectors accounts for the minus sign.

T.1.a.ii) Illustrate by calculating the path integral

$$\int_{C_1} y^3\,dx + x^2 y\,dy,$$

where C_1 is the segment of the parabola $y = x^2$ starting at $\{0,0\}$ and ending at $\{2,4\}$ and then calculating the path integral

$$\int_{C_2} y^3\,dx + x^2 y\,dy$$

where C_2 is the segment of the same parabola, but starting at $\{2,4\}$ and ending at $\{0,0\}$.

Answer:

In[1]:=
```
Clear[m,n,x,y]; m[x_,y_] = y^3; n[x_,y_] = x^2 y;
```

To calculate the path integral

$$\int_{C_1} y^3\,dx + x^2 y\,dy,$$

where C_1 is the segment of the parabola $y = x^2$ starting at $\{0,0\}$ and ending at $\{2,4\}$, parameterize C_1 so that it runs from $\{0,0\}$ to $\{2,4\}$ and integrate:

In[2]:=
```
Clear[t]; x[t_] = t; y[t_] = t^2; tlow = 0; thigh = 2;
start1 = {x[tlow],y[tlow]}
```

Out[2]=
```
{0, 0}
```

In[3]:=
```
end1 = {x[thigh],y[thigh]}
```

Out[3]=
```
{2, 4}
```

In[4]:=
```
C1pathintegral = NIntegrate[m[x[t],y[t]] x'[t] +
n[x[t],y[t]] y'[t],{t,tlow,thigh}]
```

Out[4]=
39.619

To calculate the path integral $\int_{C_2} y^3\, dx + x^2\, y\, dy$, where C_2 is the segment of the parabola $y = x^2$ starting at $\{2, 4\}$ and ending at $\{0, 0\}$, parameterize C_2 so that it starts at $\{2, 4\}$ and runs to $\{0, 0\}$ and integrate:

In[5]:=
```
Clear[x,y]; x[t_] = (2 - t); y[t_] = (2 - t)^2;
tlow = 0; thigh = 2;
start2 = {x[tlow],y[tlow]}
```
Out[5]=
{2, 4}

In[6]:=
```
end2 = {x[thigh],y[thigh]}
```
Out[6]=
{0, 0}

In[7]:=
```
C2pathintegral = NIntegrate[m[x[t],y[t]] x'[t] +
  n[x[t],y[t]] y'[t],{t,tlow,thigh}]
```
Out[7]=
-39.619

Compare:

In[8]:=
```
{C1pathintegral,C2pathintegral}
```
Out[8]=
{39.619, -39.619}

Just as you knew in advance; they are negatives of each other.

T.1.b.i) This problem appears only in the electronic version.

T.1.b.ii) This problem appears only in the electronic version.

T.1.c.i) It's late and you're calculating

$$\oint_C y^2\, dx + (2\,x^2 + y)\, dy$$

where C is the ellipse

$$\left(\frac{x+1}{4}\right)^2 + \left(\frac{y}{2}\right)^2 = 1.$$

In your haste to meet your date at the local hangout, you type:

In[9]:=
```
Clear[m,n,x,y,t]
m[x_,y_] = y^2; n[x_,y_] = 2 x^2 + y;
{x[t_],y[t_]} = {-1,0} + {4 Sin[t],2 Cos[t]};
answer = NIntegrate[m[x[t],y[t]] x'[t] + n[x[t],y[t]] y'[t],{t,0, 2 Pi}]
```

Out[9]=
100.531

A friend who is looking over your shoulder says: "Good work, but your answer is wrong because your parameterization is clockwise and not counterclockwise."

After looking at a plot of some tangent vectors, you see that your parameterization is clockwise. And then you say, "The correct answer is":

In[10]:=
```
correctanswer = - answer
```
Out[10]=
−100.531

Are you right?

Answer: Yes.

■ T.2) Screwing up

T.2.a) What are the best ways of screwing up the calculation of a path integral
$$\oint_C m[x,y]\,dx + n[x,y]\,dy?$$

Answer: The best way to screw up is to give a clockwise parameterization of C instead of a counterclockwise parameterization.

The second best way to screw up is to give a counterclockwise parameterization that covers C more than once.

Case in point: Calculate
$$\oint_C -y\,dx + x\,dy,$$
given that C is the circle $x^2 + y^2 = 1$.

In[11]:=
```
Clear[m,n,x,y]; m[x_,y_] = -y; n[x_,y_] = x;
x[t_] = Cos[t]; y[t_] = Sin[t]; tlow = 0; thigh = 4 Pi;
pathintegral = NIntegrate[m[x[t],y[t]] x'[t] + n[x[t],y[t]]y'[t],{t,tlow,thigh}]
```
Out[11]=
12.5664

Check the parameterization:

Lesson 3.05 Flow Measurements by Integrals

In[12]:=
```
curveplot = ParametricPlot[{x[t],y[t]},{t,tlow,thigh},
  PlotStyle->{{Red,Thickness[0.01]}},
  AxesLabel->{"x","y"},DisplayFunction->Identity];
jump = Pi/4; tangentvectors = Table[
  Arrow[{x'[t],y'[t]},Tail->{x[t],y[t]},Red],
  {t,tlow,thigh - jump,jump}];
Show[curveplot,tangentvectors,
  DisplayFunction->$DisplayFunction];
```

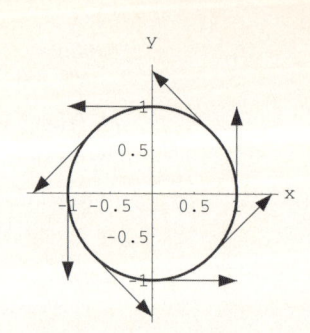

So far, so good. Everything looks fine. Now look at:

In[13]:=
```
{x[tlow],y[tlow]} == {x[thigh],y[thigh]}
```

Out[13]=
True

The starting point and the ending point are the same. So this parameterization passes the usual tests, but the calculated value of $\oint_C -y\,dx + x\,dy$ above is **dead wrong**. To see why, look at:

In[14]:=
```
ParametricPlot[{x[t],y[t]},{t,tlow,2 Pi},
  PlotStyle->{{Red,Thickness[0.01]}},
  AxesLabel->{"x","y"}];
```

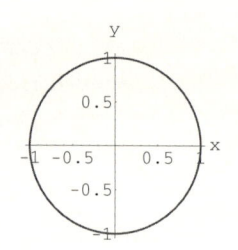

In[15]:=
```
ParametricPlot[{x[t],y[t]},{t,2 Pi,thigh},
  PlotStyle->{{Red,Thickness[0.01]}},
  AxesLabel->{"x","y"}];
```

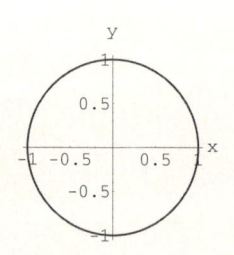

As t advances from tlow $= 0$ to thigh $= 4\pi$ as orginally specified, the parameterization goes around the curve twice, and this is not what you want. The right value for $\oint_C -y\,dx + x\,dy$ is:

In[16]:=
```
tlow = 0; correctthigh = 2 Pi;
correctpathintegral =
NIntegrate[m[x[t],y[t]] x'[t] + n[x[t],y[t]]y'[t],{t,tlow,correctthigh}]
```

Out[16]=
6.28319

■ T.3) Recognizing gradient fields: The gradient test

When you have to calculate a path integral

$$\int_C m[x,y]\,dx + n[x,y]\,dy$$

and you recognize that the vector field

$$\text{Field}[x,y] = \{m[x,y], n[x,y]\}$$

is the gradient field of a function $f[x,y]$, then a warm comfortable feeling radiates through your body. If you want to exploit the advantages you get from a gradient field, you'll have to be in a position to recognize when a given vector field is a gradient field. To this end, go with a cleared gradient field:

In[17]:=
```
Clear[f,gradf,x,y,m,n]
gradf[x_,y_] = {D[f[x,y],x],D[f[x,y],y]};
{m[x_,y_],n[x_,y_]} = gradf[x,y]
```

Out[17]=
$\{f^{(1,0)}[x, y], f^{(0,1)}[x, y]\}$

Look at this:

In[18]:=
```
gradtest = D[m[x,y],y] - D[n[x,y],x]
```

Out[18]=
0

This tells you that if you have a gradient field $\text{Field}[x,y] = \{m[x,y], n[x,y]\}$, then

$$D[m[x,y],y] - D[n[x,y],x] = 0.$$

T.3.a.i) If you are given a vector field $\text{Field}[x,y] = \{m[x,y], n[x,y]\}$ and you learn that

$$D[m[x,y],y] - D[n[x,y],x] = 0,$$

are you automatically guaranteed that $\text{Field}[x,y]$ is a gradient field?

Answer: Yes, provided that neither $m[x,y]$ nor $n[x,y]$ has a singularity (blow-up or blow-down).

T.3.a.ii) Is the vector field

$$\text{Field}[x,y] = \{e^x \cos[y], -e^x \sin[y]\}$$

a gradient field?

Answer: Look at:

In[19]:=
```
Clear[m,n,x,y]; {m[x_,y_],n[x_,y_]} = {E^x Cos[y],-E^x Sin[y]}
```

Out[19]=
$\{E^x \text{Cos}[y], -(E^x \text{Sin}[y])\}$

Good; neither $m[x,y]$ nor $n[x,y]$ has any singularities. Now go with the gradient test:

In[20]:=
```
gradtest = D[m[x,y],y] - D[n[x,y],x]
```

Out[20]=
0

Hot dog! No doubt about it, this vector field is a gradient field.

T.3.a.iii) Is the vector field

$$\text{Field}[x,y] = \{e^y \cos[x], -e^y \sin[x]\}$$

a gradient field?

Answer: Look at:

In[21]:=
```
Clear[m,n,x,y]; {m[x_,y_],n[x_,y_]} = {E^y Cos[x],-E^y Sin[x]}
```

Out[21]=
$\{E^y \text{Cos}[x], -(E^y \text{Sin}[x])\}$

Good; neither $m[x,y]$ nor $n[x,y]$ has any singularities. Now go with the gradient test:

In[22]:=
```
gradtest = D[m[x,y],y] - D[n[x,y],x]
```

Out[22]=
$2 E^y \text{Cos}[x]$

This is not 0. Absolutely no doubt about it, this vector field is **not** a gradient field.

T.3.a.iv) Is the vector field

$$\text{Field}[x,y] = \left\{\frac{-y}{x^2+y^2}, \frac{x}{x^2+y^2}\right\}$$

a gradient field?

Answer: Look at:

In[23]:=
```
Clear[m,n,Field,x,y]
{m[x_,y_],n[x_,y_]} = {-y/(x^2 + y^2),x/(x^2 + y^2)};
Field[x_,y_] = {m[x,y],n[x,y]}
```

Out[23]=
$$\left\{-\left(\frac{y}{x^2+y^2}\right), \frac{x}{x^2+y^2}\right\}$$

A big fat singularity at $\{x,y\} = \{0,0\}$. You can see this by plotting.

In[24]:=
```
Plot[m[0,y],{y,-1,1},PlotStyle->Red,AspectRatio->1];
```

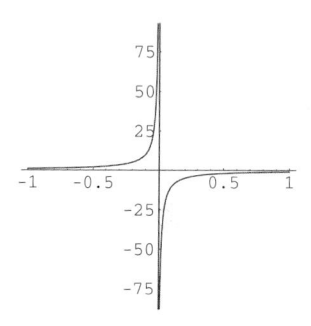

In[25]:=
```
Plot[n[x,0],{x,-1,1},PlotStyle->Red,AspectRatio->1];
```

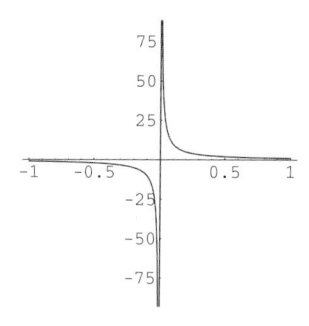

This vector field fails the first part of the gradient test. Now look at the second part of the gradient test:

In[26]:=
```
gradtest = Simplify[D[n[x,y],x] - D[m[x,y],y]]
```

Out[26]=
0

This field passes the second part of the gradient test.

Whether this vector field is a genuine gradient field is in doubt.

Try something else by looking at

$$\oint_C m[x,y]\,dx + n[x,y]\,dy$$

where C is the circle of radius 1 centered at the singularity at $\{0,0\}$:

In[27]:=
```
Clear[t]
{x[t_],y[t_]} = {Cos[t],Sin[t]};
tlow = 0; thigh = 2 Pi;
NIntegrate[m[x[t],y[t]] x'[t] + n[x[t],y[t]] y'[t],{t,tlow,thigh}]
```

Out[27]=
6.28319

Not zero. This tells you for sure that this vector field is not a gradient field.

T.3.a.v) Is every vector field a gradient field?

Answer: Hell no.

T.3.b) Here's a vector field:

In[28]:=
```
Clear[Field,m,n,x,y]
{m[x_,y_],n[x_,y_]} = {x^2 + 2 x Sin[y],Sin[5 y] + x^2 Cos[y]};
Field[x_,y_] = {m[x,y],n[x,y]}
```

Out[28]=
$\{x^2 + 2\ x\ Sin[y],\ x^2\ Cos[y] + Sin[5\ y]\}$

No singularities. Give it the second part of the gradient test:

In[29]:=
```
gradtest = D[m[x,y],y] - D[n[x,y],x]
```

Out[29]=
0

Good. Field$[x,y]$ is definitely a gradient field.

Try to come up with a function $f[x,y]$ so that grad$f[x,y]$ = Field$[x,y]$.

Answer: To do this, fix any point $\{a,b\}$ you like and parameterize a line C (or other curve) running from $\{a,b\}$ to the variable point $\{x,y\}$: $\{0,0\}$ is usually a good choice for $\{a,b\}$.

In[30]:=
```
Clear[t]; {a,b} = {0,0};
fixedpoint = {a,b}; variablepoint = {x,y};
tlow = 0; thigh = 1;
{x[t_],y[t_]} = fixedpoint + t(variablepoint - fixedpoint)
```

Out[30]=
{t x, t y}

To get a function $f[x,y]$ with $\text{gradf}[x,y] = \{m[x,y], n[x,y]\}$, all you gotta do is set
$$f[x,y] = \int_C m[x,y]\, dx + n[x,y]\, dy$$
where C is the line (or other curve) running from the fixed point $\{a,b\}$ to the variable point $\{x,y\}$:

In[31]:=
```
Clear[f]
f[x_,y_] = Integrate[m[x[t],y[t]] x'[t] + n[x[t],y[t]] y'[t],{t,tlow,thigh}]
```

Out[31]=
$$\frac{1}{5} + \frac{x^3}{3} - \frac{\cos[5y]}{5} + x^2 \sin[y]$$

Try it out:

In[32]:=
```
Clear[gradf]; gradf[x_,y_] = {D[f[x,y],x],D[f[x,y],y]}
```

Out[32]=
$$\{x^2 + 2x \sin[y], x^2 \cos[y] + \sin[5y]\}$$

Compare:

In[33]:=
```
{m[x,y],n[x,y]}
```

Out[33]=
$$\{x^2 + 2x \sin[y], x^2 \cos[y] + \sin[5y]\}$$

Great. This tells you that $\text{gradf}[x,y] = \{m[x,y], n[x,y]\}$, just as you wanted.

See what happens when you go with $\{a,b\} = \{1, \pi/2\}$:

In[34]:=
```
Clear[x,y,t]; {a,b} = {1,Pi/2};
fixedpoint = {a,b}; variablepoint = {x,y};
tlow = 0; thigh = 1;
{x[t_],y[t_]} = fixedpoint + t(variablepoint - fixedpoint)
```

Out[34]=
$$\{1 + t(-1+x),\ \frac{\text{Pi}}{2} + t(\frac{-\text{Pi}}{2} + y)\}$$

Set $f[x,y] = \int_C m[x,y]\, dx + n[x,y]\, dy$ where C is the line running from $\{a,b\}$ to $\{x,y\}$:

In[35]:=
```
Clear[f]
f[x_,y_] = Integrate[m[x[t],y[t]] x'[t] +
n[x[t],y[t]] y'[t],{t,tlow,thigh}]
```

Lesson 3.05 Flow Measurements by Integrals

Out[35]=

$$-1 + \frac{(-1+x)(1+x+x^2)}{3} + x^2 \cos\left[\frac{-Pi+2y}{2}\right] + \frac{5(-Pi+2y)\sin\left[\frac{5(-Pi+2y)}{2}\right]}{5}$$

Looks bad; check whether it feels good:

In[36]:=
```
Clear[gradf]; gradf[x_,y_] = {D[f[x,y],x],D[f[x,y],y]}
```

Out[36]=

$$\left\{\frac{(-1+x)(1+2x)}{3} + \frac{1+x+x^2}{3} + 2x\cos\left[\frac{-Pi+2y}{2}\right],\right.$$

$$\left.\cos\left[\frac{5(-Pi+2y)}{2}\right] - x^2\sin\left[\frac{-Pi+2y}{2}\right]\right\}$$

Compare:

In[37]:=
```
Expand[gradf[x,y] - {m[x,y],n[x,y]},Trig->True]
```

Out[37]=
```
{0, 0}
```

If feels great! Each time you change the fixed point $\{a,b\}$, you make a different function $f[x,y]$ whose gradient is $\{m[x,y],n[x,y]\}$.

T.3.c) What is the value of

$$\int_C e^{-5xy}(3\cos[3x] - 5y\sin[3x])\,dx - 5x\,e^{-5xy}\sin[3x]\,dy$$

for any curve C running from $\{-0.7, 0\}$ to $\{1.1, 0.4\}$?

Answer: Here's the vector field:

In[38]:=
```
Clear[Field,m,n,x,y]
{m[x_,y_],n[x_,y_]} = {E^(-5 x y) (3 Cos[3 x] - 5 y Sin[3 x]),
    -5 x E^(-5 x y) Sin[3 x]};
Field[x_,y_] = {m[x,y],n[x,y]}
```

Out[38]=

$$\left\{\frac{3\cos[3x] - 5y\sin[3x]}{E^{5xy}}, \frac{-5x\sin[3x]}{E^{5xy}}\right\}$$

No singularities (because $e^s > 0$ no matter what s is). Give it the second part of the gradient test:

In[39]:=
```
gradtest = Together[D[m[x,y],y] - D[n[x,y],x]]
```

Out[39]=
0

Good. Now you know that Field$[x, y] = \{m[x, y], n[x, y]\}$ is a gradient field. This is really good news because this tells you that

$$\int_C e^{-5xy}\,(3\cos[3\,x] - 5\,y\,\sin[3\,x])\ dx - 5\,x\,e^{-5xy}\sin[3\,x]\,dy$$

calculates out to the same value no matter what curve C you go with as long as C starts at $\{-0.7, 0\}$ and stops at $\{1.1, 0.4\}$.

This information is quite a relief because now you know that you can calculate this integral by using any cheap curve C running from $\{-0.7, 0\}$ to $\{1.1, 0.4\}$. One really cheap curve is the straight line parameterized by:

In[40]:=
```
Clear[x,y,t]; start = {-0.7,0}; end = {1.1,0.4};
tlow = 0; thigh = 1; {x[t_],y[t_]} = start + t (end - start)
```

Out[40]=
{-0.7 + 1.8 t, 0.4 t}

Here comes the calculation of

$$\int_C e^{-5xy}\,(3\cos[3\,x] - 5\,y\,\sin[3\,x])\ dx - 5\,x\,e^{-5xy}\sin[3\,x]\,dy$$

for any curve C running from $\{-0.7, 0\}$ to $\{1.1, 0.4\}$:

In[41]:=
```
NIntegrate[m[x[t],y[t]] x'[t] + n[x[t],y[t]] y'[t],{t,tlow,thigh}]
```

Out[41]=
0.845731

Not hard at all.

■ T.4) Line integrals

T.4.a) What do folks mean when they talk about line integrals?

Answer: A line integral is the same thing as a path integral. This alternate terminology is in common use. This is unfortunate because many path integrals involve curves that are not lines.

■ T.5) Summary of main ideas

This problem appears only in the electronic version.

Give It a Try

Experience with the starred (★) problems will be useful for understanding developments later in the course.

■ G.1) Flow along and flow across★

G.1.a) Here's a vector field:

In[1]:=
```
Clear[x,y,m,n,Field]
{m[x_,y_],n[x_,y_]} = {x^2 - 2 y,-y^2 + x}; Field[x_,y_] = {m[x,y],n[x,y]}
```

Out[1]=
$\{x^2 - 2y, x - y^2\}$

Here's the circle C of radius 3 centered at $\{1,2\}$ parameterized in the counterclockwise way:

In[2]:=
```
Clear[t]
{x[t_],y[t_]} = {1,2} + 3 {Cos[t],Sin[t]};
tlow = 0; thigh = 2 Pi;
curveplot = ParametricPlot[{x[t],y[t]},
  {t,tlow,thigh},PlotStyle->{{Thickness[0.01],Red}},
  AspectRatio->Automatic,AxesLabel->{"x","y"}];
```

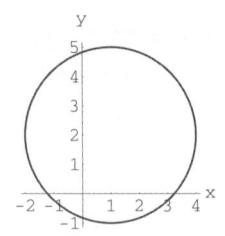

Calculate

$$\oint_C -n[x,y]\,dx + m[x,y]\,dy$$

and use your result to determine whether the net flow of this vector field across this curve is from inside to outside, outside to inside, or 0.

Calculate

$$\oint_C m[x,y]\,dx + n[x,y]\,dy$$

and use your result to determine whether the net flow of this vector field along this curve is clockwise, counterclockwise, or 0.

G.1.b) Here's a vector field:

In[3]:=
```
Clear[x,y,m,n,Field]
{m[x_,y_],n[x_,y_]} = {0.5 x - 1.2 y,1}; Field[x_,y_] = {m[x,y],n[x,y]}
```

Out[3]=
{0.5 x - 1.2 y, 1}

Here's a parameterization and a plot of a closed curve C:

In[4]:=
```
Clear[t]
{x[t_],y[t_]} = {3 t (3 - t),t (t - 3)^2};
tlow = 0; thigh = 3;
curveplot = ParametricPlot[{x[t],y[t]},
 {t,tlow,thigh},PlotStyle->
 {{Thickness[0.01],Red}},PlotLabel->
 "The closed curve C",AspectRatio->Automatic,
 AxesLabel->{"x","y"}];
```

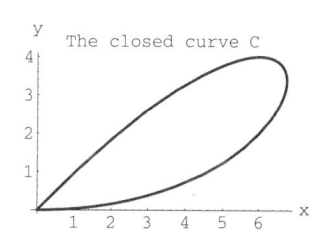

Is C parameterized in the counterclockwise or clockwise way?

Use a path integral to determine whether the net flow of this vector field across C is from outside to inside, inside to outside, or 0.

Use a path integral to determine whether the net flow of this vector field along C is clockwise, counterclockwise, or 0.

G.1.c) Here's a vector field:

In[5]:=
```
Clear[x,y,m,n,Field]
{m[x_,y_],n[x_,y_]} = {x^3 - 3 x y^2,3 x^2 y - y^3};
Field[x_,y_] = {m[x,y],n[x,y]}
```

Out[5]=
{x^3 - 3 x y^2, 3 x^2 y - y^3}

Here's a parameterization and a plot of a closed curve C.

In[6]:=
```
Clear[t]
{x[t_],y[t_]} =
{3 Sin[t] Cos[t],Sin[t]^2 + Cos[6 t]/6 + 2};
tlow = 0; thigh = Pi;
curveplot = ParametricPlot[{x[t],y[t]},
 {t,tlow,thigh},PlotStyle->{{Thickness[0.01],Red}},
 AspectRatio->Automatic,AxesLabel->{"x","y"}];
```

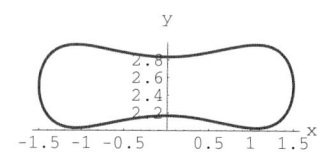

Is the curve parameterized in the counterclockwise or clockwise way?

Use a path integral to determine whether the net flow of this vector field across this curve is from outside to inside, inside to outside, or 0.

Use a path integral to determine whether the net flow of this vector field along this curve is clockwise, counterclockwise, or 0.

G.1.d.i) Here's the gradient field of the function $f[x,y] = e^{2x-y}$:

In[7]:=
```
Clear[f,gradf,x,y,m,n,Field]; f[x_,y_] = E^(2 x - y);
gradf[x_,y_] ={D[f[x,y],x],D[f[x,y],y]};
{m[x_,y_],n[x_,y_]} = gradf[x,y];
Field[x_,y_] = {m[x,y],n[x,y]}
```

Out[7]=
$$\{2 E^{2x-y}, -E^{2x-y}\}$$

Here's a parameterization and a plot of a closed curve C.

In[8]:=
```
Clear[t]
{x[t_],y[t_]} = {6 Sin[t] Cos[t](1 - 0.7 Sin[4 t]),
3 Sin[t]^2 + 2}; tlow = 0; thigh = Pi;
curveplot = ParametricPlot[{x[t],y[t]},
{t,tlow,thigh},PlotStyle->{{Thickness[0.01],Red}},
AspectRatio->Automatic,AxesLabel->{"x","y"}];
```

> Explain how you know in advance that the net flow of this vector field along this curve is 0.
>
> Use a path integral to determine whether the net flow of this vector field across this curve is from outside to inside, inside to outside, or 0.

G.1.d.ii)
> You already know that the net flow of a gradient field along a closed curve is guaranteed to be 0.
>
> Is it true that the net flow of a gradient field across a closed curve is guaranteed to be 0?

■ G.2) Path integrals: Backward and forward★

G.2.a.i) Suppose C_1 and C_2 are the same non-closed physical curve, but the starting point of C_1 is the ending point of C_2 and the ending point of C_1 is the starting point of C_2.

> Express $\int_{C_2} m[x,y]\,dx + n[x,y]\,dy$ in terms of $\int_{C_1} m[x,y]\,dx + n[x,y]\,dy$.

G.2.a.ii)
> Calculate the path integral
> $$\int_{C_1} x^2 y\,dx - 3xy\,dy,$$

> where C_1 starts at $\{-1,3\}$, runs to $\{1,0\}$ on a straight line, and then follows the parabola $y = 3(x-1)^2$ to $\{2,3\}$ where it stops.
>
> Then calculate the path integral
>
> $$\int_{C_2} x^2 y \, dx - 3xy \, dy,$$
>
> where C_2 starts at $\{2,3\}$, runs to $\{1,0\}$ on the parabola $y = 3(x-1)^2$ and then follows the straight line from $\{1,0\}$ to $\{-1,3\}$ where it stops.

G.2.b) Here's a parameterization of the ellipse

$$\left(\frac{x-1}{3}\right)^2 + \left(\frac{y+2}{2}\right)^2 = 1:$$

In[9]:=
```
Clear[x,y,t]
tlow = 0; thigh = 2 Pi;
{x[t_],y[t_]} = {1,-2} + {3 Sin[t],2 Cos[t]}
```

Out[9]=
```
{1 + 3 Sin[t], -2 + 2 Cos[t]}
```

Call this ellipse C. Now look at this calculation:

In[10]:=
```
NIntegrate[y[t]^2 x'[t] + (x[t] y[t] + 1) y'[t],{t,tlow,thigh}]
```

Out[10]=
```
-37.6991
```

> Does this result calculate
>
> $$\oint_C y^2 \, dx + (xy + 1) \, dy?$$
>
> If not, how do you modify this result to get the value of
>
> $$\oint_C y^2 \, dx + (xy + 1) \, dy?$$

■ G.3) Calculations and interpretations*

Many different notations for path integrals are in regular use in science. In this problem, you will meet some of them.

Go with a given vector field

$$\text{Field}[x, y] = \{m[x, y], n[x, y]\},$$

a curve C, and a direction for C via a parameterization
$$P[t] = \{x[t], y[t]\}, \qquad \text{for tlow} \leq t \leq \text{thigh}.$$
All the next four integrals calculate out to the same value, and all measure the flow of
$$\text{Field}[x, y] = \{m[x, y], n[x, y]\}$$
along C:

i) $\displaystyle\int_C m[x, y] \, dx + n[x, y] \, dy;$

ii) $\displaystyle\int_{\text{tlow}}^{\text{thigh}} \text{Field}[x[t], y[t]] \bullet \{x'[t], y'[t]\} \, dt;$

iii) $\displaystyle\int_C \text{Field} \bullet dP, \text{where } dP = \{x'[t], y'[t]\} \, dt;$

iv) $\displaystyle\int_C \text{Field} \bullet \text{unittan} \, ds.$

This last integral is with respect to length measured on the curve from the start of the curve. The way to see that the last integral is the same as the others is to notice that
$$ds = \sqrt{x'[t]^2 + y'[t]^2} \, dt$$
and
$$\text{unittan}[t] = \frac{\{x'[t], y'[t]\}}{\sqrt{x'[t]^2 + y'[t]^2}}.$$
When you transform to the t variable, you get
$$\text{Field} \bullet \text{unittan} \, ds \longleftrightarrow \text{Field}[x[t], y[t]] \bullet \{x'[t], y'[t]\} \, dt.$$
All of the next four integrals calculate out to the same value, and all measure the flow of
$$\text{Field}[x, y] = \{m[x, y], n[x, y]\}$$
across C:

i) $\displaystyle\int_C -n[x, y] \, dx + m[x, y] \, dy;$

ii) $\displaystyle\int_{\text{tlow}}^{\text{thigh}} \text{Field}[x[t], y[t]] \bullet \{y'[t], -x'[t]\} \, dt;$

iii) $\displaystyle\int_C \text{Field} \bullet \text{unitnormal} \, ds.$

This last integral is with respect to length measured on the curve from the start of the curve. The way to see that the last integral is the same as the others is to

notice that
$$ds = \sqrt{x'[t]^2 + y'[t]^2}\, dt$$
and
$$\text{unitnormal}[t] = \frac{\{y'[t], -x'[t]\}}{\sqrt{x'[t]^2 + y'[t]^2}}.$$

When you transform to the t variable, you get
$$\text{Field} \bullet \text{unitnormal}\, ds \longleftrightarrow \text{Field}[x[t], y[t]] \bullet \{y'[t], -x'[t]\}\, dt.$$

G.3.a.i) Calculate
$$\oint_C \text{Field} \bullet \text{unittan}\, ds$$
in the case in which
$$\text{Field}[x, y] = \{x^2 y^2, x y^2\}$$
and the curve C is the ellipse
$$x^2 + 2 y^2 = 1.$$
Give an interpretation of the result as a flow-along measurement, and illustrate with a plot.

G.3.a.ii) Calculate
$$\oint_C \text{Field} \bullet dP$$
in the case in which
$$\text{Field}[x, y] = \{x^2 y^2, x y^2\}$$
and the curve C is the ellipse
$$x^2 + 2 y^2 = 1.$$
Give an interpretation of the result.

G.3.a.iii) Calculate
$$\oint_C \text{Field} \bullet \text{unitnormal}\, ds$$
in the case in which
$$\text{Field}[x, y] = \{x^2 y^2, x y^2\}$$

and the curve C is the ellipse
$$x^2 + 2y^2 = 1.$$
Give an interpretation of the result.

G.3.a.iv) Calculate
$$\int_C \text{Field} \bullet \text{unitnormal} \, ds$$
in the case in which
$$\text{Field}[x, y] = \{x^2 y^2, x y^2\}$$
and the curve C is the top half of the ellipse
$$x^2 + 2y^2 = 1$$
starting on the far right and ending on the far left.
Give an interpretation of the result.

G.3.b) Calculate
$$\oint_C (-5y) \, dx + x \, dy$$
where C is the circle
$$x^2 + (y-2)^2 = 9$$
and interpret the result in two ways:
→ As a flow-along C measurement of a certain vector field, and
→ As a flow-across C measurment of another vector field.

G.3.c) Calculate
$$\oint_C \sin[y] \, dx + \cos[x] \, dy$$
where C is the circle
$$x^2 + (y-1)^2 = 0.8$$
and give two interpretations of the measurement.

■ G.4) Water*

This problem appears only in the electronic version.

■ G.5) Sources and sinks*

The simplest way to spot a source of new fluid or a drain of old fluid at a point $\{a, b\}$ is to center a circle $C[r]$ of very small radius r at $\{a, b\}$ and then to calculate the flow-across $C[r]$ measurement

$$\oint_{C[r]} -n[x,y]\, dx + m[x,y]\, dy.$$

If this measurement is positive for **all** very small r's, then you can be sure that the point $\{a, b\}$ is a source of new fluid.

If this measurement is negative for **all** very small r's, then you can be sure that the point $\{a, b\}$ is a drain of old fluid.

Try this out on Field$[x, y] = \{3\,x^2, 4\,y^4\}$:

In[11]:=
```
Clear[Field,m,n,x,y]; {m[x_,y_],n[x_,y_]} = {3 x^2, 4 y^4};
Field[x_,y_] = {m[x,y],n[x,y]};
```

Center a circle $C[r]$ of radius r at $\{a, b\} = \{2, 1\}$ and calculate the flow-across $C[r]$ measurement

$$\oint_{C[r]} -n[x,y]\, dx + m[x,y]\, dy :$$

In[12]:=
```
{a,b} = {2,1}; Clear[r,t]
tlow = 0; thigh = 2 Pi;
{x[t_],y[t_]} = {a,b} + r {Cos[t],Sin[t]};
Integrate[-n[x[t],y[t]] x'[t] + m[x[t],y[t]] y'[t],{t,tlow,thigh}]
```
Out[12]=
```
            2         2
     4 Pi r  (7 + 3 r )
```

This is positive no matter what r you go with.

The upshot: $\{a, b\} = \{2, 1\}$ is a source for new fluid.

Now go with cleared values of $\{a, b\}$ and center a circle $C[r]$ of very small radius r at $\{a, b\}$ and calculate the flow-across $C[r]$ measurement

$$\oint_{C[r]} -n[x,y]\, dx + m[x,y]\, dy :$$

Lesson 3.05 Flow Measurements by Integrals

In[13]:=
```
Clear[a,b,r,t]
tlow = 0; thigh = 2 Pi;
{x[t_],y[t_]} = {a,b} + r {Cos[t],Sin[t]};
Integrate[-n[x[t],y[t]] x'[t] + m[x[t],y[t]] y'[t],{t,tlow,thigh}]
```

Out[13]=
$$2\,Pi\,r^2\,(3\,a + 8\,b^3 + 6\,b\,r^2)$$

This tells you that:

→ $\{a, b\}$ is a source of new fluid if $(3a + 8b^3) > 0$, and

→ $\{a, b\}$ is a sink (drain) for old fluid if $(3a + 8b^3) < 0$.

Here's a sample plot of some of the sources and sinks in this vector field:

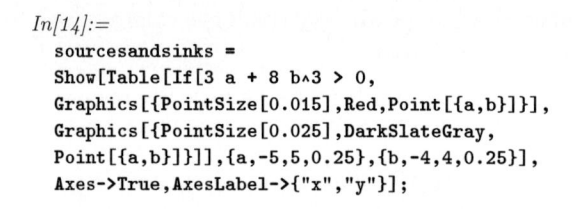

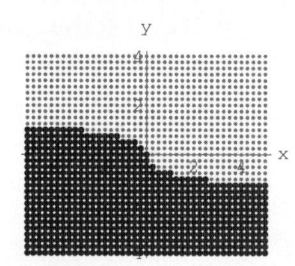

In[14]:=
```
sourcesandsinks =
Show[Table[If[3 a + 8 b^3 > 0,
Graphics[{PointSize[0.015],Red,Point[{a,b}]}],
Graphics[{PointSize[0.025],DarkSlateGray,
Point[{a,b}]}]],{a,-5,5,0.25},{b,-4,4,0.25}],
Axes->True,AxesLabel->{"x","y"}];
```

The larger points are sinks; the smaller points are sources.

The sinks are in the lower part of the plot. Think of the sources as little individual springs feeding the flow. Think of the sinks as little holes through which fluid seeps out as the flow goes by.

G.5.a.i) Give a sample plot of some of the sources and sinks in the vector field
$$\text{Field}[x, y] = \{x^2, y^3\}.$$

G.5.a.ii) Give a sample plot of some of the sources and sinks in the vector field
$$\text{Field}[x, y] = \{3x, -x^2 y^3\}.$$

G.5.b) Here's a look at the vector field
$$\text{Field}[x, y] = 3\left\{\frac{x}{x^2 + y^2}, \frac{y}{x^2 + y^2}\right\}:$$

Note the singularity at $\{0, 0\}$.

In[15]:=
```
Clear[Field,m,n,x,y]
{m[x_,y_],n[x_,y_]} = 3 {x/(x^2 + y^2),y/(x^2 + y^2)};
Field[x_,y_] = {m[x,y],n[x,y]};
fieldplot = Table[Arrow[Field[x,y],Tail->{x,y},Blue],
  {x,-5,5,2},{y,-5,5,1}]; singularity = {0,0};
singularityplot =
Graphics[{Red,PointSize[0.03],Point[singularity]}];
Show[fieldplot,singularityplot,
Axes->True,AxesLabel->{"x","y"}];
```

Go with the circle C_2 of radius 2 centered at the singularity at $\{0,0\}$ and look at the calculation

$$\oint_{C_2} -n[x,y]\,dx + m[x,y]\,dy$$

of the flow of this vector field across C_2:

In[16]:=
```
Clear[x,y,t]
r = 2; {x[t_],y[t_]} = r {Cos[t],Sin[t]}; tlow = 0; thigh = 2 Pi;
Integrate[-n[x[t],y[t]] x'[t] + m[x[t],y[t]] y'[t],{t,tlow,thigh}]
```
Out[16]=
6 Pi

Now go with the circle C_1 of radius 1 centered at the singularity at $\{0,0\}$ and look at this calculation

$$\oint_{C_1} -n[x,y]\,dx + m[x,y]\,dy$$

of the flow of this vector field across C_1:

In[17]:=
```
Clear[x,y,t]
r = 1; {x[t_],y[t_]} = r {Cos[t],Sin[t]}; tlow = 0; thigh = 2 Pi;
Integrate[-n[x[t],y[t]] x'[t] + m[x[t],y[t]] y'[t],{t,tlow,thigh}]
```
Out[17]=
6 Pi

Now go with any circle C_r of radius r centered at the singularity at $\{0,0\}$ and look at this calculation

$$\oint_{C_r} -n[x,y]\,dx + m[x,y]\,dy$$

of the flow of this vector field across C_r:

In[18]:=
```
Clear[r,x,y,t]; {x[t_],y[t_]} = r {Cos[t],Sin[t]};
tlow = 0; thigh = 2 Pi;
Integrate[-n[x[t],y[t]] x'[t] + m[x[t],y[t]] y'[t],{t,tlow,thigh}]
```
Out[18]=
6 Pi

No matter what positive radius you go with, the flow of this vector field across the circle of radius r centered at the singularity is 6π.

> What clue does this give you about the location of the only source of new fluid in this flow?

■ G.6) Gradient fields are where the mathematical action is★

This problem appears only in the electronic version.

■ G.7) Work and how physicists measure it

What's work for some folks is fun for other folks. Trig identities come to mind; they always seem to be work to the math student but fun to the math teacher.

Physicists have their own technical notion of work. They envision a vector field

$$\text{Field}[x,y] = \{m[x,y], n[x,y]\}$$

to represent the force (push) on an object positioned at $\{x,y\}$. In this interpretation, the vector field Field$[x,y]$ is called a force field. Next, physicists say that if you push an object along a curve C parameterized by $\{x[t], y[t]\}$ with tlow $\leq t \leq$ thigh, then the work done by the force field Field$[x,y]$ for you for the duration of the trip is measured by

$$\int_C m[x,y]\,dx + n[x,y]\,dy = \int_{\text{tlow}}^{\text{thigh}} (m[x[t], y[t]]\,x'[t] + n[x[t], y[t]]\,y'[t])\,dt$$

$$= \int_{\text{tlow}}^{\text{thigh}} \text{Field}[[x[t], y[t]] \bullet \{x'[t], y'[t]\}\,dt.$$

This might not be what your own notion of work is, but the physicists have a pretty good reason for using that word for this measurement. Think of it this way: If, at a point on the trip,

→ Field$[x[t], y[t]] \bullet \{x'[t], y'[t]\} > 0$, then at this point the force field Field$[x,y]$ is working this much to push the object and you do no work at all.

But if

→ Field$[x[t], y[t]] \bullet \{x'[t], y'[t]\} < 0$, then at this point the force field Field$[x,y]$ is against your efforts to advance the object; you are working this much and the force field Field$[x,y]$ does no work at all.

With this in mind, you can think of

$$\int_C m[x,y]\,dx + n[x,y]\,dy = \int_{\text{tlow}}^{\text{thigh}} \text{Field}[[x[t], y[t]] \bullet \{x'[t], y'[t]\}\,dt.$$

as a measurement of

the force field's work − your work.

→ If
$$\int_C m[x,y]\,dx + n[x,y]\,dy = \int_{tlow}^{thigh} \text{Field}[[x[t],y[t]] \bullet \{x'[t],y'[t]\}\,dt = 0,$$
then the force field did the same amount of work that you did during the trip.

→ If
$$\int_C m[x,y]\,dx + n[x,y]\,dy = \int_{tlow}^{thigh} \text{Field}[[x[t],y[t]] \bullet \{x'[t],y'[t]\}\,dt > 0,$$
then the force field did most of the work during the object's trip.

→ If $\int_C m[x,y]\,dx + n[x,y]\,dy = \int_{tlow}^{thigh} \text{Field}[[x[t],y[t]] \bullet \{x'[t],y'[t]\}\,dt < 0$, then you did most of the work during the object's trip.

G.7.a) Is there a difference between flow along the curve measurements and work?

Answer: Mathematically there is no difference because they are both measured by the same formula
$$\int_C m[x,y]\,dx + n[x,y]\,dy.$$
The difference is in the interpretation.

When you are talking about flow along the curve measurements, then you envision
$$\text{Field}[x,y] = \{m[x,y], n[x,y]\}$$
as the velocity vector at $\{x,y\}$ of a fluid flow. The fluid is flowing and the curve is just sitting there.

When you are talking about work, then you envision $\text{Field}[x,y] = \{m[x,y], n[x,y]\}$ as the force on an object at $\{x,y\}$ moving on a curve. This time the force field is just sitting there and the object is moving on the curve.

G.7.b) Here is a force field:

```
In[19]:=
  Clear[Field,m,n,x,y]
  m[x_,y_] = x/4; n[x_,y_] = (x - y)/5;
  Field[x_,y_] = {m[x,y],n[x,y]};
  forcefieldplot =
  Table[Arrow[Field[x,y],Tail->{x,y},Blue],
  {x,-2,4},{y,-3,3}];
  Show[forcefieldplot,Axes->Automatic,
  AxesLabel->{"x","y"}];
```

Lesson 3.05 Flow Measurements by Integrals

An object starts at $\{3.3, 0\}$ and moves through this force field one time around the ellipse

$$\left(\frac{x-1}{2.3}\right)^2 + \left(\frac{y}{1.3}\right)^2 = 1.$$

> Which way should the object go (counterclockwise or clockwise) to make the force field do most of the work?

G.7.c) Comment on the statement:

> "If a given force field $\text{Field}[x, y] = \{m[x,y], n[x,y]\}$ is a gradient field, and you are pushing an object from a start point $\{x_0, y_0\}$ to an end point $\{x_1, y_1\}$, you might as well push it on the line C starting at $\{x_0, y_0\}$ and ending at $\{x_1, y_1\}$, because if C_1 is any other curve starting at $\{x_0, y_0\}$ and ending at $\{x_1, y_1\}$, then
>
> $$\int_C m[x,y]\,dx + n[x,y]\,dy = \int_{C_1} m[x,y]\,dx + n[x,y]\,dy.\text{"}$$

G.7.d) This problem appears only in the electronic version.

G.7.e) You are moving along the x-axis, starting at $\{0, 0\}$ and ending at $\{s, 0\}$, under a constant force $F = \{d_1, d_2\}$ at all points.

> Explain the formula:
>
> $$\text{work done by the force field} = Fs.$$
>
> Given that physicists measure force in newtons and distance in meters, say why physicists measure work in newton-meters (which they call joules).

G.7.f) Write a few words on what you think is the difference between the everyday English language definition of work and the technical definition of work as used by physicists.

> To get started, think about this: According to physicists, if there is no change of position, then there is no work. If you must hold a heavy old computer in your arms while you stand in place for one hour, physicists would say that you did no work. Do you agree?

■ **G.8) Spin fields**

This problem appears only in the electronic version.

■ **G.9) "Calculus Cal" screws up again**

This problem appears only in the electronic version.

■ **G.10) Force fields and their trajectories**

This problem appears only in the electronic version.

LESSON 3.06

Sources, Sinks, Swirls, and Singularities

Basics

■ **B.1) Using a 2D integral to measure flow across closed curves**

B.1.a) Explain this:

To calculate the net flow of a vector field

$$\text{Field}[x, y] = \{m[x, y], n[x, y]\}$$

across the boundary C of a region R, you have your choice:

$\to$ You can go through the labor of parameterizing C, and then calculate

$$\oint_C -n[x, y]\, dx + m[x, y]\, dy$$

or

$\to$ If the field has no singularities inside R, you can put

$$\text{divField}[x, y] = D[m[x, y], x] + D[n[x, y], y]$$

and calculate the 2D integral

$$\iint_R \text{divField}[x, y]\, dx\, dy.$$

Answer: Back in the lesson on 2D integrals, you met up with the Gauss-Green formula. The Gauss-Green formula says that if C is the boundary curve of a region R, then you are guaranteed that

$$\oint_C -n[x,y]\,dx + m[x,y]\,dy = \iint_R D[m[x,y],x] + D[n[x,y],y]\,dx\,dy.$$

Go with Field$[x,y] = \{m[x,y], n[x,y]\}$ and put

$$\text{divField}[x,y] = D[m[x,y],x] + D[n[x,y],y]$$

and read off

$$\oint_C -n[x,y]\,dx + m[x,y]\,dy = \iint_R \text{divField}[x,y]\,dx\,dy.$$

Because

$$\oint_C -n[x,y]\,dx + m[x,y]\,dy$$

measures the net flow of the vector field

$$\text{Field}[x,y] = \{m[x,y], n[x,y]\}$$

across C, you are guaranteed that

$$\iint_R \text{divField}[x,y]\,dx\,dy$$

makes the same measurement.

B.1.b) Here is the rectangle R with corners at $\{-3,-1\}, \{3,-1\}, \{3,1\}$, and $\{-3,1\}$:

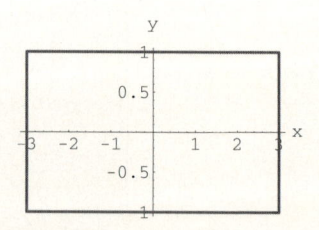

In[1]:=
```
Rplot = Show[Graphics[{Red,Thickness[0.01],
Line[{{-3,-1},{3,-1},{3,1},{-3,1},{-3,-1}}]}],
Axes->True,AxesLabel->{"x","y"},
AspectRatio->1/GoldenRatio];
```

Use a 2D integral to measure the net flow of the vector field

$$\text{Field}[x,y] = \{x^3 + y, x - y\}$$

across the boundary curve C of this rectangle.

Answer: Enter the vector field:

In[2]:=
```
Clear[x,y,m,n,Field]; {m[x_,y_],n[x_,y_]} = {x^3 + y,x - y};
Field[x_,y_] = {m[x,y],n[x,y]};
```

Calculate the divergence, divField[x, y]:

In[3]:=
```
Clear[divField]; divField[x_,y_] = D[m[x,y],x] + D[n[x,y],y]
```

Out[3]=
$$-1 + 3x^2$$

Take another look at R:

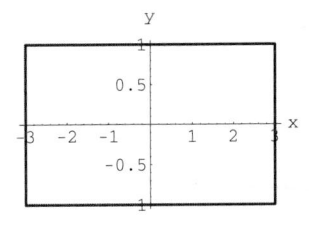

In[4]:=
```
Show[Rplot];
```

Calculate

$$\iint_R \text{divField}[x, y] \, dx \, dy :$$

In[5]:=
```
Integrate[divField[x,y],{y,-1,1},{x,-3,3}]
```

Out[5]=
96

Big-time positive.

This means that the net flow of this vector field across the boundary of this rectangle is from the inside to the outside. There must be a lot of sources of new fluid inside the rectangle.

B.1.c.i) Take a look at divField[x, y] for the vector field

$$\text{Field}[x, y] = \{\sin[y] - x, \cos[x] - y\} :$$

In[6]:=
```
Clear[x,y,m,n,Field,divField]
{m[x_,y_],n[x_,y_]} = {Sin[y] - x,Cos[x] - y};
Field[x_,y_] = {m[x,y],n[x,y]};
divField[x_,y_] = D[m[x,y],x] + D[n[x,y],y]
```

Out[6]=
−2

You look at this and note that divField[x, y] < 0 no matter what $\{x, y\}$ is. Then you say: "Good, this tells me that the flow of this vector field across any closed curve is from outside to inside." You are right.

> Why are you right?

Answer: Take any closed curve C, and call R the region C encloses. You can calculate the net flow

$$\oint_C -n[x,y]\,dx + m[x,y]\,dy$$

of

$$\text{Field}[x,y] = \{m[x,y], n[x,y]\}$$

across C by calculating the 2D integral

$$\iint_R \text{divField}[x,y]\,dx\,dy.$$

Because $\text{divField}[x,y] < 0$ no matter what $\{x,y\}$ you go with, you are guaranteed that

$$\iint_R \text{divField}[x,y]\,dx\,dy < 0.$$

So:

$$\oint_C -n[x,y]\,dx + m[x,y]\,dy = \iint_R \text{divField}[x,y]\,dx\,dy < 0$$

no matter what closed curve C you go with.

This tells you that the flow of this vector field across any closed curve is from outside to inside.

B.1.c.ii) Take a look at $\text{divField}[x,y]$ for the vector field

$$\text{Field}[x,y] = \{\sin[y] + x^5, \cos[x] + y^3\}:$$

In[7]:=
```
Clear[x,y,m,n,Field,divField]
{m[x_,y_],n[x_,y_]} = {Sin[y] + x^5,Cos[x] + y^3};
Field[x_,y_] = {m[x,y],n[x,y]};
divField[x_,y_] = D[m[x,y],x] + D[n[x,y],y]
```
Out[7]=
$$5x^4 + 3y^2$$

You look at this and note that $\text{divField}[x,y] > 0$ unless $\{x,y\} = \{0,0\}$. Then you say: "Good, this tells me that the flow of this vector field across any closed curve is from inside to outside." You are right.

> **Why are you right?**

Answer: Take any closed curve C and call R the region C encloses. You can calculate the flow

$$\oint_C -n[x,y]\,dx + m[x,y]\,dy$$

of
$$\text{Field}[x, y] = \{m[x, y], n[x, y]\}$$
across C by calculating the 2D integral
$$\iint_R \text{divField}[x, y] \, dx \, dy.$$
Because divField$[x, y] > 0$ except at one point, you are guaranteed that
$$\iint_R \text{divField}[x, y] \, dx \, dy > 0.$$
So:
$$\oint_C -n[x, y] \, dx + m[x, y] dy = \iint_R \text{divField}[x, y] \, dx \, dy > 0$$
no matter what closed curve C you go with.

This tells you that the flow of this vector field across any closed curve is from inside to outside.

■ B.2) Sources, sinks, and the divergence of a vector field

B.2.a) Given a vector field
$$\text{Field}[x, y] = \{m[x, y], n[x, y]\},$$
you calculate
$$\text{divField}[x, y] = D[m[x, y], x] + D[n[x, y], y].$$

> How does the sign of divField$[x, y] = D[m[x, y], x] + D[n[x, y], y]$ tell you whether $\{x, y\}$ is a source of new fluid or a sink (drain) for old fluid?

Answer: If divField$[x_0, y_0] > 0$, then the point $\{x_0, y_0\}$ is a source of new fluid. If divField$[x_0, y_0] < 0$, then the point $\{x_0, y_0\}$ is a sink for old fluid.

Here's why: Take a small circle C centered at $\{x_0, y_0\}$. Calculate the flow of Field$[x, y]$ across C by calculating
$$\oint_C -n[x, y] \, dx + m[x, y] \, dy = \iint_R \text{divField}[x, y] \, dx \, dy.$$

Here's the kicker: If divField$[x_0, y_0] > 0$, then it is positive for all $\{x, y\}$'s close to $\{x_0, y_0\}$. So if C is so small that divField$[x, y] > 0$ at all $\{x, y\}$'s inside C, then you see that
$$\oint_C -n[x, y] \, dx + m[x, y] \, dy = \iint_R \text{divField}[x, y] \, dx \, dy > 0.$$

This means that if divField$[x_0, y_0] > 0$, then the net flow of Field$[x, y]$ across small circles centered at $\{x_0, y_0\}$ is from inside to outside.

The upshot: If divField$[x_0, y_0] > 0$, then the point $\{x_0, y_0\}$ is a source of new fluid.

Similarly, if divField$[x_0, y_0] < 0$, then the net flow of Field$[x, y]$ across small circles centered at $\{x_0, y_0\}$ is from outside to inside.

So if divField$[x_0, y_0] < 0$, then the point $\{x_0, y_0\}$ is a sink for old fluid.

Check this out: Go with Field$[x, y] = \{3\,x\,y, y\}$ and look at divField$[x, y]$:

In[8]:=
```
Clear[x,y,Field,m,n,divField]
{m[x_,y_],n[x_,y_]} = {3 x y, y};
Field[x_,y_] = {m[x,y],n[x,y]};
divField[x_,y_] = D[m[x,y],x] + D[n[x,y],y]
```

Out[8]=
1 + 3 y

See whether the point $\{0, 2\}$ is a source or a sink:

In[9]:=
```
divField[0,2]
```

Out[9]=
7

Positive. This tells you that the point $\{0, 2\}$ is a source.

Take a look at this vector field on a small circle centered at $\{0, 2\}$ to see whether this calculation agrees with reality:

In[10]:=
```
Clear[x,y,t]
point = {0,2}; radius = 0.1; tlow = 0; thigh = 2 Pi;
pointplot = Graphics[{PointSize[0.04],Point[point]}];
{x[t_],y[t_]} = point + radius {Cos[t],Sin[t]};
smallcircle = ParametricPlot[{x[t],y[t]},{t,tlow,thigh},
PlotStyle->{{Red,Thickness[0.01]}},
DisplayFunction->Identity]; scalefactor = 0.05;
fieldplot = Table[Arrow[scalefactor Field[x[t],y[t]],
Tail->{x[t],y[t]},Blue],{t, tlow, thigh,(thigh - tlow)/16}];
Show[pointplot,smallcircle,fieldplot,Axes->True,
AxesLabel->{"x","y"},AspectRatio->Automatic,
DisplayFunction->$DisplayFunction];
```

Confirm by looking at the normal components of the field vectors on the curve:

Basics (B.2)

In[11]:=
```
Clear[normal]; normal[t_] = {y'[t],- x'[t]};
Clear[normalcomponent]; normalcomponent[t_] =
  (((Field[x[t],y[t]].normal[t])/
  (normal[t].normal[t])) normal[t]); jump = Pi/5;
actualflowacross = Table[Arrow[
  scalefactor normalcomponent[t],
  Tail->{x[t],y[t]}],{t,0, 2 Pi - jump,jump}];
Show[pointplot,smallcircle,actualflowacross,
  Axes->True,AxesLabel->{"x","y"},AspectRatio->Automatic,
  DisplayFunction->$DisplayFunction];
```

Yessiree, Bob. The plot shows a lot more flow from inside to outside than from outside to inside, just as you would expect on a small circle centered at a source.

B.2.b) Here's a vector field:

In[12]:=
```
Clear[Field,m,n,x,y]
{m[x_,y_],n[x_,y_]} = {Sin[x] Cos[y], Sin[y] Cos[x]};
Field[x_,y_] = {m[x,y],n[x,y]};
```

> Give a sample plot of some of the sources and sinks in this vector field.

Answer: Here's divField[x, y]:

In[13]:=
```
Clear[divField]
divField[x_,y_] = D[m[x,y],x] + D[n[x,y],y]
```
Out[13]=
```
2 Cos[x] Cos[y]
```

A point $\{x, y\}$ is a source if divField[x, y] > 0 and $\{x, y\}$ is a sink if divField[x, y] < 0. Here comes the plot:

In[14]:=
```
sourcesandsinks = Show[
  Table[If[N[divField[x,y]] > 0,
  Graphics[{PointSize[0.015],Red,Point[{x,y}]}],
  Graphics[{PointSize[0.025],DarkSlateGray,
  Point[{x,y}]}]],{x,-5,5,0.25},{y,-4,4,0.25}],
  Axes->True,AxesLabel->{"x","y"}];
```

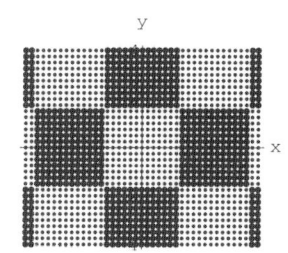

The larger points are sinks; the smaller points are sources.

Alternate squares of sources and sinks. Think of the sources as little individual springs feeding the flow. Think of the sinks as little holes through which fluid seeps out as the flow goes by.

B.2.c.i)

If every point inside a closed curve (like a deformed circle) C is a source of a given vector field, and if the vector field has no singularities inside C, then how do you know that the net flow of the given vector field across C is from inside to outside?

Answer: If every point inside a closed curve C is a source of a given vector field, then

→ new fluid is oozing out of each point inside C, and

→ there there is no place within C to absorb excess outside-to-inside flow.

The result: If every point inside a closed curve C is a source of a given vector field, then the flow of this vector field across C is automatically from inside to outside. For example, look at:

In[15]:=
```
Clear[Field,m,n,x,y,divField]
{m[x_,y_],n[x_,y_]} = {x^3 - y,y^3 + x};
Field[x_,y_] = {m[x,y],n[x,y]};
divField[x_,y_] = D[m[x,y],x] + D[n[x,y],y]
```

Out[15]=
$$3x^2 + 3y^2$$

Unless $\{x,y\} = \{0,0\}$, divField$[x,y] > 0$. This tells you that all points $\{x,y\}$ except $\{0,0\}$ are sources for Field$[x,y]$, and the lone exception is not a sink.

This vector field has no singularities.

On the basis of this information you can say with confidence and authority that the flow of this vector field across any closed curve is from inside to outside.

B.2.c.ii)

If every point inside a closed curve (like a deformed circle) C is a sink of a given vector field, and if the vector field has no singularities inside C, then how do you know that the net flow of the given vector field across C is automatically from outside to inside?

Answer: If every point inside a closed curve C is a sink of a given vector field, then

→ old fluid is soaking into each point inside C, and

→ there there is no place within C to generate excess inside-to-outside flow.

The result: If every point inside a closed curve C is a sink of a given vector field, then the flow of this vector field across C is automatically from outside to inside. For example, look at:

In[16]:=
```
Clear[Field,m,n,x,y,divField]
{m[x_,y_],n[x_,y_]} = { y - x^3 ,x - y^7};
Field[x_,y_] = {m[x,y],n[x,y]};
divField[x_,y_] = D[m[x,y],x] + D[n[x,y],y]
```
Out[16]=
$-3 x^2 - 7 y^6$

Unless $\{x,y\} = \{0,0\}$, divField$[x,y] < 0$. This tells you that all points $\{x,y\}$ except $\{0,0\}$ are sinks for Field$[x,y]$ and the lone exception is not a source. This vector field has no singularities.

On the basis of this information you can say with confidence and authority that the flow of this vector field across any closed curve is from outside to inside.

B.2.c.iii) If there are no sinks and there are no sources of a given vector field inside a closed curve (like a deformed circle) C and if the vector field has no singularities inside C, then how do you know that the net flow of the given vector field across C is 0?

Answer: If there are no sinks and there are no sources of a given vector field inside C and if there are no singularities inside C, then no new fluid is injected and no old fluid is sucked up inside C. So:

→ What flows from inside to outside must be replaced by equal outside-to-inside flow.

→ What flows from outside to inside must be replaced by equal inside-to-outside flow.

The result: If there are no sinks and there are no sources of a given vector field inside C and there are no singularities inside C, then the net flow of the vector field across C is 0. For example, look at:

In[17]:=
```
Clear[Field,m,n,x,y,divField]
{m[x_,y_],n[x_,y_]} = {Cos[x] Cosh[y], Sin[x] Sinh[y]};
Field[x_,y_] = {m[x,y],n[x,y]};
divField[x_,y_] = D[m[x,y],x] + D[n[x,y],y]
```
Out[17]=
0

This tells you that this vector field has no sources or sinks. This vector field has no singularities, so on the basis of this information you can say with confidence and authority that the net flow of this vector field across any closed curve is 0.

B.2.d) Why do most folks call divField$[x,y]$ the divergence of a vector field Field$[x,y]$?

Answer: The name fits.

If divField$[x,y] > 0$, then new fluid is oozing out of the point $\{x,y\}$ and diverges elsewhere. If divField$[x,y] < 0$, then old fluid is sucked into the point $\{x,y\}$ and converges onto this point. If divField$[x,y] = 0$, then no new fluid is introduced, and no old fluid is sucked off as the flow passes by $\{x,y\}$.

■ B.3) Flow-across-the-curve measurements in the presence of singularities

The presentation of this problem was heavily influenced by *The Feynman Lectures on Physics*, by Richard P. Feynman, Robert B. Leighton, and Matthew Sands, Addison-Wesley, 1964.

B.3.a) Here are two closed curves, each parameterized correctly in the counterclockwise way:

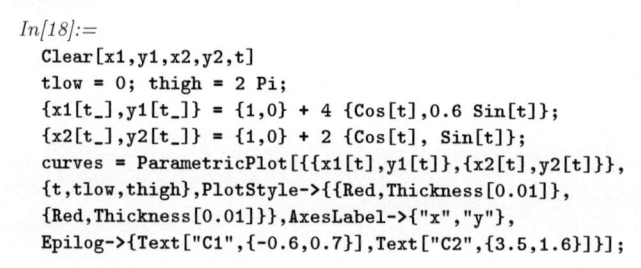

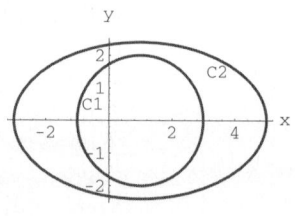

In[18]:=
```
Clear[x1,y1,x2,y2,t]
tlow = 0; thigh = 2 Pi;
{x1[t_],y1[t_]} = {1,0} + 4 {Cos[t],0.6 Sin[t]};
{x2[t_],y2[t_]} = {1,0} + 2 {Cos[t], Sin[t]};
curves = ParametricPlot[{{x1[t],y1[t]},{x2[t],y2[t]}},
{t,tlow,thigh},PlotStyle->{{Red,Thickness[0.01]},
{Red,Thickness[0.01]}},AxesLabel->{"x","y"},
Epilog->{Text["C1",{-0.6,0.7}],Text["C2",{3.5,1.6}]}];
```

Go with
$$\text{Field}[x,y] = \{m[x,y], n[x,y]\} = \left\{\frac{6x}{x^2+y^2}, \frac{6y}{x^2+y^2}\right\},$$

and look at the measurements
$$\oint_{C_1} -n[x,y]\,dx + m[x,y]\,dy \text{ and } \oint_{C_2} -n[x,y]\,dx + m[x,y]\,dy$$

of the flow of this vector field across each curve:

In[19]:=
```
Clear[Field,m,n,x,y]
{m[x_,y_],n[x_,y_]} = {6 x/(x^2 + y^2),6 y/(x^2 + y^2)};
Field[x_,y_] = {m[x,y],n[x,y]};
flowacrossC1 = NIntegrate[
-n[x1[t],y1[t]] x1'[t] + m[x1[t],y1[t]] y1'[t],{t,tlow,thigh}]
```

Out[19]=
37.6991

In[20]:=
```
flowacrossC2 = NIntegrate[
-n[x2[t],y2[t]] x2'[t] + m[x2[t],y2[t]] y2'[t],{t,tlow,thigh}]
```

Out[20]=
37.6991

The curves are different, but the flow of this vector field across the one curve is the same as the flow of this vector field across the other.

> **Was this an accident?**

Answer: You got it right. In mathematics, there are no accidents.

Take a look at the vector field.

In[21]:=
```
Field[x,y]
```
Out[21]=
$$\left\{\frac{6x}{x^2+y^2}, \frac{6y}{x^2+y^2}\right\}$$

Note the nasty singularity at $\{0,0\}$.

In[22]:=
```
Field[0,0]
```
Power::infy: Infinite expression $\frac{1}{0}$ encountered.

Infinity::indet:
    Indeterminate expression 0 6 ComplexInfinity encountered.

Power::infy: Infinite expression $\frac{1}{0}$ encountered.

Infinity::indet:
    Indeterminate expression 0 6 ComplexInfinity encountered.

Out[22]=
{Indeterminate, Indeterminate}

Look at divField$[x, y]$:

In[23]:=
```
Clear[divField]
divField[x_,y_] = Together[D[m[x,y],x] + D[n[x,y],y]]
```
Out[23]=
0

Ah-ha! The vector field has no sources or sinks other than at the singularity at $\{0,0\}$. Check out the position of the singularity relative to the curves:

In[24]:=
```
singularity = {0,0};
singularityplot =
Graphics[{PointSize[0.03],Point[singularity]}];
Show[curves,singularityplot];
```

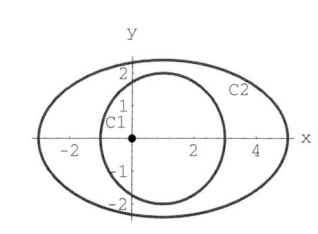

The singularity is not between the curves. And because there are no sources or sinks anywhere but at the singularity, the same amount of fluid that flows across the inner curve also flows across the outer curve. This is why when you measure the flow of the vector field

$$\text{Field}[x,y] = \{m[x,y], n[x,y]\} = \left\{\frac{6x}{x^2+y^2}, \frac{6y}{x^2+y^2}\right\},$$

as above, you get

$$\oint_{C_1} -n[x,y]dx + m[x,y]dy = \oint_{C_2} -n[x,y]dx + m[x,y]dy.$$

B.3.b) Look at the following curve C:

In[25]:=
```
Cplot = Graphics[{Thickness[0.01],Red,
  Line[{{7,0},{4,4},{5,5},{-1,6},{-2,3},{-5,2},
  {-3,-4},{1,-2},{4,-3},{7,0}}]}];
label= Graphics[Text["C",{2.5,5}]];
Show[Cplot,label,Axes->True,AxesLabel->{"x","y"},
  AspectRatio->Automatic];
```

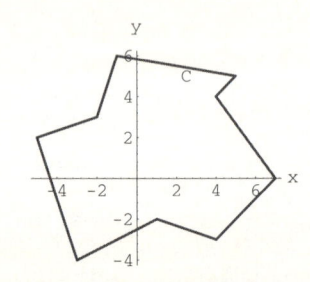

Calculate the flow of

$$\text{Field}[x,y] = \left\{\frac{2(x-2)}{(x-2)^2+(y-1)^2}, \frac{2(y-1)}{(x-2)^2+(y-1)^2}\right\}$$

across C without going to all the bother of parameterizing C.

Answer: This is definitely a case in which a little knowledge can save a lot of work. Only a bean-counting dweeb would find pleasure in parameterizing that silly polygonal curve C. Here you go:

In[26]:=
```
Clear[Field,m,n,x,y]
m[x_,y_] = 2 (x - 2)/((x - 2)^2 + (y - 1)^2);
n[x_,y_] = 2 (y - 1)/((x - 2)^2 + (y - 1)^2);
Field[x_,y_] = {m[x,y],n[x,y]}
```

Out[26]=

$$\left\{\frac{2(-2+x)}{(-2+x)^2+(-1+y)^2}, \frac{2(-1+y)}{(-2+x)^2+(-1+y)^2}\right\}$$

Note the singularity at $\{2,1\}$.

In[27]:=
```
Field[2,1]
```

Out[27]=
{Indeterminate, Indeterminate}

Look at divField$[x, y]$:

In[28]:=
```
Together[D[m[x,y],x] + D[n[x,y],y]]
```

Out[28]=
0

Good. If there were no singularities inside C, then this information would be enough to tell you that the flow of this field across C is 0. But the nasty singularity (blow-up) at $\{2, 1\}$ inside the curve C might be a sink or source.

This won't hold you back because you can pull off a pretty neat stunt: Encapsulate the singularity by centering a little circle C_1 around the singularity at $\{2, 1\}$, taking care that the circle lies completely within C. Here's one:

In[29]:=
```
singularity = {2,1};
Clear[x1,y1,t]
{x1[t_],y1[t_]} = singularity + 0.5 {Cos[t],Sin[t]};
C1plot = ParametricPlot[{x1[t],y1[t]},
{t,0,2 Pi},PlotStyle->{{Thickness[0.01],Red}},
DisplayFunction->Identity];
singularityplot = Graphics[{Blue,
PointSize[0.03],Point[singularity]}];
extralabel = Graphics[Text["C1",{1.3,0.6}]];
Show[Cplot,C1plot,singularityplot,label,extralabel,
PlotRange->All,Axes->True,AxesLabel->{"x","y"},
AspectRatio->Automatic,
DisplayFunction->$DisplayFunction];
```

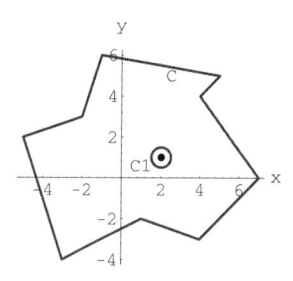

Because
$$\text{divField}[x, y] = 0$$
at all points between the circle C_1 and the ugly curve C and because there are no singularities between the two curves, you can be certain that the flow of this vector field across the little circle C_1 is the same as the flow of this vector field across C. In other words,

$$\oint_C -n[x,y]\,dx + m[x,y]\,dy = \oint_{C1} -n[x,y]\,dx + m[x,y]\,dy.$$

The beauty of this is that you can calculate $\oint_{C1} -n[x,y]\,dx + m[x,y]\,dy$ with a couple of flicks of your fingers:

In[30]:=
```
flowacrossC1 =NIntegrate[-n[x1[t],
y1[t]]x1'[t] + m[x1[t],y1[t]]y1'[t],{t,0,2 Pi}]
```

Out[30]=
12.5664

Lesson 3.06 Sources, Sinks, Swirls, and Singularities

The net flow across the little circle C_1 is from inside to outside. There was a gushing source at the singularity. The flow of the given vector field across C is the same as the flow of this field across C_1:

In[31]:=
```
flowacrossC = flowacrossC1
```
Out[31]=
```
12.5664
```

And you did this without going to the trouble of parameterizing C.

A little knowledge saved a lot of work.

B.3.c) Stay with the same curve C as in part B.3.b):

In[32]:=
```
Cplot = Graphics[{Thickness[0.01],Red,
Line[{{7,0},{4,4},{5,5},{-1,6},{-2,3},{-5,2},
{-3,-4},{1,-2},{4,-3},{7,0}}]}];
label= Graphics[Text["C",{2.5,5}]];
Show[Cplot,label,Axes->True,AxesLabel->{"x","y"},
AspectRatio->Automatic];
```

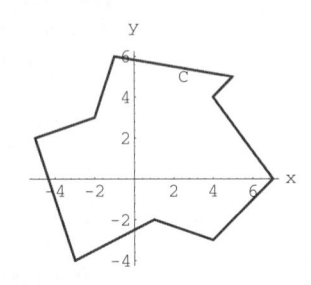

Go with:

In[33]:=
```
Clear[Field,m,n,x,y]
m[x_,y_] = (x - 2)/((x - 2)^2 + (y - 1)^2) - 2 (x + 3)/((x + 3)^2 + (y + 2)^2);
n[x_,y_] = (y - 1)/((x - 2)^2 + (y - 1)^2) - 2 (y + 2)/((x + 3)^2 + (y + 2)^2);
Field[x_,y_] = {m[x,y],n[x,y]}
```

Out[33]=
$$\left\{\frac{-2+x}{(-2+x)^2+(-1+y)^2} - \frac{2(3+x)}{(3+x)^2+(2+y)^2},\frac{-1+y}{(-2+x)^2+(-1+y)^2} - \frac{2(2+y)}{(3+x)^2+(2+y)^2}\right\}$$

> Calculate the flow of this vector field across C without breaking into a heavy sweat.

Answer: Look at the vector field again. Note the singularities at $\{2,1\}$ and at $\{-3,-2\}$.

In[34]:=
```
{Field[2,1],Field[-3,-2]}
```

Out[34]=
```
{{Indeterminate, Indeterminate}, {Indeterminate, Indeterminate}}
```

Now look at:

In[35]:=
```
Together[D[m[x,y],x] + D[n[x,y],y]]
```

Out[35]=
0

Good. The vector field has no sources or sinks aside from the singularities. Encapsulate the singularities by centering a little circle C_1 at the singularity at $\{2,1\}$ and centering another little circle C_2 at the singularity at $\{-3,-2\}$ taking care that both little circles lie completely within C and taking care that they do not invade each other's territory. Here are two very acceptable circles:

In[36]:=
```
singularity1 = {2,1}; singularity2 = {-3,-2};
Clear[x1,y1,x2,y2,t]
{x1[t_],y1[t_]} = singularity1 + 0.8 {Cos[t],Sin[t]};
{x2[t_],y2[t_]} = singularity2 + 0.4 {Cos[t],Sin[t]};
littlecircles =
ParametricPlot[{{x1[t],y1[t]},{x2[t],y2[t]}},
{t,0,2 Pi},PlotStyle->{{Thickness[0.01],Red},
{Thickness[0.01],Red}},DisplayFunction->Identity];
singularityplot = {Graphics[{Blue,PointSize[0.02],
Point[singularity1]}],Graphics[{Blue,PointSize[0.02],
Point[singularity2]}]}; extralabels =
{Graphics[Text["C1",{1,0.5}]],
Graphics[Text["C2",{-3,-1.3}]]};
Show[Cplot,littlecircles,singularityplot,label,extralabels,
PlotRange->All,Axes->True,AxesLabel->{"x","y"},
AspectRatio->Automatic,DisplayFunction->$DisplayFunction];
```

Because

$$\text{divField}[x,y] = 0$$

at all points between the circles and the ugly curve C and since there are no singularities between the circles and C, you can be certain that the net flow across C is given by

$$\oint_C -n[x,y]dx + m[x,y]dy = \oint_{C1} -n[x,y]dx + m[x,y]dy + \oint_{C2} -n[x,y]dx + m[x,y]dy.$$

Here you go:

In[37]:=
```
flowacrossC1 = NIntegrate[-n[x1[t],
y1[t]]x1'[t] + m[x1[t],y1[t]]y1'[t],{t,0,2 Pi}];
flowacrossC2 = NIntegrate[-n[x2[t],
y2[t]]x2'[t] + m[x2[t],y2[t]]y2'[t],{t,0,2 Pi}];
flowacrossC = flowacrossC1 + flowacrossC2
```

Out[37]=
-6.28319

Lesson 3.06 Sources, Sinks, Swirls, and Singularities

The net flow of this vector field across C is from outside to inside. At least one of the singularities must have been a big-time sink.

Tutorials

■ **T.1) The pleasure of calculating path integrals**

$$\oint_C m[x,y]\,dx + n[x,y]\,dy$$

when $D[n[x,y],x] - D[m[x,y],y] = 0$

T.1.a.i) Here is a curve:

In[1]:=
```
Cplot = Graphics[{Thickness[0.01],Red,
Line[{{7,0},{2,2},{5,5},{-2,2},{-4,1},{-5,-3},
{-3,-4},{1,-2},{4,-3},{7,0}}]}];
label = Graphics[Text["C",{2.5,4.5}]];
Show[Cplot,label,Axes->True,AxesLabel->{"x","y"},
AspectRatio->Automatic];
```

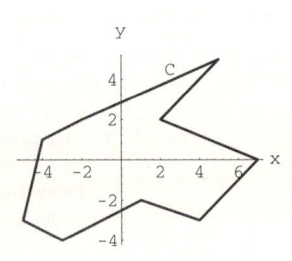

Here are two functions $m[x,y]$ and $n[x,y]$:

In[2]:=
```
Clear[m,n,x,y]
m[x_,y_] = x^2 + 2(x - 6)/((x - 6)^2 + (y - 1)^4);
n[x_,y_] = y^2 + 4 ((y - 1)^3)/((x - 6)^2 + (y - 1)^4);
```

Note the singularity at $\{6,1\}$ and plot it:

In[3]:=
```
singularity = {6,1};
singularityplot =
Graphics[{PointSize[0.03],Point[singularity]}];
Show[Cplot,singularityplot,label,Axes->True,
AxesLabel->{"x","y"},AspectRatio->Automatic];
```

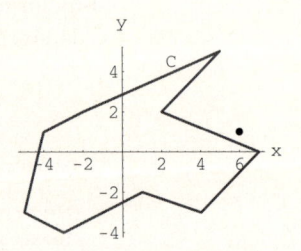

Good. The singularity is not inside the region enclosed by C. Now check

$D[n[x,y],x] - D[m[x,y],y]$:

In[4]:=
```
Together[D[n[x,y],x]- D[m[x,y],y]]
```

Out[4]=
0

Good. Now, without further calculation, you can be sure that
$$\oint_C m[x,y]\, dx + n[x,y]\, dy = 0.$$

> **Why can you be sure of this?**

Answer: Go with the $m[x,y]$ and $n[x,y]$ used above.
$$\oint_C m[x,y]\, dx + n[x,y]\, dy$$
measures the net flow of the vector field
$$\text{Field}[x,y] = \{n[x,y], -m[x,y]\}$$
across C. This vector field has no singularities inside C, and because
$$\text{divField}[x,y] = D[n[x,y],x] - D[m[x,y],y] = 0,$$
this vector field has no sources or sinks within C.

And because you can't get something out of nothing, the net flow of this vector field across C is 0. So $\oint_C m[x,y]\, dx + n[x,y]\, dy = 0$.

Alternatively, you could use the Gauss-Green formula to explain this. The Gauss-Green formula says that if C is the boundary curve of a region R, then you are guaranteed that
$$\oint_C m[x,y]\, dx + n[x,y]\, dy = \iint_R D[n[x,y],x] - D[m[x,y],y]\, dx\, dy$$
$$= \iint_R 0\, dx\, dy = 0.$$

T.1.a.ii) Here is the same curve:

In[5]:=
```
Cplot = Graphics[{Thickness[0.01],Red,Line[
  {{7,0},{2,2},{5,5},{-2,2},{-4,1},
  {-5,-3},{-3,-4},{1,-2},{4,-3},{7,0}}]}];
label = Graphics[Text["C",{2.5,4.5}]];
Show[Cplot,label,Axes->True,AxesLabel->{"x","y"},
AspectRatio->Automatic];
```

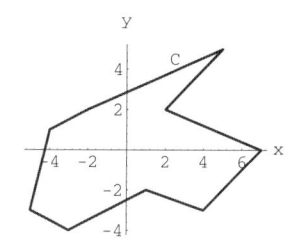

Here are two new functions $m[x,y]$ and $n[x,y]$:

In[6]:=
```
Clear[m,n,x,y]
m[x_,y_] = (1 - y)/((x + 2)^2 + (y - 1)^2);
n[x_,y_] = (x + 2)/((x + 2)^2 + (y - 1)^2);
```

Lesson 3.06 Sources, Sinks, Swirls, and Singularities

Note the singularity at $\{-2, 1\}$ and plot it:

In[7]:=
```
singularity = {-2,1};
singularityplot =
Graphics[{PointSize[0.03],Point[singularity]}];
Show[Cplot,singularityplot,label,Axes->True,
AxesLabel->{"x","y"},AspectRatio->Automatic];
```

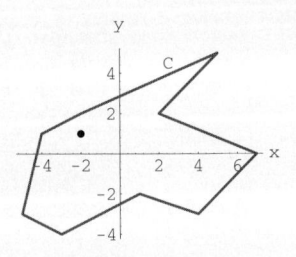

This time the singularity is **inside** the region enclosed by C. Now check

$$D[n[x,y],x] - D[m[x,y],y]:$$

In[8]:=
```
Together[D[n[x,y],x]- D[m[x,y],y]]
```

Out[8]=
0

If there were no singularities inside C, then this information would be enough to tell you that $\oint_C m[x,y]\,dx + n[x,y]\,dy = 0$, but this does not tell you that $\oint_C m[x,y]\,dx + n[x,y]\,dy = 0$ because of the nasty singularity (blow-up) inside the curve C.

Encapsulate the singularity by centering a little circle C_1 at the singularity, taking care that the circle lies completely within C. Here's one:

In[9]:=
```
Clear[x1,y1,t]
{x1[t_],y1[t_]} = singularity + 0.5 {Cos[t],Sin[t]};
C1plot = ParametricPlot[{x1[t],y1[t]},
{t,0,2 Pi},PlotStyle->{{Thickness[0.01],Red}},
DisplayFunction->Identity]; singularityplot =
Graphics[{PointSize[0.03],Point[singularity]}];
extralabel = Graphics[Text["C1",{-1.2,1}]];
Show[Cplot,C1plot,singularityplot,label,extralabel,
PlotRange->All,Axes->True,AxesLabel->{"x","y"},
AspectRatio->Automatic,DisplayFunction->$DisplayFunction];
```

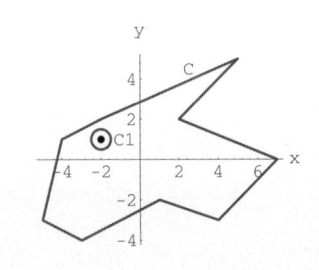

Now you can be sure that

$$\oint_C m[x,y]\,dx + n[x,y]\,dy = \oint_{C_1} m[x,y]\,dx + n[x,y]\,dy,$$

which is given by:

In[10]:=
```
NIntegrate[m[x1[t],y1[t]] x1'[t] + n[x1[t],y1[t]] y1'[t],{t,0,2 Pi}]
```

Out[10]=
6.28319

> Explain why you can be sure that this calculation is correct.

Answer: Go with the $m[x, y]$ and $n[x, y]$ used above. The integral
$$\oint_C m[x, y]\, dx + n[x, y]\, dy$$
measures the net flow of the vector field $\text{Field}[x, y] = \{n[x, y], -m[x, y]\}$ across C. This vector field has no singularities between C and C_1 and because
$$\text{divField}[x, y] = D[n[x, y], x] - D[m[x, y], y] = 0,$$
this vector field has no sources or sinks between C and C_1. And because you can't get something out of nothing, the net flow of this vector field across C is the same as the net flow of this vector field across C_1. So
$$\oint_C m[x, y]\, dx + n[x, y]\, dy = \oint_{C_1} m[x, y]\, dx + n[x, y]\, dy.$$
You've got a hard time using the Gauss-Green formula to explain this directly, because the Gauss-Green formula can fail when C encloses a singularity.

T.1.a.iii) This problem appears only in the electronic version.

T.1.b) Here are two new curves both correctly parameterized in the counterclockwise way:

In[11]:=
```
Clear[x1,y1,x2,y2,t]; tlow = 0; thigh = 2 Pi;
{x1[t_],y1[t_]} =
 2 {Cos[t],0.6 Sin[t] + 0.2 Sin[2 t]};
{x2[t_],y2[t_]} = {1,0} + 0.5 {Cos[t], Sin[t]};
curves = ParametricPlot[{{x1[t],y1[t]},{x2[t],y2[t]}},
 {t,tlow,thigh},PlotStyle->{{Red,Thickness[0.01]},
 {Red,Thickness[0.01]}},AxesLabel->{"x","y"},
 PlotRange->All,Epilog->{Text["C1",{-0.6,0.75}],
 Text["C2",{1.2,0.35}]}];
```

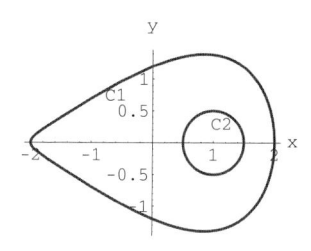

> Given functions $m[x, y]$ and $n[x, y]$, what do you check to be sure that
> $$\oint_{C_1} m[x, y]\, dx + n[x, y]\, dy = \oint_{C_2} m[x, y]\, dx + n[x, y]\, dy$$
> without going to all the trouble to make the individual calculations?

Answer: If no singularities of $\{m[x, y], n[x, y]\}$ pop up between C_1 and C_2 and if
$$D[n[x, y], x] - D[m[x, y], y] = 0$$
at all points between C_1 and C_2, then you can be sure that
$$\oint_{C_1} m[x, y]\, dx + n[x, y]\, dy = \oint_{C_2} m[x, y]\, dx + n[x, y]\, dy.$$

■ T.2) Using a 2D integral to measure flow along closed curves

This problem is a copy, paste, and edit job of B.1)

T.2.a) Explain this:

To calculate the net flow of a vector field

$$\text{Field}[x,y] = \{m[x,y], n[x,y]\}$$

along the boundary C of a region R, you have your choice:

→ You can go to the labor of parameterizing C and then calculate

$$\oint_C m[x,y]\, dx + n[x,y]\, dy$$

or

→ If the field has no singularities inside R, you can put

$$\text{rotField}[x,y] = D[n[x,y], x] - D[m[x,y], y]$$

and calculate the 2D integral

$$\iint_R \text{rotField}[x,y]\, dx\, dy.$$

Answer: Back in the lesson on 2D integrals, you met up with the Gauss-Green formula. The Gauss-Green formula says that if C is the boundary curve of a region R, then you are guaranteed that

$$\oint_C m[x,y]\, dx + n[x,y]\, dy = \iint_R D[n[x,y], x] - D[m[x,y], y]\, dx\, dy.$$

Go with $\text{Field}[x,y] = \{m[x,y], n[x,y]\}$ and put

$$\text{rotField}[x,y] = D[n[x,y], x] - D[m[x,y], y]$$

and read off

$$\oint_C m[x,y]\, dx + n[x,y]\, dy = \iint_R \text{rotField}[x,y]\, dx\, dy.$$

Because

$$\oint_C m[x,y]\, dx + n[x,y]\, dy$$

measures the flow of

$$\text{Field}[x,y] = \{m[x,y], n[x,y]\}$$

along C, you are guaranteed that

$$\iint_R \text{rotField}[x, y] \, dx \, dy$$

makes the same measurement.

T.2.b) Here is the rectangle R with corners at $\{-2, -1\}, \{2, -1\}, \{2, 1\}$ and $\{-2, 1\}$:

In[12]:=
```
Rplot = Show[Graphics[{Red,Thickness[0.01],Line[
    {{-2,-1},{2,-1},{2,1},{-2,1},{-2,-1}}]}],
    Axes->True,AxesLabel->{"x","y"}];
```

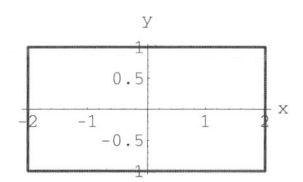

Use a 2D integral to measure the net flow of the vector field

$$\text{Field}[x, y] = \{x + y^2, x - y^2\}$$

along the boundary curve C of this rectangle.

Answer: Enter the field:

In[13]:=
```
Clear[x,y,m,n,Field]
{m[x_,y_],n[x_,y_]} = {x + y^2,x - y^2};
Field[x_,y_] = {m[x,y],n[x,y]}
```

Out[13]=
$\{x + y^2, x - y^2\}$

Calculate rotField$[x, y]$:

In[14]:=
```
Clear[rotField]
rotField[x_,y_] = D[n[x,y],x] - D[m[x,y],y]
```

Out[14]=
$1 - 2y$

Take another look at R:

In[15]:=
```
Show[Rplot];
```

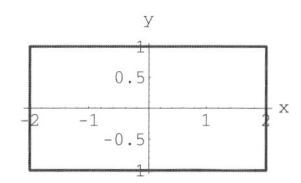

Calculate $\iint_R \text{rotField}[x, y] \, dx \, dy$:

In[16]:=
```
Integrate[rotField[x,y],{y,-1,1},{x,-2,2}]
```

Lesson 3.06 Sources, Sinks, Swirls, and Singularities

Out[16]=
8

Positive. This tells you that the net flow of this vector field along the boundary of this rectangle is counterclockwise.

T.2.c.i) Take a look at rotField$[x, y]$ for the vector field
$$\text{Field}[x, y] = \{\sin[x] + y, \cos[y] - x\}:$$

In[17]:=
```
Clear[x,y,m,n,Field,rotField]
{m[x_,y_],n[x_,y_]} = {Sin[x] + y,Cos[y] - x};
Field[x_,y_] = {m[x,y],n[x,y]};
rotField[x_,y_] = D[n[x,y],x] - D[m[x,y],y]
```

Out[17]=
−2

You look at this and note that rotField$[x, y] < 0$ no matter what $\{x, y\}$ is. And then you say: "Good, this tells me that the flow of this vector field along any closed curve is clockwise." You are right.

> **Why are you right?**

Answer: Take any closed curve C and call R the region C encloses. You can calculate the flow
$$\oint_C m[x, y]\, dx + n[x, y]\, dy$$
of
$$\text{Field}[x, y] = \{m[x, y], n[x, y]\}$$
along C by calculating the 2D integral
$$\iint_R \text{rotField}[x, y]\, dx\, dy.$$

Because rotField$[x, y] < 0$ no matter what $\{x, y\}$ you go with, you are guaranteed that
$$\iint_R \text{rotField}[x, y]\, dx\, dy < 0.$$
So
$$\oint_C m[x, y]\, dx + n[x, y]\, dy = \iint_R \text{rotField}[x, y]\, dx\, dy < 0,$$
no matter what closed curve C you go with. This tells you that the flow of this vector field along any closed curve is clockwise.

T.2.c.ii) Take a look at rotField$[x, y]$ for the vector field
$$\text{Field}[x, y] = \{\sin[x] - y^5, \cos[y] + x^3\} :$$

In[18]:=
```
Clear[x,y,m,n,Field,rotField]
{m[x_,y_],n[x_,y_]} = {Sin[x] - y^5,Cos[y] + x^3};
Field[x_,y_] = {m[x,y],n[x,y]};
rotField[x_,y_] = D[n[x,y],x] - D[m[x,y],y]
```

Out[18]=
$$3x^2 + 5y^4$$

You look at this and note that rotField$[x, y] > 0$ unless $\{x, y\} = \{0, 0\}$. And then you say: "Good, this tells me that the net flow of this vector field along any closed curve is counterclockwise." You are right.

> **Why are you right?**

Answer: Take any closed curve C and call R the region C encloses. You can calculate the flow
$$\oint_C m[x, y]\, dx + n[x, y]\, dy$$
of
$$\text{Field}[x, y] = \{m[x, y], n[x, y]\}$$
along C by calculating the 2D integral
$$\iint_R \text{rotField}[x, y]\, dx\, dy.$$
Because rotField$[x, y] > 0$ except at one point, you are guaranteed that
$$\iint_R \text{rotField}[x, y]\, dx\, dy > 0.$$
So
$$\oint_C m[x, y]\, dx + n[x, y]\, dy = \iint_R \text{rotField}[x, y]\, dx\, dy > 0,$$
no matter what closed curve C you go with.

This tells you that the net flow of this vector field along any closed curve is counterclockwise.

■ T.3) Rotation (swirl) of a vector field

This problem is a copy, paste, and edit job of B.2).

Lesson 3.06 Sources, Sinks, Swirls, and Singularities

T.3.a) Fingering a vector field to get the meaning of the sign of
$\text{rotField}[x, y] = D[n[x, y], x] - D[m[x, y], y]$

Given a vector field

$\text{Field}[x, y] = \{m[x, y], n[x, y]\},$

lick the tip of your index finger and touch it to a point $\{x_0, y_0\}$ while the vector field is in full flow.

> What does the sign of $\text{rotField}[x_0, y_0]$ tell you about the swirl your finger feels? Illustrate with a plot.

Answer: If $\text{rotField}[x_0, y_0] > 0$, then you feel $\text{Field}[x, y]$ swirling around $\{x_0, y_0\}$ in the counterclockwise way.

If $\text{rotField}[x_0, y_0] < 0$, then you feel $\text{Field}[x, y]$ swirling around $\{x_0, y_0\}$ in the clockwise way.

Here's why: Take a small circle C centered at $\{x_0, y_0\}$. Calculate the flow of $\text{Field}[x, y]$ along C by calculating

$$\oint_C m[x, y]\, dx + n[x, y]\, dy = \iint_R \text{rotField}[x, y]\, dx\, dy.$$

Here's the kicker: If $\text{rotField}[x_0, y_0] > 0$, then it is positive for all $\{x, y\}$'s close to $\{x_0, y_0\}$, so if C is so small that $\text{rotField}[x, y] > 0$ at all $\{x, y\}$'s inside C, then you see that

$$\oint_C m[x, y]\, dx + n[x, y]\, dy = \iint_R \text{rotField}[x, y]\, dx\, dy > 0.$$

This means that if $\text{rotField}[x_0, y_0] > 0$, then the net flow of $\text{Field}[x, y]$ along small circles centered at $\{x_0, y_0\}$ is counterclockwise.

Similarly, if $\text{rotField}[x_0, y_0] < 0$, then the net flow of $\text{Field}[x, y]$ along small circles centered at $\{x_0, y_0\}$ is clockwise.

Check this out for the following vector field at $\{0, 0\}$:

In[19]:=
```
Clear[x,y,m,n,Field,rotField]
m[x_,y_] = x + 3 y^2; n[x_,y_] = y + 6 x;
Field[x_,y_] = {m[x,y],n[x,y]};
rotField[x_,y_] = D[n[x,y],x] - D[m[x,y],y]
```

Out[19]=
```
6 - 6 y
```

Look at:

In[20]:=
```
rotField[0,0]
```

Out[20]=
```
6
```

Positive. This means that Field[x, y] swirls around $\{0,0\}$ in the counterclockwise way. Take a look at what the vector field is doing near $\{0,0\}$:

In[21]:=
```
fingerpoint = {0,0}; h = 0.5; scalefactor = 0.1;
fieldplot =
  Table[Arrow[scalefactor Apply[Field,fingerpoint+{x,y}],
  Tail->fingerpoint + {x,y},Blue],
  {x,-h, h, h/3},{y,- h, h, h/3}];
fingerpointplot = Graphics[{PointSize[0.07],
  Point[fingerpoint]}];
Show[fingerpointplot,fieldplot,Axes->Automatic,
  AspectRatio->Automatic,PlotRange->All,
  AxesLabel->{"x","y"}];
```

Lick the tip of your index finger, put the tip of your finger at the point and feel the counterclockwise swirl.

T.3.b) Here's a vector field:

In[22]:=
```
Clear[Field,m,n,x,y]
{m[x_,y_],n[x_,y_]} = {Sin[x] Cos[2y],Sin[y] Cos[2x]};
Field[x_,y_] = {m[x,y],n[x,y]}
```
Out[22]=
{Cos[2 y] Sin[x], Cos[2 x] Sin[y]}

> Give a sample plot of some of the points $\{x, y\}$ about which this vector field swirls in the counterclockwise way.

Answer: Here's rotField[x, y]:

In[23]:=
```
Clear[rotField]
rotField[x_,y_] = D[n[x,y],x] - D[m[x,y],y]
```
Out[23]=
-2 Sin[2 x] Sin[y] + 2 Sin[x] Sin[2 y]

The vector field swirls around $\{x, y\}$ in the counterclockwise way if rotField[x, y] > 0. Here comes the plot:

In[24]:=
```
Show[Table[If[N[rotField[x,y]] > 0,
  Graphics[{PointSize[0.015],Red,Point[{x,y}]}],
  Graphics[{Point[{x,y}]}]],
  {x,-5,5,0.25},{y,-4,4,0.25}],
  Axes->True,AxesLabel->{"x","y"}];
```

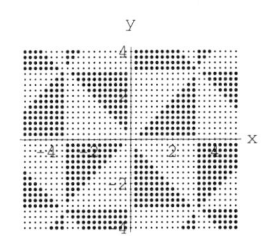

The larger points are points at which the swirl is counterclockwise; the smaller points are points at which the swirl is either 0 or clockwise.

T.3.c) Why do most folks call rotField[x, y] the rotation of a vector field Field[x, y]?

Answer: The name fits.

If rotField[x, y] > 0, then the fluid swirls around $\{x, y\}$ in the counterclockwise way.

If rotField[x, y] < 0, then the fluid swirls around $\{x, y\}$ in the clockwise way.

If rotField[x, y] = 0, then the fluid has no swirl at all as it passes by $\{x, y\}$.

■ **T.4) Summary of main ideas.**

This problem appears only in the electronic version.

Give It a Try

Experience with the starred (*) problems will be useful for understanding developments later in the course.

■ **G.1) Sources, sinks, and swirls***

G.1.a.i) Go with

$$\text{Field}[x, y] = \{e^x \sin[y], e^x \cos[y]\}.$$

Calculate divField[x, y] and use the result to say why the net flow of Field[x, y] across any closed curve is 0.

G.1.a.ii) Go with

$$\text{Field}[x, y] = \{x, y\}.$$

Calculate divField[x, y] and use the result to say why the net flow of Field[x, y] across any closed curve is from inside to outside.

G.1.b.i) Go with

$$\text{Field}[x, y] = \{e^x \sin[y], e^x \cos[y]\}.$$

Calculate rotField[x, y] and use the result to say why the net flow of Field[x, y] along any closed curve is 0.

G.1.b.ii) Go with

$$\text{Field}[x, y] = \{y, -x\}.$$

Calculate rotField[x, y] and use the result to say why the net flow of Field[x, y] along any closed curve is clockwise.

G.1.c) Here is the rectangle R with corners at $\{-1, -2\}, \{5, -2\}, \{5, 2\}$ and $\{-1, 2\}$:

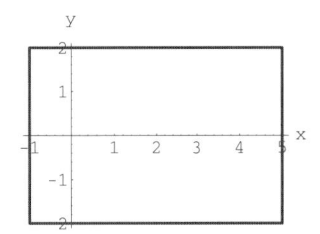

```
In[1]:=
  Rplot = Show[Graphics[{Red,Thickness[0.01],
  Line[{{-1,-2},{5,-2},{5,2},{-1,2},{-1,-2}}]}],
  Axes->True,AxesLabel->{"x","y"}];
```

Use the formula

$$\oint_C -n[x, y]\, dx + m[x, y]\, dy = \iint_R \text{divField}[x, y]\, dx\, dy$$

to measure the net flow of the vector field

$$\text{Field}[x, y] = \{x^2 + 2y^2, x^2 - 2y^2\}$$

across the boundary curve C of R.

Is the net flow of this vector field across C from outside to inside, or is it from inside to outside?

Next, use the formula

$$\oint_C m[x, y]\, dx + n[x, y]\, dy = \iint_R (D[n[x, y], x] - D[m[x.y], y])\, dx\, dy$$

to measure the net flow of the vector field

$$\text{Field}[x, y] = \{x^2 + 2y^2, x^2 - 2y^2\}$$

across the boundary curve C of R.

Is the net flow of this vector field along C counterclockwise or clockwise?

Lesson 3.06 Sources, Sinks, Swirls, and Singularities

G.1.d.i) Here's a vector field:

In[2]:=
```
Clear[Field,m,n,x,y]
{m[x_,y_],n[x_,y_]} = {Cos[x] Cos[2 y], Cos[y] Cos[2 x]};
Field[x_,y_] = {m[x,y],n[x,y]};
```

> Give a sample plot of some of the sources in this vector field.

G.1.d.ii) Go with the same vector field as in part G.1.d.i) immediately above and look at this plot:

In[3]:=
```
Clear[divField]
divField[x_,y_] = D[m[x,y],x] + D[n[x,y],y];
divplot = Plot3D[divField[x,y],{x,-5,5},{y,-5,5},
  PlotPoints->{25,25},DisplayFunction->Identity];
xyplane = Graphics3D[Polygon[
  {{-5,-5,0},{-5,5,0},{5,5,0},{5,-5,0}}]];
Show[divplot,xyplane,
  AxesLabel->{"x","y","z"},ViewPoint->CMView,
  DisplayFunction->$DisplayFunction];
```

> How is this plot related to the plot you did in part G.1.d.i) immediately above?

G.1.e.i) Here's another vector field:

In[4]:=
```
Clear[Field,m,n,x,y]; {m[x_,y_],n[x_,y_]} = {Sin[2 y], x};
Field[x_,y_] = {m[x,y],n[x,y]};
```

> Give a sample plot of some of the points at which this vector field swirls in the counterclockwise direction.

G.1.e.ii) Go with the same vector field as in part G.1.e.i) immediately above and look at this plot:

In[5]:=
```
Clear[rotField]
rotField[x_,y_] = D[n[x,y],x] - D[m[x,y],y];
rotplot = Plot3D[rotField[x,y],{x,-5,5},{y,-5,5},
  PlotPoints->{10,30},DisplayFunction->Identity];
xyplane = Graphics3D[Polygon[
  {{-5,-5,0},{-5,5,0},{5,5,0},{5,-5,0}}]];
Show[rotplot,xyplane,
  AxesLabel->{"x","y","z"},ViewPoint->CMView,
  DisplayFunction->$DisplayFunction];
```

> How is this plot related to the plot you did in part e.i) immediately above?

■ G.2) Singularity sources, sinks, and swirls*

Given a vector field

$$\text{Field}[x,y] = \{m[x,y], n[x,y]\},$$

you can calculate

$$\text{divField}[x,y] = D[m[x,y],x] + D[n[x,y],y]$$

to look for sources and sinks.

If you locate a source at $\{x,y\}$ and $\{x,y\}$ is not a singularity, you can expect new fluid to be slowly oozing into the flow at $\{x,y\}$.

If you locate a sink at $\{x,y\}$ and $\{x,y\}$ is not a singularity, you can expect old fluid to be slowly soaking out of the flow at $\{x,y\}$.

The vivid sources and sinks are often found at singularities; in fact, if you envision a sink at a singularity to be a black hole, you are thinking correctly.

To detect a source or a sink at a singularity, you center a small circle $C[r]$ of radius r at the singularity and calculate

$$\oint_{C[r]} -n[x,y]\,dx + m[x,y]\,dy.$$

→ If

$$\lim_{r \to 0} \oint_{C[r]} -n[x,y]\,dx + m[x,y]\,dy > 0,$$

you have located a singularity source (a gusher) at the singularity.

→ If

$$\lim_{r \to 0} \oint_{C[r]} -n[x,y]\,dx + m[x,y]\,dy < 0,$$

you have located a singularity sink (a black hole) at the singularity.

→ If

$$\lim_{r \to 0} \oint_{C[r]} -n[x,y]\,dx + m[x,y]\,dy = 0,$$

you have located a singularity that is neither a source nor a sink.

Try it out: The point $\{a,b\}$ is a singularity of the 2D electric field coming from a point charge of strength 2 placed at $\{a,b\}$:

Lesson 3.06 Sources, Sinks, Swirls, and Singularities

```
In[6]:=
  Clear[ElectricField,a,b,m,n,q,x,y]
  {m[x_,y_],n[x_,y_]} = q {x - a,y - b}/((x - a)^2 + (y - b)^2);
  ElectricField[x_,y_] = {m[x,y],n[x,y]}
```

Out[6]=
$$\left\{\frac{q(-a+x)}{(-a+x)^2+(-b+y)^2}, \frac{q(-b+y)}{(-a+x)^2+(-b+y)^2}\right\}$$

Center a circle $C[r]$ of radius r at $\{a,b\}$ and calculate $\oint_{C[r]} -n[x,y]\,dx + m[x,y]\,dy$:

```
In[7]:=
  singularity = {a,b};
  Clear[xr,yr,r,t]
  {xr[t_],yr[t_]} = singularity + r {Cos[t],Sin[t]};
  Integrate[-n[xr[t],yr[t]] xr'[t] + m[xr[t],yr[t]] yr'[t],{t,0,2 Pi}]
```

Out[7]=
2 Pi q

This tells you that the singularity at $\{a,b\}$ is a source of new juice if $q > 0$ (positive charge at $\{a,b\}$) and is a sink for old juice if $q < 0$ (negative charge at $\{a,b\}$).

G.2.a.i) Does the electric field above have sources or sinks other than at the singularity?

G.2.a.ii) Here's a vector field related to the electric field:

```
In[8]:=
  Clear[Field,a,b,m,n,x,y]
  {m[x_,y_],n[x_,y_]} =
    5 {x - a,y - b}/((x - a)^2 + (y - b)^2)^(1/2);
  Field[x_,y_] = {m[x,y],n[x,y]}
```

Out[8]=
$$\left\{\frac{5(-a+x)}{\text{Sqrt}[(-a+x)^2+(-b+y)^2]}, \frac{5(-b+y)}{\text{Sqrt}[(-a+x)^2+(-b+y)^2]}\right\}$$

Note the singularity at $\{a,b\}$. Now look at the following information:

```
In[9]:=
  singularity = {a,b};
  Clear[xr,yr,r,t]
  {xr[t_],yr[t_]} = singularity + r {Cos[t],Sin[t]};
  Integrate[-n[xr[t],yr[t]] xr'[t] + m[xr[t],yr[t]] yr'[t],{t,0,2 Pi}]
```

Out[9]=
$$\frac{10\,\text{Pi}\,r^2}{\text{Sqrt}[r^2]}$$

> How does this help to tell you that the singularity at $\{a, b\}$ is neither a singularity source nor a singularity sink?

G.2.a.iii) Now look at divField$[x, y]$ together with $(x - a)^2 + (y - b)^2$:

In[10]:=
```
{Together[D[m[x,y],x] + D[n[x,y],y]],Expand[(x - a)^2 + (y - b)^2]}
```

Out[10]=
$$\left\{\frac{5}{\sqrt{a^2 + b^2 - 2ax + x^2 - 2by + y^2}}, \; a^2 + b^2 - 2ax + x^2 - 2by + y^2\right\}$$

> How does this help to tell you that all points $\{x, y\}$ other than the singularity at $\{a, b\}$ are sources?
>
> How does this tell you that the big-time sources of new fluid for this field are packed near the singularity?

G.2.a.iv) Here's another vector field related to the electric field:

In[11]:=
```
Clear[Field,a,b,q,m,n,x,y]
{m[x_,y_],n[x_,y_]} = 7 {x - a,y - b}/((x - a)^2 + (y - b)^2)^(3/2);
Field[x_,y_] = {m[x,y],n[x,y]}
```

Out[11]=
$$\left\{\frac{7(-a + x)}{((-a + x)^2 + (-b + y)^2)^{3/2}}, \; \frac{7(-b + y)}{((-a + x)^2 + (-b + y)^2)^{3/2}}\right\}$$

> Determine whether $\{a, b\}$ is a singularity source, a singularity sink, or neither.
>
> Also determine whether there are sources or sinks other than at the singularity.

G.2.b.i) When you pull the plug in a bathtub, you see a good example of a singularity swirl (and a singularity sink). To detect a singularity swirl, you center a small circle $C[r]$ of radius r at the singularity and calculate a limit of a certain path integral. If this limit is positive, then you have located a counterclockwise singularity swirl. If this limit is negative, then you have located a clockwise singularity swirl. If this limit is 0, then you have located singularity that has no swirling effect at all.

> What limit do you look at?

G.2.b.ii) Here's a vector field with a singularity at $\{0, 0\}$:

In[12]:=
```
Clear[Field,m,n,x,y]
{m[x_,y_],n[x_,y_]} = {y,-x}/(x^2 + y^2);
Field[x_,y_] = {m[x,y],n[x,y]}
```

Out[12]=
$$\left\{\frac{y}{x^2 + y^2}, -\left(\frac{x}{x^2 + y^2}\right)\right\}$$

Here is rotField$[x, y]$:

In[13]:=
```
rotField[x,y] = Together[D[n[x,y],x] - D[m[x,y],y]]
```

Out[13]=
0

This field has no swirl around any point other than possibly the singularity at $\{0, 0\}$. Look at a plot of this vector field.

In[14]:=
```
scalefactor = 2;
fieldplot =
Table[Arrow[scalefactor Field[x,y],Tail->{x,y},Blue],
{x,-4.5,4.5,9/5},{y,-4.5,4.5,9/5}];
Show[fieldplot,Axes->True,AxesLabel->{"x","y"}];
```

Big-time clockwise singularity swirl around the singularity at $\{0, 0\}$.

> Test your answer to part G.2.b.i) immediately above to see whether your limit agrees with reality.

■ G.3) Agree or disagree*

> Indicate your agreement or disagreement with each of the following statements and paragraphs. When you disagree, say why you disagree. When you agree, feel free to say why you agree, but you are under no obligation to do so.

G.3.a) If
$$\text{Field}[x, y] = \{m[x, y], n[x, y]\}$$

has no sinks, sources, or singularities within a closed curve C, then the net flow of Field$[x, y]$ across C must be 0.

G.3.b) If
$$\text{Field}[x, y] = \{m[x, y], n[x, y]\}$$
has no singularities within a closed curve C and rotField$[x, y] = 0$ at all points within C, then the net flow of Field$[x, y]$ along C is 0.

G.3.c) If
$$\text{Field}[x, y] = \{m[x, y], n[x, y]\}$$
has no singularities and has the property that rotField$[x, y] = 0$ at all points $\{x, y\}$, then Field$[x, y]$ is guaranteed to be a gradient field.

G.3.d) Here is a vector field:

In[15]:=
```
Clear[m,n,x,y]; {m[x_,y_],n[x_,y_]} = {x^3,y^3};
Field[x_,y_] = {m[x,y],n[x,y]};
```

No singularities anywhere. Look at divField$[x, y]$:

In[16]:=
```
D[m[x,y],x] + D[n[x,y],y]
```

Out[16]=
$$3\,x^2 + 3\,y^2$$

This is always positive except when $\{x, y\} = \{0, 0\}$. As a result, every $\{x, y\}$ other than $\{0, 0\}$ is a source for Field$[x, y]$.

Consequently, if C is any closed curve (like a deformed circle), then the net flow of Field$[x, y]$ across C is from inside to outside.

After all, it cannot be from outside to inside because there are no sinks inside C to absorb the excess fluid.

G.3.e) Here are two curves:

In[17]:=
```
Clear[x1,y1,x2,y2,t]
{x1[t_],y1[t_]} = 0.7 {2 Cos[t],Sin[t]};
{x2[t_],y2[t_]} = 1.2 {2 Cos[t],Sin[t] + Sin[4 t]/4};
curveplots = ParametricPlot[{{x1[t],y1[t]},
{x2[t],y2[t]}},{t,0,2 Pi},PlotStyle->
{{Thickness[0.01],Red}},DisplayFunction->Identity];
label = {Graphics[Text["C1",{1,0.4}]],
Graphics[Text["C2",{2,0.75}]]};
Show[curveplots,label,PlotRange->All,
Axes->True,AxesLabel->{"x","y"},
AspectRatio->Automatic,DisplayFunction->$DisplayFunction];
```

The inner curve is C_1.

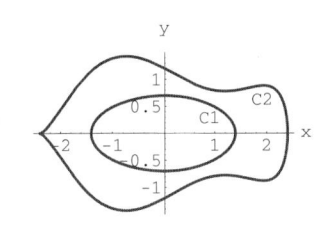

Suppose $\text{Field}[x, y] = \{m[x, y], n[x, y]\}$ has the properties that

→ all the singularities of $\text{Field}[x, y]$ are inside C_1

→ $\oint_{C_1} -n[x, y]\, dx + m[x, y]\, dy > 0$ and

→ $\text{divField}[x, y] = D[m[x, y], x] + D[n[x, y], y] > 0$ at all points $\{x, y\}$ between C_1 and C_2.

This tells you that all points $\{x, y\}$ between C_1 and C_2 are sources for the flow corresponding to $\text{Field}[x, y]$. As a result,

$$\oint_{C_1} -n[x, y]\, dx + m[x, y]\, dy < \oint_{C_2} -n[x, y]\, dx + m[x, y]\, dy,$$

so that the flow of $\text{Field}[x, y]$ across C_2 must be greater than the flow of $\text{Field}[x, y]$ across C_1.

G.3.f) Suppose $\text{Field}[x, y] = \{m[x, y], n[x, y]\}$ has exactly two singularities in the region R enclosed by a closed curve C. Also suppose that C_1 is a closed curve running totally within R.

If $\text{Field}[x, y]$ has no sinks, sources, or singularities between C and C_1, then the net flow of $\text{Field}[x, y]$ across C equals the net flow of $\text{Field}[x, y]$ across C_1.

G.3.g) Here's a closed curve C:

In[18]:=
```
Cplot = Graphics[{Thickness[0.01],Red,
Line[{{6,0},{1,1},{0,6},{-1,1},{-6,0},
{-1,-1},{0,-6},{1,-1},{6,0}}]}];
label = Graphics[Text["C",{0.8,3.5}]];
Show[Cplot,label,Axes->True,AxesLabel->{"x","y"},
AspectRatio->Automatic];
```

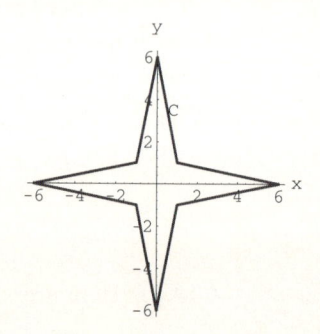

And here's a vector field:

In[19]:=
```
Clear[m,n,x,y]
{m[x_,y_],n[x_,y_]} = {(-y + (x/4 - 1))/(x^2 + y^2), x/(x^2 + y^2)};
Field[x_,y_] = {m[x,y],n[x,y]}
```

Out[19]=

$$\left\{ \frac{-1 + \dfrac{x}{4} - y}{x^2 + y^2},\ \frac{x}{x^2 + y^2} \right\}$$

Note the singularity at $\{0,0\}$. This tells you that you can calculate the flow of Field$[x,y]$ across C by calculating $\oint_{C_1} -n[x,y]\,dx + m[x,y]\,dy$ where C_1 is a small circle centered at $\{0,0\}$.

■ G.4) Flow calculations in the presence of singularities*

G.4.a) How can you tell without evaluating any path integral that the flow of the vector field
$$\text{Field}[x,y] = \left\{\frac{2x}{x^2+y^2}, \frac{2y}{x^2+y^2}\right\}$$
across the circle
$$x^2 + y^2 = r^2$$
is the same as its flow across any ellipse
$$\left(\frac{x}{a}\right)^2 + \left(\frac{y}{b}\right)^2 = 1?$$

G.4.b.i) Look at the following curve C:

```
In[20]:=
  Cplot = Graphics[{Thickness[0.01],Red,
  Line[{{2,0},{1,3},{4,3},{5,5},{-1,6},{0,4},
  {-1,0},{-3,-3},{-.3,-1},{2,-3},{2,0}}]}];
  label = Graphics[Text["C",{3.5,5}]];
  Show[Cplot,label,Axes->True,AxesLabel->{"x","y"}];
```

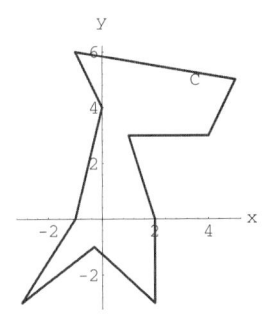

Why would only a dweeb measure the net flow of
$$\text{Field}[x,y] = \left\{\frac{-4x}{x^2+(y-2)^2}, \frac{-4(y-2)}{x^2+(y-2)^2}\right\}$$
across C by parameterizing C and calculating the path integral?

Remove yourself from the dweeb class by measuring the net flow of this vector field across C by measuring the net flow of this vector field across a convenient substitute curve.

Is the net flow of this vector field across C from inside to outside, or is it from outside to inside?

G.4.b.ii) Go with the same curve C as above:

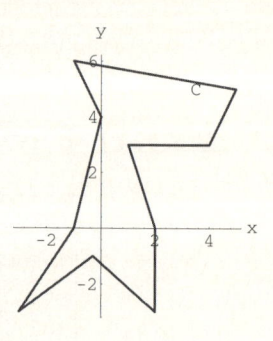

In[21]:=
```
Show[Cplot,label,Axes->True,AxesLabel->{"x","y"}];
```

You are asked to measure the net flow of

$$\text{Field}[x, y] = \left\{ \frac{4(x+2)}{(x+2)^2 + (y-2)^2}, \frac{4(y-2)}{(x+2)^2 + (y-2)^2} \right\}$$

across C. You weren't born yesterday; so you look at:

In[22]:=
```
Clear[x,y,m,n,Field,divField]
{m[x_,y_],n[x_,y_]} = {4(x + 2)/((x + 2)^2 + (y - 2)^2),
    4(y - 2)/((x + 2)^2 + (y - 2)^2)};
Field[x_,y_] = {m[x,y],n[x,y]};
divField[x_,y_] = Together[D[m[x,y],x] + D[n[x,y],y]]
```

Out[22]=
0

And then you announce that the net flow of this vector field across C is 0. You are right.

> Why are you right?

■ **G.5) 2D electric fields, dipole fields, and Gauss's law in physics**★

This problem appears only in the electronic version.

■ **G.6) The Laplacian $\partial^2 f[x,y]/\partial x^2 + \partial^2 f[x,y]/\partial y^2$ and steady-state heat**★

Here is a a cleared function and its gradient field :

In[23]:=
```
Clear[x,y,z,f,gradf,m,n,Field,divField,rotField]
gradf[x_,y_] = {D[f[x,y],x],D[f[x,y],y]};
{m[x_,y_],n[x_,y_]} = gradf[x,y];
Field[x_,y_] = {m[x,y],n[x,y]}
```

Out[23]=
$\{f^{(1,0)}[x, y], f^{(0,1)}[x, y]\}$

Here is the rotation of this gradient field:

In[24]:=
```
rotField[x_,y_] = D[n[x,y],x] - D[m[x,y],y]
```

Out[24]=
0

G.6.a.i) Is it true that all gradient fields are irrotational (have no swirls)?

G.6.a.ii) Here is the divergence of this gradient field:

In[25]:=
```
divField[x_,y_] = D[m[x,y],x] + D[n[x,y],y]
```

Out[25]=
$f^{(0,2)}[x, y] + f^{(2,0)}[x, y]$

Folks like to call

$$\frac{\partial^2 f[x,y]}{\partial x^2} + \frac{\partial^2 f[x,y]}{\partial y^2} = D[f[x,y],\{x,2\}] + D[f[x,y],\{y,2\}] = \text{divField}[x,y]$$

by the name Laplacian of $f[x, y]$.

How do you use the sign of

$$\frac{\partial^2 f[x,y]}{\partial x^2} + \frac{\partial^2 f[x,y]}{\partial y^2}$$

to determine whether a point $\{x, y\}$ is a source or a sink of the gradient field of $f[x, y]$?

How do you check the Laplacian

$$\frac{\partial^2 f[x,y]}{\partial x^2} + \frac{\partial^2 f[x,y]}{\partial y^2}$$

to learn whether a given gradient field $\text{Field}[x, y] = \text{gradf}[x, y]$ is free of sources and sinks at points other than singularities?

G.6.b.i) Here's a concrete slab:

Lesson 3.06 *Sources, Sinks, Swirls, and Singularities*

In[26]:=
```
Clear[x,y,t,r,z];
{x[t_],y[t_]} = {Cos[t],3 Sin[t]};
tlow = 0; thigh = 2Pi; zlow = 0; zhigh = 0.5;
bottom = ParametricPlot3D[{r x[t],r y[t],0},
  {r,0,1},{t,tlow,thigh},PlotPoints->{2,Automatic},
  DisplayFunction->Identity];
top = ParametricPlot3D[{r x[t],r y[t],0.5},
  {r,0,1},{t,tlow,thigh},PlotPoints->{2,Automatic},
  DisplayFunction->Identity];
side = ParametricPlot3D[{x[t],y[t],z},
  {z,zlow,zhigh},{t,tlow,thigh},PlotPoints->{2,Automatic},
  DisplayFunction->Identity]; Show[top,bottom,side,
  AxesLabel->{"x","y","z"},ViewPoint->CMView,
  BoxRatios->Automatic,Boxed->False,
  DisplayFunction->$DisplayFunction];
```

You heat this slab any way you like, and then you apply perfect insulation to the top and bottom. At the same time, you apply heating pads to the side to keep the temperature along each vertical segment through points

$$\{x[t], y[t], z\} \quad \text{for } (0 < z < 0.5)$$

at the same temperature, but your heating pad allows for the temperature to vary as t varies. In other words, different vertical line segments on the sides are kept at possibly different temperatures, but the temperature doesn't vary along any one vertical line segment on the side.

You leave for a long time and wait for the temperature inside the slab to settle into its steady-state condition. After you come back, the temperature at a point $\{x, y, z\}$ inside the slab will not vary as z varies, but the temperature at a point $\{x, y, z\}$ probably will vary as x and y vary. Put

$$\text{temp}[x, y] = \text{steady-state temperature at } \{x, y, z\}.$$

In the steady state, no point inside the slab and not on the side or the top or the bottom can be a source of new heat flow or a sink for old heat.

> Why does this tell you that
>
> $$\frac{\partial^2 \text{temp}[x,y]}{\partial x^2} + \frac{\partial^2 \text{temp}[x,y]}{\partial y^2} = 0$$
>
> at each point $\{x, y\}$ with $\{x, y, z\}$ inside but not on the outer surface of the slab?

G.6.b.ii) Assume $\{x_0, y_0\}$ is a hot spot inside the slab in the sense that

$$\text{temp}[x_0, y_0] > \text{temp}[x, y]$$

for nearby $\{x, y\}$.

> → Why do you think that the net flow of the gradient field of temp$[x,y]$ across small circles centered at $\{x_0, y_0\}$ must be from outside to inside?
>
> → What does the fact that
> $$\frac{\partial^2 \text{temp}[x,y]}{\partial x^2} + \frac{\partial^2 \text{temp}[x,y]}{\partial y^2} = 0$$
> tell you the flow of the gradient field of temp$[x,y]$ across these same small circles actually is?
>
> Say how your responses to the last two questions give you an explanation of why no such hot spot can exist.

G.6.b.iii)
> Can there be $\{x_0, y_0\}$, a cold spot, inside the slab in the sense that
> $$\text{temp}[x_0, y_0] < \text{temp}[x, y]$$
> for nearby $\{x, y\}$?

G.6.b.iv)
> How do you know that both the hottest and coldest spots of the slab will be found on the sides?

■ G.7) Calculating $\oint_C m[x,y]\,dx + n[x,y]\,dy$ in the presence of singularities

This problem appears only in the electronic version.

■ G.8) Water and electricity★

This problem appears only in the electronic version.

■ G.9) Is parallel flow always irrotational?

Some of the ideas for this problem came from Gilbert Strang's book *Calculus*, Wellesley-Cambridge Press, 1991.

G.9.a) Here is a 2D vector field consisting of equal parallel vectors:

Lesson 3.06 Sources, Sinks, Swirls, and Singularities

In[27]:=
```
Clear[Field1,m1,n1,x,y]
{m1[x_,y_],n1[x_,y_]} = 0.5 {-1,1};
Field1[x_,y_] = {m1[x,y],n1[x,y]}
```

Out[27]=
{-0.5, 0.5}

Take a look:

In[28]:=
```
field1plot = Table[Arrow[Field1[x,y],Tail->{x,y},Blue],
 {x,-2,2,4/5},{y,-2,2,4/5}];
Show[field1plot,Axes->True,AxesLabel->{"x","y"}];
```

A calming, steady flow. Check its rotation:

In[29]:=
```
D[n1[x,y],x] - D[m1[x,y],y]
```

Out[29]=
0

Just as the plot suggests, this field has no swirls at all. Here's another vector field flowing in the same direction.

In[30]:=
```
Clear[Field2,m2,n2,x,y]
{m2[x_,y_],n2[x_,y_]} = 0.2 (Sqrt[x^2 + y^2] + 1) {-1,1};
Field2[x_,y_] = {m2[x,y],n2[x,y]}
```

Out[30]=
$\{-0.2\ (1 + \text{Sqrt}[x^2 + y^2]),\ 0.2\ (1 + \text{Sqrt}[x^2 + y^2])\}$

Take a look:

In[31]:=
```
field2plot = Table[Arrow[Field2[x,y],Tail->{x,y},Blue],
 {x,-2,2,4/5},{y,-2,2,4/5}];
Show[field2plot,Axes->True,AxesLabel->{"x","y"}];
```

Another calming, steady, parallel flow. The farther $\{x, y\}$ is from $\{0, 0\}$, the faster the flow of Field2$[x, y]$.

Both Field1$[x, y]$ and Field2$[x, y]$ represent parallel flows. Field1$[x, y]$ represents flow at a steady speed, but the flow represented by Field2$[x, y]$ is of variable speed.

> Is this Field2[x, y] also rotation-free?
>
> If not, plot some sample points at which Field2[x, y] is swirling in the counterclockwise way.

G.9.b.i) One way to make a nonconstant parallel flow in the direction of $\{-1, 1\}$ that is free of all rotation is to take any nonnegative function $g[x]$ and put:

In[32]:=
```
Clear[Field,m,n,g,x,y]; {m[x_,y_],n[x_,y_]} = g[x - y] {-1,1};
Field[x_,y_] = {m[x,y],n[x,y]}
```

Out[32]=
```
{-g[x - y], g[x - y]}
```

Check rotField[x, y]:

In[33]:=
```
D[n[x,y],x] - D[m[x,y],y]
```

Out[33]=
```
0
```

Here's what you get when you go with $g[x] = \sin[x]^2$:

In[34]:=
```
Clear[Field,m,n,g,x,y]
g[x_] = Sin[x]^2;
{m[x_,y_],n[x_,y_]} = g[x - y] {-1,1};
Field[x_,y_] = {m[x,y],n[x,y]};
fieldplot = Table[Arrow[Field[x,y],Tail->{x,y},Blue],
  {x,-2,2,4/5},{y,-2,2,4/5}];
Show[fieldplot,Axes->True,AxesLabel->{"x","y"}];
```

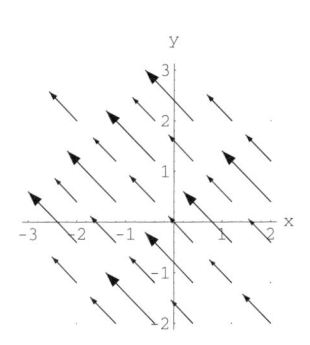

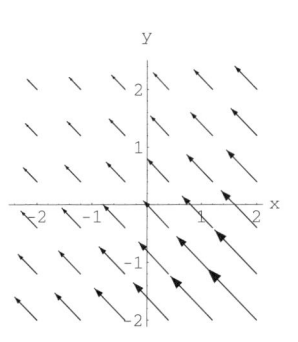

This flow is guaranteed to be free of rotation at all points because rotField[x, y] = 0 at all points $\{x, y\}$ and because there are no singularities.

Here's what you get when you go with $g[x] = 0.4 \, e^{0.2x}$:

In[35]:=
```
Clear[Field,m,n,g,x,y]
g[x_] = 0.4 E^(0.2 x);
{m[x_,y_],n[x_,y_]} = g[x - y] {-1,1};
Field[x_,y_] = {m[x,y],n[x,y]};
fieldplot = Table[Arrow[Field[x,y],Tail->{x,y},Blue],
  {x,-2,2,4/5},{y,-2,2,4/5}];
Show[fieldplot,Axes->True,AxesLabel->{"x","y"}];
```

This flow is guaranteed to be free of rotation at all points because rotField$[x, y] = 0$ at all points $\{x, y\}$ and there are no singularities.

> Given a and b with $a \neq 0$, your job is to say how to set c in terms of a and b so that when you go with a nonconstant nonnegative function $g[x]$, then the vector field,
>
> $$\text{Field}[x, y] = \{m[x, y], n[x, y]\} = g[x + c\, y]\{a, b\}$$
>
> gives you rotation-free, nonconstant, parallel flow in the direction of $\{a, b\}$.
>
> Show off your work with a plot or two.

G.9.b.ii)
> Given a and b with $a = 0$ and $b \neq 0$, is it possible to set c in terms of b so that when you go with a nonconstant nonnegative function $g[x]$, the vector field
>
> $$\text{Field}[x, y] = \{m[x, y], n[x, y]\} = g[x + c\, y]\{a, b\} = g[x + c\, y]\{0, b\}$$
>
> gives you rotation-free, nonconstant, parallel flow on the direction of $\{a, b\}$? Why?

G.9.c)
> Given definite numbers a, b, c and d with $\{a, b\} \neq \{0, 0\}$, agree or disagree with the following statement and explain why:

No matter what nonconstant function $g[x]$ you go with, when you make a nonconstant, parallel flow in the direction of $\{a, b\}$ by setting

$$\text{Field}[x, y] = \{m[x, y], n[x, y]\} = g[d\, x + c\, y]\{a, b\},$$

then either

→ Field$[x, y]$ is not free of sources or sinks or

→ Field$[x, y]$ is not rotation free.

■ G.10) Spin fields

This problem appears only in the electronic version.

LESSON 3.07

Transforming 2D Integrals

Basics

■ B.1) uv-paper and xy-paper

Here's a part of the usual xy-paper grid:

In[1]:=
```
Clear[x,y]
{xlow,xhigh} = {0,6}; {ylow,yhigh} = {0,5};
xygrid = Show[Table[Graphics[Line[
{{xlow,y},{xhigh,y}}]],{y,ylow,yhigh}],
Table[Graphics[Line[{{x,ylow},{x,yhigh}}]],
{x,xlow,xhigh}],
Axes->True,AxesLabel->{"x","y"}];
```

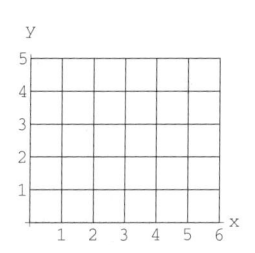

As you well know, y is constant on the horizontal lines and x is constant on the vertical lines. If you want to locate a point like $\{3, 2\}$, you can do it by going to the point at which the grid lines $x = 3$ and $y = 2$ cross each other:

In[2]:=
```
point = {3,2};
Show[xygrid,
Graphics[{Red,PointSize[0.03],Point[point]}]];
```

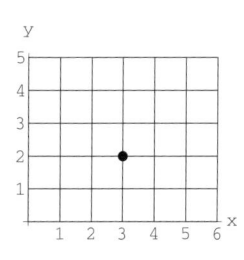

This isn't the only way to locate the point $\{3, 2\}$. You can go with:

267

In[3]:=
```
Clear[u,v,x,y]
{u[x_,y_],v[x_,y_]} = {x^2 - y^2, x y}
```
Out[3]=
$$\{x^2 - y^2, x\, y\}$$

And you can say that the point $\{3, 2\}$ is the point at which the level curves
$$u[x, y] = u[3, 2]$$
and
$$v[x, y] = v[3, 2]$$
cross each other:

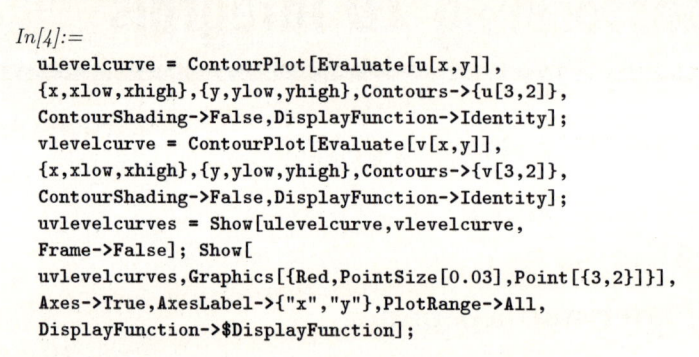

In[4]:=
```
ulevelcurve = ContourPlot[Evaluate[u[x,y]],
 {x,xlow,xhigh},{y,ylow,yhigh},Contours->{u[3,2]},
 ContourShading->False,DisplayFunction->Identity];
vlevelcurve = ContourPlot[Evaluate[v[x,y]],
 {x,xlow,xhigh},{y,ylow,yhigh},Contours->{v[3,2]},
 ContourShading->False,DisplayFunction->Identity];
uvlevelcurves = Show[ulevelcurve,vlevelcurve,
 Frame->False]; Show[
 uvlevelcurves,Graphics[{Red,PointSize[0.03],Point[{3,2}]}],
 Axes->True,AxesLabel->{"x","y"},PlotRange->All,
 DisplayFunction->$DisplayFunction];
```

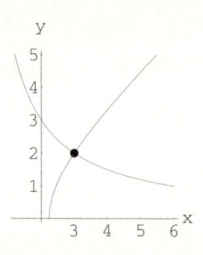

You can plot a whole grid of level curves of $u[x, y]$ and $v[x, y]$:

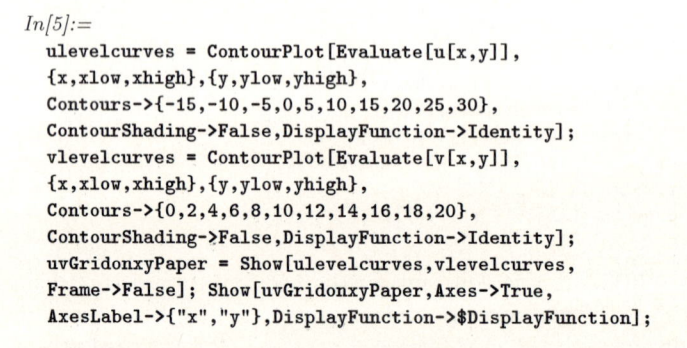

In[5]:=
```
ulevelcurves = ContourPlot[Evaluate[u[x,y]],
 {x,xlow,xhigh},{y,ylow,yhigh},
 Contours->{-15,-10,-5,0,5,10,15,20,25,30},
 ContourShading->False,DisplayFunction->Identity];
vlevelcurves = ContourPlot[Evaluate[v[x,y]],
 {x,xlow,xhigh},{y,ylow,yhigh},
 Contours->{0,2,4,6,8,10,12,14,16,18,20},
 ContourShading->False,DisplayFunction->Identity];
uvGridonxyPaper = Show[ulevelcurves,vlevelcurves,
 Frame->False]; Show[uvGridonxyPaper,Axes->True,
 AxesLabel->{"x","y"},DisplayFunction->$DisplayFunction];
```

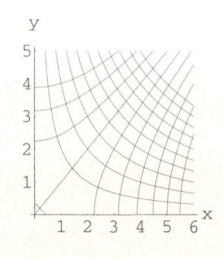

As above, you can locate the point $\{3, 2\}$ on xy-paper as the point where the level curves $u[x, y] = u[3, 2]$ and $v[x, y] = v[3, 2]$ cross each other:

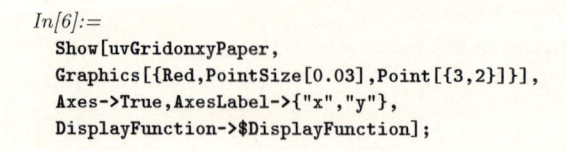

In[6]:=
```
Show[uvGridonxyPaper,
 Graphics[{Red,PointSize[0.03],Point[{3,2}]}],
 Axes->True,AxesLabel->{"x","y"},
 DisplayFunction->$DisplayFunction];
```

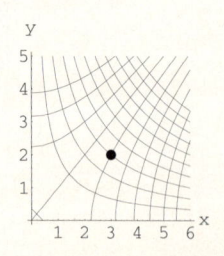

Or you can locate the point $\{4,4\}$ on xy-paper as the point where the level curves $u[x,y] = u[4,4]$ and $v[x,y] = v[4,4]$ cross each other:

In[7]:=
```
Show[uvGridonxyPaper,
  Graphics[{Red,PointSize[0.03],Point[{4,4}]}],
  Axes->True,AxesLabel->{"x","y"},
  DisplayFunction->$DisplayFunction];
```

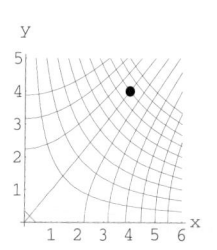

Play.

B.1.a.i) Continue to go with the $u[x,y]$ and $v[x,y]$ above.

What do folks mean when they talk about the uv-coordinates of a point with xy-coordinates $\{x,y\}$?

Answer: That's easy. The uv-coordinates of a point with xy-coordinates $\{x,y\}$ are:

In[8]:=
```
{u[x,y],v[x,y]}
```

Out[8]=
$$\{x^2 - y^2,\ x\, y\}$$

For instance, the uv-coordinates of the point with xy-coordinates $\{3,1\}$ are:

In[9]:=
```
{u[3,1],v[3,1]}
```

Out[9]=
$$\{8,\ 3\}$$

In other words, the level curves $u[x,y] = 8$ and $v[x,y] = 3$ cross at the point $\{x,y\} = \{3,1\}$. Not a whole lot to it.

B.1.a.ii) How do you make uv-paper?

Answer: With a little imagination and a good graphics system like *Mathematica*.

To see how little is involved, look at the uv-grid on xy-paper and imagine the plot to be made on a rubber sheet:

In[10]:=
```
Show[uvGridonxyPaper,Axes->True,AxesLabel->{"x","y"},
  DisplayFunction->$DisplayFunction];
```

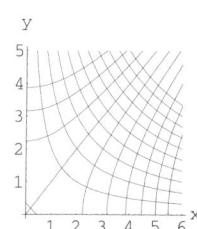

Lesson 3.07 Transforming 2D Integrals

What you see on the rubber sheet are the level curves

$$u[x,y] = k \quad \text{for } k = -15, -10, -5, 0, 5, 10, 15, 20, 25, \text{ and } 30,$$

and

$$v[x,y] = c \quad \text{for } c = 0, 2, 4, 6, 8, 10, 12, 14, 16, 18, \text{ and } 20.$$

To make uv-paper, stretch and compress the rubber sheet so that all of the curves

$$u[x,y] = k \quad \text{and} \quad v[x,y] = c$$

become straight lines crossing perpendicularly. Then apply new axes corresponding to the curves

$$v[x,y] = 0 \quad \text{and} \quad u[x,y] = 0.$$

Here's what you get:

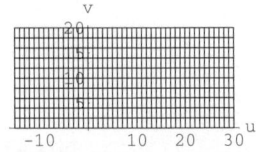

In[11]:=
```
{ulow,uhigh,ujump} = {-15,30,5};
{vlow,vhigh,vjump} = {0,20,2};
uvgrid = Show[Table[Graphics[Line[
  {{ulow,v},{uhigh,v}}]],{v,vlow,vhigh,vjump}],
  Table[Graphics[Line[{{u,vlow},{u,vhigh}}]],
  {u,ulow,uhigh}],Axes->True,AxesLabel->{"u","v"}];
```

This shows off the advantage of plotting on uv-paper; the hard-to-handle curved uv-grid on xy-paper becomes an easily-dealt-with grid of straight lines on uv-paper.

A little imagination and *Mathematica* will take you a long way.

B.1.a.iii) Continue to use $u[x,y] = x^2 - y^2$ and $v[x,y] = x\,y$.

How does the hyperbola $y = 2/x$ plot out on uv-paper?

How does the hyperbola $x^2 - y^2 = 4$ plot out on uv-paper?

How does the circle $(x-1)^2 + y^2 = 1$ plot out on uv-paper?

Answer:

In[12]:=
```
Clear[u,v,x,y]; u[x_,y_] = x^2 - y^2;
v[x_,y_] = x y;
```

The uv-paper plot of $y = 2/x$:

A parametric equation of the hyperbola $y = 2/x$ is:

In[13]:=
```
Clear[x,y,t]; {x[t_],y[t_]} = {t,2/t};
```

Here's part of this hyperbola on xy-paper:

Basics (B.1)

```
In[14]:=
    ParametricPlot[{x[t],y[t]},{t,2/5,5},
    PlotStyle->{{Thickness[0.01],Red}},
    AxesLabel->{"x","y"},
    Epilog->Text["xy-paper plot",{3,3}]];
```

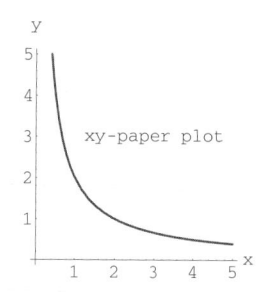

The xy-paper point

$$\{x[t], y[t]\}$$

plots out on uv-paper at the uv-paper point

$$\{u[x[t], y[t]], v[x[t], y[t]]\}.$$

Here's the uv-paper plot of the same part of the hyperbola $y = 2/x$ plotted above on xy-paper:

```
In[15]:=
    ParametricPlot[{u[x[t],y[t]],v[x[t],y[t]]},
    {t,2/5,5}, PlotStyle->{{Thickness[0.01],Red}},
    PlotRange->{1,3},AxesLabel->{"u","v"},
    AspectRatio->1,
    Epilog->Text["uv-paper plot",{10,2.5}]];
```

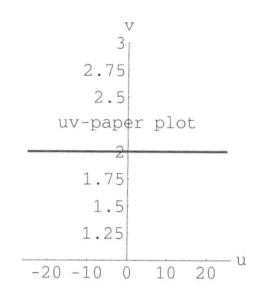

Gee whiz. The hyperbola $y = 2/x$ plots out as the line $v = 2$.

This is no accident because the hyperbola $y = 2/x$ is the level curve $xy = 2$ and this is the same as the level curve $v[x, y] = xy = 2$. When you stretched the xy-paper into uv-paper, this level curve became the line $v = 2$.

The uv-paper plot of $x^2 - y^2 = 4$:

You don't need the machine to say how this looks on uv-paper. When you remember that $u[x, y] = x^2 - y^2$, you see that the hyperbola

$$x^2 - y^2 = 4$$

plots out on uv-paper on the line

$$u = 4.$$

The uv-paper plot of $(x-1)^2 + y^2 = 1$:

A parametric equation of the circle

$$(x-1)^2 + y^2 = 1$$

is:

Lesson 3.07 Transforming 2D Integrals

In[16]:=
```
Clear[x,y,t]; {x[t_],y[t_]} = {1,0} + {Cos[t],Sin[t]};
```

Here's this circle on xy-paper in true scale:

In[17]:=
```
ParametricPlot[{x[t],y[t]},{t,0,2 Pi},
  PlotStyle->{{Thickness[0.01],Red}},
  AxesLabel->{"x","y"},AspectRatio->Automatic,
  Epilog->Text["xy-paper plot",{0.7,0.7}]];
```

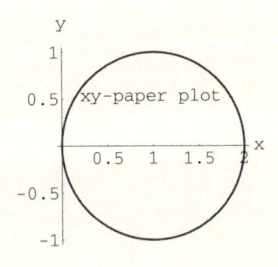

The xy-paper point

$$\{x[t], y[t]\}$$

plots out on uv-paper at the uv-paper point

$$\{u[x[t], y[t]], v[x[t], y[t]]\}.$$

Here's the true scale uv-paper plot of the circle $(x-1)^2 + y^2 = 1$:

In[18]:=
```
ParametricPlot[{u[x[t],y[t]],v[x[t],y[t]]},
  {t,0, 2 Pi}, PlotStyle->{{Thickness[0.01],Red}},
  AxesLabel->{"u","v"},AspectRatio->Automatic,
  Epilog->Text["uv-paper plot",{1.5,0.5}]];
```

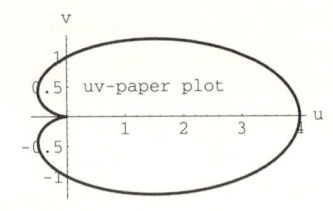

Buns. When you stretched and compressed the original rubber xy-paper to make the uv-paper, the circle got squashed.

The cusp (corner) at $\{0,0\}$ is hard to ignore; you'll find out why it happened in one of the "Give It a Try" problems.

B.1.b.i) Often you'll want to start with uv-paper and go to xy-paper.

Here's an example involving polar parameterization: Go with the uv-paper rectangle,

$$1 \leq u \leq 5 \quad \text{and} \quad 0 \leq v \leq \pi:$$

In[19]:=
```
uvpaperplot = Show[Graphics[
  {Thickness[0.01],Line[{{1,0},{5,0}}]}],
  Graphics[{Blue,Thickness[0.01],Line[{{5,0},{5,Pi}}]}],
  Graphics[{Thickness[0.01],Line[{{5,Pi},{1,Pi}}]}],
  Graphics[{Red,Thickness[0.01],Line[{{1,Pi},{1,0}}]}],
  Graphics[Text["uv-paper plot",{3,Pi/2}]],
  AspectRatio->Automatic,Axes->True,
  AxesLabel->{"u","v"}];
```

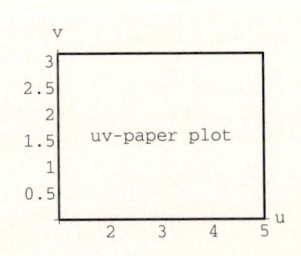

Basics (B.1) 273

> Go to xy-paper with the polar functions
> $$x[u,v] = u\cos[v] \quad \text{and} \quad y[u,v] = u\sin[v]$$
> and plot this rectangle on xy-paper.

Answer:

In[20]:=
```
Clear[x,y,u,v]; {x[u_,v_],y[u_,v_]} = {u Cos[v],u Sin[v]}
```

Out[20]=
{u Cos[v], u Sin[v]}

The bottom side of the rectangle runs with $1 \le u \le 5$ and $v = 0$:

In[21]:=
```
xybottom =
ParametricPlot[{x[u,0],y[u,0]},{u,1,5},
 PlotStyle->{{Thickness[0.01]}},
 AxesLabel->{"x","y"}];
```

The right side of the rectangle runs with $u = 5$ and $0 \le v \le \pi$:

In[22]:=
```
xyright = ParametricPlot[{x[5,v],y[5,v]},{v,0,Pi},
 PlotStyle->{{Blue,Thickness[0.01]}},
 AxesLabel->{"x","y"}];
```

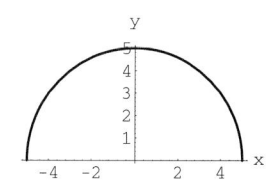

The top side of the rectangle runs with $1 \le u \le 5$ and $v = \pi$:

In[23]:=
```
xytop = ParametricPlot[{x[u,Pi],
 y[u,Pi]},{u,1,5},
 PlotStyle->{{Thickness[0.01]}},
 AxesLabel->{"x","y"}];
```

The left side of the rectangle runs with $u = 1$ and $0 \le v \le \pi$:

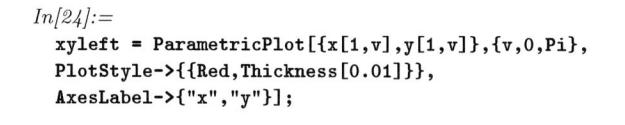

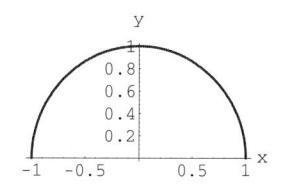

In[24]:=
```
xyleft = ParametricPlot[{x[1,v],y[1,v]},{v,0,Pi},
 PlotStyle->{{Red,Thickness[0.01]}},
 AxesLabel->{"x","y"}];
```

Here they are all assembled:

Lesson 3.07 Transforming 2D Integrals

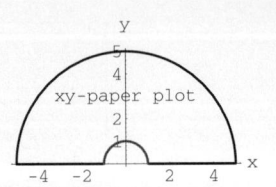

```
In[25]:=
xypaperplot =
Show[xybottom,xyright,xytop,xyleft,
Graphics[Text["xy-paper plot",{1.5,3}]],
AspectRatio->Automatic,AxesLabel->{"x","y"}];
```

Half a donut.

B.1.b.ii) How does this bear on the topic of double integrals?

Answer: It's sometimes handier than a can opener. You'll see why later, but just to whet your appetite, ask yourself this: If you are setting up a double integral over a region R, would you prefer R to look like this:

```
In[26]:=
Show[xypaperplot];
```

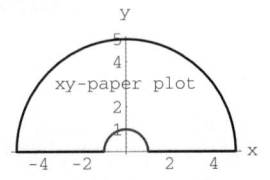

or this:

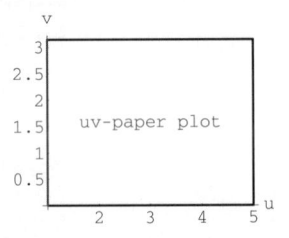

```
In[27]:=
Show[uvpaperplot];
```

This lesson is all about the art of replacing ornery regions with rectangles.

Go on and enjoy.

■ B.2) Linearizing the grids

B.2.a) Here is a microscopic plot of part of the xy-grid coming from
$$x[u,v] = u^2 - v^2 \quad \text{and} \quad y[u,v] = u\,v$$
in the vicinity of the uv-paper point $\{a,b\} = \{2.1, 1.5\}$ plotted on uv-paper:

Basics (B.2)

In[28]:=
```
{a,b} = {2.1,1.5}; jump = 0.02;
Clear[u,v,h,x,y]
{x[u_,v_],y[u_,v_]} = {u^2 - v^2, u v};
xlevelcurves = ContourPlot[Evaluate[x[u,v]],
{u,a - jump,a + jump},{v,b - 2 jump,b + 2 jump},
Contours->Table[x[a,b] + h,{h,-5 jump,5 jump,jump}],
ContourShading->False,DisplayFunction->Identity];
ylevelcurves = ContourPlot[Evaluate[y[u,v]],
{u,a - jump,a + jump},{v,b - 2 jump,b + 2 jump},
Contours->Table[y[a,b] + h,{h,-5 jump,5 jump,jump}],
ContourShading->False,DisplayFunction->Identity];
xyGridonuvPaper = Show[xlevelcurves,ylevelcurves,
Frame->False]; Show[xyGridonuvPaper,Graphics[
{Red,PointSize[0.03],Point[{a,b}]}],Axes->True,
AxesLabel->{"u","v"},DisplayFunction->$DisplayFunction];
```

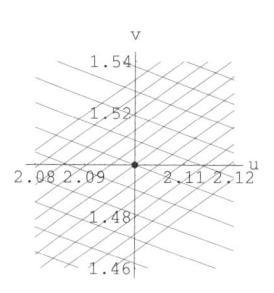

Here's the same thing for the linearizations of $x[u,v]$ and $y[u,v]$ at the same point $\{a,b\}$.

In[29]:=
```
Clear[linearx,lineary,gradx,grady]
gradx[u_,v_] = {D[x[u,v],u],D[x[u,v],v]};
grady[u_,v_] = {D[y[u,v],u],D[y[u,v],v]};
linearx[u_,v_] = Expand[x[a,b] + gradx[a,b].{u - a,v - b}]
```

Out[29]=
-2.16 + 4.2 u - 3. v

In[30]:=
```
lineary[u_,v_] = Expand[y[a,b] + grady[a,b].{u - a,v - b}]
```

Out[30]=
-3.15 + 1.5 u + 2.1 v

In[31]:=
```
linearxlevelcurves = ContourPlot[Evaluate[linearx[u,v]],
{u,a - jump,a + jump},{v,b - 2 jump,b + 2 jump},
Contours->Table[linearx[a,b] + h,{h,-5 jump,5 jump,jump}],
ContourShading->False,DisplayFunction->Identity];
linearylevelcurves = ContourPlot[Evaluate[lineary[u,v]],
{u,a - jump,a + jump},{v,b - 2 jump,b + 2 jump},
Contours->Table[lineary[a,b] + h,{h,- 5 jump,5 jump,jump}],
ContourShading->False,DisplayFunction->Identity];
linearxyGridonuvPaper = Show[linearxlevelcurves,
linearylevelcurves,Frame->False];
Show[linearxyGridonuvPaper,
Graphics[{Red,PointSize[0.03],Point[{a,b}]}],
Axes->True,AxesLabel->{"u","v"},
DisplayFunction->$DisplayFunction];
```

Compare:

Lesson 3.07 Transforming 2D Integrals

In[32]:=
```
Show[xyGridonuvPaper,linearxyGridonuvPaper,
Graphics[{Red,PointSize[0.03],Point[{a,b}]}],
Axes->True,AxesLabel->{"u","v"},
DisplayFunction->$DisplayFunction];
```

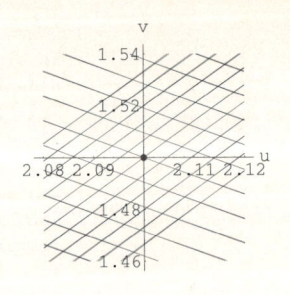

In this microscopic plot of the vicinity of $\{a, b\}$, the xy-grid on uv-paper is almost the same as the grid coming from the linearized versions of $x[u, v]$ and $y[u, v]$ at $\{a, b\}$.

> **Explain why:**
>
> You will see similar results no matter what the functions $x[u, v]$ and $y[u, v]$ are and no matter what point $\{a, b\}$ you go with (unless $\{a, b\}$ is a singularity for $x[u, v]$ or $y[u, v]$).

Answer: Maybe you remember learning way back that the linearized version of any function $x[u, v]$ (respectively $y[u, v]$) at $\{a, b\}$ is a superb approximation of $x[u, v]$ (respectively $y[u, v]$) in the vicinity of $\{a, b\}$. That's why the grid coming from the linearized versions of $x[u, v]$ and $y[u, v]$ at $\{a, b\}$ has to mimic the xy-grid so well near the point $\{a, b\}$. In fact, the closer you get to $\{a, b\}$, the better the approximation.

B.2.b)
> Summarize.

Answer:

→ Near a uv-paper point $\{a, b\}$, xy-paper plots look the same as plots on linearx-lineary paper. The closer you get to $\{a, b\}$, the better the linear grid approximates the curved grid.

→ The linearx-lineary grid on uv-paper is always a bunch of parallelograms in much the same way that the xy-grid on xy-paper is a bunch of rectangles.

■ B.3) Transforming 2D integrals: How you do it and why you do it

B.3.a) Knowing that

$$\int_{x[a]}^{x[b]} f[x]\, dx = \int_{a}^{b} f[x[u]]x'[u]\, du$$

is a mark of calculus literacy. But without the fudge function,
$$\text{fudge}[u] = x'[u],$$
the transformation fails.

In the two-variable case, when you use functions $x[u,v]$ and $y[u,v]$ to go from integrating on xy-paper to integrating on uv-paper, you've got to come up with the fudge function, $\text{fudge}[u,v]$, that makes
$$\iint_{R_{xy}} f[x,y]\, dx\, dy = \iint_{R_{uv}} f[x[u,v], y[u,v]]\, \text{fudge}[u,v]\, du\, dv$$
where R_{uv} is the uv-paper plot of the region R_{xy} originally plotted on xy-paper.

What is the physical meaning of the function $\text{fudge}[u,v]$?

Answer: The function $\text{fudge}[u,v]$ has to satisfy
$$\iint_{R_{xy}} f[x,y]\, dx\, dy = \iint_{R_{uv}} f[x[u,v], y[u,v]]\, \text{fudge}[u,v]\, du\, dv$$
no matter what $f[x,y]$ you have.

You get a pregnant clue to the physical meaning of $\text{fudge}[u,v]$ by taking $f[x,y] = 1$. For this particular $f[x,y]$, you get
$$\iint_{R_{xy}} dx\, dy = \iint_{R_{uv}} \text{fudge}[u,v]\, du\, dv.$$
Put $\text{Area}[R_{xy}]$ equal to the area of R_{xy} measured on xy-paper and realize that this equation means that
$$\text{Area}[R_{xy}] = \iint_{R_{xy}} dx\, dy = \iint_{R_{uv}} \text{fudge}[u,v]\, du\, dv.$$
If you make the measurements on uv-paper, then
$$\text{Area}[R_{uv}] = \iint_{R_{uv}} du\, dv.$$
So $\text{fudge}[u,v]$ must be the instantaneous area conversion factor that you integrate to convert uv-paper area measurements into xy-paper area measurements.

You could say that $\text{fudge}[u,v]$ is the derivative of xy-paper area with respect to uv-paper area. To dignify the $\text{fudge}[u,v]$, give it a fancy look by writing
$$A_{(x,y)}[u,v] = \text{fudge}[u,v].$$
Name it by calling $A_{(x,y)}[u,v]$ the area conversion factor because it converts uv-paper area measurements into xy-paper area measurements.

B.3.b) Now go for the throat. The formula for the area conversion factor $A_{(x,y)}[u,v]$ that you integrate to convert uv-paper area measurements to xy-paper area measurements is:

In[33]:=
```
Clear[x,y,u,v,gradx,grady,Axy]
gradx[u_,v_] = {D[x[u,v],u],D[x[u,v],v]};
grady[u_,v_] = {D[y[u,v],u],D[y[u,v],v]};
Axy[u_,v_] = Abs[Det[{gradx[u,v],grady[u,v]}]]
```

Out[33]=
$$\text{Abs}[y^{(0,1)}[u,v]\, x^{(1,0)}[u,v] - x^{(0,1)}[u,v]\, y^{(1,0)}[u,v]]$$

Some of the fancy folks call this the Jacobian determinant.

> **Where does this beauty come from?**

Answer: Ultimately, it comes from the cross product. Start with

$$x = x[u,v] \quad \text{and} \quad y = y[u,v].$$

At a fixed uv-paper point $\{a,b\}$, the linearized versions of $x[u,v]$ and $y[u,v]$ are calculated as follows:

In[34]:=
```
Clear[x,y,u,v,a,b,gradx,grady,linearx,lineary]
gradx[u_,v_] = {D[x[u,v],u],D[x[u,v],v]};
grady[u_,v_] = {D[y[u,v],u],D[y[u,v],v]};
linearx[u_,v_] = x[a,b] + gradx[a,b].{u - a,v - b}
```

Out[34]=
$$x[a,b] + (-b+v)\, x^{(0,1)}[a,b] + (-a+u)\, x^{(1,0)}[a,b]$$

In[35]:=
```
lineary[u_,v_] = y[a,b] + grady[a,b].{u - a,v - b}
```

Out[35]=
$$y[a,b] + (-b+v)\, y^{(0,1)}[a,b] + (-a+u)\, y^{(1,0)}[a,b]$$

Remember from B.2:

On uv-paper, the closer you are to $\{a,b\}$, the more spectacularly the xy-grid is approximated by the linearx-lineary grid. As a result, at the point $\{a,b\}$, the two area conversion factors are the same; in other words

$$A_{(x,y)}[a,b] = A_{(\text{linearx},\text{lineary})}[a,b].$$

Now calculate $A_{(\text{linearx},\text{lineary})}[a,b]$. To do this, remember that the linear grid is a bunch of parallelograms. For this reason, the uv-paper square with corners at $\{a,b\}, \{a+h,b\}, \{a+h,b+h\}$ and $\{a,b+h\}$ plots out on linearx-lineary paper as the parallelogram with corners at:

In[36]:=
```
Clear[h]
basepoint = {linearx[a,b],lineary[a,b]};
corner1 = {linearx[a + h,b],lineary[a + h,b]};
corner2 = {linearx[a + h,b + h],lineary[a + h,b + h]};
corner3 = {linearx[a,b + h],lineary[a,b + h]};
```

The area of this parallelogram is given by the absolute value of:

In[37]:=
```
X = corner1 - basepoint; Y = corner3 - basepoint;
X3D = Append[X,0]; Y3D = Append[Y,0];
X3DcrossY3D = Det[{{{1,0,0},{0,1,0},{0,0,1}},X3D,Y3D}];
AreaofParallelogram = Sqrt[Collect[Expand[X3DcrossY3D.X3DcrossY3D],h^4]]
```

Out[37]=

$$\text{Sqrt}[h^4 \, (y^{(0,1)}[a,b]^2 \, x^{(1,0)}[a,b]^2 - 2 \, x^{(0,1)}[a,b] \, y^{(0,1)}[a,b] \, x^{(1,0)}[a,b] \, y^{(1,0)}[a,b] + x^{(0,1)}[a,b]^2 \, y^{(1,0)}[a,b]^2)]$$

A detailed explanation of where this calculation comes from appears in the electronic version.

The original uv-paper square with corners at

$$\{a,b\}, \{a+h,b\}, \{a+h,b+h\}, \{a,b+h\}$$

has area measuring out to h^2. So the area conversion factor

$$A_{(x,y)}[a,b] = A_{(\text{linearx,lineary})}[a,b]$$

is given by the absolute value of:

In[38]:=
```
Simplify[AreaofParallelogram/h^2]
```

Out[38]=

$$\text{Sqrt}[h^4 \, (y^{(0,1)}[a,b] \, x^{(1,0)}[a,b] - x^{(0,1)}[a,b] \, y^{(1,0)}[a,b])^2] \, / \, h^2$$

Not so bad, because this is the same as the absolute value of:

In[39]:=
```
Det[{gradx[u,v],grady[u,v]}]/.{u->a,v->b}
```

Out[39]=

$$y^{(0,1)}[a,b] \, x^{(1,0)}[a,b] - x^{(0,1)}[a,b] \, y^{(1,0)}[a,b]$$

(The h's cancel out.)

And now you see why the area conversion factor $A_{(x,y)}[u,v]$ is nothing more than the absolute value of:

In[40]:=
```
Det[{gradx[u,v],grady[u,v]}]
```

Out[40]=

$$y^{(0,1)}[u,v] \, x^{(1,0)}[u,v] - x^{(0,1)}[u,v] \, y^{(1,0)}[u,v]$$

B.3.c.i) Now you get a chance to use this good stuff:

The region R_{xy} is everything inside and on the ellipse

$$\left(\frac{x}{3}\right)^2 + y^2 = 1$$

on xy-paper:

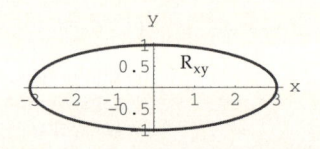

In[41]:=
```
Clear[x,y,t]
{x[t_],y[t_]} = {3 Cos[t],Sin[t]};
ParametricPlot[{x[t],y[t]},{t,0, 2Pi},
PlotStyle->{{Thickness[0.01],Red}},
AspectRatio->Automatic,AxesLabel->{"x","y"}];
```

Calculate

$$\iint_{R_{xy}} (x^2 + y^2)\, dx\, dy$$

by transforming the integral to new variables u and v.

Answer: You can describe the ellipse and everything inside it by writing

$$\{3u\cos[v], u\sin[v]\} \quad \text{with } 0 \le u \le 1 \text{ and } 0 \le v \le 2\pi.$$

Put

$$x[u,v] = 3u\cos[v] \quad \text{and} \quad y[u,v] = u\sin[v]:$$

In[42]:=
```
Clear[x,y,u,v]; Clear[gradx,grady,Axy]
{x[u_,v_],y[u_,v_]} = {3 u Cos[v],u Sin[v]};
gradx[u_,v_] = {D[x[u,v],u],D[x[u,v],v]};
grady[u_,v_] = {D[y[u,v],u],D[y[u,v],v]};
Axy[u_,v_] = Expand[Det[{gradx[u,v],grady[u,v]}],Trig->True]
```

Out[42]=
3 u

The uv-paper plot, R_{uv}, of R_{xy} is the rectangle with corners at $\{0,0\}, \{1,0\}, \{1, 2\pi\}$ and $\{0, 2\pi\}$, and everything inside it:

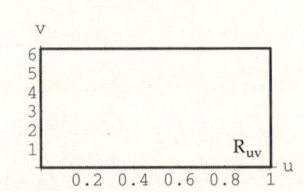

In[43]:=
```
Show[Graphics[{Red,Thickness[0.01],Line[
{{0,0},{1,0},{1,2 Pi},{0,2 Pi},{0,0}}]}],
Axes->True,AxesLabel->{"u","v"},AspectRatio->1/2];
```

This gives
$$\iint_{R_{xy}} (x^2 + y^2) \, dx \, dy = \iint_{R_{uv}} (x[u,v]^2 + y[u,v]^2) \, A_{(x,y)}[u,v] \, du \, dv$$
$$= \int_0^{2\pi} \int_0^1 (x[u,v]^2 + y[u,v]^2) \, A_{(x,y)}[u,v] \, du \, dv :$$

In[44]:=
`Integrate[(x[u,v]^2 + y[u,v]^2) Axy[u,v],{v,0,2 Pi},{u,0,1}]`

Out[44]=
$$\frac{15 \, \text{Pi}}{2}$$

So:
$$\iint_{R_{xy}} (x^2 + y^2) \, dx \, dy = \frac{15 \, \pi}{2}$$

and you're out of here without sweating and without a lot of miserable, irritating algebra.

B.3.c.ii) Look at the region R_{xy} inside the boundary described by the curves
$$y = 0.8 \, x, \quad y = 0.8 \, x + 0.5, \quad x \, y = 0.2, \quad \text{and} \quad x \, y = 0.6 :$$

In[45]:=
```
Plot[{0.8 x,0.8 x + 0.5,0.2/x ,0.6/x},
  {x,0.25,0.9},PlotStyle->{{Red,Thickness[0.01]}},
  PlotRange->{0.37,1.0},AxesLabel->{"x","y"}];
```

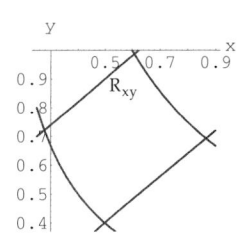

The region R_{xy} you are looking at is that stylish four-cornered figure you see above and everything inside it. Don't worry about the stray ends.

Calculate
$$\iint_{R_{xy}} x \, y^2 \, dx \, dy$$
without a lot of weeping, wailing, and gnashing of teeth by transforming the integral to new variables u and v.

Answer: This is frustrating without leaving xy-paper because the integral will have to be broken into three integrals with lots of weeping, wailing, and gnashing of teeth. But this is a great set-up for transforming to an integral easily set up on uv-paper. The R_{xy} region has its boundary formed by the curves
$$y = 0.8 \, x, \quad y = 0.8 \, x + 0.5, \quad x \, y = 0.2, \quad \text{and} \quad x \, y = 0.6.$$

Lesson 3.07 Transforming 2D Integrals

Use
$$u[x,y] = y - 0.8\,x \quad \text{and} \quad v[x,y] = x\,y.$$

In[46]:=
```
Clear[x,y,u,v]; {u[x_,y_],v[x_,y_]} = {y - 0.8 x, x y}
```
Out[46]=
{-0.8 x + y, x y}

Note
$$y = 0.8\,x \longleftrightarrow u[x,y] = 0$$
$$y = 0.8\,x + 0.5 \longleftrightarrow u[x,y] = 0.5$$
$$x\,y = 0.2 \longleftrightarrow v[x,y] = 0.2$$
$$x\,y = 0.6 \longleftrightarrow v[x,y] = 0.6.$$

After you stretch everything out so that the level curves of $u[x,y]$ and $v[x,y]$ become perpendicular straight lines on uv-paper, you see that the uv-paper plot, R_{uv}, of R_{xy} is the region inside the uv-paper rectangle with corners at $\{0, 0.2\}$, $\{0.5, 0.2\}$, $\{0.5, 0.6\}$, and $\{0, 0.6\}$.

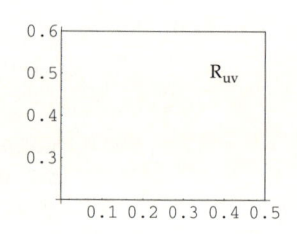

In[47]:=
```
uvboundary =
Show[Graphics[{Red,
Line[{{0,0.2},{0.5,0.2},{0.5,0.6},
{0,0.6},{0,0.2}}]}],
Axes->Automatic,Axes->Automatic];
```

Now
$$\iint_{R_{xy}} x\,y^2 \, dx\, dy = \iint_{R_{uv}} x[u,v]\, y[u,v]^2\, A_{(x,y)}[u,v]\, du\, dv$$
$$= \int_{0.2}^{0.6} \int_{0}^{0.5} x[u,v]\, y[u,v]^2\, A_{(x,y)}[u,v]\, du\, dv.$$

You can let *Mathematica* mop this up as soon as you find out what $x[u,v]$, $y[u,v]$, and $A_{(x,y)}[u,v]$ are:

In[48]:=
```
Solve[{u == u[x,y], v == v[x,y]},{x,y}]
```
Out[48]=
```
                             2
{{x -> 1.25 (-0.5 u + 0.5 Sqrt[1. u  + 3.2 v]),
                          2
  y -> 0.5 (1. u + 1. Sqrt[1. u  + 3.2 v])},
                          2
 {x -> 1.25 (-0.5 u - 0.5 Sqrt[1. u  + 3.2 v]),
                          2
  y -> 0.5 (1. u - 1. Sqrt[1. u  + 3.2 v])}}
```

The original region R_{xy} consists of $\{x, y\}$'s with positive coordinates, so you use:

In[49]:=
```
Clear[Axy,gradx,grady]
x[u_,v_] = 1.25(-0.5 u + 0.5 Sqrt[u^2 + 3.2 v]);
y[u_,v_] = 0.5 (u + Sqrt[u^2 + 3.2 v]);
gradx[u_,v_] = {D[x[u,v],u],D[x[u,v],v]};
grady[u_,v_] = {D[y[u,v],u],D[y[u,v],v]};
Axy[u_,v_] = Det[{gradx[u,v],grady[u,v]}]
```

Out[49]=
$$\frac{-1.}{2\,\text{Sqrt}[u^2 + 3.2\,v]}$$

This is negative; so throw in an extra minus sign to arrive at

$$\iint_{R_{xy}} x\,y^2\,dx\,dy = \int_{0.2}^{0.6} \int_0^{0.5} x[u,v]\,y[u,v]^2\,(-A_{(x,y)}[u,v])\,du\,dv :$$

In[50]:=
```
integral = Integrate[x[u,v] y[u,v]^2 (-Axy[u,v]),{v,0.2,0.6},{u,0,0.5}]
```

Out[50]=
0.048339

There wasn't all that much to it. Once you decided on a good uv-paper, then *Mathematica* ground it out without much trouble. The choice of uv-paper was more or less dictated by the set-up of the problem. It saved a lot of miserable, irritating algebra.

B.3.c.iii) In calculating the last two integrals, what was the decisive advantage in switching from xy-paper to uv-paper?

Answer: Setting up the integrals on xy-paper would have been frustrating. Setting up the integrals by transforming them to uv-paper was a breeze; it's the modern, up-to-date way of doing it.

Tutorials

■ T.1) Transforming $\iint_{R_{xy}} f[x,y]\,dx\,dy$ when the boundary of R_{xy} is given by parametric formulas

When the boundary of a region R_{xy} is plotted with parametric formulas, you often have a good shot at using the parametric formulas to come up with uv-paper on which R_{uv} is a rectangle. This is especially good because calculating 2D integrals over rectangles is usually very easy.

T.1.a.i) The region R_{xy} is everything inside and on the circle $(x-1)^2 + (y+2)^2 = 5$:

284   **Lesson 3.07 Transforming 2D Integrals**

In[1]:=
```
Clear[x,y,t]
{x[t_],y[t_]} = {1,-2} + Sqrt[5] {Cos[t],Sin[t]};
Rxyplot = ParametricPlot[{x[t],y[t]},{t,0,2 Pi},
PlotStyle->{{Thickness[0.01],Red}},
AspectRatio->Automatic,AxesLabel->{"x","y"}];
```

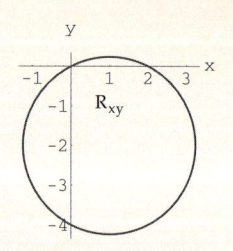

Calculate $\iint_{R_{xy}} (3x^2 + 5y^4) \, dx \, dy$.

Answer: Look at the functions used to plot the boundary of R_{xy}:

In[2]:=
```
{x[t],y[t]}
```
Out[2]=
```
{1 + Sqrt[5] Cos[t], -2 + Sqrt[5] Sin[t]}
```

Put

In[3]:=
```
Clear[x,y,u,t]
{x[u_,t_],y[u_,t_]} = {1 + u Cos[t],-2 + u Sin[t]}
```
Out[3]=
```
{1 + u Cos[t], -2 + u Sin[t]}
```

Realize this: When you run u from 0 to $\sqrt{5}$, and you run t from 0 to 2π, then $\{x[u,t], y[u,t]\}$ runs through all of R_{xy}:

In[4]:=
```
{{ulow = 0,uhigh = Sqrt[5]},{tlow = 0,thigh = 2 Pi}}
```
Out[4]=
```
{{0, Sqrt[5]}, {0, 2 Pi}}
```

The upshot: R_{ut} is the rectangle ulow $\leq u \leq$ uhigh and tlow $\leq t \leq$ thigh on ut-paper. This, and the fact that

$$\iint_{R_{xy}} f[x,y] \, dx \, dy = \iint_{R_{ut}} f[x[u,t], y[u,t]] \, A_{(x,y)}[u,t] \, du \, dt$$

$$= \int_{tlow}^{thigh} \int_{ulow}^{uhigh} f[x[u,t], y[u,t]] \, A_{(x,y)}[u,t] \, du \, dt,$$

tell you that you can turn everything over to the machine, right now:

In[5]:=
```
Clear[gradx,grady,Axy]
gradx[u_,t_] = {D[x[u,t],u],D[x[u,t],t]};
grady[u_,t_] = {D[y[u,t],u],D[y[u,t],t]};
Axy[u_,t_] = Expand[Det[{gradx[u,t],grady[u,t]}],Trig->True]
```

Out[5]=
u

Here comes the calculation of $\iint_{R_{xy}} (3x^2 + 5y^4)\, dxdy$:

In[6]:=
```
calculation = Integrate[(3 x[u,t]^2 + 5 y[u,t]^4) Axy[u,t],
   {t,tlow,thigh},{u,ulow,uhigh}]
```

Out[6]=
$$\frac{10095 \text{ Pi}}{8}$$

Nasty answer, but it wasn't hard to get.

T.1.a.ii) Could you have used the Gauss-Green formula to calculate the integral in part T.1.a.i)?

Answer: Yes.

Here is how it goes: You want to calculate
$$\iint_{R_{xy}} (3x^2 + 5y^4)\, dx\, dy,$$
and you have a counterclockwise parameterization of the boundary of R_{xy}:

In[7]:=
```
Clear[x,y,f,t]; f[x_,y_] = 3 x^2 + 5 y^4;
{x[t_],y[t_]} = {1,-2} + Sqrt[5] {Cos[t],Sin[t]};
{a,b} = {0, 2 Pi}
```

Out[7]=
{0, 2 Pi}

With $a \leq t \leq b$. The Gauss-Green formula says
$$\iint_{R_{xy}} D[n[x,y],x] - D[m[x,y],y]\, dx\, dy = \int_a^b m[x[t],y[t]]\, x'[t] + n[x[t],y[t]]\, y'[t]\, dt$$
where $a = 0$ and $b = 2\pi$.

To calculate
$$\iint_{R_{xy}} f[x,y]\, dx\, dy$$
you just say $m[x,y] = 0$ and $n[x,y] = \int_0^x f[s,y]\, ds$:

In[8]:=
```
Clear[n,m,s,y]; m[x_,y_] = 0;
n[x_,y_] = Integrate[f[s,y],{s,0,x}]
```

Out[8]=
$$5 y^4 x + x^3$$

This gives you

$$f[x,y] = D[n[x,y],x] - D[m[x,y],x]:$$

In[9]:=
```
{D[n[x,y],x] - D[m[x,y],x], f[x,y]}
```
Out[9]=
$$\{3 x^2 + 5 y^4,\ 3 x^2 + 5 y^4\}$$

Now you know that

$$\iint_{R_{xy}} f[x,y]\, dx\, dy = \iint_{R_{xy}} D[n[x,y],x] - D[m[x,y],y]\, dx\, dy$$

$$= \int_a^b m[x[t],y[t]]\, x'[t] + n[x[t],y[t]]\, y'[t]\, dt:$$

In[10]:=
```
{x[t_],y[t_]} = {1,-2} + Sqrt[5] {Cos[t],Sin[t]};
GGcalculation = Integrate[m[x[t],y[t]] x'[t] + n[x[t],y[t]] y'[t],{t,a,b}]
```
Out[10]=
$$\frac{10095\ \text{Pi}}{8}$$

As expected, this is the same nasty answer you got in part T.1.a.i).

T.1.a.iii) How do you decide whether to go with transformations as in part T.1.a.i), or with the Gauss-Green approach as in part T.1.a.ii)?

Answer: From the scientific point of view, it's a toss-up. Both methods work, so the decision is really a matter of personal choice. Most folks prefer the approach using transformations in part T.1.a.i). But when you're putting on the ritz, you might want to go with the Gauss-Green approach.

Since you're already familiar with the Gauss-Green formula, this lesson will concentrate on the approach using transformations. You should do that too, because the approach using transformations has a clear extension to three dimensions. Versions of Gauss-Green in three dimensions are complicated.

T.1.b.i) Here's a plot of the part of the surface

$$z = xy + 5$$

above the ellipse

$$\left(\frac{x}{2}\right)^2 + \left(\frac{y}{3}\right)^2 = 1$$

in the xy-plane:

In[11]:=
```
Clear[x,y,r,t]
{x[r_,t_],y[r_,t_]} = {2 r Cos[t],3 r Sin[t]};
{{rlow,rhigh},{tlow,thigh}} = {{0,1},{0, 2 Pi}};
top = ParametricPlot3D[{x[r,t],
y[r,t],x[r,t] y[r,t] + 5},{r,rlow,rhigh},
{t,tlow,thigh},DisplayFunction->Identity];
base = ParametricPlot3D[
{x[r,t],y[r,t],0},{r,rlow,rhigh},{t,tlow,thigh},
PlotPoints->{2,Automatic},DisplayFunction->Identity];
Show[top,base,PlotRange->All,AxesLabel->{"x","y","z"},
ViewPoint->CMView,DisplayFunction->$DisplayFunction];
```

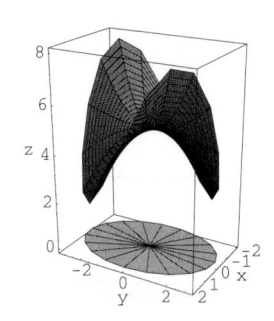

> Measure the volume under the plotted surface and above its base in the xy-plane.

Answer: It's duck soup. The volume is measured by

$$\iint_{R_{xy}} (x\,y + 5)\ dx\,dy,$$

where R_{xy} is everything inside and on the ellipse

$$\left(\frac{x}{2}\right)^2 + \left(\frac{y}{3}\right)^2 = 1$$

plotted in the xy-plane. Go with rt-paper coming from:

In[12]:=
```
{x[r,t],y[r,t]}
```

Out[12]=
{2 r Cos[t], 3 r Sin[t]}

As you run r from rlow to rhigh and as you run t from tlow to thigh, $\{x[r,t], y[r,t]\}$ sweeps out everything inside and on R_{xy}. So R_{rt} is the rectangle rlow $\leq r \leq$ rhigh and tlow $\leq t \leq$ thigh, and

$$\text{volume} = \iint_{R_{xy}} (x\,y + 5)\ dx\,dy$$

$$= \iint_{R_{rt}} (x[r,t]\,y[r,t] + 5)\ A_{(x,y)}[r,t]\ dr\,dt$$

$$= \int_{\text{tlow}}^{\text{thigh}} \int_{\text{rlow}}^{\text{rhigh}} (x[r,t]\,y[r,t] + 5)\ A_{(x,y)}[r,t]\ dr\,dt :$$

In[13]:=
```
Clear[gradx,grady,Axy]
gradx[r_,t_] = {D[x[r,t],r],D[x[r,t],t]};
grady[r_,t_] = {D[y[r,t],r],D[y[r,t],t]};
Axy[r_,t_] = Expand[Det[{gradx[r,t],grady[r,t]}],Trig->True]
```

Out[13]=
6 r

Lesson 3.07 Transforming 2D Integrals

Measure the volume:

In[14]:=
```
Integrate[(x[r,t] y[r,t] + 5)  Axy[r,t],
 {t,tlow,thigh},{r,rlow,rhigh}]
```

Out[14]=
```
30 Pi
```

Nice answer. But if you hadn't gone to *rt*-paper to calculate the integral, getting this nice answer wouldn't have been so simple.

T.1.b.ii) Could you have measured the same volume using the Gauss-Green formula?

Answer: Yes.

T.1.c.i) Here's a plot of a region R_{xy} on *xy*-paper:

In[15]:=
```
Clear[x,y,r,t]
{x[r_,t_],y[r_,t_]} = {3 r Cos[t],r Sin[t]};
{rlow,rhigh} = {1,3}; {tlow,thigh} = {Pi/4,7 Pi/4};
twosides = ParametricPlot[{{x[rlow,t],y[rlow,t]},
 {x[rhigh,t],y[rhigh,t]}},{t,tlow,thigh},
 PlotStyle->{{Thickness[0.01],Red}},
 DisplayFunction->Identity]; twomoresides =
 ParametricPlot[{{x[r,tlow],y[r,tlow]},
 {x[r,thigh],y[r,thigh]}},{r,rlow,rhigh},
 PlotStyle->{{Thickness[0.01],Red}},
 DisplayFunction->Identity]; Show[twosides,twomoresides,
 AspectRatio->Automatic,AxesLabel->{"x","y"},
 DisplayFunction->$DisplayFunction];
```

R_{xy} is everything inside and on this boundary.

Calculate $\iint_{R_{xy}} x \, dx \, dy$ with little thought, and with *Mathematica* doing the work.

Answer: Look at:

In[16]:=
```
{x[r,t],y[r,t]}
```

Out[16]=
```
{3 r Cos[t], r Sin[t]}
```

When you look at the plotting instructions above, then you see that when you run *r* from rlow to rhigh and you run *t* from tlow to thigh, then $\{x[r,t], y[r,t]\}$ describes R_{xy}. This tells you that on *rt*-paper R_{rt} is the rectangle rlow $\leq r \leq$ rhigh and tlow

$\leq t \leq$ thigh. And your thinking is almost done. Remembering that

$$\iint_{R_{xy}} x\, dx\, dy = \iint_{R_{uv}} x[r,t]\, A_{(x,y)}[r,t]\, dr\, dt,$$

turn *Mathematica* loose:

In[17]:=
```
Clear[gradx,grady,Axy]
gradx[r_,t_] = {D[x[r,t],r],D[x[r,t],t]};
grady[r_,t_] = {D[y[r,t],r],D[y[r,t],t]};
Axy[r_,t_] = Factor[Expand[Det[{gradx[r,t],grady[r,t]}],Trig->True]]
```

Out[17]=
3 r

You can see that $A_{(x,y)}[r,t]$ is never negative because $1 + \cos[2t]$ can never go negative. Now your thinking is done. Now calculate

$$\iint_{R_{xy}} x\, dx\, dy = \iint_{R_{uv}} x[r,t]\, A_{(x,y)}[r,t]\, dr\, dt$$
$$= \int_{\text{tlow}}^{\text{thigh}} \int_{\text{rlow}}^{\text{rhigh}} x[r,t]\, A_{(x,y)}[r,t]\, dr\, dt :$$

In[18]:=
```
Integrate[x[r,t] Axy[r,t],{t,tlow,thigh},{r,rlow,rhigh}]
```

Out[18]=
-78 Sqrt[2]

And you're out of here.

T.1.c.ii) | Could you have calculated the same integral using the Gauss-Green formula?

Answer: In theory, yes.

In practice, it would have been a very tedious job, because you would have to do a lot of bureaucratic work to come up with the required counterclockwise parameterization of the boundary. This would involve separate parameterizations for each of the four boundary segments. Ugh.

T.1.d.i) Here's a curve on *xy*-paper:

In[19]:=
```
Clear[x,y,r,t]
{x[t_],y[t_]} = {4 Cos[t],3 Sin[t]};
{tlow,thigh} = {1,5};
curveplot = ParametricPlot[{x[t],y[t]},{t,tlow,thigh},
PlotStyle->{{Thickness[0.01],Red}},
AspectRatio->Automatic,AxesLabel->{"x","y"}];
```

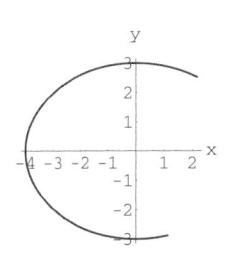

The outer normal to the curve at $\{x[t], y[t]\}$ is:

In[20]:=
```
Clear[unitnormal]
unitnormal[t_] = Expand[{y'[t],-x'[t]}/Sqrt[x'[t]^2 + y'[t]^2],Trig->True]
```
Out[20]=
$$\left\{\frac{3 \text{ Sqrt}[2] \text{ Cos}[t]}{\text{Sqrt}[25 - 7 \text{ Cos}[2\ t]]}, \frac{4 \text{ Sqrt}[2] \text{ Sin}[t]}{\text{Sqrt}[25 - 7 \text{ Cos}[2\ t]]}\right\}$$

Here's the boundary of a ribbon of constant width 1 centered on this curve:

In[21]:=
```
Clear[r,outerribbon,innerribbon,oneend,otherend]
outerribbon[t_] = {x[t],y[t]} + 0.5 unitnormal[t];
innerribbon[t_] = {x[t],y[t]} - 0.5 unitnormal[t];
oneend[r_] = {x[tlow],y[tlow]} + r unitnormal[tlow];
otherend[r_] = {x[thigh],y[thigh]} + r unitnormal[thigh];
{rlow = -0.5,rhigh = 0.5}; twosides =
ParametricPlot[{outerribbon[t],innerribbon[t]},
{t,tlow,thigh},PlotStyle->{{Thickness[0.01],Blue}},
DisplayFunction->Identity]; twomoresides =
ParametricPlot[{oneend[r],otherend[r]},{r,rlow,rhigh},
PlotStyle->{{Thickness[0.01],Blue}},
DisplayFunction->Identity];
Show[curveplot,twosides,twomoresides,
AspectRatio->Automatic,AxesLabel->{"x","y"},
DisplayFunction->$DisplayFunction];
```

Measure the area of this ribbon.

Answer: Call the inside of the ribbon R_{xy}. The integral

$$\iint_{R_{xy}} dx\, dy$$

measures the area of the ribbon. Go to *rt*-paper with:

In[22]:=
```
{rlow,rhigh} = {-0.5,0.5}; Clear[x,y,r,t]
{x[r_,t_],y[r_,t_]} = {4 Cos[t],3 Sin[t]} + r unitnormal[t]
```
Out[22]=
$$\left\{4 \text{ Cos}[t] + \frac{3 \text{ Sqrt}[2]\ r \text{ Cos}[t]}{\text{Sqrt}[25 - 7 \text{ Cos}[2\ t]]},\ 3 \text{ Sin}[t] + \frac{4 \text{ Sqrt}[2]\ r \text{ Sin}[t]}{\text{Sqrt}[25 - 7 \text{ Cos}[2\ t]]}\right\}$$

When you look at the plotting instructions above, then you see that when you run r from rlow to rhigh and you run t from tlow to thigh, then $\{x[r,t], y[r,t]\}$ describes the whole ribbon R_{xy}. This tells you that on *rt*-paper R_{rt} is the rectangle

$$\text{rlow} \le r \le \text{rhigh} \quad \text{and} \quad \text{tlow} \le t \le \text{thigh}.$$

Remembering that

$$\text{area of ribbon} = \iint_{R_{xy}} dx\, dy = \iint_{R_{rt}} A_{(x,y)}[r,t]\, dr\, dt,$$

turn *Mathematica* loose:

In[23]:=
```
Clear[gradx,grady,Axy]
gradx[r_,t_] = {D[x[r,t],r],D[x[r,t],t]}; grady[r_,t_] = {D[y[r,t],r],D[y[r,t],t]};
Axy[r_,t_] = Expand[Det[{gradx[r,t],grady[r,t]}],Trig->True]
```

Out[23]=

$$\frac{24\, r}{25 - 7\, \text{Cos}[2\, t]} + \frac{25}{\text{Sqrt}[2]\, \text{Sqrt}[25 - 7\, \text{Cos}[2\, t]]} - \frac{7\, \text{Cos}[2\, t]}{\text{Sqrt}[2]\, \text{Sqrt}[25 - 7\, \text{Cos}[2\, t]]}$$

Analyzing this, it's hard to see whether this mess can ever go negative. Take the easy way out and integrate its absolute value using NIntegrate.

In[24]:=
```
NIntegrate[Evaluate[Abs[Axy[r,t]]],{t,tlow,thigh},{r,rlow,rhigh},AccuracyGoal->2]
```

Out[24]=
14.4267

About 14.4 square units.

T.1.d.ii) This problem appears only in the electronic version.

■ T.2) Transforming $\iint_{R_{xy}} f[x,y]\, dx\, dy$ when the boundary of R_{xy} is not given with parametric formulas

When the boundary of a region R_{xy} is plotted with nonparametric formulas, things are not always as simple as they were in T.1). But even in this case, there are situations that allow you to inspect the boundary curves to help come up with favorable uv-paper. Here's one such:

T.2.a.i) R_{xy} is the region plotted below that is bounded by the curves

$$y = 0.5\, x^2 + 1, \quad y = 0.5\, x^2 - 1, \quad y = 3\, x + 2, \quad \text{and} \quad y = 3\, x - 2:$$

In[25]:=
```
Clear[x]
Rxyplot = Plot[{0.5 x^2 + 1,
  0.5 x^2 - 1,3 x + 2,3 x - 2},
  {x,-1,1.4},PlotStyle->{{Red,Thickness[0.01]}},
  AxesLabel->{"x","y"}];
```

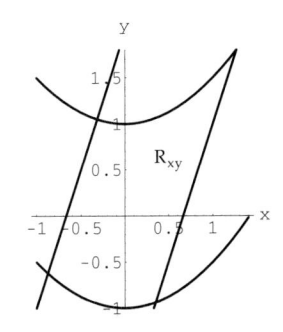

Lesson 3.07 Transforming 2D Integrals

The region R_{xy} under scrutiny is the four-sided figure you see above and everything inside it.

> Transform R_{xy} into a rectangle on uv-paper to help you to come up with a quick and easy calculation of
> $$\iint_{R_{xy}} (x^2 + y^2) \, dx \, dy.$$

Answer: Enter $f[x,y] = x^2 + y^2$:

In[26]:=
```
Clear[f,x,y]; f[x_,y_] = x^2 + y^2;
```

Look at the formulas for the functions whose plots make up the boundary of R_{xy}. They are:

$$y = 0.5\, x^2 + 1,$$
$$y = 0.5\, x^2 - 1,$$
$$y = 3\, x + 2,$$

and

$$y = 3\, x - 2:$$

Put

$$u[x,y] = y - 0.5\, x^2 \quad \text{and} \quad v[x,y] = y - 3\, x:$$

In[27]:=
```
Clear[u,v,x,y]; {u[x_,y_],v[x_,y_]} = {y - 0.5 x^2, y - 3 x};
```

The original boundary curves are level curves of $u[x,y]$ and $v[x,y]$. In fact,

→ $y = 0.5\, x^2 + 1$ is the level curve $u[x,y] = 1$
→ $y = 0.5\, x^2 - 1$ is the level curve $u[x,y] = -1$
→ $y = 3\, x - 2$ is the level curve $v[x,y] = 2$ and
→ $y = 3\, x + 2$ is the level curve $v[x,y] = -2$.

This is very good news because this tells you that R_{uv} is the rectangle

$$-1 \leq u \leq 1 \quad \text{and} \quad -2 \leq v \leq 2.$$

The upshot:

$$\iint_{R_{xy}} (x^2 + y^2) \, dx \, dy = \iint_{R_{uv}} \left(x[u,v]^2 + y[u,v]^2\right) A_{(x,y)}[u,v] \, du \, dv$$

$$= \int_{-2}^{2} \int_{-1}^{1} \left(x[u,v]^2 + y[u,v]^2\right) A_{(x,y)}[u,v] \, du \, dv.$$

First, you have to come up with formulas for $x[u,v]$ and $y[u,v]$:

In[28]:=
```
solutions = Solve[{u == u[x,y],v == v[x,y]},{x,y}]
```
Out[28]=
```
{{x -> 1. (3. + 1.41421 Sqrt[4.5 - 1. u + 1. v]),
  y -> 9. (1. + 0.111111 v + 0.471405 Sqrt[4.5 - 1. u + 1. v])},
 {x -> 1. (3. - 1.41421 Sqrt[4.5 - 1. u + 1. v]),
  y -> 9. (1. + 0.111111 v - 0.471405 Sqrt[4.5 - 1. u + 1. v])}}
```

This gives two choices.

In[29]:=
```
Clear[x1,y1,x2,y2]
{x1[u_,v_],y1[u_,v_]} =
{0.5 (6 - Sqrt[36 - 4 (2 u - 2 v)]),
 9 + v - 1.5 Sqrt[36 - 8 u + 8 v]}
```
Out[29]=
```
{0.5 (6 - Sqrt[36 - 4 (2 u - 2 v)]), 9 + v - 1.5 Sqrt[36 - 8 u + 8 v]}
```

In[30]:=
```
{x2[u_,v_],y2[u_,v_]}=
{0.5 (6 + Sqrt[36 - 4 (2 u - 2 v)]),
 9 + v + 1.5 Sqrt[36 - 8 u + 8 v]}
```
Out[30]=
```
{0.5 (6 + Sqrt[36 - 4 (2 u - 2 v)]), 9 + v + 1.5 Sqrt[36 - 8 u + 8 v]}
```

You know that R_{uv} is the rectangle

$$-1 \leq u \leq 1 \quad \text{and} \quad -2 \leq v \leq 2.$$

The uv-point $\{0.5, 1\}$ is in this rectangle.

See which solution makes uv-point $\{0.5, 1\}$ plot out inside R_{xy} on xy-paper by seeing where $\{x_1[0.5, 1], y_1[0.5, 1]\}$ and $\{x_2[0.5, 1], y_2[0.5, 1]\}$ land on xy-paper:

In[31]:=
```
Show[Rxyplot,Graphics[{Blue,PointSize[0.07],
Point[{x1[0.5,1],y1[0.5,1]}]}],
PlotRange->All];
```

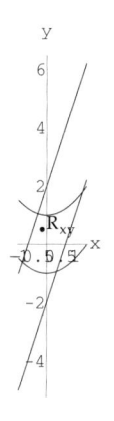

Good; this lands inside R_{xy}. Just for the heck of it, try out the other pair of solutions:

Lesson 3.07 Transforming 2D Integrals

```
In[32]:=
Show[Rxyplot,Graphics[{Blue,PointSize[0.07],
Point[{x2[0.5,1],y2[0.5,1]}]}],
PlotRange->All];
```

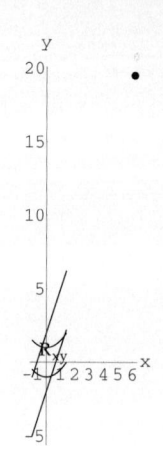

Way outside R_{xy}. This means you definitely want to go with:

```
In[33]:=
{x[u_,v_],y[u_,v_]} = {x1[u,v],y1[u,v]}
```

Out[33]=
```
{0.5 (6 - Sqrt[36 - 4 (2 u - 2 v)]), 9 + v - 1.5 Sqrt[36 - 8 u + 8 v]}
```

Calculate the area conversion factor $A_{(x,y)}[u,v]$:

```
In[34]:=
Clear[Axy,gradx,grady]
gradx[u_,v_] = {D[x[u,v],u],D[x[u,v],v]};
grady[u_,v_] = {D[y[u,v],u],D[y[u,v],v]};
Axy[u_,v_] = Chop[Det[{gradx[u,v],grady[u,v]}]]
```

Out[34]=
$$\frac{1.}{\text{Sqrt}[9 - 2\ u + 2\ v]}$$

Calculate

$$\iint_{R_{xy}} (x^2 + y^2)\, dx\, dy = \iint_{R_{uv}} (x[u,v]^2 + y[u,v]^2)\, A_{(x,y)}[u,v]\, du\, dv$$
$$= \int_{-2}^{2} \int_{-1}^{1} (x[u,v]^2 + y[u,v]^2)\, A_{(x,y)}[u,v]\, du\, dv.$$

```
In[35]:=
NIntegrate[(x[u,v]^2 + y[u,v]^2) Axy[u,v],
{v,-2,2},{u,-1,1},AccuracyGoal->2]
```

Out[35]=
1.74016

Finished.

T.2.a.ii) When you have a set-up like the problem in part T.2.a.i), what can go wrong?

Answer: In theory, nothing much can go wrong.

In practice, this technique can grind to a quick halt. The hitch is that you specify $u[x,y]$ and $v[x,y]$, and then you have to solve the simultaneous equations

$$u = u[x,y] \quad \text{and} \quad v = v[x,y]$$

for x and y to get the formulas for $x[u,v]$ and $y[u,v]$. Solving $u = u[x,y]$ and $v = v[x,y]$ for x and y is possible only in simple special situations. Samples:

In[36]:=
```
Clear[x,y,u,v]; {u[x_,y_],v[x_,y_]} = {y - Sin[x],y - x};
Solve[{u == u[x,y],v == v[x,y]},{x,y}]
```

Solve::ifun:
    Warning: Inverse functions are being used by Solve, so
      some solutions may not be found.

Solve::tdep:
    The equations appear to involve transcendental
      functions of the variables in an essentially
      non-algebraic way.

Out[36]=
```
Solve[{u == y - Sin[x], v == -x + y}, {x, y}]
```

No dice. That transcendental function $\sin[x]$ screws up the algebra.

In[37]:=
```
Clear[x,y,u,v]; {u[x_,y_],v[x_,y_]} = {y - E^(-x),y - x};
Solve[{u == u[x,y],v == v[x,y]},{x,y}]
```

Solve::ifun:
    Warning: Inverse functions are being used by Solve, so
      some solutions may not be found.

Solve::tdep:
    The equations appear to involve transcendental
      functions of the variables in an essentially
      non-algebraic way.

Out[37]=
```
Solve[{u == -E^-x + y, v == -x + y}, {x, y}]
```

No dice. That transcendental function e^{-x} screws up the algebra.

In[38]:=
```
Clear[x,y,u,v]; {u[x_,y_],v[x_,y_]} = {y - 2 x ,y + x};
Solve[{u == u[x,y],v == v[x,y]},{x,y}]
```

Out[38]=
$$\left\{\left\{x \to \frac{-u}{3} + \frac{v}{3},\ y \to \frac{u}{3} + \frac{2v}{3}\right\}\right\}$$

No sweat. This gives you

$$x[u,v] = \frac{-u+v}{3} \quad \text{and} \quad y[u,v] = \frac{u+2v}{3}.$$

Lesson 3.07 Transforming 2D Integrals

T.2.a.iii) What is a transcendental function?

This answer comes from Phillip Gillett's book *Calculus and Analytic Geometry* (2nd edition), D.C. Heath, Lexington, Massachusetts, 1984, p.335.

Answer: Transcendental functions are those that transcend the ordinary processes of algebra. The basic calculus functions $\sin[x]$, $\cos[x]$, and e^x are all transcendental. That's why you are guaranteed to fail when you try simple things like:

In[39]:=
   `Clear[x]; Solve[ x == Sin[x],x]`
Solve::ifun:
   Warning: Inverse functions are being used by Solve, so
     some solutions may not be found.
Solve::tdep:
   The equations appear to involve transcendental functions of the
     variables in an essentially non-algebraic way.

Out[39]=
   Solve[x == Sin[x], x]

In[40]:=
   `Clear[x]; Solve[ x == E^x,x]`
Solve::ifun:
   Warning: Inverse functions are being used by Solve, so
     some solutions may not be found.
Solve::tdep:
   The equations appear to involve transcendental functions of the
     variables in an essentially non-algebraic way.

Out[40]=
   Solve[x == E^x, x]

Line functions like $f[x] = 3\,x + 2$ are not transcendental:

In[41]:=
   `Clear[x]; Solve[ x == 3 x + 2,x]`

Out[41]=
   {{x -> -1}}

Determining whether a given function is transcendental is part of the stuff of advanced mathematics.

■ **T.3)** The area conversion factor $A_{(x,y)}[u, v]$

T.3.a) Polar coordinates $\{u, v\}$ are related to xy-coordinates via

$$x = u\cos[v] \quad \text{and} \quad y = u\sin[v].$$

Here is a plot of a random bunch of points $\{u, v\}$ on uv-paper with $0 \le u \le 10$ and $0 \le v \le 2\pi$:

In[42]:=
```
Clear[k]
uvpoints = Table[{Random[Real,{0,10}],
  Random[Real,{0,N[2 Pi]}]},{k,1,150}];
uvpaperplot = ListPlot[uvpoints,
  PlotStyle->{Blue,PointSize[0.02]},
  AspectRatio->Automatic,AxesLabel->{"u","v"}];
```

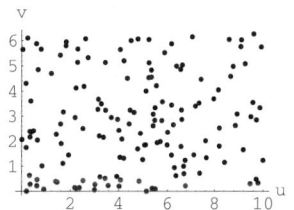

Should be fairly well scattered. If not, then rerun. Now look at the xy-paper plot of these uv-paper points:

In[43]:=
```
Clear[x,y,u,v]; Clear[uvtoxy]
x[u_,v_] = u Cos[v]; y[u_,v_] = u Sin[v];
uvtoxy[{u_,v_}] = {x[u,v],y[u,v]};
xypoints = Map[uvtoxy,uvpoints];
xypaperplot = ListPlot[xypoints,
  PlotStyle->{Blue,PointSize[0.03]},
  AspectRatio->Automatic,AxesLabel->{"x","y"}];
```

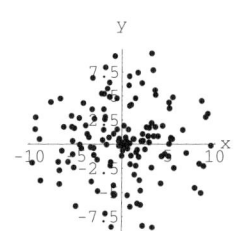

> On the xy-paper, why are the points so bunched up near the origin and sparsely scattered far from the origin?

Answer: Look at the area conversion factor:

In[44]:=
```
Clear[gradx,grady,Axy]
gradx[u_,v_] = {D[x[u,v],u],D[x[u,v],v]};
grady[u_,v_] = {D[y[u,v],u],D[y[u,v],v]};
Axy[u_,v_] = Expand[Det[{gradx[u,v],grady[u,v]}],Trig->True]
```
Out[44]=
u

At a point with uv-paper coordinates $\{u, v\}$,

$$(xy\text{-paper area measurements}) = u\,(u\,v\text{-paper area measurements}).$$

In this set-up, uv-paper coordinates are polar coordinates.

Look at:

In[45]:=
```
{x[u,v],y[u,v]}
```
Out[45]=
{u Cos[v], u Sin[v]}

The uv-paper coordinate, u, of the xy-paper point, $\{x[u,v], y[u,v]\}$, measures the distance from $\{0,0\}$ to $\{x[u,v], y[u,v]\}$ on xy-paper.

→ When
$$A_{(x,y)}[u,v] = u$$
is big, then xy-paper area measurements are a lot bigger than the corresponding uv-area measurements, so uv-paper points with u big are flung apart when they are plotted on xy-paper.

→ When
$$A_{(x,y)}[u,v] = u$$
is wee little, uv-paper points $\{u,v\}$ are compressed together when they are plotted on xy-paper. Since u is small for points near the origin, you see a pile-up near the origin on the xy-paper plot.

T.3.b.i) Take $x[u,v] = u^2 - v^2$ and $y[u,v] = 2uv$ and calculate the area conversion factor $A_{(x,y)}[u,v]$:

In[46]:=
```
Clear[x,y,u,v,gradx,grady,Axy];
x[u_,v_] = u^2 - v^2; y[u_,v_] = 2 u v;
gradx[u_,v_] = {D[x[u,v],u],D[x[u,v],v]};
grady[u_,v_] = {D[y[u,v],u],D[y[u,v],v]};
Axy[u_,v_] = Det[{gradx[u,v],grady[u,v]}]
```

Out[46]=
$$4u^2 + 4v^2$$

Here's a table of random points inside the uv-paper square with $-2 \leq u \leq 2$ and $-2 \leq v \leq 2$ plotted as dots on uv-paper:

In[47]:=
```
Clear[k]; pointcount = 150;
uvpoints = Table[{Random[Real,{-2,2}],
Random[Real,{-2,2}]},{k,1,pointcount}];
uvpointplot = Show[Table[Graphics[
{Blue,PointSize[0.02],Point[uvpoints[[k]]]}],
{k,1,pointcount},AspectRatio->Automatic,
Axes->Automatic,AxesLabel->{"u","v"}];
```

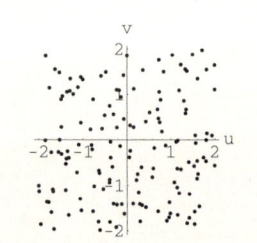

Fairly well scattered.

Here is the same plot with the size of each plotted point $\{u,v\}$ adjusted by a factor proportional to $\sqrt{A_{(x,y)}[u,v]}$.

In[48]:=
```
Clear[sizer]; sizer[u_,v_] = 0.015 Sqrt[Axy[u,v]];
sizeduvpointplot = Show[Table[Graphics[{Blue,
PointSize[Apply[sizer,uvpoints[[k]]]],
Point[uvpoints[[k]]]}],
{k,1, pointcount},AspectRatio->Automatic,
Axes->Automatic,AxesLabel->{"u","v"}];
```

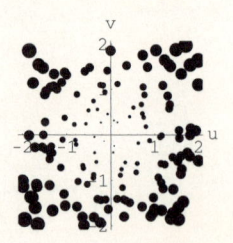

A humdinger. See the plots side by side:

In[49]:=
```
Show[GraphicsArray[
  {uvpointplot,sizeduvpointplot}]];
```

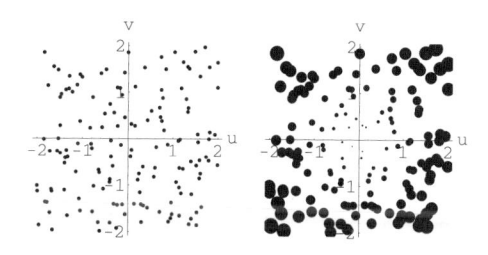

> What information is conveyed by these plots?

Answer: The area conversion factor in going from $\{u,v\}$ to $\{x[u,v], y[u,v]\}$ is $A_{(x,y)}[u,v]$. The points are sized in proportion to $\sqrt{A_{(x,y)}[u,v]}$.

Consequently, the second plot shows what the relative sizes of the plotted uv-points will be after they have been plotted on xy-paper. Evidently, as $\{u,v\}$ gets farther and farther from $\{0,0\}$, $A_{(x,y)}[u,v]$ gets bigger and bigger. This fact is suggested by the plot, and is confirmed by the formula for $A_{(x,y)}[u,v]$:

In[50]:=
```
Axy[u,v]
```
Out[50]=
$$4u^2 + 4v^2$$

If you wonder why the points are sized proportionally to $\sqrt{A_{(x,y)}[u,v]}$ instead of proportionally to $A_{(x,y)}[u,v]$, then read on.

In *Mathematica*, the PointSize instruction governs the radius of the plotted point. To make the area of the plotted point proportional to $A_{(x,y)}[u,v]$, you have to make the radius of the plotted point proportional to $\sqrt{A_{(x,y)}[u,v]}$, because the area of a circle of radius r is proportional to r^2.

T.3.b.ii) This problem appears only in the electronic version.

■ T.4) Measurements of volume, mass, and density

T.4.a) You make an object by distributing a substance over a certain region R on xy-paper.

> What does it mean to say that the density of the resulting object measures out at
>
> $p[x,y]$ grams/unit^3 at location $\{x,y\}$?

Answer: It means that the mass of the object is given by

$$\iint_R p[x,y]\,dx\,dy$$

where R is the same region that you distributed the substance over to begin with. It also means that if R_1 is a region inside R, then the total mass of the substance that was spread over R_1 is

$$\iint_{R_1} p[x,y]\,dx\,dy.$$

As a result, $p[x_0, y_0]$ is the conversion factor that converts area at $\{x_0, y_0\}$ on xy-paper to mass of the object at $\{x_0, y_0\}$.

T.4.b.i) An object is made by forming a uniform substance that weighs 2 grams/unit in the shape of the paraboloid $f[x,y] = 9 - x^2 - y^2$ over the region inside and on the circle $x^2 + y^2 = 9$ on xy-paper. Here's a look at it:

In[51]:=
```
Clear[x,y]
Plot3D[9 - x^2 - y^2,{x,-3,3},{y,-3,3},
  PlotRange->{0,9},ViewPoint->CMView,
  AxesLabel->{"x","y","z"}];
```

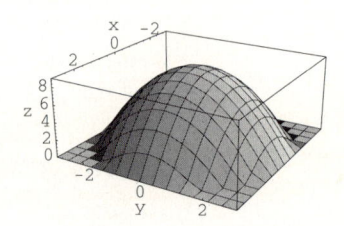

> Measure the total mass of this object. Measure the total volume of this object.

Answer: The density of the object, $p[x,y]$, at $\{x,y\}$ is $2f[x,y]$.

The total mass is given by

$$\iint_{R_{xy}} 2\,f[x,y]\,dx\,dy$$

where R is the region inside and on the circle

$$x^2 + y^2 = 9$$

on xy-paper. No one likes to integrate over circular regions, so calculate this by moving to uv-paper with

$$x[u,v] = u\cos[v] \quad \text{and} \quad y[u,v] = u\sin[v]$$

(polar coordinates). On uv-paper, R plots out as the rectangle with corners at $\{0,0\}$, $\{3,0\}$, $\{3,2\pi\}$, and $\{0,2\pi\}$ ($0 \le u \le 3$ and $0 \le v \le 2\pi$). The total mass is

$$\iint_{R_{xy}} 2\,f[x,y]\,dx\,dy = \int_0^{2\pi}\!\!\int_0^3 2\,f[x[u,v], y[u,v]]\,A_{(x,y)}[u,v]\,du\,dv:$$

In[52]:=
```
Clear[f,u,v,x,y,gradx,grady,Axy]; f[x_,y_] = 9 - x^2 - y^2;
x[u_,v_] = u Cos[v]; y[u_,v_] = u Sin[v];
gradx[u_,v_] = {D[x[u,v],u],D[x[u,v],v]};
grady[u_,v_] = {D[y[u,v],u],D[y[u,v],v]};
Axy[u_,v_] = Expand[Det[{gradx[u,v],grady[u,v]}],Trig->True]
```
Out[52]=
u

The total mass is:

In[53]:=
```
Integrate[2 f[x[u,v],y[u,v]] Axy[u,v],{v,0,2 Pi},{u,0,3}]
```
Out[53]=
81 Pi

The total volume of this object is:

In[54]:=
```
Integrate[f[x[u,v],y[u,v]] Axy[u,v],{v,0,2 Pi},{u,0,3}]
```
Out[54]=
$$\frac{81 \text{ Pi}}{2}$$

Routine stuff.

T.4.b.ii) When you rubberized the *xy*-paper and stretched it out to make the *uv*-paper, you also deformed the original shape of the object in part T.4.b.i). The deformation affected measurements on the base but had no effect on the height measurements.

> What does the resulting deformed object look like when it is plotted on *uv*-paper for the functions $x = x[u,v]$ and $y = y[u,v]$ you used above in part T.4.b.i)?
>
> What is the density of this deformed object at a point $\{u,v\}$ inside the *uv*-paper rectangle $0 \leq u \leq 3$ and $0 \leq v \leq 2\pi$?

Answer: Here is how it looks on *uv*-paper:

In[55]:=
```
Plot3D[f[x[u,v],y[u,v]],
  {v,0,2 Pi},{u,0,3},
  PlotRange->{0,9},ViewPoint->CMView,
  AxesLabel->{"x","y","z"}];
```

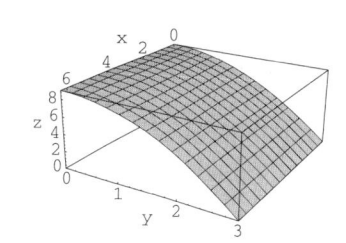

Quite a change of shape.

You can figure out what the deformed object's density is at a point $\{u,v\}$ within the uv-paper rectangle $0 \leq u \leq 3$ and $0 \leq v \leq 2\pi$. It is just what you integrate to calculate its mass:

In[56]:=
```
Expand[2 f[x[u,v],y[u,v]] Axy[u,v],Trig->True]
```

Out[56]=
$$18\,u - 2\,u^3$$

You've got to multiply by the area conversion factor because
$$2\,f[x[u,v], y[u,v]]\,A_{(x,y)}[u,v]$$
is what you integrate to calculate mass.

Give It a Try

Experience with the starred (★) problems will be useful for understanding developments later in the course.

■ G.1) Transforming 2D integrals★

G.1.a) Here's a plot of the part of the surface
$$z = e^{-(x^2 + 4y^2)}$$
above the ellipse
$$\left(\frac{x}{2}\right)^2 + y^2 = 1$$
in the xy-plane:

In[1]:=
```
Clear[x,y,r,t]
{x[r_,t_],y[r_,t_]} = {2 r Cos[t],r Sin[t]};
{{rlow,rhigh},{tlow,thigh}} = {{0,1},{0,2 Pi}};
ParametricPlot3D[
 {x[r,t],y[r,t],E^(-(x[r,t]^2 + 4 y[r,t]^2))},
 {r,rlow,rhigh},{t,tlow,thigh},
 AxesLabel->{"x","y","z"},ViewPoint->CMView];
```

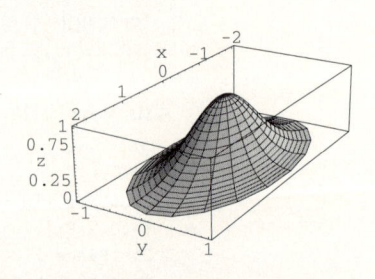

Measure the volume of the solid whose top skin is the surface plotted above and whose base is everything on the xy-plane directly below this surface.

G.1.b) Here's a plot of a region R_{xy} on xy-paper:

```
In[2]:=
Clear[x,y,r,t]
{x[r_,t_],y[r_,t_]} = {5 r Cos[t],3 r Sin[t]};
{rlow,rhigh} = {1,2}; {tlow,thigh} = {-2,2};
twosides = ParametricPlot[{{x[rlow,t],y[rlow,t]},
{x[rhigh,t],y[rhigh,t]}},{t,tlow,thigh},
PlotStyle->{{Thickness[0.01],Red}},
DisplayFunction->Identity];
twomoresides = ParametricPlot[{{x[r,tlow],y[r,tlow]},
{x[r,thigh],y[r,thigh]}},{r,rlow,rhigh},
PlotStyle->{{Thickness[0.01],Red}},
DisplayFunction->Identity];
Show[twosides,twomoresides,AspectRatio->Automatic,
AxesLabel->{"x","y"},DisplayFunction->$DisplayFunction];
```

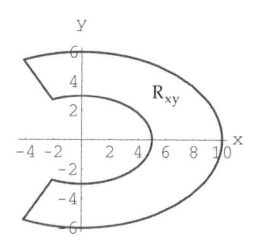

R_{xy} is everything inside and on this boundary.

> Use a transformation to favorable uv-paper to measure the area of R_{xy} with little thought and with *Mathematica* doing the work.

G.1.c.i) Here's a parallelogram plotted on xy-paper

```
In[3]:=
Plot[{0.5 x - 1,0.5 x + 2,-0.2 x,
-0.2 x + 4}, {x,-3.5,7.5},
PlotStyle->{{Magenta,Thickness[0.01]}},
PlotRange->{-0.5,3.5},AspectRatio->Automatic,
AxesLabel->{"x","y"}];
```

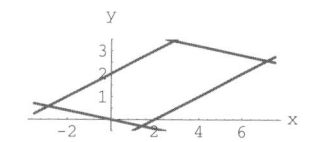

> Call R everything inside and on this parallelogram, and use a favorable transformation to help calculate
> $$\iint_R e^{y-x}\,dx\,dy.$$

G.1.c.ii) If $a \neq b$, $c < d$, and $r < s$ then you are guaranteed that the lines

$$y = a\,x + c,$$
$$y = a\,x + d,$$
$$y = b\,x + r, \text{ and}$$
$$y = b\,x + s$$

define a parallelogram on xy-paper.

Reason: Nonparallel lines cross each other.

> Assume $b < a$, $c < d$, and $r < s$, and come up with a formula that measures the area of this parallelogram in terms of a, b, c, d, r, and s.

G.1.d) Use a transformation to favorable uv-paper to calculate

$$\iint_{R_{xy}} (x+y)\, dx\, dy$$

where R_{xy} is the region with $x \geq 0$ and $y \geq 0$ bounded by the curves

$x^2 - y^2 = 1,$
$x^2 - y^2 = 4,$
$x^2 + y^2 = 4,$ and
$x^2 + y^2 = 9.$

G.1.e) Calculate

$$\iint_{R_{xy}} e^{-x^2-y^2}\, dx\, dy$$

where R_{xy} is the region on xy-paper consisting of everything within and on the circle

$x^2 + y^2 = 2.$

■ G.2) Ribbons*

G.2.a) Here's a curve on xy-paper:

```
In[4]:=
  Clear[x,y,r,t]
  {x[t_],y[t_]} = {2 t Cos[t],2 t Sin[t]};
  {tlow,thigh} = {Pi,3 Pi}; curveplot =
  ParametricPlot[{x[t],y[t]},{t,tlow,thigh},
  PlotStyle->{{Thickness[0.01],Red}},
  AspectRatio->Automatic,PlotRange->All,
  AxesLabel->{"x","y"}];
```

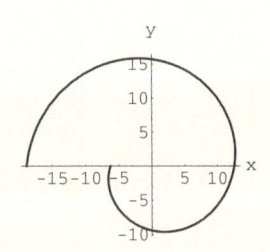

Here's the boundary of a ribbon of constant width 1 centered on this curve:

In[5]:=
```
Clear[unitnormal]
unitnormal[t_] = Expand[
  {y'[t],-x'[t]}/Sqrt[x'[t]^2 + y'[t]^2],Trig->True];
Clear[r,outerribbon,innerribbon,oneend,otherend]
outerribbon[t_] = {x[t],y[t]} + 0.5 unitnormal[t];
innerribbon[t_] = {x[t],y[t]} - 0.5 unitnormal[t];
oneend[r_] = {x[tlow],y[tlow]} + r unitnormal[tlow];
otherend[r_] = {x[thigh],y[thigh]} + r unitnormal[thigh];
{rlow = -0.5,rhigh = 0.5}; twosides =
ParametricPlot[{outerribbon[t],innerribbon[t]},{t,tlow,thigh},
PlotStyle->{{Thickness[0.01],Blue}},
DisplayFunction->Identity]; twomoresides =
ParametricPlot[{oneend[r],otherend[r]},{r,rlow,rhigh},
PlotStyle->{{Thickness[0.01],Blue}},
DisplayFunction->Identity]; Show[curveplot,twosides,
twomoresides,AspectRatio->Automatic,AxesLabel->{"x","y"},
DisplayFunction->$DisplayFunction];
```

> Change this ribbon to a new ribbon with constant width 2 centered on the given curve, and measure the area of the new ribbon.

G.2.b) Go with the same base curve as in part G.2.a).

> But this time make the width of the ribbon
> $$2 + 4\sin[2\,t]^2$$
> at $\{x[t], y[t]\}$.
> Plot the resulting ribbon and measure its area.

G.2.c) Do something artistic with ribbons. How about a real eye-catcher coming from your own mind?

■ G.3) Flow measurements*

G.3.a) To calculate the net flow of a vector field

$$\text{Field}[x, y] = \{m[x, y], n[x, y]\}$$

across the boundary C of a region R, you have your choice:

→ You can go to the labor of parameterizing C and then calculate

$$\oint_C -n[x, y]\, dx + m[x, y]\, dy.$$

→ Or, if the field has no singularities inside R, you can put
$$\text{divField}[x,y] = D[m[x,y],x] + D[n[x,y],y]$$
and calculate the 2D integral
$$\iint_R \text{divField}[x,y]\,dx\,dy.$$

Here's a vector field:

In[6]:=
```
Clear[x,y,m,n,Field]; {m[x_,y_],n[x_,y_]} = {x^3 + y ,x + y^2}
```

Out[6]=
$$\{x^3 + y,\ x + y^2\}$$

R is everything inside the parallelogram you see below:

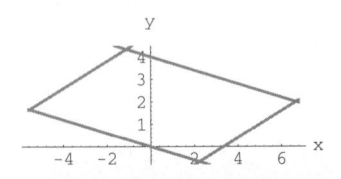

In[7]:=
```
Plot[{0.6 x - 2,0.6 x + 5,-0.3 x,
  -0.3 x + 4}, {x,-5.6,6.8},
  PlotStyle->{{Coral,Thickness[0.01]}},
  PlotRange->{-0.8,4.5},AspectRatio->Automatic,
  AxesLabel->{"x","y"}];
```

> Transform the 2D integral
> $$\iint_R \text{divField}[x,y]\,dx\,dy$$
> to favorable uv-paper to measure the net flow of this vector field across the parallelogram.
>
> Is the net flow of this vector field across this parallelogram from outside to inside or inside to outside?

G.3.b) To calculate the net flow of a vector field
$$\text{Field}[x,y] = \{m[x,y],n[x,y]\}$$
along the boundary C of a region R, you have your choice:

→ You can go to the labor of parameterizing C and then calculate
$$\oint_C m[x,y]\,dx + n[x,y]\,dy.$$

→ Or, if the field has no singularities inside R, you can put
$$\text{rotField}[x,y] = D[n[x,y],x] - D[m[x,y],y]$$
and calculate the 2D integral
$$\iint_R \text{rotField}[x,y]\,dx\,dy.$$

Here's a vector field:

In[8]:=
```
Clear[x,y,m,n,Field]
{m[x_,y_],n[x_,y_]} = {2 Sin[y] + x, x + y};
```

R is everything inside the parallelogram plotted in part G.3.a).

> Transform the 2D integral
> $$\iint_R \text{rotField}[x, y]\, dx\, dy$$
> to favorable uv-paper to measure the net flow of this vector field along the parallelogram. Is the net flow of this vector field along the parallelogram clockwise or counterclockwise?

■ G.4) Interpret the plots

G.4.a) All the plots below give information concerning the same phenomenon about what happens when you plot the region within the uv-paper square with corners at $\{1/2, -2\}$, $\{2, -2\}$, $\{2, 2\}$, and $\{1/2, 2\}$ on xy-paper coming from

$$x[u, v] = \log[u] \quad \text{and} \quad y[u, v] = \arctan[v].$$

> Interpret the information conveyed by each plot.

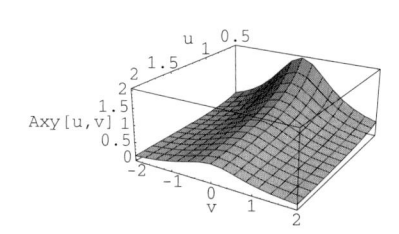

In[9]:=
```
Clear[x,y,u,v,gradx,grady,Axy]
x[u_,v_] = Log[u]; y[u_,v_] = ArcTan[v];
gradx[u_,v_] = {D[x[u,v],u],D[x[u,v],v]};
grady[u_,v_] = {D[y[u,v],u],D[y[u,v],v]};
Axy[u_,v_] = Det[{gradx[u,v],grady[u,v]}]
```

Out[9]=
$$\frac{1}{u\,(1 + v^2)}$$

In[10]:=
```
Plot3D[Axy[u,v],{u,1/2,2},{v,-2,2},
AxesLabel->{"u","v","Axy[u,v]"},
ViewPoint->CMView];
```

308 Lesson 3.07 *Transforming 2D Integrals*

In[11]:=
```
Clear[gradAxy]
gradAxy[u_,v_] = {D[Axy[u,v],u],D[Axy[u,v],v]};
scalefactor = 0.3;
Show[Table[Arrow[scalefactor gradAxy[u,v],
Tail->{u,v},Red],{u,1/2,2,1/4},{v,-2,2,1/4}],
Axes->Automatic];
```

In[12]:=
```
Clear[sizer]; scalefactor = 0.08;
sizer[u_,v_] = scalefactor Sqrt[Axy[u,v]];
Clear[k]; uvpoints = Table[
{Random[Real,{1/2,2}],Random[Real,{-2,2}]},{k,1,150}];
Show[Table[Graphics[{PointSize[Apply[sizer,uvpoints[[k]]]],
Red, Point[uvpoints[[k]]]}],{k,1, 150}],
AspectRatio->1,Axes->Automatic,AxesLabel->{"u","v"}];
```

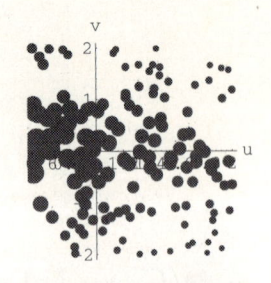

In[13]:=
```
ContourPlot[Axy[u,v],{u,1/2,2},{v,-2,2}];
```

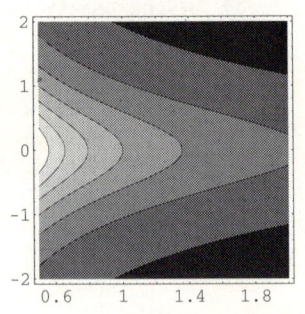

■ G.5) Semi-log paper and log-log paper

Semi-log and log paper are friends of every scientist because they make analysis of exponential and power functions very easy. Back at the beginning of Calculus&*Mathematica*, you used semi-log paper to some advantage.

Here are the lines

$y = e^{-2}$,

$y = e^{-1}$,

$y = e^0$,

$y = e$,

$y = e^2$, and

$y = e^3$

plotted on xy-paper for $-5 \leq x \leq 15$:

In[14]:=
```
Clear[t]
xylines = ParametricPlot[{{t,E^(-2)},
{t,E^(-1)},{t,E^(0)},{t,E^(1)},{t,E^(2)},
{t,E^(3)}},{t,-5,15},PlotStyle->{{Blue}},
AxesLabel->{"x","y"},AspectRatio->Automatic];
```

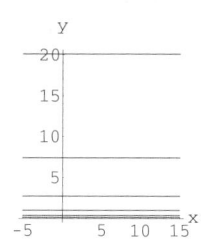

Semi-log paper is uv-paper for

$$u[x,y] = x \quad \text{and} \quad v[x,y] = \log[y].$$

Here are the same lines plotted on semi-log paper:

In[15]:=
```
Clear[u,v,x,y]
u[x_,y_] = x; v[x_,y_] = Log[y];
uvlines = ParametricPlot[{{u[t,E^(-2)],v[t,E^(-2)]},
{u[t,E^(-1)],v[t,E^(-1)]},{u[t,E^(0)],v[t,E^(0)]},
{u[t,E^(1)],v[t,E^(1)]},{u[t,E^(2)],v[t,E^(2)]},
{u[t,E^(3)],v[t,E^(3)]}},{t,-5,15},PlotStyle->{Blue},
AxesLabel->{"u","v"},AspectRatio->Automatic];
```

G.5.a) Plot the xy-curves $y = 3\,e^{-0.75x}$ and $y = 2\,e^{1.5x}$ on semi-log paper. Describe what you see, and try to explain why you see it.

Why is it a good idea to plot xy-data on semi-log paper to reveal exponential relationships between the x-coordinate and the y-coordinate?

G.5.b) Log-log paper is uv-paper for

$$u[x,y] = \log[x] \quad \text{and} \quad v[x,y] = \log[y].$$

Plot the power curves $y = 4\,x^{-0.6}$ and $y = 5\,x^{1.4}$ on xy-paper and then plot them on log-log paper. Why is the result not terribly surprising?

Why is it a good idea to plot xy-data on log-log paper to reveal power relationships between the x-coordinate and the y-coordinate?

■ **G.6)** What can happen when $A_{(x,y)}[u,v]$ is 0: What does the sign of $\text{Det}[\{\text{gradx}[u,v], \text{grady}[u,v]\}]$ tell you?

This problem appears only in the electronic version.

■ G.7) Volume, mass, and density

G.7.a.i) An object is made by forming a uniform substance that has density 1.71 grams/unit^3 in the shape of the surface
$$f[x,y] = \frac{3.14}{1 + x^2 + y^2}$$
over the region inside and on the circle
$$x^2 + y^2 = 8$$
on xy-paper.

> Use polar coordinate paper with
> $$x[u,v] = u\cos[v] \quad \text{and} \quad y[u,v] = u\sin[v]$$
> to measure the total mass and volume of this object.

G.7.a.ii) When you rubberized the xy-paper and stretched it out to make the uv-paper, you also deformed the original shape of the object in part G.7.a.i).

> What does the resulting deformed object look like when it is plotted on uv-paper for the functions
> $$x[u,v] = u\cos[v] \quad \text{and} \quad y[u,v] = u\sin[v]$$
> used above in part G.7.a.i)?
>
> What is the density of this deformed object at a point $\{u,v\}$ inside the uv-paper rectangle $0 \leq u \leq \sqrt{8}$ and $0 \leq v \leq 2\pi$?

■ G.8) Two recreational plots

This problem appears only in the electronic version.

■ G.9) What went wrong?*

G.9.a) Let R_{xy} be the region on xy-paper consisting of everything inside and on the circle
$$x^2 + y^2 = 4.$$
Polar cordinate paper is handy for calculating
$$\iint_{R_{xy}} x^2 + y^2 \, dx \, dy :$$

Go to polar coordinates via:

In[16]:=
```
Clear[x,y,u,v,gradx,grady,Axy]
x[u_,v_] = u Cos[v]; y[u_,v_] = u Sin[v];
gradx[u_,v_] = {D[x[u,v],u],D[x[u,v],v]};
grady[u_,v_] = {D[y[u,v],u],D[y[u,v],v]};
Axy[u_,v_] = Expand[Det[{gradx[u,v],grady[u,v]}],Trig->True]
```
Out[16]=
u

In[17]:=
```
Clear[f]; f[x_,y_] = x^2 + y^2
```
Out[17]=
$x^2 + y^2$

Here are six attempts at calculations of
$$\iint_{R_{xy}} x^2 + y^2 \, dx \, dy.$$
Identify the correct calculations and determine what went wrong in the incorrect calculations. There is always a possibility that a calculation produces a correct answer, but the method is wrong. Identify these as well.

Calculation 1: The uv-paper rectangle

$$0 \leq u \leq 4 \quad \text{and} \quad 0 \leq v \leq 2\pi$$

plots out on xy-paper as the circle $x^2 + y^2 = 4$ and everything inside it. So

$$\iint_{R_{xy}} x^2 + y^2 \, dx \, dy$$

is given by:

In[18]:=
```
Integrate[f[x[u,v],y[u,v]] Axy[u,v],{v,0,2 Pi},{u,0,4}]
```
Out[18]=
128 Pi

Calculation 2: The uv-paper rectangle

$$0 \leq u \leq 2 \quad \text{and} \quad 0 \leq v \leq 2\pi$$

plots out on xy-paper as the circle $x^2 + y^2 = 4$ and everything inside it. So

$$\iint_{R_{xy}} x^2 + y^2 \, dx \, dy$$

is given by:

In[19]:=
```
Integrate[f[x[u,v],y[u,v]] Axy[u,v],{v,0,2 Pi},{u,0,2}]
```

Out[19]=
8 Pi

Calculation 3: The uv-paper rectangle
$$0 \leq u \leq 2 \quad \text{and} \quad 0 \leq v \leq 4\pi$$
plots out on xy-paper as the circle $x^2 + y^2 = 4$ and everything inside it. So
$$\iint_{R_{xy}} x^2 + y^2 \, dx \, dy$$
is given by:

In[20]:=
  Integrate[f[x[u,v],y[u,v]] Axy[u,v],{v,0,4 Pi},{u,0,2}]

Out[20]=
16 Pi

Calculation 4: The uv-paper rectangle
$$-2 \leq u \leq 0 \quad \text{and} \quad 0 \leq v \leq 2\pi$$
plots out on xy-paper as the circle $x^2 + y^2 = 4$ and everything inside it. So
$$\iint_{R_{xy}} x^2 + y^2 \, dx \, dy$$
is given by:

In[21]:=
  Integrate[f[x[u,v],y[u,v]] Axy[u,v],{u,-2,0},{v,0,2 Pi}]

Out[21]=
-8 Pi

Calculation 5: The uv-paper rectangle
$$0 \leq u \leq 2 \quad \text{and} \quad -\pi \leq v \leq \pi$$
plots out on xy-paper as the circle $x^2 + y^2 = 4$ and everything inside it. So
$$\iint_{R_{xy}} x^2 + y^2 \, dx \, dy$$
is given by:

In[22]:=
  Integrate[f[x[u,v],y[u,v]] Axy[u,v],{u,0,2},{v,-Pi,Pi}]

Out[22]=
8 Pi

Calculation 6: The uv-paper rectangle
$$0 \leq u \leq 2 \quad \text{and} \quad 10\pi \leq v \leq 12\pi$$
plots out on xy-paper as the circle $x^2 + y^2 = 4$ and everything inside it. Therefore
$$\iint_{R_{xy}} x^2 + y^2 \, dx \, dy$$

is given by:

In[23]:=
```
Integrate[f[x[u,v],y[u,v]] Axy[u,v],{u,0,2},{v,10 Pi,12 Pi}]
```

Out[23]=
```
8 Pi
```

■ G.10) Linear equations and area conversion factors

This problem appears only in the electronic version.

■ G.11) Eigenvalues and eigenvectors

This problem appears only in the electronic version.

LESSON 3.08

Transforming 3D Integrals

Basics

■ B.1) 3D integrals

B.1.a.i) Here's a plot of the solid R whose top skin is the surface

$$z = 4 - x^2 - \frac{y^2}{4}$$

and whose bottom skin lies on the plane $z = 1$.

In[1]:=
```
topskin = Plot3D[4 - x^2 - (y^2)/4,
  {x,-1.8,1.8},{y,-3.5,3.5},PlotRange->{0.9,5},
  ClipFill->None,DisplayFunction->Identity];
base = Graphics3D[
  Polygon[{{-2,-4,1},{2,-4,1},{2,4,1},{-2,4,1}}]];
Show[topskin,base,Boxed->False,
  BoxRatios->Automatic,AxesLabel->{"x","y","z"},
  ViewPoint->CMView,DisplayFunction->$DisplayFunction];
```

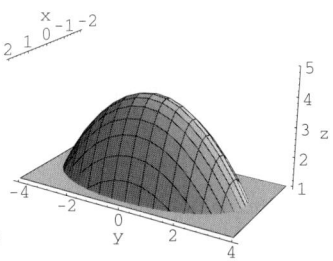

> You could use a 2D integral to measure the volume of this solid, but just to get experience, use a 3D integral to measure the volume of this solid.

Answer: Just as $\iint_R dx\, dy$ measures the area of a region R in two dimensions,

$$\iiint_R dx\, dy\, dz$$

315

Lesson 3.08 Transforming 3D Integrals

measures the volume of a solid R in three dimensions. You are free to choose any order you like for the variables so that

$$\iiint_R dx\, dy\, dz = \iiint_R dz\, dy\, dx = \iiint_R dz\, dx\, dy.$$

In the current situation, the top is $z = 4 - x^2 - y^2/4$ and the bottom is $z = 1$. With z so conveniently displayed, it's natural to integrate with respect to z first. To set the limits for the first integral, fix x and y and enter the lowest and the highest values of z for this fixed x and y:

$$\int_{zlow[x,y]}^{zhigh[x,y]} dz.$$

In[2]:=
```
Clear[zhigh,zlow,x,y,z]
{zlow[x_,y_],zhigh[x_,y_]} = {1,4 - x^2 - (y^2)/4};
firstintegral = Integrate[1,{z,zlow[x,y],zhigh[x,y]}]
```

Out[2]=
$$3 - x^2 - \frac{y^2}{4}$$

For the second integral, you've got your choice of whether to integrate with respect to y or x next. Go with y. The main problem is deciding what limits to insert for yhigh[x] and ylow[x] in the second integral

$$\int_{ylow[x]}^{yhigh[x]} \int_{zlow[x,y]}^{zhigh[x,y]} dz\, dy.$$

The key to deciding how to set yhigh[x] and ylow[x] is to ask: For a fixed x, what are the lowest and the highest values y can have no matter what z is? The simplest way of answering this is to plot the shadow of this solid on xy-paper:

In[3]:=
```
shadoweqn = Eliminate[{z == zhigh[x,y], z == zlow[x,y]},z]
```

Out[3]=
$$4 x^2 == 12 - y^2$$

In[4]:=
```
shadowcurves = Solve[shadoweqn,y]
```

Out[4]=
```
{{y -> Sqrt[12 - 4 x^2]}, {y -> -Sqrt[12 - 4 x^2]}}
```

In[5]:=
```
Clear[ylow,yhigh]
ylow[x_] = -Sqrt[12 - 4 x^2];
yhigh[x_] = Sqrt[12 - 4 x^2];
shadowplot = Plot[{ylow[x],yhigh[x]},
{x,-Sqrt[3],Sqrt[3]},
PlotStyle->{Blue,Red},
AxesLabel->{"x","y"}];
```

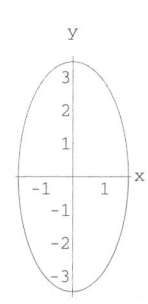

The second integral

$$\int_{\text{ylow}[x]}^{\text{yhigh}[x]} \int_{\text{zlow}[x,y]}^{\text{zhigh}[x,y]} dz\, dy$$

is:

In[6]:=
```
secondintegral = Integrate[firstintegral,{y,ylow[x],yhigh[x]}]
```

Out[6]=

$$-\frac{(12 - 4x^2)^{3/2}}{6} + 2\,\text{Sqrt}[12 - 4x^2]\,(3 - x^2)$$

For the third integral, you have no choice but to integrate with respect to x. The main problem is deciding what limits to insert for xhigh and xlow in the third integral

$$\int_{\text{xlow}}^{\text{xhigh}} \int_{\text{ylow}[x]}^{\text{yhigh}[x]} \int_{\text{zlow}[x,y]}^{\text{zhigh}[x,y]} dz\, dy\, dx.$$

Note that xlow and xhigh are numbers and are not functions. The key to deciding how to set xlow and xhigh is to ask: What are the lowest and the highest values x can have no matter what z and y are? The simplest way of getting a hold on xlow and xhigh is to look at the shadow plot again:

In[7]:=
```
Show[shadowplot];
```

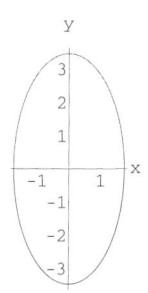

In this problem, xlow and xhigh happen where ylow[x] and yhigh[x] meet:

In[8]:=
```
ends = Solve[yhigh[x] == ylow[x]]
```

Lesson 3.08 Transforming 3D Integrals

Out[8]=
`{{x -> Sqrt[3]}, {x -> -Sqrt[3]}}`

In[9]:=
`{xlow,xhigh} = {-Sqrt[3],Sqrt[3]};`

The volume is measured by the third integral

$$\int_{xlow}^{xhigh} \int_{ylow[x]}^{yhigh[x]} \int_{zlow[x,y]}^{zhigh[x,y]} dz\, dy\, dx :$$

In[10]:=
`volume = Integrate[secondintegral,{x,xlow,xhigh}]`

Out[10]=
`9 Pi`

You can get this with a single 3D integral instruction for the volume measurement

$$\int_{xlow}^{xhigh} \int_{ylow[x]}^{yhigh[x]} \int_{zlow[x,y]}^{zhigh[x,y]} dz\, dy\, dx :$$

In[11]:=
```
threeDintegral =
Integrate[1,{x,xlow,xhigh},{y,ylow[x],yhigh[x]},{z,zlow[x,y],zhigh[x,y]}]
```

Out[11]=
`9 Pi`

Good. However, doing this in one step does not relieve you of the experience of setting zhigh[x, y], zlow[x, y], yhigh[x], ylow[x], xhigh and xlow.

Tough break.

B.1.a.ii) Discuss the physical meaning of the three integrals that were calculated in part B.1.a.i):

$$\text{first integral} = \int_{zlow[x,y]}^{zhigh[x,y]} dz,$$

$$\text{second integral} = \int_{ylow[x]}^{yhigh[x]} \int_{zlow[x,y]}^{zhigh[x,y]} dz\, dy$$

and

$$\text{third integral} = \int_{xlow}^{xhigh} \int_{ylow[x]}^{yhigh[x]} \int_{zlow[x,y]}^{zhigh[x,y]} dz\, dy\, dx.$$

Answer: The first integral,

$$\int_{zlow[x,y]}^{zhigh[x,y]} dz,$$

fixes x and y and measures the length of the stick that runs from $\{x, y, \text{zlow}[x, y]\}$ to $\{x, y, \text{zhigh}[x, y]\}$. For example:

Feel free to reset $\{x_0, y_0\}$ and rerun.

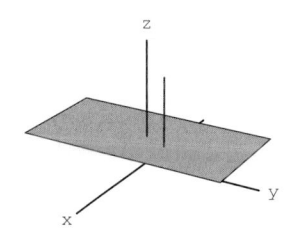

In[12]:=
```
{x0,y0} = {0.5,1};
stickinslice = Graphics3D[{DarkGreen,
Line[{{x0,y0,zlow[x0,y0]},{x0,y0,zhigh[x0,y0]}}]}];
threedims = ThreeAxes[4.5];
Show[stickinslice,base,threedims,
BoxRatios->Automatic,Boxed->False,
ViewPoint->CMView];
```

The second integral

$$\int_{\text{ylow}[x]}^{\text{yhigh}[x]} \int_{\text{zlow}[x,y]}^{\text{zhigh}[x,y]} dz\, dy$$

holds $x = x_0$ fixed but releases y and measures the accumulated area of the slice of the solid swept out by the sticks as y runs from ylow[x] to yhigh[x]:

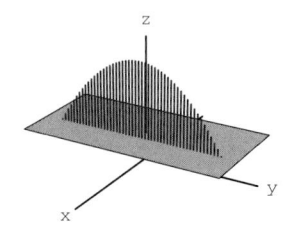

In[13]:=
```
sticksinslice =
Table[Graphics3D[{DarkGreen,
Line[{{x0,y,zlow[x0,y]},
{x0,y,zhigh[x0,y]}}]}],
{y,ylow[x0],yhigh[x0],
(yhigh[x0] - ylow[x0])/50}];
Show[sticksinslice,base,threedims,
Boxed->False,ViewPoint->CMView];
```

The third integral

$$\int_{\text{xlow}}^{\text{xhigh}} \int_{\text{ylow}[x]}^{\text{yhigh}[x]} \int_{\text{zlow}[x,y]}^{\text{zhigh}[x,y]} dz\, dy\, dx$$

releases x and measures the accumulated volume swept out by the slices as x runs from xlow to xhigh:

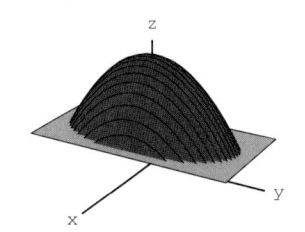

In[14]:=
```
constituentslices =
Table[Graphics3D[{DarkGreen,
Polygon[Table[{x,y,zhigh[x,y]},
{y,ylow[x] - 0.01,yhigh[x] + 0.01,
(yhigh[x] - ylow[x] + 0.02)/20}]]}],
{x,xlow,xhigh,(xhigh - xlow)/20}];
Show[constituentslices,base,threedims,
Boxed->False,ViewPoint->CMView];
```

The third integral measures the total volume.

B.1.a.iii) Calculate $\iiint_R z\, dx\, dy\, dz$ where R is the solid plotted in part B.1.a.i).

Answer: The layout is essentially the same as the calculation of the volume of R done in part B.1.a.i) above.

In[15]:=
```
Clear[f]; f[x_,y_,z_] = z;
firstintegral = Integrate[f[x,y,z],{z,zlow[x,y],zhigh[x,y]}]
```
Out[15]=
$$-\left(\frac{1}{2}\right) + \frac{\left(4 - x^2 - \frac{y^2}{4}\right)^2}{2}$$

In[16]:=
```
secondintegral = Integrate[firstintegral,{y,ylow[x],yhigh[x]}]
```
Out[16]=
$$\frac{(12 - 4x^2)^{5/2}}{80} + \frac{(12 - 4x^2)^{3/2}(-4 + x^2)}{6} + \text{Sqrt}[12 - 4x^2]\,(15 - 8x^2 + x^4)$$

In[17]:=
```
thirdintegral = Integrate[secondintegral,{x,xlow,xhigh}]
```
Out[17]=
18 Pi

You can get this with one instruction for calculating

$$\iiint_R f[x,y,z]\, dx\, dy\, dz = \int_{\text{xlow}}^{\text{xhigh}} \int_{\text{ylow}[x]}^{\text{yhigh}[x]} \int_{\text{zlow}[x,y]}^{\text{zhigh}[x,y]} f[x,y,z]\, dz\, dy\, dx :$$

In[18]:=
```
onestep = Integrate[f[x,y,z],
   {x,xlow,xhigh},{y,ylow[x],yhigh[x]},{z,zlow[x,y],zhigh[x,y]}]
```
Out[18]=
18 Pi

Good. Doing this in one step does not relieve you of the burden of setting zhigh$[x,y]$, zlow$[x,y]$, yhigh$[x]$, ylow$[x]$, xhigh and xlow. You set them up for calculating $\iiint_R f[x,y,z]\, dx\, dy\, dz$ the same way you set them up for calculating the volume of R by calculating $\iiint_R dx\, dy\, dz$ in part B.1.a.i).

■ B.2) Transforming 3D integrals

B.2.a) In two dimensions, you know that

$$\iint_{R_{xy}} f[x,y]\, dx\, dy = \iint_{R_{uv}} f[x[u,v], y[u,v]] |A_{(x,y)}[u,v]|\, du\, dv$$

where the area conversion factor $A_{(x,y)}[u,v]$ is given by:

In[19]:=
```
Clear[x,y,u,v,gradx,grady,Axy]
gradx[u_,v_] = {D[x[u,v],u],D[x[u,v],v]};
grady[u_,v_] = {D[y[u,v],u],D[y[u,v],v]};
Axy[u_,v_] = Det[{gradx[u,v],grady[u,v]}]
```

Out[19]=
$$y^{(0,1)}[u,v]\, x^{(1,0)}[u,v] - x^{(0,1)}[u,v]\, y^{(1,0)}[u,v]$$

In three dimensions, what is the volume conversion factor $V_{(x,y,z)}[u,v,w]$ that makes

$$\iiint_{R_{xyz}} f[x,y,z]\, dx\, dy\, dz$$
$$= \iiint_{R_{uvw}} f[x[u,v,w], y[u,v,w], z[u,v,w]]\, V_{(x,y,z)}[u,v,w]\, du\, dv\, dw\,?$$

Answer: In mathematics, there is nothing like the power of analogy, and analogy works beautifully here. In two dimensions the area conversion factor $A_{(x,y)}[u,v]$ is given by:

In[20]:=
```
Clear[x,y,u,v,gradx,grady,Axy]
gradx[u_,v_] = {D[x[u,v],u],D[x[u,v],v]};
grady[u_,v_] = {D[y[u,v],u],D[y[u,v],v]};
Axy[u_,v_] = Abs[Det[{gradx[u,v],grady[u,v]}]]
```

Out[20]=
$$\mathrm{Abs}[y^{(0,1)}[u,v]\, x^{(1,0)}[u,v] - x^{(0,1)}[u,v]\, y^{(1,0)}[u,v]]$$

Analogously, in three dimensions the volume conversion factor $V_{(x,y,z)}[u,v,w]$ is given by:

In[21]:=
```
Clear[x,y,z,u,v,gradx,grady,gradz,Vxyz]
gradx[u_,v_,w_] = {D[x[u,v,w],u],D[x[u,v,w],v],D[x[u,v,w],w]};
grady[u_,v_,w_] = {D[y[u,v,w],u],D[y[u,v,w],v],D[y[u,v,w],w]};
gradz[u_,v_,w_] = {D[z[u,v,w],u],D[z[u,v,w],v],D[z[u,v,w],w]};
Vxyz[u_,v_,w_] = Abs[Det[{gradx[u,v,w],grady[u,v,w],gradz[u,v,w]}]]
```

Out[21]=
$$\mathrm{Abs}[z^{(0,0,1)}[u,v,w]\, y^{(0,1,0)}[u,v,w]\, x^{(1,0,0)}[u,v,w] -$$
$$y^{(0,0,1)}[u,v,w]\, z^{(0,1,0)}[u,v,w]\, x^{(1,0,0)}[u,v,w] -$$
$$z^{(0,0,1)}[u,v,w]\, x^{(0,1,0)}[u,v,w]\, y^{(1,0,0)}[u,v,w] +$$

$$x^{(0,0,1)}[u,v,w]\, z^{(0,1,0)}[u,v,w]\, y^{(1,0,0)}[u,v,w] +$$
$$y^{(0,0,1)}[u,v,w]\, x^{(0,1,0)}[u,v,w]\, z^{(1,0,0)}[u,v,w] -$$
$$x^{(0,0,1)}[u,v,w]\, y^{(0,1,0)}[u,v,w]\, z^{(1,0,0)}[u,v,w]]$$

You just arrange the three gradients as (horizontal) rows of a matrix which most folks call the Jacobian matrix.

In[22]:=
gradmatrix = {gradx[u,v,w],grady[u,v,w],gradz[u,v,w]}

Out[22]=
$$\{\{x^{(1,0,0)}[u,v,w],\, x^{(0,1,0)}[u,v,w],\, x^{(0,0,1)}[u,v,w]\},$$
$$\{y^{(1,0,0)}[u,v,w],\, y^{(0,1,0)}[u,v,w],\, y^{(0,0,1)}[u,v,w]\},$$
$$\{z^{(1,0,0)}[u,v,w],\, z^{(0,1,0)}[u,v,w],\, z^{(0,0,1)}[u,v,w]\}\}$$

Look at it in matrix form:

In[23]:=
MatrixForm[gradmatrix]

Out[23]=
$$\begin{pmatrix} x^{(1,0,0)}[u,v,w] & x^{(0,1,0)}[u,v,w] & x^{(0,0,1)}[u,v,w] \\ y^{(1,0,0)}[u,v,w] & y^{(0,1,0)}[u,v,w] & y^{(0,0,1)}[u,v,w] \\ z^{(1,0,0)}[u,v,w] & z^{(0,1,0)}[u,v,w] & z^{(0,0,1)}[u,v,w] \end{pmatrix}$$

Take the absolute value of its determinant:

In[24]:=
Abs[Det[gradmatrix]]

Out[24]=
$$\mathrm{Abs}[z^{(0,0,1)}[u,v,w]\, y^{(0,1,0)}[u,v,w]\, x^{(1,0,0)}[u,v,w] -$$
$$y^{(0,0,1)}[u,v,w]\, z^{(0,1,0)}[u,v,w]\, x^{(1,0,0)}[u,v,w] -$$
$$z^{(0,0,1)}[u,v,w]\, x^{(0,1,0)}[u,v,w]\, y^{(1,0,0)}[u,v,w] +$$
$$x^{(0,0,1)}[u,v,w]\, z^{(0,1,0)}[u,v,w]\, y^{(1,0,0)}[u,v,w] +$$
$$y^{(0,0,1)}[u,v,w]\, x^{(0,1,0)}[u,v,w]\, z^{(1,0,0)}[u,v,w] -$$
$$x^{(0,0,1)}[u,v,w]\, y^{(0,1,0)}[u,v,w]\, z^{(1,0,0)}[u,v,w]]$$

This is the same as the volume conversion factor $V_{(x,y,z)}[u,v,w]$.

Quite a pill to swallow, but who cares? The computer is the one that has to swallow it.

B.2.b.i) Check out the formula for the volume conversion factor

$$V_{(x,y,z)}[u,v,w] = \text{Det}[\{\text{gradx}[u,v,w],\text{grady}[u,v,w],\text{gradz}[u,v,w]\}]$$

by using it to relate the volume of the ellipsoid

$$\left(\frac{x}{a}\right)^2 + \left(\frac{y}{b}\right)^2 + \left(\frac{z}{c}\right)^2 = r^2$$

to the known volume $4\pi r^3/3$ of the sphere of radius r,

$$u^2 + v^2 + w^2 = r^2.$$

Answer:

In[25]:=
```
Clear[x,y,z,a,b,c]
ellipsoideqn = ((x/a)^2 + (y/b)^2 + (z/c)^2 == r^2)
```

Out[25]=
$$\frac{x^2}{a^2} + \frac{y^2}{b^2} + \frac{z^2}{c^2} == r^2$$

Go to *uvw*-space with:

In[26]:=
```
Clear[u,v,w]
{x[u_,v_,w_],y[u_,v_,w_],z[u_,v_,w_]} = {a u, b v, c w};
```

Here is the *uvw*-equation of the original ellipsoid:

In[27]:=
```
ellipsoideqn/.{x->x[u,v,w],y->y[u,v,w],z->z[u,v,w]}
```

Out[27]=
$$u^2 + v^2 + w^2 == r^2$$

This tells you that if R_{xyz} is everything inside and on the ellipsoid

$$\left(\frac{x}{a}\right)^2 + \left(\frac{y}{b}\right)^2 + \left(\frac{z}{c}\right)^2 = r^2,$$

then R_{uvw} is everything inside and on the sphere

$$u^2 + v^2 + w^2 = r^2.$$

The volume of the sphere R_{uvw} is $4\pi r^3/3$.

The *uvw*-to-*xyz*-volume conversion factor is:

In[28]:=
```
Clear[gradx,grady,gradz,Vxyz]
gradx[u_,v_,w_] = {D[x[u,v,w],u],D[x[u,v,w],v],D[x[u,v,w],w]};
grady[u_,v_,w_] = {D[y[u,v,w],u],D[y[u,v,w],v],D[y[u,v,w],w]};
gradz[u_,v_,w_] = {D[z[u,v,w],u],D[z[u,v,w],v],D[z[u,v,w],w]};
Vxyz[u_,v_,w_] = Det[{gradx[u,v,w],grady[u,v,w],gradz[u,v,w]}]
```

Out[28]=
a b c

So the xyz-volume measurement of R_{xyz} is

$a\,b\,c$ times the uvw-volume measurement of R_{uvw}.

In short, the volume measurement of the ellipsoid

$$\left(\frac{x}{a}\right)^2 + \left(\frac{y}{b}\right)^2 + \left(\frac{z}{c}\right)^2 = r^2$$

is

$$\frac{a\,b\,c\,4\,\pi\,r^3}{3}.$$

Makes sense when you think about it.

B.2.b.ii) Now get a bit more serious.

> Calculate
>
> $$\iiint_{R_{xyz}} e^{-x} \sin[y]\, dx\, dy\, dz$$
>
> where R_{xyz} is the region inside and on the box bounded by the planes
>
> $z = 2x - 1$ (bottom),
> $z = 2x + 3$ (top),
> $y = x$ (side),
> $y = x + 4$ (side),
> $y = -2x$ (side), and
> $y = -2x + 5$ (side).

Answer: A direct calculation would be a revolting task. Even the task of plotting R_{xyz} wouldn't be pleasant. But after you transform the integral to a favorable uvw-space, then everything is fairly easy.

Look at the formulas for the boundary surfaces:

$z = 2x - 1$ (bottom),
$z = 2x + 3$ (top),
$y = x$ (side),
$y = x + 4$ (side),
$y = -2x$ (side),
$y = -2x + 5$ (side)

and make the assignments:

In[29]:=
```
Clear[u,v,w,x,y,z]
{u[x_,y_,z_],v[x_,y_,z_],w[x_,y_,z_]} = {z - 2 x,y - x,y + 2 x}
```
Out[29]=
```
{-2 x + z, -x + y, 2 x + y}
```

Note:
$$z = 2x - 1 \longleftrightarrow u[x,y,z] = -1,$$
$$z = 2x + 3 \longleftrightarrow u[x,y,z] = 3,$$
$$y = x \longleftrightarrow v[x,y,z] = 0,$$
$$y = x + 4 \longleftrightarrow v[x,y,z] = 4,$$
$$y = -2x \longleftrightarrow w[x,y,z] = 0, \text{ and}$$
$$y = -2x + 5 \longleftrightarrow w[x,y,z] = 5.$$

In uvw-space, R_{uvw} is everything inside and on the box bounded by the planes $u = -1$, $u = 3$, $v = 0$, $v = 4$, $w = 0$, and $w = 5$. Also

$$\iiint_{R_{xyz}} e^{-x} \sin[y] \, dx \, dy \, dz$$
$$= \iiint_{R_{uvw}} e^{-x(u,v,w)} \sin[y[u,v,w]] \, V_{(x,y,z)}[u,v,w] \, du \, dv \, dw$$
$$= \int_0^5 \int_0^4 \int_{-1}^3 e^{-x(u,v,w)} \sin[y[u,v,w]] \, V_{(x,y,z)}[u,v,w] \, du \, dv \, dw.$$

To calculate $x[u,v,w]$, $y[u,v,w]$, $z[u,v,w]$, and $V_{(x,y,z)}[u,v,w]$, use:

In[30]:=
```
xyzsolved = Solve[{u == u[x,y,z],v == v[x,y,z], w == w[x,y,z]},{x,y,z}]
```
Out[30]=
$$\{\{z \to u - \frac{2(v - w)}{3}, y \to v + \frac{-v + w}{3}, x \to \frac{-v + w}{3}\}\}$$

In[31]:=
```
{x[u_,v_,w_],y[u_,v_,w_],z[u_,v_,w_]} = {x,y,z}/.xyzsolved[[1]]
```
Out[31]=
$$\{\frac{-v + w}{3}, v + \frac{-v + w}{3}, u - \frac{2(v - w)}{3}\}$$

Now turn everything over to *Mathematica*:

In[32]:=
```
Clear[gradx,grady,gradz,Vxyz]
gradx[u_,v_,w_] = {D[x[u,v,w],u],D[x[u,v,w],v],D[x[u,v,w],w]};
grady[u_,v_,w_] = {D[y[u,v,w],u],D[y[u,v,w],v],D[y[u,v,w],w]};
gradz[u_,v_,w_] = {D[z[u,v,w],u],D[z[u,v,w],v],D[z[u,v,w],w]};
Vxyz[u_,v_,w_] = Abs[Det[{gradx[u,v,w],grady[u,v,w],gradz[u,v,w]}]]
```

Lesson 3.08 Transforming 3D Integrals

Out[32]=
$$\frac{1}{3}$$

So

$$\iiint_{R_{xyz}} e^{-x} \sin[y] \, dx \, dy \, dz$$
$$= \iiint_{R_{uvw}} e^{-x[u,v,w]} \sin[y[u,v,w]] \, V_{[x,y,z]}[u,v,w] \, du \, dv \, dw$$
$$= \int_0^5 \int_0^4 \int_{-1}^3 e^{-x[u,v,w]} \sin[y[u,v,w]] \left(\frac{1}{3}\right) du \, dv \, dw.$$

The first integral

$$\int_{-1}^3 e^{-x[u,v,w]} \sin[y[u,v,w]] \left(\frac{1}{3}\right) du$$

is:

In[33]:=
```
firstintegral = Integrate[E^(-x[u,v,w]) Sin[y[u,v,w]] (1/3),{u,-1,3}]
```

Out[33]=
$$\frac{4 E^{v/3 - w/3} \sin[v + \frac{-v + w}{3}]}{3}$$

The second integral

$$\int_0^4 \int_{-1}^3 e^{-x[u,v,w]} \sin[y[u,v,w]] \left(\frac{1}{3}\right) du \, dv$$

is:

In[34]:=
```
secondintegral = Integrate[firstintegral,{v,0,4}]
```

Out[34]=
$$\frac{-4 \left(-2 \cos[\frac{w}{3}] + \sin[\frac{w}{3}]\right)}{5 E^{w/3}} + \frac{4 \left(-2 E^{4/3} \cos[\frac{8+w}{3}] + E^{4/3} \sin[\frac{8+w}{3}]\right)}{5 E^{w/3}}$$

The third and final integral

$$\int_0^5 \int_0^4 \int_{-1}^3 e^{-x[u,v,w]} \sin[y[u,v,w]] \left(\frac{1}{3}\right) du \, dv \, dw$$

is:

In[35]:=
```
thirdintegral = Integrate[secondintegral,{w,0,5}]
```

Out[35]=
$$\frac{-6\left(-1 + E^{4/3}\,\text{Cos}\left[\frac{8}{3}\right] - 3\,E^{4/3}\,\text{Sin}\left[\frac{8}{3}\right]\right)}{5} +$$
$$\frac{6\left(-\text{Cos}\left[\frac{5}{3}\right] + E^{4/3}\,\text{Cos}\left[\frac{13}{3}\right] + 3\,\text{Sin}\left[\frac{5}{3}\right] - 3\,E^{4/3}\,\text{Sin}\left[\frac{13}{3}\right]\right)}{5\,E^{5/3}}$$

Or, in decimals:

In[36]:=
N[thirdintegral]

Out[36]=
14.2704

You can get

$$\iiint_{R_{xyz}} e^{-x}\sin[y]\,dx\,dy\,dz = \int_0^5 \int_0^4 \int_{-1}^3 e^{-x[u,v,w]}\sin[y[u,v,w]]\left(\frac{1}{3}\right)du\,dv\,dw$$

in one step:

In[37]:=
quickway = Integrate[E^(-x[u,v,w]) Sin[y[u,v,w]] (1/3),{w,0,5},{v,0,4},{u,-1,3}]

Out[37]=
$$\frac{-6\left(-1 + E^{4/3}\,\text{Cos}\left[\frac{8}{3}\right] - 3\,E^{4/3}\,\text{Sin}\left[\frac{8}{3}\right]\right)}{5} +$$
$$\frac{6\left(-\text{Cos}\left[\frac{5}{3}\right] + E^{4/3}\,\text{Cos}\left[\frac{13}{3}\right] + 3\,\text{Sin}\left[\frac{5}{3}\right] - 3\,E^{4/3}\,\text{Sin}\left[\frac{13}{3}\right]\right)}{5\,E^{5/3}}$$

Or in decimals:

In[38]:=
N[quickway]

Out[38]=
14.2704

Went through that one like a hot knife going through butter.

B.2.b.iii) The solid R_{xyz} consists of everything under the surface

$$z = 0.2\,e^{-y}$$

and over the region inside the ellipse

$$x^2 + \left(\frac{y}{3}\right)^2 = 1$$

in the xy-plane. Here's a look:

328 Lesson 3.08 Transforming 3D Integrals

In[39]:=
```
Clear[x,y,z,r,t,s]
{x[r_,t_],y[r_,t_]} = {r Cos[t],3 r Sin[t]};
z[r_,t_] = 0.2 E^(-y[r,t]);
{{rlow,rhigh},{tlow,thigh}} = {{0,1},{0,2 Pi}};
topskin = ParametricPlot3D[{x[r,t],y[r,t],z[r,t]},
{r,rlow,rhigh},{t,tlow,thigh},DisplayFunction->Identity];
baseskin = ParametricPlot3D[{x[r,t],y[r,t],0},
{r,rlow,rhigh},{t,tlow,thigh},PlotPoints->{2,Automatic},
DisplayFunction->Identity];
sideskin = ParametricPlot3D[{x[rhigh,t],y[rhigh,t],
s z[rhigh,t]},{s,0,1},{t,tlow,thigh},
PlotPoints->{2,Automatic},DisplayFunction->Identity];
```

In[40]:=
```
Show[topskin,baseskin,sideskin,
BoxRatios->Automatic,PlotRange->All,
AxesLabel->{"x","y","z"},ViewPoint->CMView,
DisplayFunction->$DisplayFunction];
```

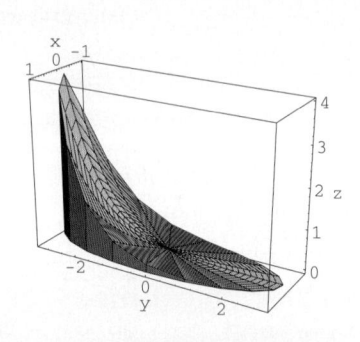

Calculate
$$\iiint_{R_{xyz}} e^{-y} \, dx \, dy \, dz$$
by transforming to an easy 3D integral.

Answer: Look at the plotting functions above:

In[41]:=
```
{x[r,t],y[r,t]}
```
Out[41]=
{r Cos[t], 3 r Sin[t]}

In[42]:=
```
z[r,t]
```
Out[42]=
$$\frac{0.2}{E^{3 r \, \text{Sin}[t]}}$$

Go to *rst*-space with:

In[43]:=
```
Clear[x,y,z,r,s,t]
{x[r_,s_,t_],y[r_,s_,t_],z[r_,s_,t_]} = {r Cos[t],3 r Sin[t],0.2 s E^(3 r Sin[t])}
```

Out[43]=
$$\{r\,\text{Cos}[t],\ 3\,r\,\text{Sin}[t],\ 0.2\,E^{3\,r\,\text{Sin}[t]}\,s\}$$

As you run r from rlow to rhigh, t from tlow to thigh and s from 0 to 1,
$$\{x[r,s,t],\ y[r,s,t],\ z[r,s,t]\}$$
sweeps out all of R_{xyz}; so R_{rst} is the box
$$\text{rlow} \le r \le \text{rhigh}, \quad 0 \le s \le 1, \quad \text{tlow} \le t \le \text{thigh},$$
and
$$\iiint_{R_{xyz}} e^{-y}\,dx\,dy\,dz = \iiint_{R_{rst}} e^{-y[r,s,t]}\,V_{(x,y,z)}[r,s,t]\,dr\,ds\,dt$$
$$= \int_{\text{tlow}}^{\text{thigh}} \int_0^1 \int_{\text{rlow}}^{\text{rhigh}} e^{-y[r,s,t]}\,V_{(x,y,z)}[r,s,t]\,dr\,ds\,dt.$$

In[44]:=
```
Clear[gradx,grady,gradz,Vxyz]
gradx[r_,s_,t_] = {D[x[r,s,t],r],D[x[r,s,t],s],D[x[r,s,t],t]};
grady[r_,s_,t_] = {D[y[r,s,t],r],D[y[r,s,t],s],D[y[r,s,t],t]};
gradz[r_,s_,t_] = {D[z[r,s,t],r],D[z[r,s,t],s],D[z[r,s,t],t]};
Vxyz[r_,s_,t_] = Det[{gradx[r,s,t],grady[r,s,t],gradz[r,s,t]}]
```

Out[44]=
$$-0.6\,E^{3\,r\,\text{Sin}[t]}\,r\,\text{Cos}[t]^2 - 0.6\,E^{3\,r\,\text{Sin}[t]}\,r\,\text{Sin}[t]^2$$

This is negative no matter what r, s, and t are in this box. Reset it:

In[45]:=
```
Clear[Vxyz]
Vxyz[r_,s_,t_] = -Det[{gradx[r,s,t],grady[r,s,t],gradz[r,s,t]}]
```

Out[45]=
$$0.6\,E^{3\,r\,\text{Sin}[t]}\,r\,\text{Cos}[t]^2 + 0.6\,E^{3\,r\,\text{Sin}[t]}\,r\,\text{Sin}[t]^2$$

So the integral
$$\iiint_{R_{rst}} e^{-y}\,dx\,dy\,dz = \iiint_{R_{rst}} e^{-y[r,s,t]}\,V_{(x,y,z)}[r,s,t]\,dr\,ds\,dt$$
$$= \int_{\text{rlow}}^{\text{rhigh}} \int_0^1 \int_{\text{tlow}}^{\text{thigh}} e^{-y[r,s,t]}\,V_{(x,y,z)}[r,s,t]\,dr\,ds\,dt$$
is given by:

In[46]:=
```
Integrate[E^(-y[r,s,t]) Vxyz[r,s,t],{r,rlow,rhigh},{s,0,1},{t,tlow,thigh}]
```

Out[46]=
0.6 Pi

Nice answer. But if you hadn't gone to *rst*-space to calculate the integral, getting this nice answer wouldn't have been nearly so simple.

■ B.3) Mass and density

This problem appears only in the electronic version.

■ B.4) Average value of a function

B.4.a.i) Given a function $f[x]$ and an interval $a \leq x \leq b$, folks like to say that the average value of $f[x]$ and this interval is given by

$$\frac{\int_a^b f[x]\,dx}{b-a}.$$

> Why do they like to say this?

Answer: They like to say this because it works.

To see it in action, go with $f[x] = \sin[x]$ for $a = 0 \leq x \leq \pi = b$:

In[47]:=
```
Clear[f,x,k]
f[x_] = Sin[x]; {a,b} = {0, Pi};
theoreticalaverage = N[Integrate[f[x],{x,a,b}]/(b - a)];
pointcount = 100;
experimentalaverage = Sum[N[f[Random[Real,{N[a],N[b]}]]],
{k,1,pointcount}]/pointcount;
{theoreticalaverage,experimentalaverage}
```

Out[47]=
{0.63662, 0.644709}

See what happens when you use more points:

In[48]:=
```
pointcount = 400;
experimentalaverage = Sum[N[f[Random[Real,{N[a],N[b]}]]],
{k,1,pointcount}]/pointcount;
{theoreticalaverage,experimentalaverage}
```

Out[48]=
{0.63662, 0.636959}

Uh-huh.

B.4.a.ii)
> Wait a minute! This is supposed to be a mathematics course.
>
> Where do you get off saying that something is correct because it works?

Answer: Good point. You can explain why it works another way by looking at plots like this:

In[49]:=
```
Clear[f,x]
f[x_] = Sin[x]; {a,b} = {0, Pi};
fplot = Plot[f[x],{x,a,b},
PlotStyle->{{Thickness[0.01],Red}},
DisplayFunction->Identity];
average = Integrate[f[x],{x,a,b}]/(b - a);
averagefplot = Graphics[{Blue,
Polygon[{{a,0},{b,0},{b,average},{a,average}}]}];
Show[averagefplot,fplot,AxesLabel->{"x","y"},
PlotRange->{0,1},AspectRatio->Automatic,
DisplayFunction->$DisplayFunction];
```

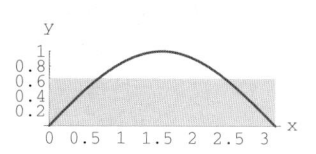

The height of the rectangle is

$$\frac{\int_a^b f[x]\,dx}{b-a}.$$

The width of the rectangle is $(b-a)$. The area inside the rectangle measures out to

$$\left(\frac{\int_a^b f[x]\,dx}{b-a}\right)(b-a) = \int_a^b f[x]\,dx \text{ square units.}$$

This tells you that the area of the flat rectangle measures the signed area under the curve and over the x-axis $y = f[x]$ for $a \leq x \leq b$.

This is why folks say that the average value of $f[x]$ for $a \leq x \leq b$ is the height of the rectangle, $\int_a^b f[x]\,dx / (b-a)$. Try it again:

In[50]:=
```
Clear[f,x]
f[x_] = E^(-x); {a,b} = {-1, 2};
fplot = Plot[f[x],{x,a,b},
PlotStyle->{{Thickness[0.01],Red}},
DisplayFunction->Identity];
average = Integrate[f[x],{x,a,b}]/(b - a);
averagefplot = Graphics[{Blue,
Polygon[{{a,0},{b,0},{b,average},{a,average}}]}];
Show[averagefplot,fplot,AxesLabel->{"x","y"},
AspectRatio->Automatic,
DisplayFunction->$DisplayFunction];
```

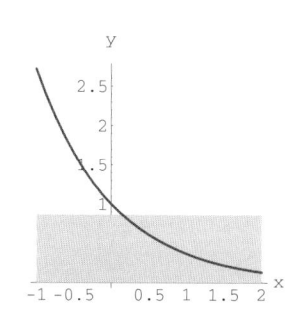

And experimentally several times:

In[51]:=
```
theoreticalaverage = N[Integrate[f[x],{x,a,b}]/(b - a)];
pointcount = 200;
experimentalaverage = Sum[N[f[Random[Real,{N[a],N[b]}]]],
{k,1,pointcount}]/pointcount;
{theoreticalaverage,experimentalaverage}
```

Out[51]=
{0.860982, 0.795809}

See what happens when you use more points:

Lesson 3.08 Transforming 3D Integrals

In[52]:=
```
pointcount = 200;
experimentalaverage = Sum[N[f[Random[Real,{N[a],N[b]}]]],
  {k,1,pointcount}]/pointcount; {theoreticalaverage,experimentalaverage}
```
Out[52]=
{0.860982, 0.850199}

Usually not far off.

B.4.b) Given a function $f[x,y]$ and a region R in xy-paper, folks like to say that the average value of $f[x,y]$ on R is given by

$$\frac{\iint_R f[x,y]\, dx\, dy}{\iint_R dx\, dy}.$$

> Why do they like to say this?

Answer: Look at these:

In[53]:=
```
{{a,b},{c,d}} = {{0,2},{-1,4}};
Clear[f,x,y]; f[x_,y_] = Sin[2 x + y] + 1;
fplot = Plot3D[f[x,y],{x,a,b},{y,c,d},Axes->True,AxesLabel->{"x","y","z"},
  ViewPoint->CMView,DisplayFunction->Identity];
theoreticalaverage = Integrate[f[x,y],{x,a,b},{y,c,d}]/Integrate[1,{x,a,b},{y,c,d}];
averagefplot = Plot3D[theoreticalaverage,{x,a,b},{y,c,d},
  Axes->True,AxesLabel->{"x","y","z"},ViewPoint->CMView,DisplayFunction->Identity];
Show[GraphicsArray[{fplot,averagefplot}],DisplayFunction->$DisplayFunction];
```

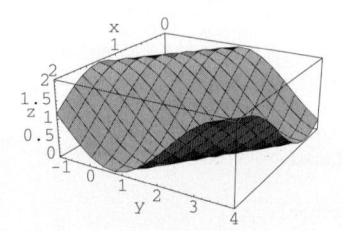

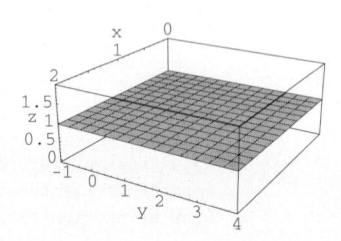

The volume measurements of the solids between the xy-plane and the plotted surfaces are the same. Reason:

$$\text{volume left} = \iint_R f[x,y]\, dx\, dy$$

$$\text{volume right} = \text{base} \times \text{height} = \iint_R dx\, dy \left(\frac{\iint_R f[x,y]\, dx\, dy}{\iint_R dx\, dy} \right)$$

$$= \iint_R f[x,y]\, dx\, dy.$$

Both solids have the same base, so the height of the solid on the right must be the average value of $f[x, y]$ for $\{x, y\}$ in R. See what happens experimentally:

In[54]:=
```
pointcount = 200;
experimentalaverage1 = Sum[N[f[Random[Real,{N[a],N[b]}],
Random[Real,{N[c],N[d]}]]],{k,1,pointcount}]/pointcount;
{N[theoreticalaverage],experimentalaverage1}
```

Out[54]=
{0.961822, 0.972281}

In[55]:=
```
pointcount = 200;
experimentalaverage2 = Sum[N[f[Random[Real,{N[a],N[b]}],
Random[Real,{N[c],N[d]}]]],{k,1,pointcount}]/pointcount;
{N[theoreticalaverage],experimentalaverage2}
```

Out[55]=
{0.961822, 0.999339}

In[56]:=
```
{N[theoreticalaverage],(experimentalaverage1 + experimentalaverage2)/2}
```

Out[56]=
{0.961822, 0.98581}

Lookin' good and feeling good.

B.4.c) Given a function $f[x, y, z]$ and a solid R in xyz-space, folks like to say that the average value of $f[x, y, z]$ on R is given by

$$\frac{\iiint_R f[x, y, z]\, dx\, dy\, dz}{\iiint_R dx\, dy\, dz}.$$

Why do they like to say this?

Answer: This time, a plot is not possible. Try an experiment with R as the box
$0 \le x \le 2, -1 \le y \le 4,$ and $-2 \le z \le 3$:

In[57]:=
```
Clear[f,x,y,z];
{{xlow,xhigh},{ylow,yhigh},{zlow,zhigh}} = {{0,2},{-1,4},{-2,3}};
f[x_,y_,z_] = Sin[2 x + y - z] + 1;
theoreticalaverage = Integrate[f[x,y,z],{x,xlow,xhigh},
{y,ylow,yhigh},{z,zlow,zhigh}]/
Integrate[1,{x,xlow,xhigh},{y,ylow,yhigh},{z,zlow,zhigh}];
```

```
In[58]:=
    pointcount = 200;
    experimentalaverage1 = Sum[N[f[Random[Real,{N[xlow],N[xhigh]}],
    Random[Real,{N[ylow],N[yhigh]}],Random[Real,{N[zlow],N[zhigh]}]]],
    {k,1,pointcount}]/pointcount; {N[theoreticalaverage],experimentalaverage1}
```

Out[58]=
    {1.00368, 0.972838}

```
In[59]:=
    pointcount = 200;
    experimentalaverage2 = Sum[N[f[Random[Real,{N[xlow],N[xhigh]}],
    Random[Real,{N[ylow],N[yhigh]}],
    Random[Real,{N[zlow],N[zhigh]}]]],{k,1,pointcount}]/pointcount;
    {N[theoreticalaverage],experimentalaverage2}
```

Out[59]=
    {1.00368, 1.00403}

```
In[60]:=
    {N[theoreticalaverage],(experimentalaverage1 + experimentalaverage2)/2}
```

Out[60]=
    {1.00368, 0.988436}

Nice. Another way to look at this is to note that when you go with

$$\text{average} = \frac{\iiint_R f[x,y,z]\, dx\, dy\, dz}{\iiint_R dx\, dy\, dz},$$

then you get

$$\iiint_R f[x,y,z]\, dx\, dy\, dz = \iiint_R \text{average}\, dx\, dy\, dz.$$

Tutorials

■ T.1) Cylinders, spheres, and tubes: Plotting them and integrating on them

T.1.a.i) Here's a cylinder:

```
In[1]:=
    Clear[x,y,z,r,s,t]
    {x[r_,s_,t_],y[r_,s_,t_],z[r_,s_,t_]} = {r Cos[t],r Sin[t],s};
    {{rlow,rhigh},{slow,shigh},{tlow,thigh}} = {{0,2},{0,3},{0, 2 Pi}};
    sideplot = ParametricPlot3D[{x[rhigh,s,t],y[rhigh,s,t],z[rhigh,s,t]},
    {s,slow,shigh},{t,tlow,thigh},PlotPoints->{2,Automatic},DisplayFunction->Identity];
    topplot = ParametricPlot3D[{x[r,shigh,t],y[r,shigh,t],z[r,shigh,t]},
    {r,rlow,rhigh},{t,tlow,thigh},PlotPoints->{2,Automatic},DisplayFunction->Identity];
    baseplot = ParametricPlot3D[{x[r,slow,t],y[r,slow,t],z[r,slow,t]},
    {r,rlow,rhigh},{t,tlow,thigh},PlotPoints->{2,Automatic},DisplayFunction->Identity];
```

In[2]:=
```
Show[sideplot,topplot,baseplot,
AxesLabel->{"x","y","z"},
ViewPoint->CMView,
BoxRatios->Automatic,
DisplayFunction->$DisplayFunction];
```

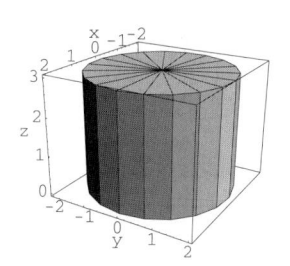

Measure the volume contained in this cylinder.

Answer: You don't need calculus for this one. The base of this cylinder is a circle of radius 2; its area measures out to:

In[3]:=
```
basearea = 2^2 Pi
```
Out[3]=
```
4 Pi
```

The height of the cylinder is:

In[4]:=
```
height = shigh
```
Out[4]=
```
3
```

The volume of the cylinder measures out to:

In[5]:=
```
volume = basearea height
```
Out[5]=
```
12 Pi
```

Now try it using calculus:

In[6]:=
```
Clear[gradx,grady,gradz,Vxyz]
gradx[r_,s_,t_] = {D[x[r,s,t],r],D[x[r,s,t],s],D[x[r,s,t],t]};
grady[r_,s_,t_] = {D[y[r,s,t],r],D[y[r,s,t],s],D[y[r,s,t],t]};
gradz[r_,s_,t_] = {D[z[r,s,t],r],D[z[r,s,t],s],D[z[r,s,t],t]};
Vxyz[r_,s_,t_] = Expand[Det[{gradx[r,s,t],grady[r,s,t],gradz[r,s,t]}],Trig->True]
```
Out[6]=
```
-r
```

The volume of this cylinder measures out to

$$\int_{tlow}^{thigh} \int_{slow}^{shigh} \int_{rlow}^{rhigh} \left(-V_{(x,y,z)}[r,s,t]\right) \, dr \, ds \, dt :$$

In[7]:=
```
volume = -Integrate[Vxyz[r,s,t],
{t,tlow,thigh},{s,slow,shigh},{r,rlow,rhigh}]
```

T.1.a.ii) This time, go with a cylinder whose top is on the plane:

In[8]:=
```
Clear[top,x,y]; top[x_,y_] = -x - y + 6;
```

and whose bottom is on the plane:

In[9]:=
```
Clear[bottom,x,y]; bottom[x_,y_] = x - y - 4;
```

Put:

In[10]:=
```
Clear[x,y,z,s]; {slow,shigh} = {0,1};
z[x_,y_,s_] = bottom[x,y] + s(top[x,y] - bottom[x,y])
```
Out[10]=
```
-4 + s (10 - 2 x) + x - y
```

And note that:

In[11]:=
```
{z[x,y,slow] == bottom[x,y],z[x,y,shigh] == top[x,y]}
```
Out[11]=
```
{True, True}
```

This gives you an easy way of plotting the cockeyed cylinder

→ whose top skin runs with the plane $z = \text{top}[x, y]$

→ whose bottom skin runs with the plane $z = \text{bottom}[x, y]$ and

→ whose sides run with circles of radius 2 perpendicular to the z-axis and centered on the z-axis:

In[12]:=
```
Clear[x,y,newz,r,s,t]
{{rlow,rhigh},{tlow,thigh}} = {{0,2},{0,2 Pi}};
{x[r_,s_,t_],y[r_,s_,t_]} = {r Cos[t],r Sin[t]};
newz[r_,s_,t_] = z[x[r,s,t],y[r,s,t],s];
sideplot = ParametricPlot3D[{x[rhigh,s,t],y[rhigh,s,t],
  newz[rhigh,s,t]},{s,slow,shigh},{t,tlow,thigh},
  PlotPoints->{2,Automatic},DisplayFunction->Identity];
topplot = ParametricPlot3D[{x[r,shigh,t],y[r,shigh,t],
  newz[r,shigh,t]},{r,rlow,rhigh},{t,tlow,thigh},
  PlotPoints->{2,Automatic},DisplayFunction->Identity];
baseplot = ParametricPlot3D[{x[r,slow,t],y[r,slow,t],
  newz[r,slow,t]},{r,rlow,rhigh},{t,tlow,thigh},
  PlotPoints->{2,Automatic},DisplayFunction->Identity];
Show[sideplot,topplot,baseplot,AxesLabel->{"x","y","z"},
  ViewPoint->CMView,BoxRatios->Automatic,
  DisplayFunction->$DisplayFunction];
```

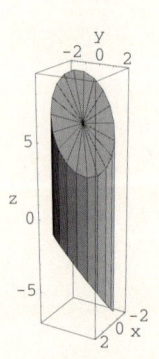

> Measure the volume contained in this weirdo cylinder.

Answer: You **do** want to use calculus for this one.

In[13]:=
```
Clear[gradx,grady,gradz,Vxyz]
gradx[r_,s_,t_] = {D[x[r,s,t],r],D[x[r,s,t],s],D[x[r,s,t],t]};
grady[r_,s_,t_] = {D[y[r,s,t],r],D[y[r,s,t],s],D[y[r,s,t],t]};
gradnewz[r_,s_,t_] = {D[newz[r,s,t],r],D[newz[r,s,t],s],D[newz[r,s,t],t]};
Vxyz[r_,s_,t_] = Abs[Expand[Det[{gradx[r,s,t],grady[r,s,t],gradnewz[r,s,t]}],
  Trig->True]]
```

Out[13]=
Abs[-10 r + 2 r² Cos[t]]

The volume inside this weirdo measures out to

$$\int_{tlow}^{thigh} \int_{slow}^{shigh} \int_{slow}^{shigh} Abs[V_{(x,y,z)}[r,s,t]] \, dr \, ds \, dt :$$

In[14]:=
```
volume = NIntegrate[Vxyz[r,s,t],{t,tlow,thigh},{s,slow,shigh},{r,rlow,rhigh}]
```

Out[14]=
125.664

Looking bad, but feeling good.

T.1.b) One way to see how to plot the sphere

$$x^2 + y^2 + z^2 = r^2$$

is to slice it with the plane $z = c$ where $-r \leq c \leq r$ and then to think about what you get. What you get is the circle

$$x^2 + y^2 = r^2 - c^2$$

plotted in the plane

$$z = c.$$

Now, when you go with a given radius r, you can plot the sphere $x^2 + y^2 + z^2 = r^2$. Take a look in the case that $r = 3$:

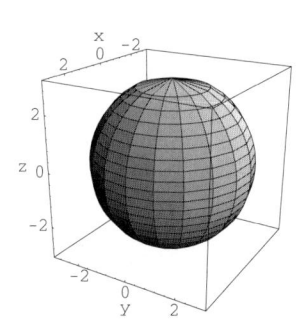

In[15]:=
```
Clear[c,t]; radius = 3;
{clow,chigh} = {-radius,radius};
{tlow,thigh} = {0,2 Pi};
ParametricPlot3D[{Sqrt[radius^2 - c^2] Cos[t],
Sqrt[radius^2 - c^2] Sin[t],c},
{c,clow,chigh},{t,tlow,thigh},
ViewPoint->CMView,Axes->True,
AxesLabel->{"x","y","z"}];
```

Lesson 3.08 *Transforming 3D Integrals*

> Agree that R is the region consisting of everything on and inside the sphere
>
> $$x^2 + y^2 + z^2 = 9$$
>
> as plotted above, and calculate the integral
>
> $$\iiint_R x^2 \, dx \, dy \, dz.$$

Answer: Look at the plotting code:

```
ParametricPlot3D[{Sqrt[radius^2 - c^2] Cos[t],
Sqrt[radius^2 - c^2] Sin[t],c},
{c,clow,chigh},{t,tlow,thigh},
```

And go with:

In[16]:=
```
Clear[x,y,z,c,r,t]
{x[c_,r_,t_],y[c_,r_,t_],z[c_,r_,t_]} = {r Cos[t],r Sin[t],c};
```

with

$$0 \le r \le \sqrt{\text{radius}^2 - c^2},$$

$$-\text{radius} \le c \le \text{radius}, \text{ and}$$

$$0 \le t \le 2\pi.$$

This gives you

$$\iiint_R x^2 \, dx \, dy \, dz = \iiint_{R_{crt}} x[c,r,t]^2 \, V_{(x,y,z)}[c,r,t] \, dc \, dr \, dt$$

$$= \iiint_{R_{crt}} x[c,r,t]^2 \, V_{(x,y,z)}[c,r,t] \, dr \, dc \, dt$$

$$= \int_0^{2\pi} \int_{-\text{radius}}^{\text{radius}} \int_0^{\sqrt{\text{radius}^2 - c^2}} x[c,r,t]^2 \, V_{(x,y,z)}[c,r,t] \, dr \, dc \, dt :$$

You gotta integrate with respect to r before you integrate with respect to c, because the limit of integration with respect to r changes as c changes.

Now calculate the volume conversion factor:

In[17]:=
```
Clear[gradx,grady,gradz,Vxyz]
gradx[c_,r_,t_] = {D[x[c,r,t],c],D[x[c,r,t],r],D[x[c,r,t],t]};
grady[c_,r_,t_] = {D[y[c,r,t],c],D[y[c,r,t],r],D[y[c,r,t],t]};
gradz[c_,r_,t_] = {D[z[c,r,t],c],D[z[c,r,t],r],D[z[c,r,t],t]};
Vxyz[c_,r_,t_] = Expand[Det[{gradx[c,r,t],grady[c,r,t],gradz[c,r,t]}],Trig->True]
```

Out[17]=
r

Calculate
$$\iiint_R x^2 \, dx \, dy \, dz = \int_0^{2\pi} \int_{-\text{radius}}^{\text{radius}} \int_0^{\sqrt{\text{radius}^2 - c^2}} x[c,r,t]^2 \, V_{(x,y,z)}[c,r,t] \, dr \, dc \, dt :$$

In[18]:=
```
Integrate[x[c,r,t]^2 Vxyz[c,r,t],{t,0,2 Pi},
  {c,-radius,radius},{r,0,Sqrt[radius^2 - c^2]}]
```

Out[18]=
$$\frac{324 \text{ Pi}}{5}$$

Done.

T.1.c) Remember the main unit normal and the binormal from the lesson on perpendicularity:

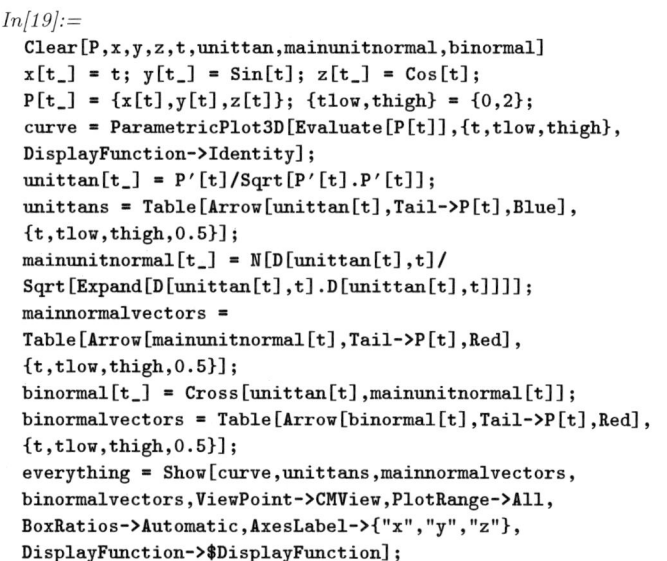

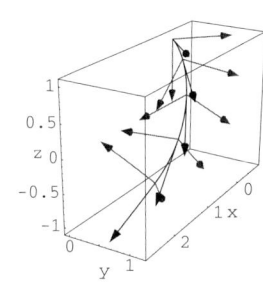

In[19]:=
```
Clear[P,x,y,z,t,unittan,mainunitnormal,binormal]
x[t_] = t; y[t_] = Sin[t]; z[t_] = Cos[t];
P[t_] = {x[t],y[t],z[t]}; {tlow,thigh} = {0,2};
curve = ParametricPlot3D[Evaluate[P[t]],{t,tlow,thigh},
  DisplayFunction->Identity];
unittan[t_] = P'[t]/Sqrt[P'[t].P'[t]];
unittans = Table[Arrow[unittan[t],Tail->P[t],Blue],
  {t,tlow,thigh,0.5}];
mainunitnormal[t_] = N[D[unittan[t],t]/
  Sqrt[Expand[D[unittan[t],t].D[unittan[t],t]]]];
mainnormalvectors =
Table[Arrow[mainunitnormal[t],Tail->P[t],Red],
  {t,tlow,thigh,0.5}];
binormal[t_] = Cross[unittan[t],mainunitnormal[t]];
binormalvectors = Table[Arrow[binormal[t],Tail->P[t],Red],
  {t,tlow,thigh,0.5}];
everything = Show[curve,unittans,mainnormalvectors,
  binormalvectors,ViewPoint->CMView,PlotRange->All,
  BoxRatios->Automatic,AxesLabel->{"x","y","z"},
  DisplayFunction->$DisplayFunction];
```

Remember how much fun it was to make a tube consisting of all circles of a fixed radius centered on the curve and lying in planes perpendicular to the curve:

In[20]:=
```
radius = 0.2; {slow,shigh} = {0,2 Pi};
ParametricPlot3D[Evaluate[P[t] +
  radius Cos[s] mainunitnormal[t] +
  radius Sin[s] binormal[t]],
  {t,tlow,thigh},{s,slow,shigh},
  ViewPoint->CMView,BoxRatios->Automatic,
  AxesLabel->{"x","y","z"}];
```

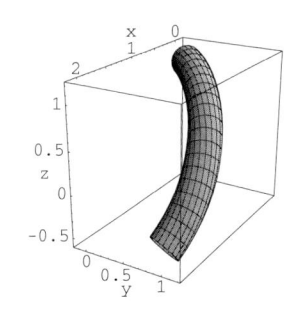

> Put flat caps on each end of this tube, and then measure the volume of the resulting container.

Answer: To plot the skin of the container, you plotted

$$P[t] + \text{radius} \cos[s] \, \text{mainunitnormal}[t] + \text{radius} \sin[s] \, \text{binormal}[t]],$$

where radius = 0.2, slow ≤ s ≤ shigh, and tlow ≤ t ≤ thigh. To describe the whole container and everything inside it, go with:

In[21]:=
```
Clear[x,y,z,r,s,t]
{x[r_,s_,t_],y[r_,s_,t_],z[r_,s_,t_]} = Chop[ExpandAll[P[t] +
  r Cos[s] mainunitnormal[t] + r Sin[s] binormal[t],Trig->True]]
```

Out[21]=
{t - 0.707107 r Sin[s], 0.853553 r Sin[s - t] + Sin[t] - 0.146447 r Sin[s + t],
 -0.853553 r Cos[s - t] + Cos[t] - 0.146447 r Cos[s + t]}

with $0 \le r \le 0.2$, slow ≤ s ≤ shigh, and tlow ≤ t ≤ thigh. Call the container and everything inside it R_{xyz} and measure its volume by calculating

$$\iiint_{R_{xyz}} dx\,dy\,dz = \iiint_{R_{rst}} V_{(x,y,z)}[r,s,t]\,dr\,ds\,dt$$

$$= \int_{tlow}^{thigh} \int_{slow}^{shigh} \int_{0}^{0.2} V_{(x,y,z)}[r,s,t]\,dr\,ds\,dt.$$

Now if *Mathematica* can hack this, you're out of here.

In[22]:=
```
Clear[gradx,grady,gradz,Vxyz]
gradx[r_,s_,t_] = {D[x[r,s,t],r],D[x[r,s,t],s],D[x[r,s,t],t]};
grady[r_,s_,t_] = {D[y[r,s,t],r],D[y[r,s,t],s],D[y[r,s,t],t]};
gradz[r_,s_,t_] = {D[z[r,s,t],r],D[z[r,s,t],s],D[z[r,s,t],t]};
Vxyz[r_,s_,t_] = Expand[Det[{gradx[r,s,t],grady[r,s,t],gradz[r,s,t]}],Trig->True]
```

Out[22]=
$1.41421\,r - 0.707107\,r^2 \cos[s]$

Haw,

$$\text{the volume of container} = \int_{tlow}^{thigh} \int_{slow}^{shigh} \int_{0}^{0.2} V_{(x,y,z)}[r,s,t]\,dr\,ds\,dt$$

is measured by:

In[23]:=
```
volume = NIntegrate[Abs[Vxyz[r,s,t]],{t,tlow,thigh},{s,slow,shigh},{r,0,0.2}]
```

Out[23]=
0.355431

Done.

Tutorials (T.2)

■ T.2) Integrating on solids bounded by sets of surfaces

T.2.a) A solid R is given as all points that are simultaneously between:

$\rightarrow$ the planes $\quad x+y+z=-2 \quad$ and $\quad x+y+z=7$
$\rightarrow$ the planes $\quad x-y-z=-4 \quad$ and $\quad x-y-z=8$

and

$\rightarrow$ the surfaces $\quad x^3-y+2z=-2 \quad$ and $\quad x^3-y+2z=6$.

> Measure the volume of this solid.

Answer: Going in blind is the thing to do. Look again at the description of the surface:

The solid R was given as all points that are simultaneously between:

$\rightarrow$ the planes $\quad x+y+z=-2 \quad$ and $\quad x+y+z=7$
$\rightarrow$ the planes $\quad x-y-z=-4 \quad$ and $\quad x-y-z=8$

and

$\rightarrow$ the surfaces $\quad x^3-y+2z=-3 \quad$ and $\quad x^3-y+2z=6$.

Put:

In[24]:=
```
Clear[x,y,z,u,v,w]
{u[x_,y_,z_],ulow,uhigh} = {x + y + z,-2,7};
{v[x_,y_,z_],vlow,vhigh} = {x - y - z,-4,8};
{w[x_,y_,z_],wlow,whigh} = {x^3 - y + 2 z,-3,6};
```

R_{uvw} is the rectangular box:

$\text{ulow} \leq u \leq \text{uhigh}, \text{vlow} \leq v \leq \text{vhigh} \quad$ and $\quad \text{wlow} \leq w \leq \text{whigh}$.

So:

$$\text{volume} = \iiint_R dx\, dy\, dz$$

$$= \iiint_{R_{uvw}} V_{(x,y,z)}[u,v,w]\, du\, dv\, dw$$

$$= \int_{\text{wlow}}^{\text{whigh}} \int_{\text{vlow}}^{\text{vhigh}} \int_{\text{ulow}}^{\text{uhigh}} V_{(x,y,z)}[u,v,w]\, du\, dv\, dw.$$

Calculate $x[u,v,w]$, $y[u,v,w]$, and $z[u,v,w]$:

In[25]:=
```
xyzsolved = Solve[{u == u[x,y,z],
  v == v[x,y,z],w == w[x,y,z]},{x,y,z}]
```

Out[25]=

$$\left\{\left\{y \to \frac{8u + u^3 - 8v + 3u^2 v + 3uv^2 + v^3 - 8w}{24},\right.\right.$$

$$\left.z \to \frac{4u - u^3 - 4v - 3u^2 v - 3uv^2 - v^3 + 8w}{24},\ x \to \frac{u+v}{2}\right\}\right\}$$

In[26]:=
`{x[u_,v_,w_],y[u_,v_,w_],z[u_,v_,w_]} = {x,y,z}/.xyzsolved[[1]]`

Out[26]=

$$\left\{\frac{u+v}{2},\ \frac{8u + u^3 - 8v + 3u^2 v + 3uv^2 + v^3 - 8w}{24},\right.$$

$$\left.\frac{4u - u^3 - 4v - 3u^2 v - 3uv^2 - v^3 + 8w}{24}\right\}$$

Calculate the volume conversion factor:

In[27]:=
```
Clear[gradx,grady,gradz,Vxyz]
gradx[u_,v_,w_] = {D[x[u,v,w],u],D[x[u,v,w],v],D[x[u,v,w],w]};
grady[u_,v_,w_] = {D[y[u,v,w],u],D[y[u,v,w],v],D[y[u,v,w],w]};
gradz[u_,v_,w_] = {D[z[u,v,w],u],D[z[u,v,w],v],D[z[u,v,w],w]};
Vxyz[u_,v_,w_] = Abs[Det[{gradx[u,v,w],grady[u,v,w],gradz[u,v,w]}]]
```

Out[27]=
$$\frac{1}{6}$$

So:

$$\text{volume of } R = \iiint_R dx\, dy\, dz$$

$$= \int_{wlow}^{whigh} \int_{vlow}^{vhigh} \int_{ulow}^{uhigh} V_{(x,y,z)}[u,v,w]\, du\, dv\, dw$$

is measured (in cubic units) by:

In[28]:=
`volume = Integrate[Vxyz[u,v,w],{w,wlow,whigh},{v,vlow,vhigh},{u,ulow,uhigh}]`

Out[28]=
162

Really easy.

■ T.3) Parallelepipeds

This problem appears only in the electronic version.

■ **T.4) Switching the order of integration**

This problem appears only in the electronic version.

Give It a Try

Experience with the starred (⋆) problems will be useful for understanding developments later in the course.

■ **G.1) 3D integrals**⋆

G.1.a) Calculate

$$\iiint_R x^2 y^2 z^2 \, dx\, dy\, dz$$

where R is the box with

$0 \leq x \leq 2$, $1 \leq y \leq 3$, and $-2 \leq z \leq 2$.

G.1.b) Calculate

$$\iiint_{R_{xyz}} \left(\frac{x}{3}\right)^2 + \left(\frac{y}{2}\right)^2 \, dx\, dy\, dz$$

where R_{xyz} is the region consisting of everything inside and on the elliptical cylinder with

sides $\left(\frac{x}{3}\right)^2 + \left(\frac{y}{2}\right)^2 = 1$

bottom $z = 0$

and

top $z = 4$.

G.1.c) Calculate

$$\iiint_R x^2 y^2 z^2 \, dx\, dy\, dz,$$

where R is the solid whose top skin is the surface
$$z = 8 - x^2 - y^2$$
and whose bottom skin is the surface
$$z = x^2 + y^2.$$

G.1.d) Calculate
$$\iiint_R x^2 y^2 z^2 \, dx \, dy \, dz,$$
where R is the solid inside the parallelepiped plotted below:

In[1]:=
```
vector1 = {3,0,2}; vector2 = {0,3,-3};
vector3 = {-3,1,4}; basepoint = {0,-2,0};
Show[Arrow[vector1,Tail->basepoint,Red],
Arrow[vector2,Tail->basepoint,Red],
Arrow[vector3,Tail->basepoint,Red],
Arrow[vector2,Tail->basepoint + vector1,Blue],
Arrow[vector3,Tail->basepoint + vector1,Blue],
Arrow[vector1,Tail->basepoint + vector2,Blue],
Arrow[vector3,Tail->basepoint + vector2,Blue],
Arrow[vector1,Tail->basepoint + vector3,Blue],
Arrow[vector2,Tail->basepoint + vector3,Blue],
Arrow[vector3,Tail->basepoint + vector1 + vector2,Blue],
Arrow[vector1,Tail->basepoint + vector2 + vector3,Blue],
Arrow[vector2,Tail->basepoint + vector1 + vector3,Blue],
Arrow[vector3,Tail->basepoint + vector1 + vector2,Blue],
Axes->True,AxesLabel->{"x","y","z"},ViewPoint->CMView];
```

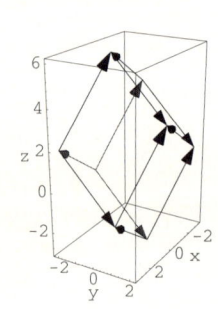

■ G.2) Volume★

G.2.a.i) Here's the plot of the outside skin of a solid:

In[2]:=
```
Clear[s,t]
skin = ParametricPlot3D[{(1 - s^2) Cos[t],
2 (1 - s^2) Sin[t],s},{s,-1,1},{t,0, 2 Pi},
ViewPoint->CMView,AxesLabel->{"x","y","z"}];
```

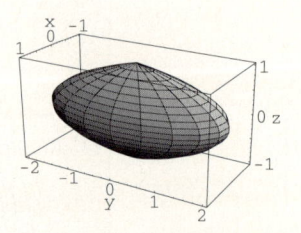

Measure the volume enclosed by this skin.

G.2.a.ii) Here's the plot of the outside skin of a related solid:

In[3]:=
```
Clear[s,t]
skin = ParametricPlot3D[{-s,-s,2s}+{(1-s^2)Cos[t],
2(1-s^2)Sin[t],0},{s,-1,1},{t,0, 2 Pi},
ViewPoint->CMView,AxesLabel->{"x","y","z"}];
```

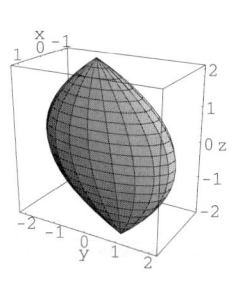

> Measure the volume enclosed by this skin.

G.2.b) A solid R is given as all points that are simultaneously between:
- → the planes $\quad x+y+z=-2\quad$ and $\quad x+y+z=5$
- → the planes $\quad x+y-z=-3\quad$ and $\quad x+y-z=7$, and
- → the surfaces $\quad x-y+2z^3=0\quad$ and $\quad x-y+2z^3=5$.

> Measure the volume of this solid.

■ G.3) Drilling and slicing spheres

One way to see how to plot the sphere
$$x^2+y^2+z^2 = \text{radius}^2$$
is to slice it with the plane $z=c$, where $-\text{radius} \leq c \leq \text{radius}$ and then to think about what you get.

What you get is the circle
$$x^2+y^2 = \text{radius}^2 - c^2$$
plotted in the plane $z=c$. Now when you go with a given radius, you can plot the sphere $x^2+y^2+z^2 = \text{radius}^2$. Take a look in the case that radius = 3:

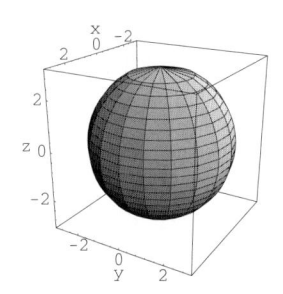

In[4]:=
```
Clear[c,t]; radius = 3;
{clow,chigh} = {-radius,radius};
{tlow,thigh} = {0,2 Pi};
ParametricPlot3D[{Sqrt[radius^2 - c^2] Cos[t],
Sqrt[radius^2 - c^2] Sin[t],c},{c,clow,chigh},
{t,tlow,thigh},ViewPoint->CMView,
Axes->True,BoxRatios->Automatic,
AxesLabel->{"x","y","z"}];
```

G.3.a.i) Use a 3D integral to measure the volume of this sphere.

G.3.a.ii) Use a 3D integral to come up with a formula for a function

volume[radius]

that measures the volume of a sphere of a given radius.

G.3.b.i) Here's a sphere of radius 2 shown in true scale with an axis running from top to bottom:

In[5]:=
```
Clear[c,t]; radius = 2; {clow,chigh} = {-radius,radius};
{tlow,thigh} = {0,2 Pi}; sphereplot =
ParametricPlot3D[{Sqrt[radius^2 - c^2] Cos[t],
Sqrt[radius^2 - c^2] Sin[t],c},{c,clow,chigh},
{t,tlow,thigh},DisplayFunction->Identity]; Show[sphereplot,
Graphics3D[{Thickness[0.01],Blue,Line[{{0,0,-2.5},{0,0,2.5}}]}],
ViewPoint->CMView,Axes->True,BoxRatios->Automatic,
AxesLabel->{"x","y","z"},DisplayFunction->$DisplayFunction];
```

Plot what you get after you take a circular drill of radius 1 and bore a straight circular hole through the sphere, centering the drill on the plotted axis.

G.3.b.ii) Go back to the set-up in part G.3.b.i) immediately above and imagine that the sphere is the skin of a solid ball made of redwood. Measure the volume of the redwood that remains after you take a circular drill of radius 1 and bore a straight circular hole through the sphere, centering the drill on the plotted axis.

G.3.c) Here's a sphere whose top has been sliced off:

In[6]:=
```
Clear[c,t]; radius = 2;
{clow,chigh} = {-radius,1}; {tlow,thigh} = {0,2 Pi};
sphereplot = ParametricPlot3D[
{Sqrt[radius^2 - c^2] Cos[t],
Sqrt[radius^2 - c^2] Sin[t],c},{c,clow,chigh},
{t,tlow,thigh},ViewPoint->CMView,Axes->True,
BoxRatios->Automatic,AxesLabel->{"x","y","z"}];
```

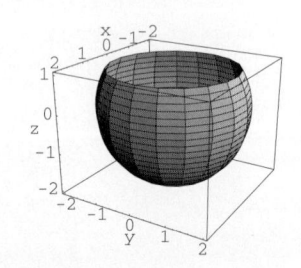

Think of this as a container for your favorite cold liquid refreshment and measure the volume of liquid this container can hold.

G.3.d) This problem was adapted from the manual *Uses of Technology in the Mathematics Curriculum*, by Benny Evans and Jasper Johnson in 1990.

A cubic foot of water weighs 62.4 pounds. A cubic foot of seasoned balsa wood weighs 7.3 pounds.

> You carve a perfect sphere 2 feet in diameter from a block of balsa wood, paint it with waterproof paint, let the paint dry, throw it into a pond and wait until it stops bobbing. How deeply will it be submerged?

■ G.4) Tubes, horns, and squashed doughnuts*

G.4.a.i) Here's a base curve:

```
In[7]:=
  Clear[s,P]
  P[s_] = {-s,2 s,s^2};
  curve = ParametricPlot3D[Evaluate[P[s]],{s,0,2},
  AxesLabel->{"x","y","z"},ViewPoint->CMView];
```

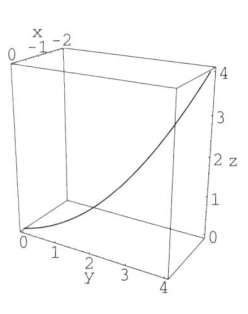

> Plot the tube whose skin consists of each circle of a radius 1.4 centered on the curve and lying in the plane that cuts the curve perpendicularly at the center of the circle.

G.4.a.ii) Imagine that flat caps have been attached to the open ends of the tube in part G.4.a.i) above, and measure the volume of the resulting container.

G.4.b) Look at this horn:

```
In[8]:=
  Clear[s,P]; P[s_] = {-s,2 s,s^2};
  unittan[s_] = P'[s]/Sqrt[P'[s].P'[s]];
  mainunitnormal[s_] = N[D[unittan[s],s]/
  Sqrt[Expand[D[unittan[s],s].D[unittan[s],s]]]];
  binormal[s_] = N[Det[{{{1,0,0},{0,1,0},{0,0,1}},
  unittan[s],mainunitnormal[s]}]]; Clear[hornskin,r];
  r[s_] = 0.25 s; hornskin[s_,t_] = Together[P[s] +
  r[s] Cos[t] mainunitnormal[s] +
  r[s] Sin[t] binormal[s]]; ParametricPlot3D[Evaluate[
  hornskin[s,t]],{s,0, 2},{t,0,2 Pi},
  AxesLabel->{"x","y","z"},ViewPoint->CMView];
```

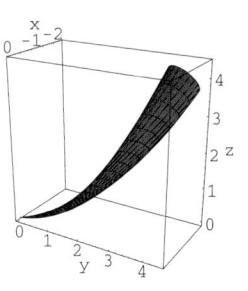

> Imagine that a flat cap has been attached to the open end, and measure the volume of the resulting container.

G.4.c) Here is a squashed doughnut:

In[9]:=
```
Clear[P,s,unittan,mainunitnormal,binormal]
P[s_] := {6 Cos[s],6 Sin[s],0};
unittan[s_] = P'[s]/
Sqrt[Expand[P'[s].P'[s],Trig->True]];
mainunitnormal[s_] = N[D[unittan[s],s]/
Sqrt[Expand[D[unittan[s],s].D[unittan[s],s],
Trig->True]]]; binormal[s_] = N[Expand[Cross[
unittan[s],mainunitnormal[s]],Trig->True]];
Clear[donutskin,rad,t]; rad[s] = Sin[s]^2 + 0.5;
donutskin[s_,t_] = P[s] +
rad[s] Cos[t] mainunitnormal[s] +
rad[s] Sin[t] binormal[s];
ParametricPlot3D[Evaluate[donutskin[s,t]],
{s,0,2 Pi},{t,0,2 Pi},PlotPoints->{30,Automatic},
AxesLabel->{"x","y","z"},ViewPoint->CMView];
```

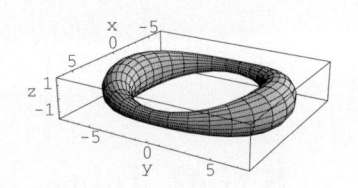

> Try to explain why the plot came out the way it did, and then measure the volume inside this squashed treat.

■ G.5) Measuring when you can't see

G.5.a.i) Here is a modest curve in three dimensions:

In[10]:=
```
Clear[curveplotter,t]; {tlow,thigh} = {0,4};
curveplotter[t_] = {3 Cos[2 t],3 t Sin[t],3 t};
curve = ParametricPlot3D[Evaluate[curveplotter[t]],
{t,tlow,thigh},DisplayFunction->Identity];
Show[curve,ViewPoint->CMView,
Axes->True,AxesLabel->{"x","y","z"},
PlotRange->All,DisplayFunction->$DisplayFunction];
```

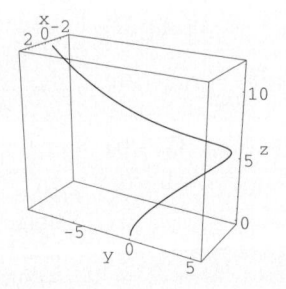

Here is the same curve shown with a couple of sample circles parallel to the xy-plane and centered on the curve:

Give It a Try (G.5) 349

In[11]:=
```
Clear[s]
circle1 = ParametricPlot3D[
Evaluate[curveplotter[1] + 2 {Cos[s],Sin[s],0}],
{s,0,2 Pi},DisplayFunction->Identity];
circle2= ParametricPlot3D[
Evaluate[curveplotter[2] + 2 {Cos[s],Sin[s],0}],
{s,0,2 Pi},DisplayFunction->Identity];
Show[curve,circle1,circle2,
ViewPoint->CMView,Axes->True,AxesLabel->{"x","y","z"},
PlotRange->All,DisplayFunction->$DisplayFunction];
```

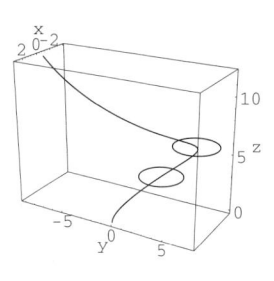

Here's what you get when you make the tube consisting of all such circles:

In[12]:=
```
Clear[tubeplotter,s,t]
tubeplotter[s_,t_] = curveplotter[t] +
2 {Cos[s],Sin[s],0}; tube = ParametricPlot3D[
Evaluate[tubeplotter[s,t]],{t,tlow,thigh},{s,0,2 Pi},
DisplayFunction->Identity];
tubeplot = Show[tube,curve,ViewPoint->CMView,
Axes->True,AxesLabel->{"x","y","z"},
PlotRange->All,DisplayFunction->$DisplayFunction];
```

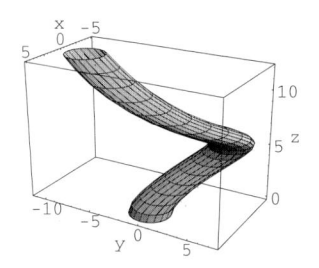

> Imagine that the ends of this tube are covered with flat plates and measure the volume of the resulting container.

G.5.a.ii) Calculus&Mathematica thanks 1991 C&M student Jonathan Paetsch for suggesting this part.

In part G.5.a.i), the core curve was plotted by

$$\text{curveplotter}[t] = \{3\cos[2\,t], 3\,t\sin[t], 3\,t\}$$

with $0 \leq t \leq 4$. This time, suppose the core curve is given by

$$\text{curveplotter}[t] = \{f[t], g[t], 3\,t\}$$

with $0 \leq t \leq 4$, where $f[t]$ and $g[t]$ are unspecified functions. When you center circles of radius 2 parallel to the xy-plane as above, you can't plot the tube because you don't have concrete formulas for $f[t]$ and $g[t]$. Nevertheless, you can imagine that flat plates are put on the open ends and you can measure the volume of the resulting container.

> Do it.

G.5.b) A solid R is given as all points that are simultaneously between:
- → the planes $\quad x+y+2z=-1 \quad$ and $\quad x+y+2z=5$
- → the planes $\quad x-y-2z=-3 \quad$ and $\quad x-y-2z=6$, and
- → the surfaces $\quad \sinh[x]-y+2z=-1 \quad$ and $\quad \sinh[x]-y+2z=2$.

Lesson 3.08 Transforming 3D Integrals

When you try to measure the volume of this solid, you go with:

In[13]:=
```
Clear[x,y,z,u,v,w]
{u[x_,y_,z_],ulow,uhigh} = {x + y + 2 z,-1,5};
{v[x_,y_,z_],vlow,vhigh} = {x - y - 2 z,-3,6};
{w[x_,y_,z_],wlow,whigh} = {Sinh[x] - y + 2 z,-1,2};
```

Then you observe that R_{uvw} is the rectangular box

$$\text{ulow} \leq u \leq \text{uhigh},$$
$$\text{vlow} \leq v \leq \text{vhigh, and}$$
$$\text{wlow} \leq w \leq \text{whigh}.$$

So:

$$\text{volume} = \iiint_R dx\, dy\, dz$$
$$= \iiint_{R_{uvw}} V_{(x,y,z)}[u,v,w]\, du\, dv\, dw$$
$$= \int_{\text{wlow}}^{\text{whigh}} \int_{\text{vlow}}^{\text{vhigh}} \int_{\text{ulow}}^{\text{uhigh}} V_{(x,y,z)}[u,v,w]\, du\, dv\, dw.$$

But when you try to get formulas for $x[u,v,w]$, $y[u,v,w]$, and $z[u,v,w]$, you run into trouble:

In[14]:=
```
xyzsolved = Solve[{u == u[x,y,z],v == v[x,y,z],w == w[x,y,z]},{x,y,z}]
```
Solve::ifun:
 Warning: Inverse functions are being used
  by Solve, so some solutions may not be
  found.

Out[14]=
{}

Reason: $\sinh[x] = (e^x - e^{-x})/2$ is a transcendental function which resists algebra. Because the formulas for functions $x[u,v,w]$, $y[u,v,w]$, and $z[u,v,w]$ are not available, it may seem impossible to calculate the volume conversion factor $V_{(x,y,z)}[u,v,w]$ for converting uvw-volume measurements into xyz-volume measurements.

On the other hand, you can calculate $V_{(u,v,w)}[x,y,z]$ which converts xyz-volume measurements into uvw-volume measurements:

In[15]:=
```
Clear[gradu,gradv,gradw]
gradu[x_,y_,z_] = {D[u[x,y,z],x],D[u[x,y,z],y],D[u[x,y,z],z]};
gradv[x_,y_,z_] = {D[v[x,y,z],x],D[v[x,y,z],y],D[v[x,y,z],z]};
gradw[x_,y_,z_] = {D[w[x,y,z],x],D[w[x,y,z],y],D[w[x,y,z],z]};
Vuvw[x_,y_,z_] = Abs[Det[{gradu[x,y,z],gradv[x,y,z],gradw[x,y,z]}]]
```

Out[15]=
8

> Look at this and use just a smidgen of common sense to figure out what $V_{(x,y,z)}[u,v,w]$ has got to be.
>
> Once you have $V_{(x,y,z)}[u,v,w]$ nailed down, use it to help measure the volume of the solid described at the beginning.

■ G.6) Average values, centroids, and centers of mass

G.6.a.i) Agree that avf_R stands for the average value of $f[x,y,z]$ as $\{x,y,z\}$ varies through a specified solid region R in xyz-space. The formula for avf_R is

$$avf_R = \frac{\iiint_R f[x,y,z]\,dx\,dy\,dz}{\iiint_R dx\,dy\,dz}.$$

> How do you know that avf_R is a number that falls between the minimum value of $f[x,y,z]$ on R and the maximum value of $f[x,y,z]$ on R?

G.6.a.ii) Suppose $f[x,y,z]$ is not constant on R.

> Why do you expect that there are points $\{x_1,y_1,z_1\}$ and $\{x_2,y_2,z_2\}$ in R with $f[x_1,y_1,z_1] < avf_R$ and $f[x_2,y_2,z_2] > avf_R$?

G.6.b) Calculate the average value of
$$f[x,y,z] = x^2 + y^2 + z^2$$
on the trapezoidal box $3 \le x \le 7$, $5 \le y \le 9$, and $0 \le z \le 6+6x+6y$.

G.6.c) You get the centroid $\{x_c, y_c, z_c\}$ of a solid R by setting:

$x_c =$ the average value of $f[x,y,z] = x$ on R,

$y_c =$ the average value of $g[x,y,z] = y$ on R,

and

$z_c =$ the average value of $h[x,y,z] = z$ on R.

Lesson 3.08 Transforming 3D Integrals

> Calculate the centroid of the trapezoidal box, $3 \le x \le 7$, $5 \le y \le 9$, and $0 \le z \le 6 + 6x + 6y$.

G.6.d.i) Take a nonnegative function $p[x, y, z]$ and think of it this way: The bigger $p[x, y, z]$ is, the more weight the point $\{x, y, z\}$ has. The weighted average value of a function $f[x, y, z]$ with respect to $p[x, y, z]$ on a region R in three dimensions is the number

$$avf_{pR} = \frac{\iiint_R f[x,y,z] p[x,y,z] \, dx \, dy \, dz}{\iiint_R p[x,y,z] \, dx \, dy \, dz}.$$

G.6.d.ii) > Given a weight function $p[x, y, z]$, how do you know that avf_{pR} is a number that falls between the minimum value of $f[x, y, z]$ on R and the maximum value of $f[x, y, z]$ on R?

G.6.d.iii) Take $p[x, y, z] = e^{-x-y-z}$.

> Calculate the weighted average of
> $$f[x, y, z] = x^2 + y^2 + z^2$$
> with respect to $p[x, y, z]$ of the trapezoidal box
> $$0 \le x \le 1,$$
> $$0 \le y \le 2,$$
> and
> $$0 \le z \le 1 + 2x + y.$$

G.6.e) A solid R has density

$$p[x, y, z] \text{ grams/cm}^3$$

at a point $\{x, y, z\}$. You get the center of mass $\{x_m, y_m, z_m\}$ of a solid R by setting:

$x_m =$ the average value of $f[x, y, z] = x$ with respect to $p[x, y, z]$ on R,

$y_m =$ the average value of $g[x, y, z] = y$ with respect to $p[x, y, z]$ on R

and

$z_m =$ the average value of $h[x, y, z] = z$ with respect to $p[x, y, z]$ on R.

The top of a solid is the domed surface

$$z = 8 - x^2 - y^2$$

and the bottom is the paraboloid
$$z = x^2 + y^2.$$
The density $p[x, y, z]$ at a point $\{x, y, z\}$ in the solid is proportional to the distance of $\{x, y, z\}$ from the plane $z = 0$, so for $z \geq 0$,
$$p[x, y, z] = c\, z,$$
where c is a positive constant of proportionality.

Calculate the center of mass $\{x_m, y_m, z_m\}$ of the solid.

G.7) Switching the order of integration

G.7.a) Calculate the triple integral
$$\int_0^1 \int_0^{1-x} \int_0^{x+y} \sin[\pi(x + y + z)]\, dz\, dy\, dx.$$
Then rewrite the integral in the order
$$\int_-^- \int_-^- \int_-^- \sin[\pi(x + y + z)]\, dy\, dx\, dz,$$
filling in the blanks with the appropriate limits.
Confirm your answer by calculating both integrals.

G.7.b) Calculate the triple integral
$$\int_0^1 \int_0^1 \int_0^{x^2+y^2} (x + y)\, dz\, dx\, dy.$$
Then rewrite the integral in the order
$$\int_-^- \int_-^- \int_-^- (x + y)\, dy\, dx\, dz,$$
filling in the blanks with the appropriate limits.
Confirm your answer by calculating it.

G.8) Linear equations and volume

This problem appears only in the electronic version.

354 Lesson 3.08 Transforming 3D Integrals

■ G.9) The volume conversion factor $V_{[x,y,z]}[u,v,w]$

G.9.a) Here's a cylinder plotted in uvw-space:

In[16]:=
```
Clear[s,t]
{{slow,shigh},{tlow,thigh}} = {{0,1},{0,2 Pi}};
uvwplot = ParametricPlot3D[{Cos[t],Sin[t],s},
  {s,slow,shigh},{t,tlow,thigh},
  BoxRatios->Automatic,
  Boxed->False,ViewPoint->CMView,
  AxesLabel->{"u","v","w"}];
```

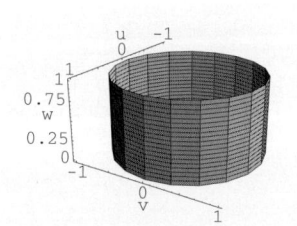

Go from uvw-space to xyz-space with:

In[17]:=
```
Clear[x,y,z]
{x[u_,v_,w_],y[u_,v_,w_],z[u_,v_,w_]} = {u,v,u v w};
```

Here's how that cylinder plots out in xyz-space:

In[18]:=
```
ParametricPlot3D[Evaluate[
  {x[Cos[t],Sin[t],s],y[Cos[t],
  Sin[t],s],z[Cos[t],Sin[t],s]}],
  {s,slow,shigh},{t,tlow,thigh},
  AxesLabel->{"x","y","z"}];
```

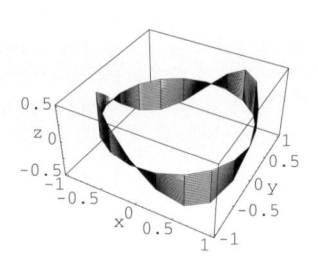

Damaged goods.

Look at the volume conversion factor $V_{(x,y,z)}[u,v,w]$:

In[19]:=
```
Clear[gradx,grady,gradz,Vxyz]
gradx[u_,v_,w_] = {D[x[u,v,w],u],D[x[u,v,w],v],D[x[u,v,w],w]};
grady[u_,v_,w_] = {D[y[u,v,w],u],D[y[u,v,w],v],D[y[u,v,w],w]};
gradz[u_,v_,w_] = {D[z[u,v,w],u],D[z[u,v,w],v],D[z[u,v,w],w]};
Vxyz[u_,v_,w_] = Abs[Det[{gradx[u,v,w],grady[u,v,w],gradz[u,v,w]}]]
```

Out[19]=
```
Abs[u v]
```

> Use the volume conversion factor to help explain why the xyz-plot of the original uvw-cylinder turned out the way it did.

G.9.b) Take any cleared functions $f[u]$ and $g[u,v]$.

Someone said that if you start in uvw-space with a solid R_{uvw} and go to xyz-space via:

$$x[u, v, w] = u,$$
$$y[u, v, w] = v + f[u]$$

and

$$z[u, v, w] = w + g[u, v],$$

then the xyz-plot, R_{xyz}, of R_{uvw} is likely to be a different shape from the original shape of R_{uvw}, but the xyz-volume measurement of R_{xyz} is likely to be the same as the uvw-volume measurement of R_{uvw}.

> What's your opinion?

LESSON 3.09

Spherical Coordinates

Basics

■ **B.1) Spherical coordinates**

You can specify a point in three dimensions as usual by its three coordinates $\{x, y, z\}$.

Another way to specify a point is to plant the x-, y-, and z-axes at $\{0, 0, 0\}$ and then run a stick from the $\{0, 0, 0\}$ to the point:

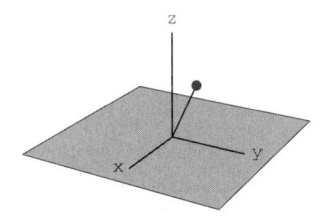

Then you measure the length r of this stick:

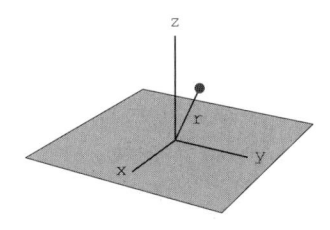

Next measure the angle s that this stick makes with the z-axis:

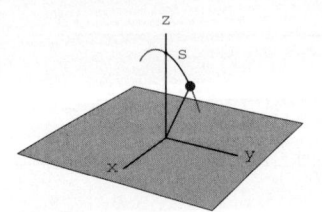

Finally, measure the angle t that the shadow of this ray on the xy-plane makes with the x-axis:

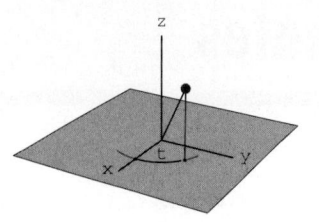

If you know the r, s, and t measurements for a point, then you can recover the $\{x, y, z\}$ coordinates via:

$$z = r\cos[s],$$

$$x = r\sin[s]\cos[t],$$

and

$$y = r\sin[s]\sin[t].$$

Most folks call the measurements $\{r, s, t\}$ the spherical coordinates of the point $\{x, y, z\}$.

B.1.a.i) Describe the sphere

$$x^2 + y^2 + z^2 = 9$$

in spherical coordinates.

Answer: Remember that the spherical coordinates r, s, and t reflect the following measurements:

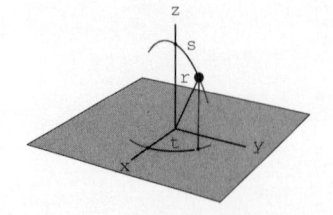

You get the whole sphere

$$x^2 + y^2 + z^2 = 9$$

by setting $r = 3$, running s from 0 to π and running t from 0 to 2π. So the description of the sphere in spherical coordinates is

$$r = 3, \qquad 0 \le s \le \pi \text{ and } 0 \le t \le 2\pi.$$

Check by plotting in xyz-space:

In[1]:=
```
Clear[x,y,z,r,s,t]
{x[r_,s_,t_],y[r_,s_,t_],z[r_,s_,t_]} =
{r Sin[s] Cos[t],r Sin[s] Sin[t],r Cos[s]};
{{slow,shigh},{tlow,thigh}} = {{0,Pi},{0,2 Pi}};
Clear[sphereplotter]; sphereplotter[s_,t_] =
{x[3,s,t],y[3,s,t],z[3,s,t]};
sphereplot = ParametricPlot3D[Evaluate[
sphereplotter[s,t]],{s,slow,shigh},{t,tlow,thigh},
Boxed->False,BoxRatios->Automatic,ViewPoint->CMView,
AxesLabel->{"x","y","z"}];
```

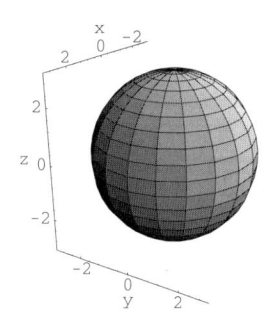

Looks good and feels great.

B.1.a.ii) Describe the part of the sphere

$$x^2 + y^2 + z^2 = 16$$

consisting of those points with $z \ge 0$ in spherical coordinates.

Answer: Remember that the spherical coordinates r, s, and t reflect the following measurements:

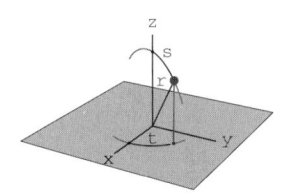

To get the description of the part of the sphere $x^2 + y^2 + z^2 = 16$ consisting of those points with $z \ge 0$ in spherical coordinates, you set $r = 4$, run s from 0 to $\pi/2$ and run t from 0 to 2π. This results in:

$$r = 4, \qquad 0 \le s \le \pi/2 \text{ and } 0 \le t \le 2\pi.$$

Lesson 3.09 Spherical Coordinates

Check:

In[2]:=
```
Clear[x,y,z,r,s,t]
{x[r_,s_,t_],y[r_,s_,t_],z[r_,s_,t_]} =
{r Sin[s] Cos[t],r Sin[s] Sin[t],r Cos[s]};
{{slow,shigh},{tlow,thigh}} = {{0,Pi/2},{0,2 Pi}};
Clear[sphereplotter]; sphereplotter[s_,t_] =
{x[4,s,t],y[4,s,t],z[4,s,t]};
hemisphereplot = ParametricPlot3D[Evaluate[
sphereplotter[s,t]],{s,slow,shigh},{t,tlow,thigh},
Boxed->False,BoxRatios->Automatic,ViewPoint->CMView,
AxesLabel->{"x","y","z"}];
```

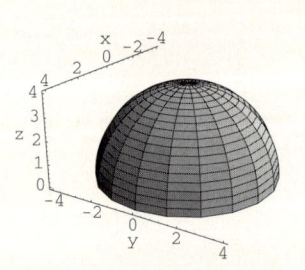

No problem-o.

B.1.b.i) Look at:

In[3]:=
```
Clear[x,y,z,r,s,t]
{x[r_,s_,t_],y[r_,s_,t_],z[r_,s_,t_]} =
{r Sin[s] Cos[t],r Sin[s] Sin[t],r Cos[s]};
{{slow,shigh},{tlow,thigh}} = {{0,Pi/4},{0,2 Pi}};
Clear[plotter];
plotter[s_,t_] = {x[2,s,t],y[2,s,t],z[2,s,t]};
ParametricPlot3D[Evaluate[plotter[s,t]],
{s,slow,shigh},{t,tlow,thigh},
Boxed->False,BoxRatios->Automatic,ViewPoint->CMView,
AxesLabel->{"x","y","z"}];
```

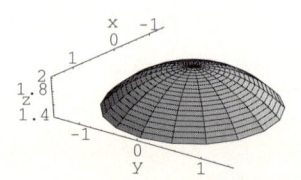

Explain why the plot turned out the way it did.

Answer: Note that s is allowed to run only from 0 to $\pi/4$. Remember that s is the indicated angle:

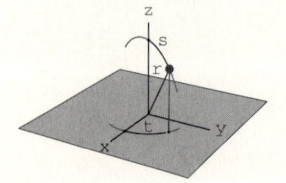

Limiting s to the interval $0 \leq s \leq \pi/4$ in the plot results in a plot of the top cap of the sphere

$$x^2 + y^2 + z^2 = 2^2.$$

Play with this by plotting with other intervals like $\pi/4 \leq s \leq \pi/2$.

B.1.b.ii) Look at:

In[4]:=
```
Clear[x,y,z,r,s,t]
{x[r_,s_,t_],y[r_,s_,t_],z[r_,s_,t_]} =
{r Sin[s] Cos[t],r Sin[s] Sin[t],r Cos[s]};
{{rlow,rhigh},{tlow,thigh}} = {{0,3},{0,2 Pi}};
Clear[plotter];
plotter[r_,t_] = {x[r,Pi/6,t],y[r,Pi/6,t],z[r,Pi/6,t]};
ParametricPlot3D[Evaluate[plotter[r,t]],
{r,rlow,rhigh},{t,tlow,thigh},
Boxed->False,BoxRatios->Automatic,ViewPoint->CMView,
AxesLabel->{"x","y","z"}];
```

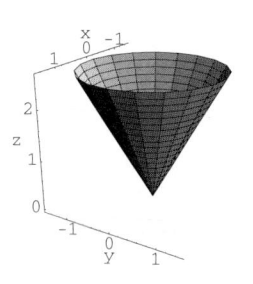

Pack in some good Baskin Robbins Ice Cream.

> Explain why the plot turned out the way it did.

Answer: This time s is held at $\pi/6$, but t and r are allowed to roam. Remember that s is the indicated angle:

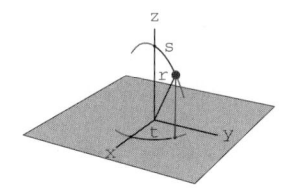

Holding $s = \pi/6$ results in a cone whose sides make an angle of $\pi/6$ radians with the z-axis.

For some variations, try:

In[5]:=
```
Clear[x,y,z,r,s,t]
{x[r_,s_,t_],y[r_,s_,t_],z[r_,s_,t_]} =
{r Sin[s] Cos[t],r Sin[s] Sin[t],r Cos[s]};
{{rlow,rhigh},{tlow,thigh}} = {{-2,3},{0,2 Pi}};
Clear[plotter];
plotter[r_,t_] = {x[r,Pi/6,t],y[r,Pi/6,t],z[r,Pi/6,t]};
ParametricPlot3D[Evaluate[plotter[r,t]],
{r,rlow,rhigh},{t,tlow,thigh},
Boxed->False,BoxRatios->Automatic,ViewPoint->CMView,
AxesLabel->{"x","y","z"}];
```

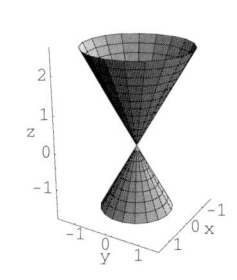

And:

In[6]:=
```
Clear[x,y,z,r,s,t]; Clear[plotter];
{x[r_,s_,t_],y[r_,s_,t_],z[r_,s_,t_]} =
 {r Sin[s] Cos[t],r Sin[s] Sin[t],r Cos[s]};
{{rlow,rhigh},{tlow,thigh}} = {{-2,3},{0,2 Pi/3}};
plotter[r_,t_] = {x[r,Pi/6,t],y[r,Pi/6,t],z[r,Pi/6,t]};
ParametricPlot3D[Evaluate[plotter[r,t]],
 {r,rlow,rhigh},{t,tlow,thigh},
 Boxed->False,BoxRatios->Automatic,ViewPoint->CMView,
 AxesLabel->{"x","y","z"}];
```

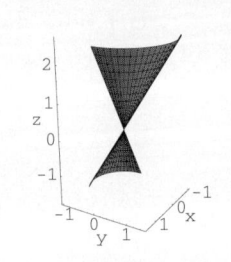

And you can go a little nuts:

In[7]:=
```
Clear[x,y,z,r,s,t,rad]
{x[r_,s_,t_],y[r_,s_,t_],z[r_,s_,t_]} =
 {r Sin[s] Cos[t],r Sin[s] Sin[t],r Cos[s]};
{tlow,thigh} = {0,2 Pi}; rad[t_] = 2 + 0.4 Sin[4 t];
Clear[plotter]; plotter[r_,t_] =
 {x[r rad[t],Pi/6,t],y[r rad[t],Pi/6,t],z[r rad[t],Pi/6,t]};
ParametricPlot3D[Evaluate[plotter[r,t]],
 {t,tlow,thigh},{r,0,1},PlotPoints->{35,2},
 Boxed->False,BoxRatios->Automatic,ViewPoint->CMView,
 AxesLabel->{"x","y","z"}];
```

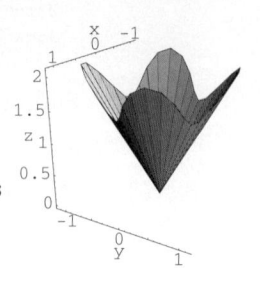

Play.

■ B.2) Integrating with spherical coordinates

Here is the volume conversion factor $V_{(x,y,z)}[r,s,t]$ for converting spherical coordinate rst-volume measurements into standard xyz-volume measurements:

In[8]:=
```
Clear[x,y,z,r,s,t]; Clear[gradx,grady,gradz,Vxyz]
x[r_,s_,t_] = r Sin[s] Cos[t];
y[r_,s_,t_] = r Sin[s] Sin[t]; z[r_,s_,t_] = r Cos[s];
gradx[r_,s_,t_] = {D[x[r,s,t],r],D[x[r,s,t],s],D[x[r,s,t],t]};
grady[r_,s_,t_] = {D[y[r,s,t],r],D[y[r,s,t],s],D[y[r,s,t],t]};
gradz[r_,s_,t_] = {D[z[r,s,t],r],D[z[r,s,t],s],D[z[r,s,t],t]};
Vxyz[r_,s_,t_] = Abs[Expand[Det[{gradx[r,s,t],grady[r,s,t],gradz[r,s,t]}],Trig->True]]
```

Out[8]=
$$\text{Abs}[r^2 \, \text{Sin}[s]]$$

In most situations you can get by with $0 \leq s \leq \pi$, so you usually don't have to multiply by -1, because $\sin[s] \geq 0$ for $0 \leq s \leq \pi$.

With these ground rules, $V_{(x,y,z)}[r,s,t]$ is given by:

In[9]:=
```
Clear[Vxyz]; Vxyz[r_,s_,t_] = r^2 Sin[s]
```

Out[9]=
$$r^2 \, \text{Sin}[s]$$

B.2.a) Use spherical coordinates to get a quick evaluation of

$$\iiint_{R_{xyz}} x^2 \, dx \, dy \, dz$$

where R_{xyz} is everything inside and on the sphere

$$x^2 + y^2 + z^2 = a^2.$$

Answer: If you have an integral

$$\iiint_{R_{xyz}} f[x, y, z] \, dx \, dy \, dz$$

and R_{xyz} is a solid sphere, then spherical coordinates are the way to go.

Enter $f[x, y, z] = x^2$:

In[10]:=
```
Clear[a,x,y,z,r,s,t,f,Vxyz]; f[x_,y_,z_] = x^2;
```

Remember that R_{xyz} can be described by $0 \leq r \leq a$, $0 \leq s \leq \pi$, and $0 \leq t \leq 2\pi$.

When you set $r = a$, all you get is the skin of the sphere. But when you run r from 0 to a, then you get the sphere and all that good stuff inside.

So

$$\iiint_{R_{xyz}} f[x, y, z] \, dx \, dy \, dz$$

$$= \int_0^{2\pi} \int_0^{\pi} \int_0^{a} f[x[r, s, t], y[r, s, t], z[r, s, t]] \, V_{(x,y,z)}[r, s, t] \, dr \, ds \, dt.$$

Put in the spherical coordinate change of variables:

In[11]:=
```
x[r_,s_,t_] = r Sin[s] Cos[t];
y[r_,s_,t_] = r Sin[s] Sin[t];
z[r_,s_,t_] = r Cos[s];
Vxyz[r_,s_,t_] = r^2 Sin[s];
```

Calculate

$$\iiint_{R_{xyz}} f[x, y, z] \, dx \, dy \, dz$$

$$= \int_0^{2\pi} \int_0^{\pi} \int_0^{a} f[x[r, s, t], y[r, s, t], z[r, s, t]] \, V_{(x,y,z)}[r, s, t] \, dr \, ds \, dt :$$

In[12]:=
```
Clear[a]
Integrate[f[x[r,s,t],y[r,s,t],z[r,s,t]] Vxyz[r,s,t],
{t,0,2 Pi},{s,0,Pi},{r,0,a}]
```

Out[12]=
$$\frac{5}{4\,a\,\text{Pi}}{15}$$

$$\frac{4\,a\,\text{Pi}^5}{15}$$

The computer did almost all the work, just the way it should. After all, what's a computer for?

Tutorials

■ T.1) Balls and eggs

T.1.a) Here's a plot of the sphere of radius 1 centered at $\{0,0,0\}$:

In[1]:=
```
Clear[x,y,z,r,s,t]; Clear[firstsphereplotter];
{x[r_,s_,t_],y[r_,s_,t_],z[r_,s_,t_]} =
{r Sin[s] Cos[t],r Sin[s] Sin[t],r Cos[s]};
{{slow,shigh},{tlow,thigh}} = {{0,Pi},{0,2 Pi}};
firstsphereplotter[s_,t_] =
{x[1,s,t],y[1,s,t],z[1,s,t]}; firstsphereplot =
ParametricPlot3D[Evaluate[firstsphereplotter[s,t]],
{s,slow,shigh},{t,tlow,thigh},Boxed->False,
BoxRatios->Automatic,ViewPoint->CMView,
AxesLabel->{"x","y","z"}];
```

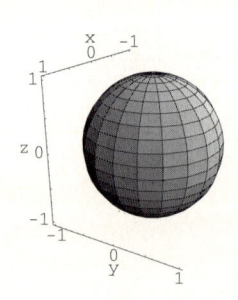

> Add the plot of the sphere of radius 0.2 centered at $\{1,1,1.5\}$ to this plot.

Answer: That's easy. Just copy, paste, and edit the code that made the original plot:

In[2]:=
```
center = {1,1,1.5}; Clear[x,y,z,r,s,t]
{x[r_,s_,t_],y[r_,s_,t_],z[r_,s_,t_]} =
center + {r Sin[s] Cos[t],r Sin[s] Sin[t],r Cos[s]};
r = 0.2; {{slow,shigh},{tlow,thigh}} =
{{0,Pi},{0,2 Pi}}; Clear[secondsphereplotter];
secondsphereplotter[s_,t_] = {x[r,s,t],y[r,s,t],z[r,s,t]};
secondsphereplot = ParametricPlot3D[Evaluate[
secondsphereplotter[s,t]],{s,slow,shigh},{t,tlow,thigh},
DisplayFunction->Identity];
Show[firstsphereplot,secondsphereplot,Boxed->False,
BoxRatios->Automatic,ViewPoint->CMView,
PlotRange->All,AxesLabel->{"x","y","z"},
DisplayFunction->$DisplayFunction];
```

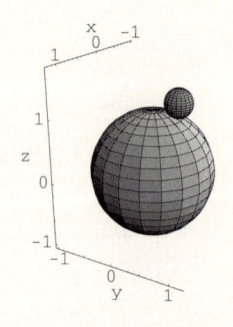

This plants the seeds for some interesting movies.

> **T.1.b)** Use spherical coordinates to get a quick evaluation of
> $$\iiint_{R_{xyz}} (x^2 + y^2)\, dx\, dy\, dz$$
> where R_{xyz} is the sphere of radius 0.1 centered on $\{2, 5, 0\}$.

Answer: Go with spherical coordinates based at $\{2, 5, 0\}$ and calculate the volume conversion factor:

In[3]:=
```
Clear[x,y,z,r,s,t]
center = {2,5,0};
{x[r_,s_,t_],y[r_,s_,t_],z[r_,s_,t_]} =
center + {r Sin[s] Cos[t],r Sin[s] Sin[t],r Cos[s]};
{{rlow,rhigh},{slow,shigh},{tlow,thigh}} =
{{0,0.1},{0,Pi},{0,2 Pi}};
Clear[gradx,grady,gradz,Vxyz]
gradx[r_,s_,t_] = {D[x[r,s,t],r],D[x[r,s,t],s],D[x[r,s,t],t]};
grady[r_,s_,t_] = {D[y[r,s,t],r],D[y[r,s,t],s],D[y[r,s,t],t]};
gradz[r_,s_,t_] = {D[z[r,s,t],r],D[z[r,s,t],s],D[z[r,s,t],t]};
Vxyz[r_,s_,t_] = Expand[Det[{gradx[r,s,t],
grady[r,s,t],gradz[r,s,t]}],Trig->True]
```

Out[3]=
$$r^2\, \text{Sin}[s]$$

Because you are going with $0 \le s \le \pi$, you know that $r^2 \sin[s]$ can never go negative, so you don't have to multiply by -1. Now you are ready to calculate

$$\iiint_{R_{xyz}} x^2 + y^2\, dx\, dy\, dz =$$

$$\iiint_{R_{rst}} \left(x[r,s,t]^2 + y[r,s,t]^2\right) V_{(x,y,z)}[r,s,t]\, dr\, ds\, dt$$

$$= \int_{tlow}^{thigh} \int_{slow}^{shigh} \int_{rlow}^{rhigh} \left(x[r,s,t]^2 + y[r,s,t]^2\right) V_{(x,y,z)}[r,s,t]\, dr\, ds\, dt :$$

In[4]:=
```
Integrate[(x[r,s,t]^2 + y[r,s,t]^2) Vxyz[r,s,t],
{t,tlow,thigh},{s,slow,shigh},{r,rlow,rhigh}]
```

Out[4]=
0.038672 Pi

Routine stuff.

T.1.c.i) Look at this:

Lesson 3.09 Spherical Coordinates

In[5]:=
```
Clear[x,y,z,r,s,t]; Clear[plotter];
{x[r_,s_,t_],y[r_,s_,t_],z[r_,s_,t_]} =
{2 r Sin[s] Cos[t],5 r Sin[s] Sin[t],3 r Cos[s]};
{{slow,shigh},{tlow,thigh}} = {{0,Pi},{0,2 Pi}};
plotter[s_,t_] = {x[1,s,t],y[1,s,t],z[1,s,t]};
ParametricPlot3D[Evaluate[plotter[s,t]],
{s,slow,shigh},{t,tlow,thigh},Boxed->False,
BoxRatios->Automatic,ViewPoint->CMView,
AxesLabel->{"x","y","z"}];
```

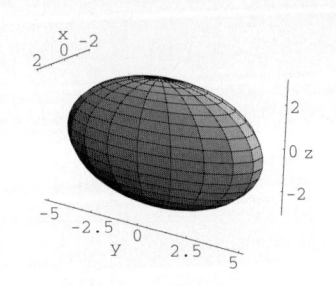

What do folks call this skin?

Answer: It all depends on who you are talking to. Common-sense folks call it an egg. Fancy folks call it an ellipsoid.

T.1.c.ii) Agree that R_{xyz} is everything inside and on the skin plotted in part T.1.c.i) above. Calculate
$$\iiint_{R_{xyz}} (x^2 + y^2 + z^2) \, dx \, dy \, dz.$$

Answer: You plotted this skin by going with
$$\{x[r,s,t], y[r,s,t], z[r,s,t]\} = \{2r\sin[s]\cos[t], 5r\sin[s]\sin[t], 3r\cos[s]\}$$
and keeping $r = 1$, running s from slow to shigh and running t from tlow to thigh. Remembering that R_{xyz} is everything inside and on this skin, you see that you can describe R_{xyz} with
$$\{x[r,s,t], y[r,s,t], z[r,s,t]\} = \{2r\sin[s]\cos[t], 5r\sin[s]\sin[t], 3r\cos[s]\}$$
by running r from 0 to 1, running s from slow to shigh, and running t from tlow to thigh. In short,
$$\iiint_{R_{xyz}} (x^2 + y^2 + z^2) \, dx \, dy \, dz$$
$$= \iiint_{R_{rst}} \left(x[r,s,t]^2 + y[r,s,t]^2 + z[r,s,t]^2 \right) V_{(x,y,z)}[r,s,t] \, dr \, ds \, dt$$
$$= \int_{tlow}^{thigh} \int_{slow}^{shigh} \int_0^1 \left(x[r,s,t]^2 + y[r,s,t]^2 + z[r,s,t]^2 \right) V_{(x,y,z)}[r,s,t] \, dr \, ds \, dt.$$

Calculate the volume conversion factor:

In[6]:=
```
Clear[gradx,grady,gradz,Vxyz]
gradx[r_,s_,t_] = {D[x[r,s,t],r],D[x[r,s,t],s],D[x[r,s,t],t]};
grady[r_,s_,t_] = {D[y[r,s,t],r],D[y[r,s,t],s],D[y[r,s,t],t]};
gradz[r_,s_,t_] = {D[z[r,s,t],r],D[z[r,s,t],s],D[z[r,s,t],t]};
Vxyz[r_,s_,t_] = Expand[Det[{gradx[r,s,t],grady[r,s,t],gradz[r,s,t]}],Trig->True]
```

Out[6]=
$$30\,r^2\,\text{Sin}[s]$$

Here comes the calculation of

$$\iiint_{R_{xyz}} \left(x^2 + y^2 + z^2\right)\,dx\,dy\,dz$$

$$= \int_{\text{tlow}}^{\text{thigh}} \int_{\text{slow}}^{\text{shigh}} \int_0^1 \left(x[r,s,t]^2 + y[r,s,t]^2 + z[r,s,t]^2\right) V_{(x,y,z)}[r,s,t]\,dr\,ds\,dt :$$

In[7]:=
```
Integrate[(x[r,s,t]^2 + y[r,s,t]^2 + z[r,s,t]^2) Vxyz[r,s,t],
{t,tlow,thigh},{s,slow,shigh},{r,0,1}]
```

Out[7]=
304 Pi

■ T.2) Ice cream cones and tops

T.2.a.i) Here's an ice cream cone plotted in two steps:

In[8]:=
```
Clear[x,y,z,r,s,t,radial]; Clear[sideplotter]
x[r_,s_,t_] = r Sin[s] Cos[t];
y[r_,s_,t_] = r Sin[s] Sin[t]; z[r_,s_,t_] = r Cos[s];
sideplotter[r_,t_] = {x[r,Pi/6,t],y[r,Pi/6,t],z[r,Pi/6,t]};
side = ParametricPlot3D[Evaluate[sideplotter[r,t]],
{r,0,3},{t,0,2 Pi},Boxed->False,BoxRatios->Automatic,
ViewPoint->CMView,AxesLabel->{"x","y","z"}];
```

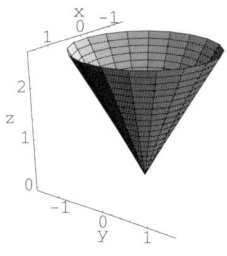

In[9]:=
```
Clear[capplotter]; capplotter[s_,t_] =
{x[3,s,t],y[3,s,t],z[3,s,t]};
cap = ParametricPlot3D[Evaluate[capplotter[s,t]],
{s,0,Pi/6},{t,0,2 Pi},Boxed->False,
BoxRatios->Automatic,
ViewPoint->CMView,AxesLabel->{"x","y","z"}];
```

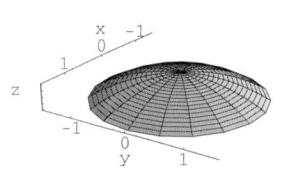

Here's the whole ice cream cone:

In[10]:=
```
Show[cap,side,Boxed->False,BoxRatios->Automatic,
ViewPoint->CMView,AxesLabel->{"x","y","z"}];
```

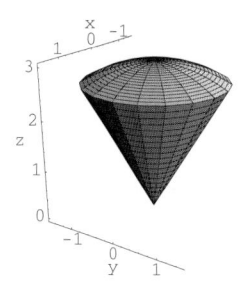

Stingy serving.

Lesson 3.09 Spherical Coordinates

> Imagine that the ice cream is packed solidly all the way to the bottom, and measure the volume of the ice cream.

Answer: When you review the plotting instructions above, you find that you plotted the side by plotting

$$\left\{x\left[r,\frac{\pi}{6},t\right], y\left[r,\frac{\pi}{6},t\right], z\left[r,\frac{\pi}{6},t\right]\right\}$$

with $0 \le r \le 3$ and $0 \le t \le 2\pi$. You also find that you plotted the cap by plotting

$$\{x[3, s, t], y[3, s, t], z[3, s, t]\}$$

with $0 \le s \le \pi/6$ and $0 \le t \le 2\pi$.

When you go with

$$0 \le r \le 3, \quad 0 \le s \le \frac{\pi}{6} \quad \text{and} \quad 0 \le t \le 2\pi,$$

the points $\{x[r, s, t], y[r, s, t], z[r, s, t]\}$ will sweep out the skin of the top and everything inside. Call the skin of the top and everything inside by the name R_{xyz}. To measure the volume of R, all you gotta do is calculate

$$\text{volume} = \iiint_{R_{xyz}} dx\, dy\, dz$$

$$= \iiint_{R_{rst}} V_{(xx,y,z)}[r, s, t]\, dr\, ds\, dt$$

$$= \int_0^{2\pi} \int_0^{\pi/6} \int_0^3 V_{(xx,y,z)}[r, s, t]\, dr\, ds\, dt.$$

In[11]:=
```
Clear[gradx,grady,gradz,Vxyz]
gradx[r_,s_,t_] = {D[x[r,s,t],r],D[x[r,s,t],s],D[x[r,s,t],t]};
grady[r_,s_,t_] = {D[y[r,s,t],r],D[y[r,s,t],s],D[y[r,s,t],t]};
gradz[r_,s_,t_] = {D[z[r,s,t],r],D[z[r,s,t],s],D[z[r,s,t],t]};
Vxyz[r_,s_,t_] = Expand[Det[{gradx[r,s,t],grady[r,s,t],gradz[r,s,t]}],Trig->True]
```

Out[11]=
$$r^2 \, \text{Sin}[s]$$

Because $0 \le s \le \pi/6$, this can never be negative; so

$$\text{volume} = \iiint_{R_{xyz}} dx\, dy\, dz$$

$$= \iiint_{R_{rst}} V_{(xx,y,z)}[r, s, t]\, dr\, ds\, dt$$

$$= \int_0^{2\pi} \int_0^{\pi/6} \int_0^3 V_{(xx,y,z)}[r, s, t]\, dr\, ds\, dt$$

measures out to:

In[12]:=
```
volume = Integrate[Vxyz[r,s,t],{t,0,2 Pi},{s,0,Pi/6},{r,0,3}]
```

Out[12]=

$$2\left(9 - \frac{3^{5/2}}{2}\right) Pi$$

That measurement is a bit too esoteric; check it out in plain decimal cubic units:

In[13]:=
```
N[volume]
```

Out[13]=
7.57608

Very, very easy.

Once again, if you can plot something, you've got a good chance of setting up the integral to measure its volume.

T.2.a.ii) You can use spherical coordinates to come up with your own custom-made coordinate systems. Here's a sample involving a top. It's only a small variant of what you did in part T.2.a.i) above.

In[14]:=
```
Clear[x,xx,y,yy,z,zz,r,s,t,radial]
xx[r_,s_,t_] = r Sin[s] Cos[t];
yy[r_,s_,t_] = r Sin[s] Sin[t];
zz[r_,s_,t_] = r Cos[s];
radial[s_] = 3 + E^(-5 s);
{x[r_,s_,t_],y[r_,s_,t_],z[r_,s_,t_]} =
{xx[r radial[s],s,t],yy[r radial[s],s,t],
zz[r radial[s],s,t]}; Clear[sideplotter,radial];
radial[s_] = 3 + E^(-5 s);
sideplotter[r_,t_] = {x[r,Pi/6,t],y[r,Pi/6,t],z[r,Pi/6,t]};
side = ParametricPlot3D[Evaluate[sideplotter[r,t]],
{r,0,1},{t,0,2 Pi},Boxed->False,BoxRatios->Automatic,
ViewPoint->CMView,AxesLabel->{"x","y","z"}];
```

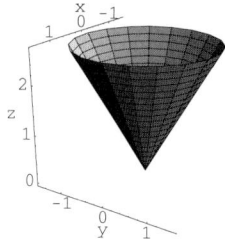

In[15]:=
```
Clear[capplotter]
capplotter[s_,t_] = {x[1,s,t],y[1,s,t],z[1,s,t]};
cap = ParametricPlot3D[Evaluate[capplotter[s,t]],
{s,0,Pi/6},{t,0,2 Pi},Boxed->False,
BoxRatios->Automatic,ViewPoint->CMView,
AxesLabel->{"x","y","z"}];
```

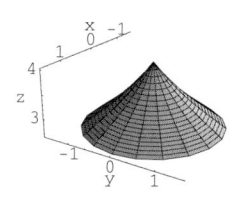

Here's the whole top:

Lesson 3.09 Spherical Coordinates

In[16]:=
```
Show[cap,side,Boxed->False,BoxRatios->Automatic,
ViewPoint->CMView,AxesLabel->{"x","y","z"}];
```

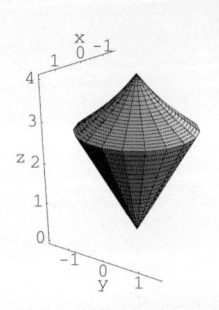

Spiffy.

> Measure the volume inside this spinner.

Answer: When you review the plotting instructions above, you find that you plotted the side by plotting

$$\left\{x\left[r,\frac{\pi}{6},t\right],y\left[r,\frac{\pi}{6},t\right],z\left[r,\frac{\pi}{6},t\right]\right\}$$

with $0 \le r \le 1$ and $0 \le t \le 2\pi$. You also find that you plotted the cap by plotting

$$\{x[1,s,t], y[1,s,t], z[1,s,t]\}$$

with $0 \le s \le \pi/6$ and $0 \le t \le 2\pi$. When you go with $0 \le r \le 1$, $0 \le s \le \pi/6$, and $0 \le t \le 2\pi$, the points $\{x[r,s,t], y[r,s,t], z[r,s,t]\}$ will sweep out the skin of the top and everything inside. Call the skin of the top and everything inside R_{xyz}. To measure the volume of R_{xyz}, all you have to do is calculate

$$\text{volume} = \iiint_{R_{xyz}} dx\, dy\, dz$$

$$= \iiint_{R_{rst}} V_{(xx,y,z)}[r,s,t]\, dr\, ds\, dt$$

$$= \int_0^{2\pi} \int_0^{\pi/6} \int_0^1 V_{(xx,y,z)}[r,s,t]\, dr\, ds\, dt :$$

In[17]:=
```
Clear[gradx,grady,gradz,Vxyz]
gradx[r_,s_,t_] = {D[x[r,s,t],r],D[x[r,s,t],s],D[x[r,s,t],t]};
grady[r_,s_,t_] = {D[y[r,s,t],r],D[y[r,s,t],s],D[y[r,s,t],t]};
gradz[r_,s_,t_] = {D[z[r,s,t],r],D[z[r,s,t],s],D[z[r,s,t],t]};
Vxyz[r_,s_,t_] = Expand[Det[{gradx[r,s,t],grady[r,s,t],gradz[r,s,t]}],Trig->True]
```

Out[17]=

$$27\, r^2\, \text{Sin}[s] + \frac{r^2\, \text{Sin}[s]}{15\, s\, E} + \frac{9\, r^2\, \text{Sin}[s]}{10\, s\, E} + \frac{27\, r^2\, \text{Sin}[s]}{5\, s\, E}$$

Because $0 \le s \le \pi/6$, this can never be negative, so

$$\text{volume} = \iiint_{R_{xyz}} dx\, dy\, dz$$
$$= \iiint_{R_{rst}} V_{(xx,y,z)}[r,s,t]\, dr\, ds\, dt$$
$$= \int_0^{2\pi} \int_0^{\pi/6} \int_0^1 V_{(xx,y,z)}[r,s,t]\, dr\, ds\, dt$$

measures out to:

In[18]:=
  volume = Integrate[Vxyz[r,s,t],{t,0,2 Pi},{s,0,Pi/6},{r,0,1}]

Out[18]=

$$2\left(\dfrac{4173916}{445107} - \dfrac{5\left(102717 + \dfrac{1313}{E^{(5\,Pi)/3}} + \dfrac{17628}{E^{(5\,Pi)/6}}\right)}{593476\, E^{(5\,Pi)/6}} - \dfrac{8011926 + \dfrac{1313}{E^{(5\,Pi)/2}} + \dfrac{26442}{E^{(5\,Pi)/3}} + \dfrac{308151}{E^{(5\,Pi)/6}}}{593476\, \text{Sqrt}[3]}\right) Pi$$

That measurement is way too abstruse; check it out in plain decimal cubic units:

In[19]:=
  N[volume]

Out[19]=
9.40702

Easy. Once again, if you can plot something, you've got a good chance of setting up the integral that measures its volume.

■ T.3) Jennifer Phillips's flower

T.3.a) When she was asked to do some art with spherical coordinates, 1992 Calculus&Mathematica student Jennifer Phillips responded with the flower you are about to see. She commented, "I only did one, but I like it a lot." So will you.

> Activate, enjoy, read her code, and think about what you too can do with spherical coordinates and tubes.

The bloom:

In[20]:=
```
Clear[x,y,z,r,s,t]
x[r_,s_,t_] = r Sin[s] Cos[t];
y[r_,s_,t_] = r Sin[s] Sin[t];
z[r_,s_,t_] = r Cos[s];
r[s_,t_] = 1.5(1 - Sin[16 s]) + 0.5(1 - Cos[4 t]);
bloom = ParametricPlot3D[{x[r[s,t],s,t],
y[r[s,t],s,t],z[r[s,t],s,t]},{s,0,Pi/2},{t,0,2 Pi},
Axes->None,ViewPoint->CMView,
Boxed->False,PlotRange->All];
```

The stem:

In[21]:=
```
Clear[s,P]
P[s_] = {-s,-2 s,s^2};
Clear[unittan,mainnormal,mainunitnormal,
binormal,secondunitnormal]
unittan[s_] = P'[s]/Sqrt[Expand[P'[s].P'[s]]];
mainnormal[s_] = Together[D[unittan[s],s]];
mainunitnormal[s_] = Simplify[mainnormal[s]/
Sqrt[mainnormal[s].mainnormal[s]]];
binormal[s_] = Simplify[
Det[{{{1,0,0},{0,1,0},{0,0,1}},unittan[s],
mainunitnormal[s]}]];
r = 0.3;Clear[stem,t]
stem[s_,t_] = Together[P[s] +
r Cos[t] mainunitnormal[s] +
r Sin[t] binormal[s]] + {2,4,-4};
stemplot = ParametricPlot3D[stem[s,t],
{s,0,2},{t,0,2 Pi},Axes->None,ViewPoint->CMView,
Boxed->False,PlotRange->All];
```

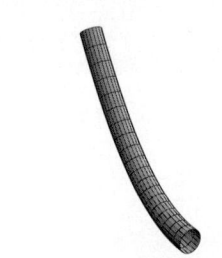

A leaf:

In[22]:=
```
Clear[x,y,z,r,s,t]
x[r_,s_,t_] = r Sin[s] Cos[t] + 2;
y[r_,s_,t_] = r Sin[s] Sin[t] - 2;
z[r_,s_,t_] = r Cos[s] + 2;
r[s_] = 3.3(1 - Sin[2 s]);
leaf = ParametricPlot3D[{z[r[s],s,t],
x[r[s],s,t],y[r[s],s,t]},{s,0,Pi/4},{t,0,Pi},
Axes->None,ViewPoint->CMView,Boxed->False,
PlotRange->All];
```

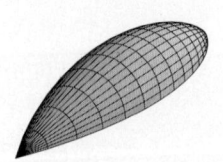

All together:

In[23]:=
```
Show[bloom,stemplot,leaf,Axes->None,ViewPoint->CMView,
  Boxed->False,PlotRange->All];
```

Exquisite! Way to go, Jennifer!

Give It a Try

Experience with the starred (★) problems will be useful for understanding developments later in the course.

■ G.1) Plotting with spherical coordinates★

G.1.a) Here are the formulas for spherical coordinates:

In[1]:=
```
Clear[x,y,z,r,s,t]
x[r_,s_,t_] = r Sin[s] Cos[t];
y[r_,s_,t_] = r Sin[s] Sin[t];
z[r_,s_,t_] = r Cos[s];
```

> Account for why the following plots turn out the way they do:

In[2]:=
```
ParametricPlot3D[{x[4,s,t],y[4,s,t],z[4,s,t]},
  {s,0,Pi/4},{t,0,Pi},Axes->Automatic];
```

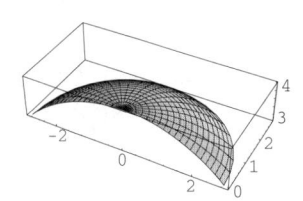

In[3]:=
```
ParametricPlot3D[
  {x[r,s,Pi/3],y[r,s,Pi/3],z[r,s,Pi/3]},
  {s,Pi/3,2 Pi/3},{r,0,3},Axes->Automatic];
```

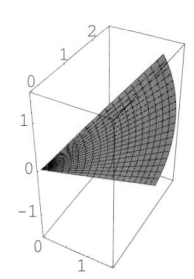

```
In[4]:=
ParametricPlot3D[
{x[r,Pi/6,t],y[r,Pi/6,t],z[r,Pi/6,t]},
{r,1,2},{t,0, Pi},Axes->Automatic];
```

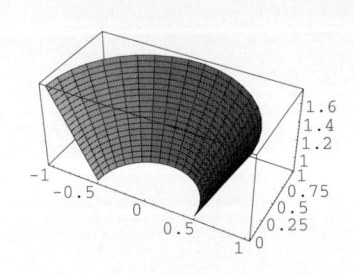

G.1.b.i) Describe the part of the sphere
$$x^2 + y^2 + z^2 = 4$$
consisting of those points with $y \geq 0$ in spherical coordinates.

G.1.b.ii) Describe the cone
$$z = \sqrt{x^2 + y^2}$$
in spherical coordinates.

■ G.2) Integrating*

G.2.a) A sphere of radius rad contains a volume of $4\pi \text{rad}^3/3$ cubic units.

Let R be the region inside and on the sphere
$$x^2 + y^2 + z^2 = \text{rad}^2$$
and use spherical coordinates to measure
$$\text{volume of } R = \iiint_R dx\, dy\, dz$$
to give a quick explanation of how this formula comes to be.

G.2.b) Calculate
$$\iiint_{R_{xyz}} (x^2 + y^2 + z^2)\, dx\, dy\, dz$$
where R_{xyz} is the region consisting of everything between and on the spheres $x^2 + y^2 + z^2 = a^2$ and $x^2 + y^2 + z^2 = b^2$. Assume $0 < a < b$.

G.2.c) A spherical shell whose inner radius is 2 cm and whose outer radius is 8 cm is made of material of varying density. At a point $\{x, y, z\}$, the density of the material is 0.39 times the square of the distance to the center of the sphere. Assume all air has been pumped out of the hollow interior.

> Does the shell sink or float in water?

G.2.d) Measure the volume of the solid region that is inside the sphere
$$x^2 + y^2 + z^2 = 4,$$
but is outside the cylinder
$$x^2 + y^2 = 1.$$

G.2.e) Calculate
$$\iiint_{R_{xyz}} \left(\frac{x}{a}\right)^2 + \left(\frac{y}{b}\right)^2 + \left(\frac{z}{c}\right)^2 \, dx \, dy \, dz,$$
where R_{xyz} is the region consisting of everything inside and on the ellipsoid
$$\left(\frac{x}{a}\right)^2 + \left(\frac{y}{b}\right)^2 + \left(\frac{z}{c}\right)^2 = 1.$$

■ G.3) Ice cream cones

This problem appears only in the electronic version.

■ G.4) Skins

G.4.a) Here's the skin of a mutant solid:

```
In[5]:=
    Clear[x,xx,y,yy,y,zz,r,radial,s,t]
    {xx[r_,s_,t_],yy[r_,s_,t_],zz[r_,s_,t_]} =
    {r Sin[s] Cos[t],r Sin[s] Sin[t],r Cos[s]};
    radial[s_] = 1.5(1 - Sin[2 s]);
    {x[r_,s_,t_],y[r_,s_,t_],z[r_,s_,t_]} =
    {xx[r radial[s],s,t],yy[r radial[s],s,t],
    zz[r radial[s],s,t]}; mutant =
    ParametricPlot3D[{x[1,s,t],y[1,s,t],z[1,s,t]},
    {s,0,Pi/2},{t,0,2 Pi},Axes->Automatic,
    ViewPoint->CMView,PlotRange->All];
```

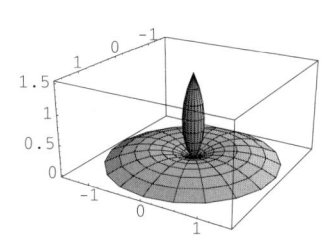

376   Lesson 3.09 Spherical Coordinates

> Measure the volume of the whole mutant solid.
>
> Then measure the volume of the base and the centerpiece, and check your first measurement by adding the second and third measurements.

G.4.b) Here is the skin of the shell of a young snail:

In[6]:=
```
Clear[x,y,z,radial,s,t]
x[r_,s_,t_] = r Sin[s] Cos[t];
y[r_,s_,t_] = r Sin[s] Sin[t];
z[r_,s_,t_] = r Cos[s];
radial[t_] = 1.5 (1 - Cos[t/4]);
ParametricPlot3D[{x[radial[t],s,t],
y[radial[t],s,t],z[radial[t],s,t]},
{s,0,Pi/2},{t,0,2 Pi},AxesLabel->{"x","y","z"},
ViewPoint->CMView,PlotRange->All];
```

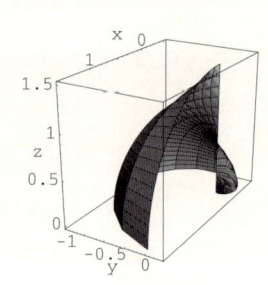

> Measure the volume under this shell and over the xy-floor.

G.4.c) Here's the skin of another snail shell:

In[7]:=
```
Clear[x,y,z,r,s,t]
x[r_,s_,t_] = r Sin[s] Cos[t];
y[r_,s_,t_] = r Sin[s] Sin[t];
z[r_,s_,t_] = r Cos[s];
r[t_] = 1.5 (1 - Cos[t/8]);
ParametricPlot3D[{x[r[t],s,t],y[r[t],s,t],
z[r[t],s,t]},{s,0,Pi/2},{t,0,4 Pi},
AxesLabel->{"x","y","z"},
ViewPoint->CMView,PlotRange->All];
```

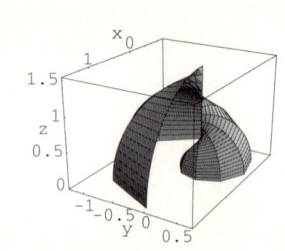

> Measure the volume under this shell and over the xy-floor.

■ **G.5) Floaters**

G.5.a) All measurements here are in centimeters.

A solid floater of radius r is to be made in the shape of the sphere

$$x^2 + y^2 + z^2 = r^2.$$

The density$[x, y, z]$ of the floater is to be proportional to the distance of the point $\{x, y, z\}$ from $\{0, 0, 0\}$; this means

$$\text{density}[x, y, z] = k\sqrt{x^2 + y^2 + z^2} \text{ grams/cm}^3.$$

> Give a formula for the function $k[r]$ such that this beast of radius r will float in water if $0 < k < k[r]$ and will sink if $k > k[r]$.
>
> Plot $k[r]$ as a function of r.
>
> Does $k[r]$ have a limiting value
>
> $$\lim_{r \to \infty} k[r]?$$

G.5.b) A cubic centimeter of platinum weighs 21.4 grams. A cubic centimeter of ethyl alcohol weighs 0.81 grams.

> Your job is to build a hollow platinum sphere 100 centimeters in diameter with the outer wall h centimeters thick so that the hollow sphere will float in the ethyl alcohol. How small does h have to be to pull this off?

■ G.6) Earth and moon

G.6.a.i) The radius of the earth is approximately 6400 km. The diameter of the moon is approximately 1750 km. The moon rotates about the earth in an approximately circular orbit, and the distance from the center of the earth to the center of the moon is approximately 384,000 km.

> Make a true scale plot showing the earth and moon in position as appropriately sized spheres.

G.6.a.ii) Some scientists believe that the moon formed as a result of a huge collision between the earth and a large interplanetary body.

See William K. Hartman's book, *Astronomy: The Cosmic Journey*, Wadsworth, Belmont, California, 1991, p. 163.

Many of these same scientists believe that at some time long, long ago, the moon was in circular orbit about the earth, but the distance between the center of the earth and the center of the moon was only about 70,000 km.

> Make a true scale plot showing the earth and moon in this position as appropriately sized spheres.

■ G.7) Star Wars

Calculus&*Mathematica* thanks space engineer Lynn Purser of Teledyne-Brown Engineering, Huntsville, Alabama, for suggesting a problem like this in his talk at the First Annual *Mathematica* Conference in 1990.

Lesson 3.09 Spherical Coordinates

Here's a curve and an ellipsoid:

In[8]:=
```
Clear[h,P]; Clear[x,y,z,r,s,t]; Clear[plotter];
P[h_] = {h - 12,2 h,0.1 h(h - 10) + h/2};
{hlow,hhigh} = {1,15};
curveplot = ParametricPlot3D[P[h],{h,hlow,hhigh},
DisplayFunction->Identity];
{x[r_,s_,t_],y[r_,s_,t_],z[r_,s_,t_]} =
{r Sin[s] Cos[t],r Sin[s] Sin[t],2 r Cos[s]};
{{slow,shigh},{tlow,thigh}} = {{0,Pi},{0,2 Pi}};
plotter[s_,t_] = {-9,5,3} + {x[1,s,t],y[1,s,t],z[1,s,t]};
ellipseplot = ParametricPlot3D[Evaluate[plotter[s,t]],
{s,slow,shigh},{t,tlow,thigh},DisplayFunction->Identity];
Show[ellipseplot,curveplot,BoxRatios->Automatic,
ViewPoint->CMView,PlotRange->All,
AxesLabel->{"x","y","z"},
DisplayFunction->$DisplayFunction];
```

Here's how you can pick up the ellipsoid and move it so that the center of the ellipsoid is at $P[h]$ on the curve, and aligned so that the long axis of the ellipsoid shares ink with the unit tangent vector to the curve at the point $P[h]$.

Here's a plot in the case $h = 8$:

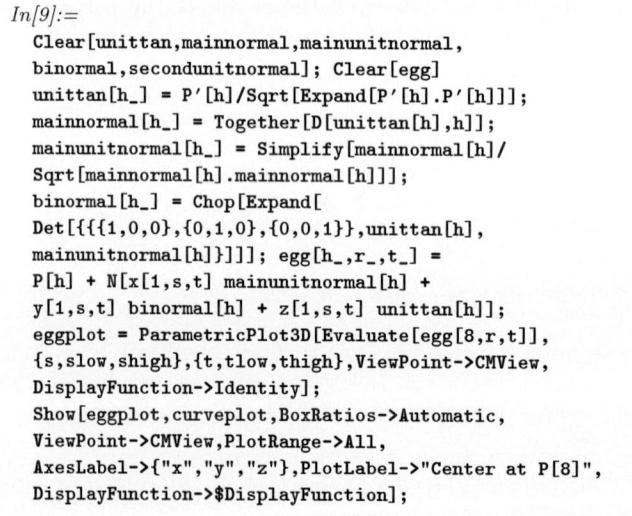

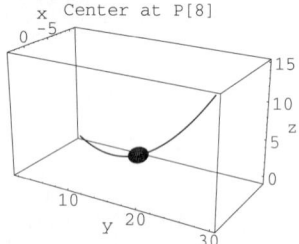

In[9]:=
```
Clear[unittan,mainnormal,mainunitnormal,
binormal,secondunitnormal]; Clear[egg]
unittan[h_] = P'[h]/Sqrt[Expand[P'[h].P'[h]]];
mainnormal[h_] = Together[D[unittan[h],h]];
mainunitnormal[h_] = Simplify[mainnormal[h]/
Sqrt[mainnormal[h].mainnormal[h]]];
binormal[h_] = Chop[Expand[
Det[{{{1,0,0},{0,1,0},{0,0,1}},unittan[h],
mainunitnormal[h]}]]]; egg[h_,r_,t_] =
P[h] + N[x[1,s,t] mainunitnormal[h] +
y[1,s,t] binormal[h] + z[1,s,t] unittan[h]];
eggplot = ParametricPlot3D[Evaluate[egg[8,r,t]],
{s,slow,shigh},{t,tlow,thigh},ViewPoint->CMView,
DisplayFunction->Identity];
Show[eggplot,curveplot,BoxRatios->Automatic,
ViewPoint->CMView,PlotRange->All,
AxesLabel->{"x","y","z"},PlotLabel->"Center at P[8]",
DisplayFunction->$DisplayFunction];
```

Play with other values of h.

Now stay with the same curve, but look at this cone:

```
In[10]:=
  Clear[x,y,z,r,s,t]; Clear[plotter];
  {x[r_,s_,t_],y[r_,s_,t_],z[r_,s_,t_]} =
  {r Sin[s] Cos[t],r Sin[s] Sin[t],r Cos[s]};
  {{rlow,rhigh},{tlow,thigh}} = {{0,5},{0,2 Pi}};
  plotter[s_,t_] = {-9,5,3} +
  {x[r,Pi/8,t],y[r,Pi/8,t],z[r,Pi/8,t]};
  coneplot = ParametricPlot3D[Evaluate[plotter[s,t]],
  {r,rlow,rhigh},{t,tlow,thigh},PlotPoints->{2,Automatic},
  DisplayFunction->Identity]; Show[coneplot,curveplot,
  BoxRatios->Automatic,ViewPoint->CMView,
  PlotRange->All,AxesLabel->{"x","y","z"},
  DisplayFunction->$DisplayFunction];
```

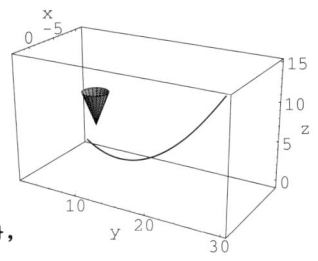

At the tip of this cone is a phaser (zapper). The cone itself represents the range of the phaser in the sense that any object that finds itself in the cone will be immediately disintegrated by the phaser. Any object outside the cone is immune to attack by the phaser.

G.7.a.i) Your mission is to take the tip of the cone, and move it to a point $P[h]$ on the curve so that the central axis of the cone is aligned with the unit tangent to the curve at $P[h]$, and directed so that the open end points in the direction of the parameterization of the curve.

Plot in the cases $h = 4, 9,$ and 15.

G.7.a.ii) Stay with the same curve as in part G.7.a.i) above. A hostile space station is anchored at the point $\{-3.2, 17.6, 3.5\}$.

If you move your phaser as above along the given curve, will your phaser be able to zap this space station?

Back up your answer with a decisive plot showing the cone eating the point or showing that the cone cannot eat the point.

G.7.a.iii) Stay with the same curve as in parts G.7.a.i) and G.7.a.ii) above. When you think about it, you see that there are two ways for an object to become vulnerable to the phaser. The most obvious way is for the object to find itself at a point sitting in the opening at the wide end of the cone.

Give a plot of the skin of the solid consisting of all points that are vulnerable this way.

G.7.a.iv) For some curves, there is another way for an object to become vulnerable to the phaser.

Say what this way is and discuss the problems you might encounter in plotting these additional vulnerable points.

■ G.8) Planes between spheres

Here are two spheres:

In[11]:=
```
Clear[x,y,z,r,s,t]; center1 = {2,-1,0};
{x[r_,s_,t_],y[r_,s_,t_],z[r_,s_,t_]} =
center1 + {r Sin[s] Cos[t],r Sin[s] Sin[t],r Cos[s]};
r1 = 1; {{slow,shigh},{tlow,thigh}} =
{{0,Pi},{0,2 Pi}}; Clear[sphere1plotter];
sphere1plotter[s_,t_] = {x[r1,s,t],y[r1,s,t],z[r1,s,t]};
sphere1plot = ParametricPlot3D[Evaluate[
sphere1plotter[s,t]],{s,slow,shigh},{t,tlow,thigh},
DisplayFunction->Identity]; center2 = {1,0,1};
Clear[x,y,z,r,s,t]
{x[r_,s_,t_],y[r_,s_,t_],z[r_,s_,t_]} =
center2 + {r Sin[s] Cos[t],r Sin[s] Sin[t],
r Cos[s]}; r2 = 0.6; {{slow,shigh},{tlow,thigh}} =
{{0,Pi},{0,2 Pi}}; Clear[sphere2plotter];
sphere2plotter[s_,t_] = {x[r2,s,t],y[r2,s,t],z[r2,s,t]};
sphere2plot = ParametricPlot3D[Evaluate[
sphere2plotter[s,t]],{s,slow,shigh},{t,tlow,thigh},
DisplayFunction->Identity];
Show[sphere1plot,sphere2plot,Boxed->False,
BoxRatios->Automatic,ViewPoint->CMView,
PlotRange->All,AxesLabel->{"x","y","z"},
DisplayFunction->$DisplayFunction];
```

G.8.a) As you can see, there is not a lot of room between these spheres, but there is room enough to insert a plane between them so that the two spheres are on different sides of the plane. Insert such a plane into this plot.

■ G.9) Art

G.9.a) Use spherical coordinates, tubes, and anything else you want to mix in to make a gallery of eye-catching surfaces of your own design. Show off.

■ G.10) Four dimensions: Is it just science fiction?

G.10.a.i) The equation
$$x^2 + y^2 + z^2 + t^2 = 3^2$$
defines a (hyper)sphere of radius 3 in four dimensions.

> If you set $t = 2$, you get a cross-section of the four-dimensional sphere. Identify this cross-section and plot it.

G.10.a.ii) The equation
$$x^2 + y^2 + z^2 + t^2 = \text{rad}^2$$
defines a (hyper)sphere of radius rad in four dimensions.

> Use a system of double polar coordinates and the appropriate volume conversion factor to give a formula that measures the four-dimensional volume of this (hyper)sphere of radius rad.

G.10.b) Conjure up the equation of a (hyper)sphere of radius r in five dimensions.

> What is the five-dimensional volume of this sphere?

■ G.11) Eigenvalues and eigenvectors

This problem appears only in the electronic version.

LESSON 3.10

3D Surface Measurements

Basics

■ B.1) Sources, sinks, and Gauss's formula in 3D

B.1.a) Given a three-dimensional vector field
$$\text{Field}[x, y, z] = \{m[x, y, z], n[x, y, z], p[x, y, z]\}:$$

In[1]:=
```
Clear[Field,m,n,p,x,y,z]
Field[x_,y_,z_] = {m[x,y,z],n[x,y,z],p[x,y,z]};
```

You calculate the divergence, divField$[x, y, z]$, of Field$[x, y, z]$ via the formula:

In[2]:=
```
Clear[divField]
divField[x_,y_,z_] = D[m[x,y,z],x] + D[n[x,y,z],y] + D[p[x,y,z],z]
```

Out[2]=
$$p^{(0,0,1)}[x, y, z] + n^{(0,1,0)}[x, y, z] + m^{(1,0,0)}[x, y, z]$$

Not much of a surprise.

> What do you do with this formula?

Answer: You use it the same way you use its 2D counterpart.

B.1.b.i) Gauss's 3D formula says that if you go with a 3D vector field
$$\text{Field}[x, y, z] = \{m[x, y, z], n[x, y, z], p[x, y, z]\}$$

383

and if R is a solid body in three dimensions with boundary surface (skin) C, then

$$\iiint_R \text{divField}[x,y,z]\,dx\,dy\,dz = \iint_C \text{Field} \bullet \text{outerunitnormal}\,dA$$
$$= \text{flow of Field}[x,y,z]\text{ across }C.$$

> Use Gauss's formula to measure the flow of the vector field
> $$\text{Field}[x,y,z] = \{y - 0.5, x\,y, z^2\}$$
> across the surface of the box consisting of all points $\{x,y,z\}$ with $-1 \le x \le 3$, $0 \le y \le 2$, and $-2 \le z \le 4$.

Answer:

In[3]:=
```
Clear[x,y,z,m,n,p,Field]
{m[x_,y_,z_],n[x_,y_,z_],p[x_,y_,z_]} = {y - 0.5,x y,z^2};
Field[x_,y_,z_] = {m[x,y,z],n[x,y,z],p[x,y,z]};
```

Gauss's formula tells you that the net flow of Field$[x,y,z]$ across the surface of the box with $-1 \le x \le 3$, $0 \le y \le 2$, and $-2 \le z \le 4$ is measured by

$$\int_{-1}^{3} \int_{0}^{2} \int_{-2}^{4} \text{divField}[x,y,z]\,dx\,dy\,dz.$$

Here you go:

In[4]:=
```
Clear[divF]
divField[x_,y_,z_] = D[m[x,y,z],x] + D[n[x,y,z],y] + D[p[x,y,z],z];
Integrate[divField[x,y,z],{x,-2,4},{y,0,2},{z,-1,3}]
```

Out[4]=
144

Very positive.

Strong flow from inside the box to outside the box. There must be lots of spigots turned on inside this box.

B.1.b.ii) In three dimensions, if

$$\text{Field}[x,y,z] = \{m[x,y,z], n[x,y,z], p[x,y,z]\},$$

then

$$\text{divField}[x,y,z] = D[m[x,y,z],x] + D[n[x,y,z],y] + D[p[x,y,z],z]$$

tells you about sources and sinks in the same way that divField$[x,y]$ tells you about sources and sinks in two dimensions:

→ The points $\{x,y,z\}$ with divField$[x,y,z] > 0$ are sources of new fluid.
→ The points $\{x,y,z\}$ with divField$[x,y,z] < 0$ are sinks for old fluid.

> Use Gauss's formula to explain why this interpretation is legitimate.

Answer: Here's how to see why: Take a small sphere C centered at $\{x_0, y_0, z_0\}$. Calculate the flow across C, which, according to Gauss's formula, is

$$\iiint_R \text{divField}[x, y, z] \, dx \, dy \, dz$$

where R is the solid region consisting of the sphere C and everything inside C.

Here's the kicker: If you start out with $\text{divField}[x_0, y_0, z_0] > 0$, then $\text{divField}[x, y, z]$ is positive for all $\{x, y, z\}$'s close to $\{x_0, y_0, z_0\}$, so, if C is small enough, then $\text{divField}[x, y, z] > 0$ for all $\{x, y, z\}$'s inside C. This tells you that for small spheres C centered at $\{x_0, y_0, z_0\}$,

$$\text{flow-across-}C = \iiint_R \text{divField}[x, y, z] \, dx \, dy \, dz > 0.$$

And this tells you that if $\text{divField}[x_0, y_0, z_0] > 0$, then the net flow of $\text{Field}[x, y, z]$ across small spheres centered at $\{x_0, y_0, z_0\}$ is from inside to outside.

The upshot: If $\text{divField}[x_0, y_0, z_0] > 0$, then the point $\{x_0, y_0, z_0\}$ is a source of new fluid.

Similarly, if $\text{divField}[x_0, y_0, z_0] < 0$, then the net flow of $\text{Field}[x, y, z]$ across small spheres centered at $\{x_0, y_0, z_0\}$ is from outside to inside. Consequently, if $\text{divField}[x_0, y_0, z_0] < 0$, then the point $\{x_0, y_0, z_0\}$ is a sink (or drain) for old fluid.

B.1.b.iii)
> Given
> $$\text{Field}[x, y, z] = \{y - 0.5, x\,y, z^2\},$$
> say how to identify the points $\{x, y, z\}$ that are sources, and the points $\{x, y, z\}$ that are sinks.

Answer:

In[5]:=
```
Clear[x,y,z,m,n,p,Field]
{m[x_,y_,z_],n[x_,y_,z_],p[x_,y_,z_]} = {y - 0.5,x y,z^2}
Field[x_,y_,z_] = {m[x,y,z],n[x,y,z],p[x,y,z]};
```

Out[5]=
$$\{-0.5 + y,\ x\ y,\ z^2\}$$

Calculate the divergence:

In[6]:=
```
Clear[divField]
divField[x_,y_,z_] = D[m[x,y,z],x] + D[n[x,y,z],y] + D[p[x,y,z],z]
```

Out[6]=
x + 2 z

Find out where divField$[x, y, z]$ is zero:

In[7]:=
```
Solve[divField[x,y,z] == 0]
```
Out[7]=
```
{{x -> -2 z}}
```

This says that the points $\{x, y, z\}$ with $x = -2z$ are neither sources nor sinks. Now look at:

In[8]:=
```
Clear[a]; divField[x,y,z]/.x->(-2 z + a)
```
Out[8]=
```
a
```

Food for thought. Upon reflection, this tells you that

if $x > -2z$, then divField$[x, y, z] > 0$

and if $x < -2z$, then $divField[x, y, z] < 0$.

As a result, the points $\{x, y, z\}$ that are sources are the points with $x > -2z$, and the points $\{x, y, z\}$ that are sinks are the points with $x < -2z$.

B.1.b.iv) Given

$$\text{Field}[x, y, z] = \{x - z, y - x, z - y\},$$

say how to identify the points $\{x, y, z\}$ that are sources, and the points $\{x, y, z\}$ that are sinks.

Use your answer to determine whether the net flow of this vector field across the sphere of radius 2 centered at $\{4, 2, 1\}$ is from inside to outside or from outside to inside.

Answer:

In[9]:=
```
Clear[m,n,p,Field,x,y,z]
{m[x_,y_,z_],n[x_,y_,z_],p[x_,y_,z_]} = { x - z, y - x, z - y};
Field[x_,y_,z_] = {m[x,y,z],n[x,y,z],p[x,y,z]}
```
Out[9]=
```
{x - z, -x + y, -y + z}
```

divField$[x, y, z]$ is given by:

In[10]:=
```
Clear[divField]
divField[x_,y_,z_] = D[m[x,y,z],x] + D[n[x,y,z],y] + D[p[x,y,z],z]
```
Out[10]=
```
3
```

No matter what $\{x, y, z\}$ happens to be, divField$[x, y, z] = 3$. So all points $\{x, y, z\}$ are sources.

And now, without further calculation, you know that the net flow of this vector field across sphere of radius 2 centered at $\{4, 2, 1\}$ is from inside to outside. Reason: There are no sinks inside the sphere to absorb excess outside to inside flow.

B.1.c) Explain the reasoning behind Gauss's formula

Answer: A heavy, long-winded explanation is possible, but leave it at this:

Once you have a good feeling for the ideas in B.2), B.3), and B.4), you can build your own explanation of Gauss's formula by copying and pasting the explanation of the two-dimensional Gauss-Green formula from an earlier lesson and making technical adjustments. If you cannot find satisfaction without seeing some more details, see: W. Kaplan, *Advanced Calculus*, Addison-Wesley, Reading, Massachusetts, 1972, p. 338.

■ B.2) Measuring area on surfaces

B.2.a) Here's a parallelogram in three dimensions:

In[11]:=
```
basepoint = {1,2,0}; X = {-2,1,1}; Y = {1,-2,1};
parallelogram = Show[Graphics3D[Polygon[
  {basepoint,basepoint + X,
  basepoint + X + Y,basepoint + Y}]],
  ViewPoint->CMView,
  Axes->True,AxesLabel->{"x","y","z"}];
```

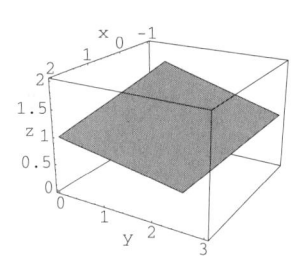

Because of its unfortunate position, it might seem hard to measure the area of this parallelogram, but the cross product can bail you out.

How?

Answer: Throw in the vectors that define this parallelogram:

In[12]:=
```
Show[parallelogram,Arrow[X,Tail->basepoint,Blue],
  Arrow[Y,Tail->basepoint,Blue],
  Graphics3D[{PointSize[0.03],Point[basepoint]}],
  Graphics3D[Text["X",basepoint + X/2 + {0,0.2,0}]],
  Graphics3D[Text["Y",basepoint + Y/2 +{0.2,0,0}]],
  ViewPoint->CMView,PlotRange->All,
  Axes->True,AxesLabel->{"x","y","z"}];
```

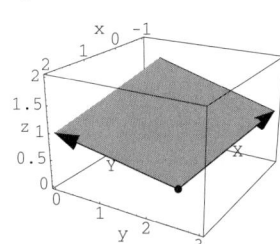

As you probably remember,

$$\|X \times Y\| = \|X\| \|Y\| |\sin[\text{angle between}]|.$$

This quantity is also equal to the area of this parallelogram; so the area in square units of this unfortunately positioned parallelogram is given by:

In[13]:=
```
XcrossY = Cross[X,Y]; Sqrt[XcrossY.XcrossY]
```

Out[13]=

$3^{3/2}$

Easy. Rerun for other choices of X and Y to get the hang of it.

B.2.b.i) Here is a look at part of the plane $x + y + z = 2$:

In[14]:=
```
Clear[x,y,z,u,v]
x[u_,v_] = u; y[u_,v_] = v;
z[u_,v_] = 2 - u - v;
surface = ParametricPlot3D[{x[u,v],y[u,v],z[u,v]},
{u,0,1},{v,0,1},PlotPoints->{2,2},
Axes->Automatic,AxesLabel->{"x","y","z"},
ViewPoint->CMView];
```

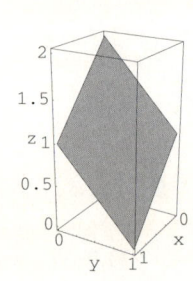

Here is another way of getting a plot of the same surface. This plot also includes the vectors that generate the parallelogram.

In[15]:=
```
{a,b} = {0,0}; h = 1;
Show[Graphics3D[Polygon[{{x[a,b],y[a,b],z[a,b]},
{x[a + h,b],y[a + h,b],z[a + h,b]},
{x[a + h,b + h],y[a + h,b + h],z[a + h,b + h]},
{x[a,b + h],y[a,b + h],z[a,b + h]}}]],
Arrow[{x[a + h,b],y[a + h,b],z[a + h,b]} -
{x[a,b],y[a,b],z[a,b]},
Tail->{x[a,b],y[a,b],z[a,b]},Blue],
Arrow[{x[a,b + h],y[a,b + h],z[a,b + h]} -
{x[a,b],y[a,b],z[a,b]},
Tail->{x[a,b],y[a,b],z[a,b]},Blue],
PlotRange->All,Axes->Automatic,
AxesLabel->{"x","y","z"},ViewPoint->CMView];
```

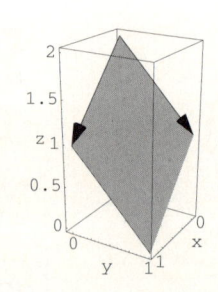

For this choice of a, b, and h, the corresponding parallelogram has area measurement:

In[16]:=
```
X = {x[a + h,b],y[a + h,b],z[a + h,b]} - {x[a,b],y[a,b],z[a,b]};
Y = {x[a,b + h],y[a,b + h],z[a,b + h]} - {x[a,b],y[a,b],z[a,b]};
XcrossY = Cross[X,Y]; Sqrt[XcrossY.XcrossY]
```

Basics (B.2)

Out[16]=
```
Sqrt[3]
```

For unspecified choices of a, b, and h, the corresponding parallelogram has its area measured by:

In[17]:=
```
Clear[a,b,h]
X = {x[a + h,b],y[a + h,b],z[a + h,b]} - {x[a,b],y[a,b],z[a,b]};
Y = {x[a,b + h],y[a,b + h],z[a,b + h]} - {x[a,b],y[a,b],z[a,b]};
XcrossY = Cross[X,Y]; Sqrt[XcrossY.XcrossY]
```

Out[17]=
```
Sqrt[3] Sqrt[h^4]
```

> What is the area conversion factor $SA_{(x,y,z)}[u,v]$ that you use to convert uv-paper area measurement into xyz-surface area measured on the plane $x + y + z = 2$?

Answer: The uv-paper rectangle with corners at $\{a, b\}$, $\{a+h, b\}$, $\{a+h, b+h\}$, and $\{a, b+h\}$ has uv-paper area given by:

In[18]:=
```
Clear[h]; uvarea = h^2
```

Out[18]=
```
h^2
```

This uv-paper rectangle plots out in xyz-coordinates as the parallelogram with corners at

$\{x[a, b], y[a, b], z[a, b]\}$,
$\{x[a + h, b], y[a + h, b], z[a + h, b]\}$,
$\{x[a + h, b + h], y[a + h, b + h], z[a + h, b + h]\}$,

and

$\{x[a, b + h], y[a, b + h], z[a, b + h]\}$.

The area of this parallelogram is given by:

In[19]:=
```
Clear[a,b,h]
X = {x[a + h,b],y[a + h,b],z[a + h,b]} - {x[a,b],y[a,b],z[a,b]};
Y = {x[a,b + h],y[a,b + h],z[a,b + h]} - {x[a,b],y[a,b],z[a,b]};
XcrossY = Cross[X,Y]; planararea = Sqrt[XcrossY.XcrossY]
```

Out[19]=
```
Sqrt[3] Sqrt[h^4]
```

The area conversion factor $SA_{(x,y,z)}[u,v]$ that you integrate to convert uv-paper area into xyz-surface area measured on the plane $x + y + z = 2$ is:

Lesson 3.10 3D Surface Measurements

In[20]:=
```
Clear[SAxyz]; SAxyz[u_,v_] = Expand[planararea/uvarea]
```
Out[20]=
$$\frac{\text{Sqrt}[3]\ \text{Sqrt}[h^4]}{h^2}$$

And this is the same as $\sqrt{3}$ because the h's cancel out:

In[21]:=
```
SAxyz[u,v]/.Sqrt[h^4]->h^2
```
Out[21]=
```
Sqrt[3]
```

Still pretty easy.

B.2.b.ii) Here is a plot of the uv-paper circle
$$u^2 + v^2 = 4$$
on the plane
$$x + y + z = 2$$
under the parameterization
$$x[u,v] = u, y[u,v] = v, \text{ and } z[u,v] = 2 - u - v$$
used above.

In[22]:=
```
Clear[x,y,z,u,v,t]; x[u_,v_] = u;
y[u_,v_] = v; z[u_,v_] = 2 - u - v;
u[t_] = 2 Cos[t]; v[t_] = 2 Sin[t];
curve = ParametricPlot3D[{x[u[t],v[t]],
y[u[t],v[t]],z[u[t],v[t]]},{t,0,2 Pi},
DisplayFunction->Identity]; {a,b} = {-4,-4};
h = 8; plane = Graphics3D[Polygon[{{x[a,b],y[a,b],z[a,b]},
{x[a+h,b],y[a+h,b],z[a+h,b]},{x[a+h,b+h],
y[a+h,b+h],z[a+h,b+h]},{x[a,b+h],y[a,b+h],z[a,b+h]}}]];
Show[curve,plane,Axes->Automatic,
AxesLabel->{"x","y","z"},ViewPoint->CMView,
DisplayFunction->$DisplayFunction];
```

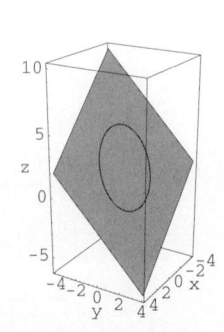

> Come up with a measurement of the area enclosed by this curve as measured on the plane $x + y + z = 2$.

Answer: Area on the plane is $SA_{(x,y,z)}[u,v]$ times uv-paper area as calculated above in part B.2.b.i) where $SA_{(x,y,z)}[u,v]$ is:

In[23]:=
```
SAxyz[u,v]
```

Out[23]=
```
Sqrt[3]
```

The circle
$$u^2 + v^2 = 4$$
is a circle of radius 2 on uv-paper. It encloses an area of $2^2 \pi = 4\pi$ square units measured on uv-paper. Its plot on the plane $x + y + z = 2$ encloses a total of:

In[24]:=
```
SAxyz[u,v] 4 Pi
```

Out[24]=
```
4 Sqrt[3] Pi
```

square units.

B.2.c.i) Here is a portion of the surface whose parametric equations are:

In[25]:=
```
Clear[x,y,z,u,v]; x[u_,v_] = u;
y[u_,v_] = v; z[u_,v_] = u v;
{ulow,uhigh} = {0,1}; {vlow,vhigh} = {0,2};
surface = ParametricPlot3D[Evaluate[
{x[u,v],y[u,v],z[u,v]}],
{u,ulow,uhigh},{v,vlow,vhigh},
ViewPoint->CMView,BoxRatios->Automatic,
Axes->Automatic,AxesLabel->{"x","y","z"}];
```

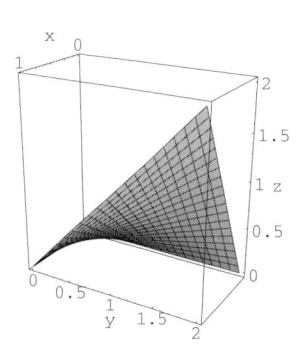

Measure the surface area of the plotted portion of this surface.

Answer: *Mathematica* plots this surface by approximating the true surface with a whole herd of little patches that resemble parallelograms. The message is clear. You can use what you know about measuring the area of parallelograms to help measure the area of curved surfaces. See what some parallelograms will do:

In[26]:=
```
Clear[h,basepoint,leg1,leg2,parallelogram]
basepoint[u_,v_] = {x[u,v],y[u,v],z[u,v]};
leg1[u_,v_,h_] = {x[u + h,v],y[u + h,v],z[u + h,v]}
- basepoint[u,v]; leg2[u_,v_,h_] = {x[u,v + h],
y[u,v + h],z[u,v + h]} - basepoint[u,v];
parallelogram[u_,v_,h_] = Graphics3D[{Black,Polygon[{
basepoint[u,v],basepoint[u,v] + leg1[u,v,h],
basepoint[u,v] + leg1[u,v,h] + leg2[u,v,h],
basepoint[u,v] + leg2[u,v,h]}]}]; jump = 0.3;
h = 0.25; parallelograms = Table[parallelogram[u,v,h],
{u,ulow,uhigh - jump,jump},{v,vlow,vhigh - jump,jump}];
Show[surface,parallelograms,ViewPoint->CMView,
Axes->Automatic,AxesLabel->{"x","y","z"}];
```

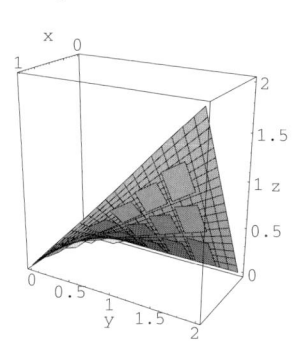

Lesson 3.10 3D Surface Measurements

Hug that surface, mama! The smart money says: The smaller the parallelograms are, the closer they hug the surface.

Check it out:

In[27]:=
```
jump = 0.2; h = 0.1;
parallelograms = Table[parallelogram[u,v,h],
  {u,ulow,uhigh - jump,jump},{v,vlow,vhigh - jump,jump}];
Show[surface,parallelograms,ViewPoint->CMView,
  Axes->Automatic,AxesLabel->{"x","y","z"}];
```

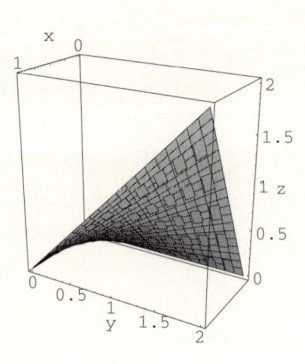

The smart money wins. These smaller parallelograms are really glued to the surface. Even smaller parallelograms will share even more ink with the surface.

Here's how to get the area conversion factor $SA_{(x,y,z)}[u,v]$ for going from uv-paper area to xyz-area on the surface at a uv-point $\{u,v\}$. Notice that for small positive h, the area conversion factors at a uv-point $\{u,v\}$ are nearly the same for area on the surface and for area on the parallelograms plotted above which are determined by the vectors:

In[28]:=
```
Clear[h]; leg1[u,v,h]
```
Out[28]=
$\{h, 0, -(u\ v) + (h + u)\ v\}$

In[29]:=
```
leg2[u,v,h]
```
Out[29]=
$\{0, h, -(u\ v) + u\ (h + v)\}$

For fixed $\{u,v\}$ and h, the area conversion factor for converting uv-area to xyz-area on the parallelogram determined by $\mathrm{leg1}[u,v,h]$ and $\mathrm{leg2}[u,v,h]$ is given by:

In[30]:=
```
cross = Cross[leg1[u,v,h],leg2[u,v,h]];
Clear[paraareafactor]
paraareafactor[u_,v_,h_] = Sqrt[(cross.cross)]/h^2
```
Out[30]=
$$\frac{\sqrt{h^4 + h^2 (-(u\ v) + (h + u)\ v)^2 + h^2 (-(u\ v) + u\ (h + v))^2}}{h^2}$$

The area conversion factor $SA_{(x,y,z)}[u,v]$ for the surface at $\{u,v\}$ is the limiting case of the above as $h \to 0$:

In[31]:=
```
Clear[SAxyz];
SAxyz[u_,v_] = Limit[paraareafactor[u,v,h],h->0]
```

Out[31]=
$$\text{Sqrt}[1 + u^2 + v^2]$$

The total surface area of the surface plotted above is

$$\int_{vlow}^{vhigh} \int_{ulow}^{uhigh} SA_{(x,y,z)}[u,v] \, du \, dv :$$

In[32]:=
```
surfarea = NIntegrate[SAxyz[u,v],{u,ulow,uhigh},{v,vlow,vhigh}]
```

Out[32]=
3.18041

And once you've done one of these, you've done them all.

B.2.c.ii) Here is a direct formula for the area conversion factor $SA_{(x,y,z)}[u,v]$ that converts uv-paper area measurements into area measurements on the surface whose parametric formulas are $\{x[u,v], y[u,v], z[u,v]\}$:

In[33]:=
```
Clear[x,y,z,u,v,cross,SAxyz]
cross[u_,v_] = Cross[D[{x[u,v],y[u,v],z[u,v]},u],
   D[{x[u,v],y[u,v],z[u,v]},v]];
SAxyz[u_,v_] = Sqrt[cross[u,v].cross[u,v]]
```

Out[33]=
$$\text{Sqrt}[(y^{(0,1)}[u,v] \, x^{(1,0)}[u,v] - x^{(0,1)}[u,v] \, y^{(1,0)}[u,v])^2 +$$
$$\text{Power}[-(z^{(0,1)}[u,v] \, x^{(1,0)}[u,v]) + x^{(0,1)}[u,v] \, z^{(1,0)}[u,v], 2] +$$
$$(z^{(0,1)}[u,v] \, y^{(1,0)}[u,v] - y^{(0,1)}[u,v] \, z^{(1,0)}[u,v])^2 \,]$$

Nasty looking, but useful.

> Explain where this formula for $SA_{(x,y,z)}[u,v]$ comes from.

Answer: This answer appears only in the electronic version.

B.2.c.iii)
> Measure the surface area of the surface of a sphere of radius r.

Answer: Go to spherical coordinates.

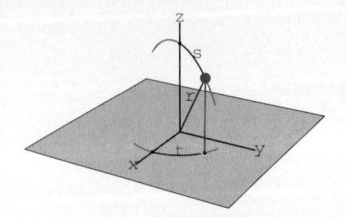

The surface of the sphere of radius r is described by
$$\{x[s,t], y[s,t], z[s,t]\} = \{r\sin[s]\cos[t], r\sin[s]\sin[t], r\cos[s]\}$$
with s running from 0 to π and t running from 0 to 2π.

In[34]:=
```
Clear[x,y,z,r,s,t]; x[s_,t_] = r Sin[s] Cos[t];
y[s_,t_] = r Sin[s] Sin[t];
z[s_,t_] = r Cos[s];
```

To measure the area of the surface of the sphere of radius r, all you have to do is calculate
$$\int_0^{2\pi}\int_0^{\pi} SA_{(x,y,z)}[s,t]\,ds\,dt:$$

In[35]:=
```
Clear[SAxyz,cross]
cross[s_,t_] = cross[D[{x[s,t],y[s,t],z[s,t]},s]
  D[{x[s,t],y[s,t],z[s,t]},t];
SAxyz[s_,t_] = Sqrt[Expand[cross[s,t].cross[s,t],Trig->True]]
```

Out[35]=
$$\sqrt{\frac{r^4}{2} - \frac{r^4 \cos[2s]}{2}}$$

In[36]:=
```
Integrate[SAxyz[s,t],{s,0,Pi},{t,0,2 Pi}]
```

Out[36]=
```
4 Pi Sqrt[r⁴]
```

You can get the same answer from:

In[37]:=
```
2 Integrate[SAxyz[s,t],{s,0,Pi/2},{t,0,2 Pi}]
```

Out[37]=
```
4 Pi Sqrt[r⁴]
```

■ B.3) Surface integrals

Just as you can integrate functions on curves with respect to arc length, you can integrate functions on surfaces with respect to surface area. And you don't have to go to a lot of trouble to do it.

B.3.a) Here is a parameterized surface C:

In[38]:=
```
Clear[x,y,z,u,v]; x[u_,v_] = u Cos[v];
y[u_,v_] = -u Sin[v]; z[u_,v_] = v;
{{ulow,uhigh},{vlow,vhigh}} = {{1,3},{0,3 Pi/2}};
surface = ParametricPlot3D[{x[u,v],y[u,v],z[u,v]},
{u,ulow,uhigh},{v,vlow,vhigh},ViewPoint->CMView,
Boxed->False,AxesLabel->{"x","y","z"}];
```

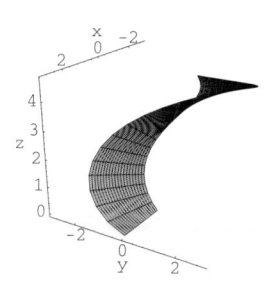

Calculate

$$\iint_C (x^2 + y^2) \, dA$$

where the integral is taken with respect to surface area.

Answer: $\iint_C (x^2 + y^2) \, dA$ is calculated via the formula

$$\int_{vlow}^{vhigh} \int_{ulow}^{uhigh} \left(x[u,v]^2 + y[u,v]^2 \right) SA_{(x,y,z)}[u,v] \, du \, dv.$$

Calculating this is duck soup.

First you need the area conversion factor $SA_{(x,y,z)}[u,v]$:

In[39]:=
```
Clear[SAxyz]
cross = Cross[D[{x[u,v],y[u,v],z[u,v]},u],
D[{x[u,v],y[u,v],z[u,v]},v]];
SAxyz[u_,v_] = Sqrt[Expand[cross.cross,Trig->True]]
```

Out[39]=
```
Sqrt[1 + u^2]
```

Now turn the integral

$$\int_{vlow}^{vhigh} \int_{ulow}^{uhigh} \left(x[u,v]^2 + y[u,v]^2 \right) SA_{(x,y,z)}[u,v] \, du \, dv$$

over to the machine:

In[40]:=
```
integral = NIntegrate[(x[u,v]^2 + y[u,v]^2) SAxyz[u,v],
{u,ulow,uhigh},{v,vlow,vhigh}];
```

Out[40]=
103.125

Not exciting, but not hard.

■ B.4) Surface integrals for measuring flow across surfaces

B.4.a) Here's a surface:

```
In[41]:=
  Clear[x,y,z,u,v]; x[u_,v_] = u;
  y[u_,v_] = v; z[u_,v_] = (10 - u^2 - v^2)/3;
  {ulow,uhigh} = {-2,2}; {vlow,vhigh} = {-2,2};
  surface = ParametricPlot3D[
  Evaluate[{x[u,v],y[u,v],z[u,v]}],
  {u,ulow,uhigh},{v,vlow,vhigh},
  AxesLabel->{"x","y","z"},ViewPoint->CMView];
```

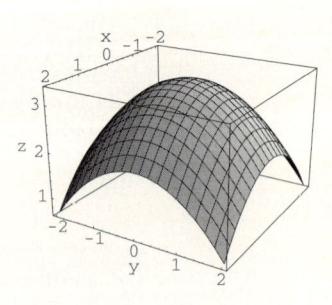

Take $\{a, b\} = \{0.5, 1.0\}$ and look at this plot of the two curves

$$P_1[u] = \{x[u,b], y[u,b], z[u,b]\} \text{ with ulow} \leq u \leq \text{uhigh}$$

and

$$P_2[v] = \{x[a,v], y[a,v], z[a,v]\} \text{ with vlow} \leq v \leq \text{vhigh}$$

together with the surface plotted above:

```
In[42]:=
  {a,b} = {0.5,1.0}; Clear[P1,P2]
  P1[u_] = {x[u,b],y[u,b],z[u,b]};
  P2[v_] = {x[a,v],y[a,v],z[a,v]};
  curve1 = ParametricPlot3D[Evaluate[
  P1[u]],{u,ulow,uhigh},DisplayFunction->Identity];
  curve2 = ParametricPlot3D[Evaluate[
  P2[v]],{v,vlow,vhigh},DisplayFunction->Identity];
  surface = Insert[surface,EdgeForm[],{1,1}];
  curves = Show[surface,curve1,curve2,Boxed->False,
  ViewPoint->CMView,BoxRatios->Automatic,
  DisplayFunction->$DisplayFunction];
```

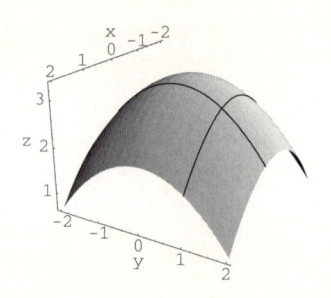

Two intersecting curves running on the surface.

Now calculate the area conversion factor $SA_{(x,y,z)}[u,v]$

```
In[43]:=
  Clear[SAxyz,cross]
  cross[u_,v_] = Cross[D[{x[u,v],y[u,v],z[u,v]},u],
  D[{x[u,v],y[u,v],z[u,v]},v]];
  SAxyz[u_,v_] = Sqrt[cross[u,v].cross[u,v]]
Out[43]=
  Sqrt[1 + (4 u^2)/9 + (4 v^2)/9]
```

Keeping the same $\{a, b\}$ as above, add to the last plot the three vectors:

$$D[\{x[u,v], y[u,v], z[u,v]\}, u]/.\{u \to a, v \to b\},$$
$$D[\{x[u,v], y[u,v], z[u,v],\}, v\}]/.\{u \to a, v \to b\},$$

and

$$\text{cross}[a, b],$$

all with their tails at the point of intersection:

In[44]:=
```
tail = {x[a,b],y[a,b],z[a,b]};
vector1 = D[{x[u,v],y[u,v],z[u,v]},u]/.{u->a ,v->b};
vector2 = D[{x[u,v],y[u,v],z[u,v]},v]/.{u->a ,v->b};
vector3 = cross[a,b];
Show[curves,Arrow[vector1,Tail->tail],
Arrow[vector2,Tail->tail],
Arrow[vector3,Tail->tail,Blue]];
```

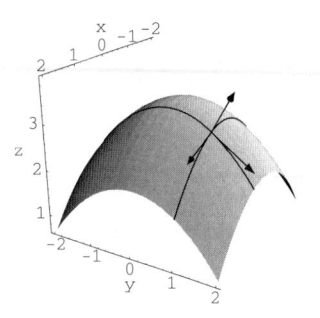

Totally radical or what!

> Explain why this happened and how it relates to the area conversion factor $SA_{(x,y,z)}[u, v]$.

Answer: Both $D[\{x[u,v], y[u,v], z[u,v]\}, u]$ and $D[\{x[u,v], y[u,v], z[u,v]\}, v]$ are tangent to the surface at $\{x[u,v], y[u,v], z[u,v]\}$, so their cross product, $\text{cross}[u, v]$, is automatically normal (perpendicular) to the surface at $\{x[u,v], y[u,v], z[u,v]\}$.

The upshot: The area conversion factor

$$SA_{(x,y,z)}[u, v] = \sqrt{\text{cross}[u, v] \bullet \text{cross}[u, v]}$$

is the same as the length of the normal vector $\text{cross}[u, v]$.

B.4.b) Given a 3D vector field

$$\text{Field}[x, y, z] = \{m[x, y, z], n[x, y, z], p[x, y, z]\}$$

and given a surface parameterized by $\{x[u,v], y[u,v], z[u,v]\}$ with $\text{ulow} \leq u \leq \text{uhigh}$ and $\text{vlow} \leq v \leq \text{vhigh}$, you can measure the flow of $\text{Field}[x, y, z]$ across the surface by calculating

$$\int_{\text{vlow}}^{\text{vhigh}} \int_{\text{ulow}}^{\text{uhigh}} \text{Field}[x[u,v], y[u,v], z[u,v]] \bullet \text{normal}[u,v] \, du \, dv$$

where $\text{normal}[u, v] = \text{cross}[u, v]$ is given by:

In[45]:=
```
Clear[x,y,z,u,v,normal]
normal[u_,v_] = Cross[D[{x[u,v],y[u,v],z[u,v]},u],
D[{x[u,v],y[u,v],z[u,v]},v]]
```

Lesson 3.10 3D Surface Measurements

Out[45]=

$\{z^{(0,1)}[u,v]\, y^{(1,0)}[u,v] - y^{(0,1)}[u,v]\, z^{(1,0)}[u,v],$

$-(z^{(0,1)}[u,v]\, x^{(1,0)}[u,v]) + x^{(0,1)}[u,v]\, z^{(1,0)}[u,v],$

$y^{(0,1)}[u,v]\, x^{(1,0)}[u,v] - x^{(0,1)}[u,v]\, y^{(1,0)}[u,v]\}$

Don't try to memorize this formula.

> Where does the flow-across formula
> $$\int_{vlow}^{vhigh} \int_{ulow}^{uhigh} \text{Field}[x[u,v], y[u,v], z[u,v]] \bullet \text{normal}[u,v]\, du\, dv$$
> come from?

Answer: Put

$$\text{normcompField}[x[u,v], y[u,v], z[u,v]]$$

equal to the component of

$$\text{Field}[x[u,v], y[u,v], z[u,v]]$$

in the direction perpendicular to the surface at $\{x[u,v], y[u,v], z[u,v]\}$.

You measure the flow of $\text{Field}[x, y, z]$ across the surface by calculating

$$\text{flow across} = \iint_{\text{surface}} \text{normcompField}[x,y,z]\, dA$$

$$= \int_{vlow}^{vhigh} \int_{ulow}^{uhigh} \text{normcompField}[x[u,v], y[u,v], z[u,v]]\, SA_{(x,y,z)}[u,v]\, du\, dv.$$

Put this in your pocket for a minute.

Remember that $\text{normal}[u,v]$ as calculated above is perpendicular to the surface at the point $\{x[u,v], y[u,v], z[u,v]\}$. At this point, the component of

$$\text{Field}[x[u,v], y[u,v], z[u,v]]$$

in the direction perpendicular to the surface is

$$\text{normcompField}[x[u,v], y[u,v], z[u,v]]$$

$$= \text{Field}[x[u,v], y[u,v], z[u,v]] \bullet \frac{\text{normal}[u,v]}{\sqrt{\text{normal}[u,v] \bullet \text{normal}[u,v]}}$$

because $SA_{(x,y,z)}[u,v] = \sqrt{\text{normal}[u,v] \bullet \text{normal}[u,v]}$. So flow across $= \iint_{\text{surface}} \text{normcompField}[x,y,z]\, dA$

$= \int_{vlow}^{vhigh} \int_{ulow}^{uhigh} \text{normcompField}[x[u,v], y[u,v], z[u,v]]\, SA_{(x,y,z)}[u,v]\, du\, dv$

$= \int_{vlow}^{vhigh} \int_{ulow}^{uhigh} \text{Field}[x[u,v], y[u,v], z[u,v]] \bullet \left(\frac{\text{normal}[u,v]}{SA_{(x,y,z)}[u,v]} \right) SA_{(x,y,z)}[u,v]\, du\, dv$

$= \int_{vlow}^{vhigh} \int_{ulow}^{uhigh} \text{Field}[x[u,v], y[u,v], z[u,v]] \bullet \text{normal}[u,v]\, du\, dv$

because the $SA_{(x,y,z)}[u,v]$ terms miraculously cancel out.

Explanation finished.

B.4.c) Here's a new surface:

In[46]:=
```
Clear[x,y,z,u,v]; x[u_,v_] = v;
y[u_,v_] = 2 E^(-v) Sin[Pi u]; z[u_,v_] = u;
{ulow,uhigh} = {0,2}; {vlow,vhigh} = {0.5,3.5};
surface = ParametricPlot3D[
Evaluate[{x[u,v],y[u,v],z[u,v]}],
{u,ulow,uhigh},{v,vlow,vhigh},
ViewPoint->CMView,AxesLabel->{"x"," y","z"}];
```

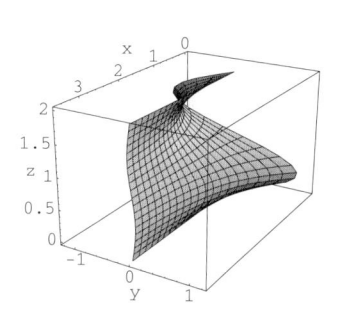

Here is this surface together with some of the normal vectors whose lengths measure the area conversion factor $SA_{(x,y,z)}[u,v]$, which converts uv-paper area measurements into xyz-area measurements on the surface:

In[47]:=
```
Clear[normal]
normal[u_,v_] = Cross[D[{x[u,v],y[u,v],z[u,v]},u],
D[{x[u,v],y[u,v],z[u,v]},v]];
normals = Table[Arrow[normal[u,v],
Tail->{x[u,v],y[u,v],z[u,v]}],
{u,0.5,1.5,0.5},{v,1.5,2.5}];
surface = Insert[surface,EdgeForm[],{1,1}];
Show[surface,normals,ViewPoint->CMView,
Boxed->False,AxesLabel->{"x","y","z"}];
```

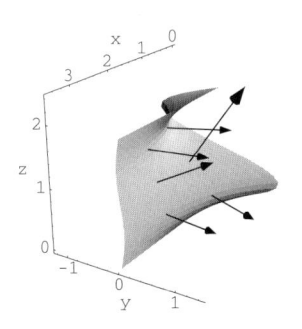

The normals are pointing out from the side of the surface you are sitting on.

Determine whether the net flow of the vector field

$\text{Field}[x,y,z] = \{y\,z, x\,z, x\,y\}$

across this surface is with the plotted normals or is against the plotted normals.

Lesson 3.10 3D Surface Measurements

Answer: Enter the vector field:

In[48]:=
```
Clear[x,y,z,m,n,p,Field]
{m[x_,y_,z_],n[x_,y_,z_],p[x_,y_,z_]} = {y z, x z, x y};
Field[x_,y_,z_] = {m[x,y,z],n[x,y,z],p[x,y,z]}
```

Out[48]=
$\{y\, z,\ x\, z,\ x\, y\}$

Reenter the parameterization of the surface and calculate normal[u, v]:

In[49]:=
```
Clear[u,v]; x[u_,v_] = v;
y[u_,v_] = 2 E^(-v) Sin[Pi u]; z[u_,v_] = u;
{ulow,uhigh} = {0,2}; {vlow,vhigh} = {0.5,3.5};
Clear[normal]
normal[u_,v_] = Cross[D[{x[u,v],y[u,v],z[u,v]},u],D[{x[u,v],y[u,v],z[u,v]},v]]
```

Out[49]=
$$\left\{\frac{2\,\text{Sin}[\text{Pi}\ u]}{E^v},\ 1,\ \frac{-2\,\text{Pi}\,\text{Cos}[\text{Pi}\ u]}{E^v}\right\}$$

Calculate

$$\int_{vlow}^{vhigh}\int_{ulow}^{uhigh} \text{Field}[x[u,v], y[u,v], z[u,v]] \bullet \text{normal}[u,v]\ du\ dv :$$

In[50]:=
```
NIntegrate[Field[x[u,v],y[u,v],z[u,v]].normal[u,v],
    {u,ulow,uhigh},{v,vlow,vhigh}]
```

Out[50]=
12.7339

Positive. This tells you that the net flow of Field[x, y, z] across this surface is in the direction of the plotted normals.

Take another look:

In[51]:=
```
Show[surface,normals,ViewPoint->CMView,
    Boxed->False,AxesLabel->{"x","y","z"}];
```

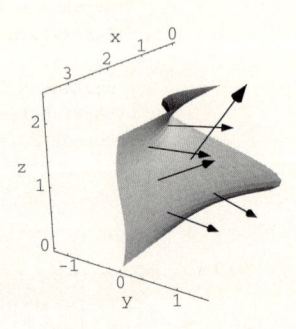

In short, the net flow of Field[x, y, z] = $\{y\, z, x\, z, x\, y\}$ across this surface is from behind the surface to the front of the surface.

■ B.5) Flux of the electric field

This problem appears only in the electronic version.

Tutorials

■ T.1) Measuring flow across surfaces: Gauss's 3D formula versus calculation by surface integrals

T.1.a.i) Here's a surface:

In[1]:=
```
Clear[x,y,z,r,s,t]
{x[r_,s_,t_],y[r_,s_,t_],z[r_,s_,t_]} =
 {0.6 r Sin[s] Cos[t],r Sin[s] Sin[t],2 r Cos[s]};
{{slow,shigh},{tlow,thigh}} = {{0,Pi},{0,2 Pi}};
Clear[surfaceplotter]; surfaceplotter[s_,t_] =
 {0,1,0} + {x[3,s,t],y[3,s,t],z[3,s,t]};
surface = ParametricPlot3D[Evaluate[surfaceplotter[s,t]],
 {s,slow,shigh},{t,tlow,thigh},
 Boxed->False,BoxRatios->Automatic,
 ViewPoint->CMView,AxesLabel->{"x","y","z"}];
```

> The surface is the skin of a solid egg. Measure the flow of
>
> $$\text{Field}[x, y, z] = \{-4x, y^2, 2z\}$$
>
> across this surface. Determine whether the net flow is from inside to outside or from outside to inside.

Answer: This is a natural for Gauss's 3D formula because the surface is all of the outside skin of a solid region. Enter the vector field:

In[2]:=
```
Clear[x,y,z,m,n,p,Field]
{m[x_,y_,z_],n[x_,y_,z_],p[x_,y_,z_]} = {-4 x,y^2,2 z};
Field[x_,y_,z_] = {m[x,y,z],n[x,y,z],p[x,y,z]}
```

Out[2]=
$$\{-4\,x,\ y^2,\ 2\,z\}$$

Calculate divField$[x, y, z]$:

In[3]:=
```
Clear[divField]
divField[x_,y_,z_] = D[m[x,y,z],x] + D[n[x,y,z],y] + D[p[x,y,z],z]
```

Lesson 3.10 3D Surface Measurements

Out[3]=
```
-2 + 2 y
```

Call the surface you see above C and call the solid consisting of all points inside and on the surface R. Gauss's formula tells you that flow of Field$[x, y, z]$ across C is

$$\iiint_R \text{divField}[x, y, z]\, dx\, dy\, dz.$$

To calculate this integral, take advantage of the parameterization of the surface. Checking the plotting instructions above, you see that R is described by

In[4]:=
```
Clear[x,y,z,r,s,t]
{x[r_,s_,t_],y[r_,s_,t_],z[r_,s_,t_]} =
  {0.6 r Sin[s] Cos[t],r Sin[s] Sin[t],2 r Cos[s]}
```

Out[4]=
```
{0.6 r Cos[t] Sin[s], r Sin[s] Sin[t], 2 r Cos[s]}
```

With

In[5]:=
```
{{rlow,rhigh},{slow,shigh},{tlow,thigh}} = {{0,3},{0,Pi},{0,2 Pi}};
```

you know that the flow of Field$[x, y, z]$ across C is measured by

$$\iiint_R \text{divField}[x, y, z]\, dx\, dy\, dz$$

$$= \int_{\text{rlow}}^{\text{rhigh}} \int_{\text{slow}}^{\text{shigh}} \int_{\text{tlow}}^{\text{thigh}} \text{divField}[x[r,s,t], y[r,s,t], z[r,s,t]]\, V_{(x,y,z)}[r,s,t]\, dr\, ds\, dt.$$

Turn *Mathematica* loose:

In[6]:=
```
Clear[gradx,grady,gradz,Vxyz]
gradx[r_,s_,t_] = {D[x[r,s,t],r],D[x[r,s,t],s],D[x[r,s,t],t]};
grady[r_,s_,t_] = {D[y[r,s,t],r],D[y[r,s,t],s],D[y[r,s,t],t]};
gradz[r_,s_,t_] = {D[z[r,s,t],r],D[z[r,s,t],s],D[z[r,s,t],t]};
Vxyz[r_,s_,t_] = Expand[Det[{gradx[r,s,t],grady[r,s,t],gradz[r,s,t]}],Trig->True]
```

Out[6]=
```
      2
1.2 r  Sin[s]
```

Good, this never goes negative for the r, s, and t used here. Here comes the measurement of the net flow of Field$[x, y, z]$ across C:

In[7]:=
```
Integrate[divField[x[r,s,t],y[r,s,t],z[r,s,t]] Vxyz[r,s,t],
  {r,rlow,rhigh},{s,slow,shigh},{t,tlow,thigh}]
```

Out[7]=
```
-86.4 Pi
```

Negative. Because Gauss's 3D formula says
$$\iiint_R \text{divField}[x, y, z] \, dx \, dy \, dz = \iint_C \text{Field} \bullet \text{outerunitnormal} \, dA$$
$$= \text{flow of Field}[x, y, z] \text{ across } C,$$
this negative result tells you that the net flow of this vector field is against the outer normals.

The bottom line: The net flow of this vector field across this surface is from outside to inside. There must be lots of sinks inside the surface.

T.1.a.ii) Go with the same surface as in part T.1.a.i) above, but this time measure the flow of the 3D vector field
$$\text{Field}[x, y, z] = \{x + \sin[y], y + 4 e^{-z}, -2z + x^3\}$$
across this surface. Determine whether the net flow across the surface is from inside to outside or from outside to inside.

Answer: Enter the vector field:

In[8]:=
```
Clear[x,y,z,m,n,p,Field]
{m[x_,y_,z_],n[x_,y_,z_],p[x_,y_,z_]} = {x + Sin[y],y + 4 E^(-z),-2 z + x^3};
Field[x_,y_,z_] = {m[x,y,z],n[x,y,z],p[x,y,z]}
```

Out[8]=
$$\{x + \text{Sin}[y], \frac{4}{E^z} + y, x^3 - 2z\}$$

Calculate divField[x,y,z]:

In[9]:=
```
Clear[divField]
divField[x_,y_,z_] = D[m[x,y,z],x] + D[n[x,y,z],y] + D[p[x,y,z],z]
```

Out[9]=
0

No sources or sinks. The net flow of this vector field across this surface is 0. In fact, the net flow of this vector field across any surface that is the whole skin of a solid region is 0. The outside-to-inside flow is exactly balanced by the inside-to-outside flow.

You could've done this one by hand.

T.1.a.iii) Go with the same surface as in part T.1.a.i) above, but this time measure the flow of the 3D vector field
$$\text{Field}[x, y, z] = \{x + \sin[y], y + 4 e^{-z}, -5z + x^3\}$$

Lesson 3.10 3D Surface Measurements

> across this surface. Determine whether the net flow is from inside to outside or from outside to inside.

Answer: Enter the vector field:

In[10]:=
```
Clear[x,y,z,m,n,p,Field]
{m[x_,y_,z_],n[x_,y_,z_],p[x_,y_,z_]} = {x + Sin[y],y + 4 E^(-z),-5 z + x^3};
Field[x_,y_,z_] = {m[x,y,z],n[x,y,z],p[x,y,z]}
```

Out[10]=
$$\{x + \text{Sin}[y], \frac{4}{E^z} + y, x^3 - 5z\}$$

Calculate divField$[x, y, z]$:

In[11]:=
```
Clear[divField]
divField[x_,y_,z_] = D[m[x,y,z],x] + D[n[x,y,z],y] + D[p[x,y,z],z]
```

Out[11]=
-3

Negative; every point is a sink. The net flow of this vector field across this surface is from outside to inside. In fact, the net flow of this vector field across any surface that is the whole skin of a solid region is from outside to inside.

You could've done this one by hand.

T.1.b) Here's a surface:

In[12]:=
```
Clear[x,y,z,u,v]; x[u_,v_] = 3 v;
y[u_,v_] = -2 u v; z[u_,v_] = 3 u - v;
{ulow,uhigh} = {0,2};{vlow,vhigh} = {0,1.3};
surface = ParametricPlot3D[
Evaluate[{x[u,v],y[u,v],z[u,v]}],
{u,ulow,uhigh},{v,vlow,vhigh},
ViewPoint->CMView,AxesLabel->{"x"," y","z"}];
```

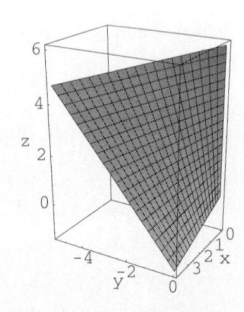

> Go with
> $$f[x, y, z] = ((x+4)(y+3)z)^2$$
> and measure the flow of the gradient field of $f[x, y, z]$ across this surface.

Answer: You can't use Gauss's formula because this surface is not the skin of a solid region, so you've got to go with a direct calculation via the surface integral.

Enter the function and its gradient field:

In[13]:=
```
Clear[f,x,y,z,gradf]; f[x_,y_,z_] = ((x + 4)(y + 3) z)^2;
gradf[x_,y_,z_] = {D[f[x,y,z],x],D[f[x,y,z],y], D[f[x,y,z],z]}
```

Out[13]=
$\{2\,(4 + x)\,(3 + y)^2\,z^2,\ 2\,(4 + x)^2\,(3 + y)\,z^2,\ 2\,(4 + x)^2\,(3 + y)^2\,z\}$

Enter the vector field:

In[14]:=
```
Clear[x,y,z,m,n,p,Field]
{m[x_,y_,z_],n[x_,y_,z_],p[x_,y_,z_]} = gradf[x,y,z];
Field[x_,y_,z_] = {m[x,y,z],n[x,y,z],p[x,y,z]};
```

Reenter the parameterization of the surface and calculate normal$[u, v]$:

In[15]:=
```
Clear[x,y,z,u,v]; x[u_,v_] = 3 v;
y[u_,v_] = -2 u v; z[u_,v_] = 3 u - v;
{ulow,uhigh} = {0,2}; {vlow,vhigh} = {0,1.3};
Clear[normal]
normal[u_,v_] = Cross[D[{x[u,v],y[u,v],z[u,v]},u],
D[{x[u,v],y[u,v],z[u,v]},v]]
```

Out[15]=
{6 u + 2 v, 9, 6 v}

Calculate

$$\int_{vlow}^{vhigh} \int_{ulow}^{uhigh} \text{Field}[x[u,v], y[u,v], z[u,v]] \bullet \text{normal}[u,v]\, du\, dv :$$

In[16]:=
```
flowacross = Integrate[Field[x[u,v],y[u,v],z[u,v]].normal[u,v],
{u,ulow,uhigh},{v,vlow,vhigh}];
N[flowacross]
```

Out[16]=
19086.8

Strongly positive.

To interpret this, look at:

In[17]:=
```
{u,ulow,uhigh}
```

Out[17]=
{u, 0, 2}

In[18]:=
```
{v,vlow,vhigh}
```

Out[18]=
{v, 0, 1.3}

Lesson 3.10 3D Surface Measurements

In[19]:=
```
scalefactor = 0.3;
normals = Table[Arrow[scalefactor normal[u,v],
  Tail->{x[u,v],y[u,v],z[u,v]}],
  {u,0.5,1.5,1},{v,0.2,1.0,0.8}];
Show[surface,normals,
ViewPoint->CMView,Boxed->False,PlotRange->All,
AxesLabel->{"x","y","z"}];
```

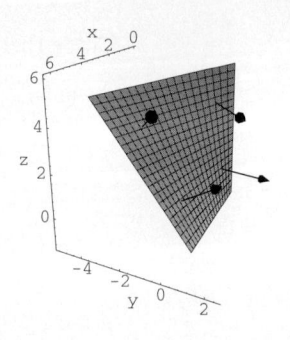

The normals are pointing out from the side of the surface you are sitting on. Because the flow measurement is hugely positive, this tells you that the net flow of gradf$[x, y, z]$ across this surface is strongly from the side you can't see to the side you can see.

■ T.2) Using Gauss's formula to avoid a calculational nightmare: Calculating flow across an oddball surface by calculating the flow across a substitute surface

C_1 is the top half of a sphere of radius 2 centered at the origin:

In[20]:=
```
Clear[x1,y1,z1,s,t]; x1[s_,t_] = 2 Sin[s] Cos[t];
y1[s_,t_] = 2 Sin[s] Sin[t]; z1[s_,t_] = 2 Cos[s];
{{slow,shigh},{tlow,thigh}} = {{0,Pi/2},{0,2 Pi}};
C1plot = ParametricPlot3D[{x1[s,t],y1[s,t],z1[s,t]},
  {s,slow,shigh},{t,tlow,thigh},Boxed->False,
ViewPoint->CMView,AxesLabel->{"x","y","z"},
PlotRange->All];
```

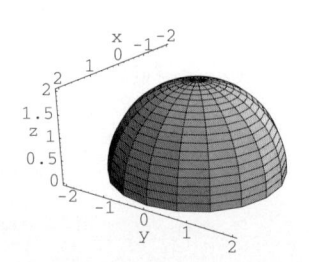

C_2 is an oddball surface that fits under C_1 but agrees with C_1 on their common boundary curve.

In[21]:=
```
Clear[x2,y2,z2,r,t]; x2[r_,t_] = r Cos[t];
y2[r_,t_] = r Sin[t];
z2[r_,t_] = 0.2 r Sin[t/2]^2 (E^Sqrt[4 - r^2] - 1);
{{rlow,rhigh},{tlow,thigh}} = {{0,2},{0,2 Pi}};
C2plot = ParametricPlot3D[{x2[r,t],y2[r,t],z2[r,t]},
  {r,rlow,rhigh},{t,tlow,thigh},Boxed->False,
ViewPoint->CMView,AxesLabel->{"x","y","z"},
PlotRange->All];
```

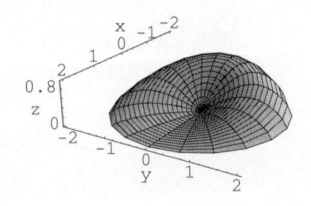

Take a look at the surfaces together from underneath:

```
In[22]:=
Show[C1plot,C2plot,ViewPoint->{1,1,-1}];
```

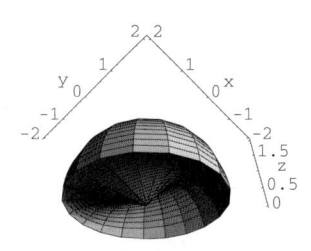

Together C_1 and C_2 make the skin of a solid region.

T.2.a.i) How do you know that when you go with a 3D vector field Field$[x, y, z]$ with divField$[x, y, z] = 0$ throughout the solid region whose top skin is C_1 and whose bottom skin is C_2, then

$$\iint_{C_1} \text{Field} \bullet \text{topunitnormal} \, dA = \iint_{C_2} \text{Field} \bullet \text{topunitnormal} \, dA,$$

so that the flow of Field$[x, y, z]$ across both surfaces is the same?

Answer: Remember that C_1 and C_2 share a common boundary curve. Make a solid region R whose top skin is C_1 and whose bottom skin is C_2. Agree that C stands for the skin of R. You are armed with the fact that divF$[x, y, z] = 0$ throughout R. Use this fact and Gauss's formula to see that

$$0 = \iiint_R 0 \, dx \, dy \, dz$$

$$= \iiint_R \text{divField}[x, y, z] \, dx \, dy \, dz$$

$$= \iint_C \text{Field} \bullet \text{outerunitnormal} \, dA$$

$$= \iint_{C_1} \text{Field} \bullet \text{topunitnormal} \, dA - \iint_{C_2} \text{Field} \bullet \text{topunitnormal} \, dA,$$

because along C_1, the outer unit normal of C agrees with the top unit normal of C_1, but along C_2 the outer unit normal of C agrees with the negative of the top unit normal of C_2. Accordingly,

$$\text{flow of Field across } C_1 = \iint_{C_1} \text{Field} \bullet \text{topunitnormal} \, dA$$

$$= \iint_{C_2} \text{Field} \bullet \text{topunitnormal} \, dA$$

$$= \text{flow of Field}[x, y, z] \text{ across } C_2.$$

T.2.a.ii) What calculational nightmare does this help you to avoid?

Answer: Because the parameterization of C involves screwy functions like

$$z_2[r,t] = 0.2\,r\sin[t/2]\left(e^{\sqrt{(4-r)^2}} - 1\right),$$

calculating

$$\iint_{C_2} \text{Field} \bullet \text{topunitnormal}\, dA$$

will probably be a nightmare. Its normals make you want to ralph. Take a look:

In[23]:=
```
Clear[normal2]
normal2[r_,t_] = Cross[D[{x2[r,t],y2[r,t],z2[r,t]},r],
D[{x2[r,t],y2[r,t],z2[r,t]},t]]
```

Out[23]=

$$\left\{-(r\,\text{Cos}[t]\,(0.2\,(-1 + E^{\text{Sqrt}[4 - r^2]})\,\text{Sin}\left[\frac{t}{2}\right]^2 - \frac{0.2\,E^{\text{Sqrt}[4 - r^2]}\,r^2\,\text{Sin}\left[\frac{t}{2}\right]^2}{\text{Sqrt}[4 - r^2]})) + \right.$$

$$0.2\,(-1 + E^{\text{Sqrt}[4 - r^2]})\,r\,\text{Cos}\left[\frac{t}{2}\right]\,\text{Sin}\left[\frac{t}{2}\right]\,\text{Sin}[t],$$

$$-0.2\,(-1 + E^{\text{Sqrt}[4 - r^2]})\,r\,\text{Cos}\left[\frac{t}{2}\right]\,\text{Cos}[t]\,\text{Sin}\left[\frac{t}{2}\right] -$$

$$r\,(0.2\,(-1 + E^{\text{Sqrt}[4 - r^2]})\,\text{Sin}\left[\frac{t}{2}\right]^2 -$$

$$\frac{0.2\,E^{\text{Sqrt}[4 - r^2]}\,r^2\,\text{Sin}\left[\frac{t}{2}\right]^2}{\text{Sqrt}[4 - r^2]})\,\text{Sin}[t],$$

$$\left. r\,\text{Cos}[t]^2 + r\,\text{Sin}[t]^2 \right\}$$

These normals will make you choke, and they will probably make *Mathematica* choke too. But you know that

$$\iint_{C_1} \text{Field} \bullet \text{topunitnormal}\, dA = \iint_{C_2} \text{Field} \bullet \text{topunitnormal}\, dA$$

so instead of confronting the nightmarish calculation of $\iint_{C_2} \text{Field} \bullet \text{topunitnormal}\, dA$, you can opt for the simpler calculation of $\iint_{C_1} \text{Field} \bullet \text{topunitnormal}\, dA$.

T.2.a.iii) Illustrate this idea by calculating the flow of the 3D vector field

$$\text{Field}[x, y, z] = \{x - y, -2y - x, z + 3\}$$

across C_2.

Answer: Enter the vector field:

In[24]:=
```
Clear[x,y,z,m,n,p,Field]
{m[x_,y_,z_],n[x_,y_,z_],p[x_,y_,z_]} = {x - y,-2 y - x, z + 3};
Field[x_,y_,z_] = {m[x,y,z],n[x,y,z],p[x,y,z]};
```

Check divField[x, y, z]:

In[25]:=
```
Clear[divField]
divField[x_,y_,z_] = D[m[x,y,z],x] + D[n[x,y,z],y] + D[p[x,y,z],z]
```
Out[25]=
0

Good. Now you know you can measure the flow of this vector field across the squirrelly surface C_2 by measuring the flow of this vector field across the clean surface C_1. Here you go:

$$\iint_{C_1} \text{Field} \bullet \text{topunitnormal} \, dA$$

$$= \int_0^{2\pi} \int_0^{\pi/2} \text{Field}[x1[s,t], y1[s,t], z1[s,t]] \bullet \text{normal1}[s,t] \, ds \, dt$$

where normal1[s, t] is:

In[26]:=
```
Clear[normal1]
normal1[s_,t_] = Expand[Cross[D[{x1[s,t],y1[s,t],z1[s,t]},s],
D[{x1[s,t],y1[s,t],z1[s,t]},t]],Trig->True]
```
Out[26]=
{-Cos[2 s - t] + 2 Cos[t] - Cos[2 s + t],
  Sin[2 s - t] + 2 Sin[t] - Sin[2 s + t], 2 Sin[2 s]}

You can see that normal1[s, t] is a top normal by looking at the third slot, which is $2\sin[2\,s]$. In the parameterization of C_1, s runs from 0 to $\pi/2$ and $2\sin[2\,s]$ remains nonnegative for $0 \leq s \leq \pi/2$.

Now you're ready to calculate:

flow of Field[x, y, z] across C_2 = flow of Field[x, y, z] across C_1

$$= \iint_{C_1} \text{Field} \bullet \text{topunitnormal} \, dA$$

$$= \int_0^{2\pi} \int_0^{\pi/2} \text{Field}[x1[s,t], y1[s,t], z1[s,t]] \bullet \text{normal1}[s,t] \, ds \, dt :$$

Lesson 3.10 3D Surface Measurements

In[27]:=
```
Integrate[Field[x1[s,t],y1[s,t],z1[s,t]].normal1[s,t],{t,0, 2 Pi},{s,0,Pi/2}]
```
Out[27]=
```
12 Pi
```

Strong flow from low to high across C_2 (as well as C_1).

The direct calculation of the flow of this field across C_2 is not practical.

■ T.3) Using Gauss's formula to take advantage of singularities: Calculating flow across the skin of a solid region by calculating the flow across a substitute sphere

Here's an elliptical cylinder with top and bottom:

In[28]:=
```
radial1 = {1,-2,-1}; radial2 = {1,1,-1}; core = {0.5,2,2};
Clear[sideplotter,topplotter,bottomplottter,r,u,t]
sideplotter[u_,t_] = Cos[t] radial1 + Sin[t] radial2 + u core;
{{ulow,uhigh},{tlow,thigh}} = {{-1,2},{0,2 Pi}};
topplotter[r_,t_] = r Cos[t] radial1 + r Sin[t] radial2 + uhigh core;
bottomplotter[r_,t_] = r Cos[t] radial1 + r Sin[t] radial2 + ulow core;
{rlow,rhigh} = {0,1};
sides = ParametricPlot3D[Evaluate[sideplotter[u,t]],
  {u,ulow,uhigh},{t,tlow,thigh},PlotPoints->{2,Automatic},DisplayFunction->Identity];
top = ParametricPlot3D[Evaluate[topplotter[r,t]],
  {r,rlow,rhigh},{t,tlow,thigh},PlotPoints->{2,Automatic},DisplayFunction->Identity];
bottom = ParametricPlot3D[Evaluate[bottomplotter[r,t]],
  {r,rlow,rhigh},{t,tlow,thigh},PlotPoints->{2,Automatic},DisplayFunction->Identity];
```

In[29]:=
```
Show[sides,top,bottom,
Boxed->False,ViewPoint->CMView,
AxesLabel->{"x","y","z"},
PlotRange->All,DisplayFunction->$DisplayFunction];
```

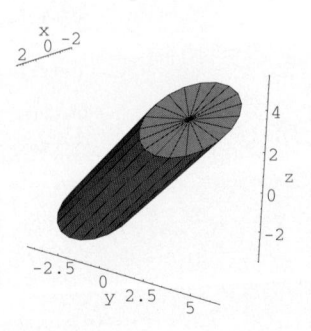

Go with the electric field

$$\text{Field}[x,y,z] = \left\{ \frac{2\,x}{(x^2+y^2+z^2)^{3/2}}, \frac{2\,y}{(x^2+y^2+z^2)^{3/2}}, \frac{2\,z}{(x^2+y^2+z^2)^{3/2}} \right\}$$

resulting from a charge of strength 2 placed at the point $\{0,0,0\}$ which is inside this skin.

T.3.a) Try to use Gauss's 3D formula to calculate the flow of this vector field across the skin of the surface plotted above.

Answer: Enter the vector field:

In[30]:=
```
Clear[x,y,z,m,n,p,Field]
{m[x_,y_,z_],n[x_,y_,z_],p[x_,y_,z_]} =
{2 x/(x^2 + y^2 + z^2)^(3/2),2 y/(x^2 + y^2 + z^2)^(3/2),
2 z/(x^2 + y^2 + z^2)^(3/2)};
Field[x_,y_,z_] = {m[x,y,z],n[x,y,z],p[x,y,z]};
```

Calculate divField$[x, y, z]$:

In[31]:=
```
Clear[divField]
divField[x_,y_,z_] = Together[D[m[x,y,z],x] + D[n[x,y,z],y] + D[p[x,y,z],z]]
```

Out[31]=
0

Good; Gauss's 3D formula tells you if you let R stand for the solid of which the plotted surface above is the outside skin, then you are guaranteed that the flow of this vector field across this surface is given by

$$\iiint_R \text{divField}[x, y, z]\, dx\, dy\, dz = \iiint_R 0\, dx\, dy\, dz = 0.$$

You happily report that the flow of Field$[x, y, z]$ across the skin plotted above is 0 and go on to the next problem.

T.3.b) Was the answer given in part T.3.a) correct?

Answer: No way.

T.3.c.i) What went wrong?

Answer: Look at:

In[32]:=
```
Field[0,0,0]
```

Out[32]=
{Indeterminate, Indeterminate, Indeterminate}

Field$[x, y, z]$ goes nuts at $\{0, 0, 0\}$.

In short, $\{0, 0, 0\}$ is a singularity of Field$[x, y, z]$, and the damage comes from the fact that $\{0, 0, 0\}$ is inside the plotted surface above. Anytime you have a singularity on or in a solid region R, Gauss's formula

$$\iiint_R \text{divField}[x, y, z]\, dx\, dy\, dz = \text{flow of Field}[x, y, z] \text{ across } C$$

has the possibility of failing. And it did fail in the "answer" given in part T.3.a).

In the absence of singularities of Field$[x, y, z]$ or (of divF$[x, y, z]$) inside or on the surface of a solid region R, Gauss's formula cannot fail.

T.3.c.ii) Are there other singularities?

Answer: Look at:

In[33]:=
```
Field[x,y,z]
```

Out[33]=

$$\left\{ \frac{2x}{(x^2 + y^2 + z^2)^{3/2}}, \frac{2y}{(x^2 + y^2 + z^2)^{3/2}}, \frac{2z}{(x^2 + y^2 + z^2)^{3/2}} \right\}$$

A short examination of this formula for Field$[x, y, z]$ shows that the only point at which the denominators are 0 is $\{0, 0, 0\}$. There are no other singularities.

T.3.d) One way to try to do the calculation of the flow of Field$[x, y, z]$ across this surface is to let C stand for the whole surface and try to do a direct calculation of

$$\iint_C \text{Field} \bullet \text{outerunitnormal}\, dA$$

by breaking it up in the form:

$$\iint_{C\text{sides}} \text{Field} \bullet \text{outerunitnormal}\, dA + \iint_{C\text{top}} \text{Field} \bullet \text{outerunitnormal}\, dA$$

$$+ \iint_{C\text{bottom}} \text{Field} \bullet \text{outerunitnormal}\, dA.$$

This is a calculational nightmare because it involves setting up three different normal vectors and then hoping that *Mathematica* can do the resulting integrals. Part of the art of mathematics is knowing how to avoid calculational nightmares.

How can you avoid this nightmare?

Answer: You center a little sphere called R_{little} at the lone singularity:

In[34]:=
```
singularity = {0,0,0};
Clear[xlittle,ylittle,zlittle,s,t,littlesphereplotter]
littleradius = 0.1;
{xlittle[s_,t_],ylittle[s_,t_],zlittle[s_,t_]} =
 singularity + {littleradius Sin[s] Cos[t],
 littleradius Sin[s] Sin[t],littleradius Cos[s]};
littlesphereplot = ParametricPlot3D[
 Evaluate[{xlittle[s,t],ylittle[s,t],
 zlittle[s,t]}],{s,0,Pi},{t,0,2Pi},
 Boxed->False,ViewPoint->CMView,PlotRange->All];
```

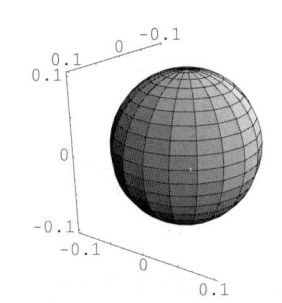

The singularity at $\{0,0,0\}$ is encapsulated inside the little sphere, and the little sphere is inside the original surface. Now you make a new hollow solid, taking everything inside the original skin C, but rejecting everything inside the skin, C_{little}, of the little sphere. Let R_{new} stand for the new hollow solid and let Cnew stand for its skin. Note that there are no singularities inside R_{new} because the lone singularity lies inside the hollow part.

Apply Gauss's formula to the new hollow solid R_{new} to see that

$$\iint_{C_{\text{new}}} \text{Field} \bullet \text{outerunitnormal} \, dA = \iiint_{R_{\text{new}}} \text{divField}[x,y,z] \, dx \, dy \, dz = 0$$

because $\text{divField}[x,y,z] = 0$ throughout R_{new}. This tells you that

$$\iint_{C_{\text{new}}} \text{Field} \bullet \text{outerunitnormal} \, dA = 0.$$

But if you let C_{little} stand for the skin of the little sphere centered at the singularity, and you take the outer unit normal from the little sphere, then you get

$$0 = \iint_{C_{\text{new}}} \text{Field} \bullet \text{outerunitnormal} \, dA$$
$$= \iint_{C} \text{Field} \bullet \text{outerunitnormal} \, dA - \iint_{C_{\text{little}}} \text{Field} \bullet \text{outerunitnormal} \, dA,$$

because the outer unit normal on the new solid points inside the little sphere.
This is big news because it tells you that

$$\iint_{C} \text{Field} \bullet \text{outerunitnormal} \, dA = \iint_{C_{\text{little}}} \text{Field} \bullet \text{outerunitnormal} \, dA.$$

As a result, you can avoid the calculation of the gruesome integral

$$\iint_{C} \text{Field} \bullet \text{outerunitnormal} \, dA$$

by calculating one single, easy integral:

$$\iint_{C_{\text{little}}} \text{Field} \bullet \text{outerunitnormal} \, dA$$
$$= \int_{0}^{2\pi} \int_{0}^{\pi} \text{Field}[\text{xlittle}[t], \text{ylittle}[t], \text{zlittle}[t]] \bullet \text{normal}[s,t] \, ds \, dt$$

where $\text{normal}[s,t]$ is given by:

In[35]:=
```
Clear[normal]
normal[s_,t_] = Expand[Cross[D[{xlittle[s,t],ylittle[s,t],
  zlittle[s,t]},s],D[{xlittle[s,t],
  ylittle[s,t],zlittle[s,t]},t]],Trig->True]
```
Out[35]=
{-0.0025 Cos[2 s - t] + 0.005 Cos[t] - 0.0025 Cos[2 s + t],
   0.0025 Sin[2 s - t] + 0.005 Sin[t] - 0.0025 Sin[2 s + t], 0.005 Sin[2 s]}

Make sure that these normals are outer normals:

In[36]:=
```
scalefactor = 5;
Show[littlesphereplot,
Table[Arrow[scalefactor normal[s,t],
Tail->{xlittle[s,t],ylittle[s,t],zlittle[s,t]}],
{s,Pi/4,3 Pi/4,Pi/4},{t,0,2 Pi,Pi/2}],
Boxed->False,ViewPoint->CMView,PlotRange->All];
```

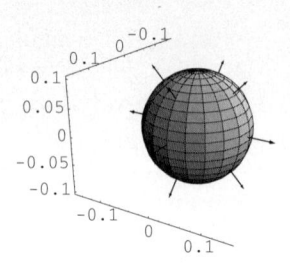

They **are** outer normals. Now you can correctly calculate the flow of Field$[x, y, z]$ across the original cylindrical surface in one short, sweet calculation:

In[37]:=
```
Integrate[Field[xlittle[s,t],ylittle[s,t],
  zlittle[s,t]].normal[s,t],{s,0,Pi},{t,0,2 Pi}]
```
Out[37]=
8. Pi

And you're out of here.

■ T.4) Combined electric fields and multiple singularities

This problem appears only in the electronic version.

■ T.5) Surface packaging: Parameterized, explicit and implicit

This problem appears only in the electronic version.

Give It a Try

Experience with the starred (*) problems will be useful for understanding developments later in the course.

■ **G.1) How do you know?**★

G.1.a) Here's a 3D vector field:

In[1]:=
```
Clear[x,y,z,m,n,p,Field]
{m[x_,y_,z_],n[x_,y_,z_],p[x_,y_,z_]} = {-6 x + 3 x Cos[z],
 4 y + 5 E^(-0.2 (x + z)),5 z - 3 Sin[z]};
Field[x_,y_,z_] = {m[x,y,z],n[x,y,z],p[x,y,z]}
```

Out[1]=

$$\{-6 x + 3 x \cos[z], \frac{5}{E^{0.2(x+z)}} + 4 y, 5 z - 3 \sin[z]\}$$

> How do you know that the net flow of this vector field across any surface that is the whole skin of a solid region is from inside to outside?

G.1.b) Here's another 3D vector field:

In[2]:=
```
Clear[x,y,z,m,n,p,Field]
{m[x_,y_,z_],n[x_,y_,z_],p[x_,y_,z_]} = {-6 x + 3 x Cos[z],
 2 y + x /(1 + x^2), z - 3 Sin[z]};
Field[x_,y_,z_] = {m[x,y,z],n[x,y,z],p[x,y,z]}
```

Out[2]=

$$\{-6 x + 3 x \cos[z], \frac{x}{1 + x^2} + 2 y, z - 3 \sin[z]\}$$

> How do you know that the net flow of this vector field across any surface that is the whole skin of a solid region is from outside to inside?

G.1.c) Here's another 3D vector field involving a parameter b:

In[3]:=
```
Clear[b,x,y,z,m,n,p,Field]
{m[x_,y_,z_],n[x_,y_,z_],p[x_,y_,z_]} = {-6 x + 3 x E^z,
 b y + 12 z,8 x + 4 z - 3 E^z};
Field[x_,y_,z_] = {m[x,y,z],n[x,y,z],p[x,y,z]}
```

Out[3]=

$$\{-6 x + 3 E^z x, b y + 12 z, -3 E^z + 8 x + 4 z\}$$

> How do you know that if $b > 2$, then the net flow of this vector field across any surface that is the whole skin of a solid region is from inside to outside?

> How do you know that if $b < 2$, then the net flow of this vector field across any surface that is the whole skin of a solid region is from outside to inside?
>
> How do you know that if $b = 2$, then the net flow of this vector field across any surface that is the whole skin of a solid region is 0?

G.1.d) Here's another 3D vector field:

In[4]:=
```
Clear[b,x,y,z,m,n,p,Field]
{m[x_,y_,z_],n[x_,y_,z_],p[x_,y_,z_]} = {6 x^2 + 3 x E^z,
 9 y + 12 z,8 x + 3 z - 3 E^z};
Field[x_,y_,z_] = {m[x,y,z],n[x,y,z],p[x,y,z]}
```

Out[4]=
$$\{3 E^z x + 6 x^2, \ 9 y + 12 z, \ -3 E^z + 8 x + 3 z\}$$

> How do you know that any point $\{x, y, z\}$ with $x > -1$ is a source for this vector field?
>
> How do you know that any point $\{x, y, z\}$ with $x < -1$ is a sink for this vector field?
>
> How do you know that the net flow of this vector field across the sphere of radius $= 1$ centered at $\{1, 0, 0\}$ is from inside to outside, but the net flow of this vector field across the sphere of radius $= 1$ centered at $\{-3, 0, 0\}$ is from outside to inside?

G.1.e) Here's another 3D vector field involving cleared functions $f[x, y], g[y, z]$ and $h[x, z]$:

In[5]:=
```
Clear[f,g,h,x,y,z,m,n,p,Field]
{m[x_,y_,z_],n[x_,y_,z_],p[x_,y_,z_]} = {g[y,z],h[x,z],f[x,y]};
Field[x_,y_,z_] = {m[x,y,z],n[x,y,z],p[x,y,z]}
```

Out[5]=
$$\{g[y, z], \ h[x, z], \ f[x, y]\}$$

This tells you that

→ $m[x, y, z] = g[y, z]$ does not change as x changes;

→ $n[x, y, z] = h[x, z]$ does not change as y changes and

→ $p[x, y, z] = f[x, y]$ does not change as z changes.

> Give a concrete example of this kind of vector field.
>
> How do you know in advance that this kind of vector field has no sources or sinks?

G.1.f) Look at this plot:

In[6]:=
```
Clear[x,y,z,s,t]; Clear[normal]
x[s_,t_] = 5/(1 + (s - Pi/2)^2) Sin[s] Cos[t];
y[s_,t_] = (4 - Sin[t])Sin[s] Sin[t];
z[s_,t_] = 4 Cos[s];
{{slow,shigh},{tlow,thigh}} = {{0,Pi/2},{0,Pi}};
surface = ParametricPlot3D[Evaluate[
{x[s,t],y[s,t],z[s,t]}],{s,slow,shigh},
{t,tlow,thigh},DisplayFunction->Identity];
normal[s_,t_] = Cross[D[{x[s,t],y[s,t],z[s,t]},s],
D[{x[s,t],y[s,t],z[s,t]},t]];
jump = Pi/8; scalefactor = 0.2;
normals = Table[Arrow[scalefactor normal[s,t],
Tail->{x[s,t],y[s,t],z[s,t]}],
{s,slow + jump,shigh - jump,jump},
{t,tlow + jump,thigh - jump,jump}];
Show[surface,normals,ViewPoint->CMView,Boxed->False,
AxesLabel->{"x","y","z"},
DisplayFunction->$DisplayFunction];
```

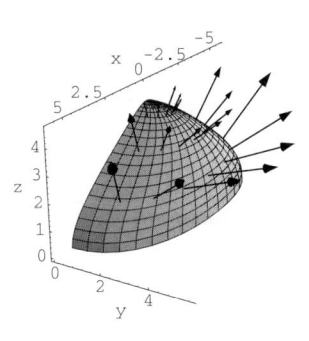

This plot shows a surface together with some of the normal vectors whose lengths measure the area conversion factor $SA_{(x,y,z)}[s,t]$, which converts st-paper area measurements into xyz-area measurements on the surface.

Notice this: The longer normal vectors have their tails on the larger plates on the *Mathematica* plot, and the shorter normal vectors have their tails on the smaller plates on the *Mathematica* plot.

> How do you know that is likely to happen no matter what surface you go with?

■ G.2) Meat-and-potatoes measurements*

G.2.a.i) Here's a vector field:

In[7]:=
```
Clear[x,y,z,m,n,p,Field]
{m[x_,y_,z_],n[x_,y_,z_],p[x_,y_,z_]} = {y^2,z^2,x^2};
Field[x_,y_,z_] = {m[x,y,z],n[x,y,z],p[x,y,z]}
```
Out[7]=
$$\{y^2, z^2, x^2\}$$

> Measure the net flow of this vector field across the surface of the three-dimensional box

$$-1 \leq x \leq 2,$$
$$-3 \leq y \leq 3,$$
and
$$0 \leq z \leq 4.$$

Is the net flow of this vector field across this skin from inside to outside, from outside to inside or 0?

G.2.a.ii) Here's a new vector field:

In[8]:=
```
Clear[x,y,z,m,n,p,Field]
{m[x_,y_,z_],n[x_,y_,z_],p[x_,y_,z_]} = {x^2,y^2,z^2};
Field[x_,y_,z_] = {m[x,y,z],n[x,y,z],p[x,y,z]}
```

Out[8]=
$\{x^2, y^2, z^2\}$

Measure the net flow of this vector field across the surface of the three-dimensional box

$$-1 \leq x \leq 2,$$
$$-3 \leq y \leq 3,$$
$$0 \leq z \leq 4.$$

Is the net flow across this skin from inside to outside, from outside to inside or 0?

G.2.b) Measure the net flow of the vector field

$$\text{Field}[x, y, z] = \{x^2, y^2, z^2\}$$

across the surface of the sphere of radius 3 centered at $\{0, 0, 2\}$.

Is the net flow across this skin from inside to outside, from outside to inside or 0?

G.2.c) Here's a new vector field:

In[9]:=
```
Clear[x,y,z,m,n,p,Field]
{m[x_,y_,z_],n[x_,y_,z_],p[x_,y_,z_]} = {x^2,y^3,z^2};
Field[x_,y_,z_] = {m[x,y,z],n[x,y,z],p[x,y,z]}
```

Out[9]=
$\{x^2, y^3, z^2\}$

Use Gauss's formula to measure the net flow of this vector field across the surface of the three-dimensional solid region between the sphere of radius 1 centered at $\{0,0,0\}$ and the sphere of radius 4 centered at $\{0,0,1\}$.

G.2.d.i) Here's a magic carpet:

In[10]:=
```
Clear[x,y,z,s,t]; {x[s_,t_],y[s_,t_],z[s_,t_]} =
  {t,0, 1 + 0.15 Cos[4 t]} + s{0,1,0};
{{slow,shigh},{tlow,thigh}} = {{0,1.5},{0,6}};
carpet = ParametricPlot3D[Evaluate[
  {x[s,t],y[s,t],z[s,t]}],{s,slow,shigh},{t,tlow,thigh},
  PlotPoints->{2,30},ViewPoint->CMView,
  BoxRatios->Automatic,Boxed->False,AxesLabel->{"x","y","z"}];
```

Look at the numbers displayed on the axes, and use them to say why it is clear that the surface area measurement of this magic carpet is a bit more than nine square units.

Then move in with a surface integral to give a more accurate measurement.

G.2.d.ii) While you're at it, take the same magic carpet as in part G.2.d.i) above and measure the flow of the 3D vector field

$$\text{Field}[x, y, z] = \{y - z, x + z, x - y\}$$

across (through) this carpet.

Determine whether the net flow of this vector field across this carpet is from under to over or from over to under.

G.2.d.iii) When that lab pest, Calculus Cal, looked over your shoulder and saw your answer to part G.2.d.ii) above, he said: "Your answer is wrong, and I'll tell you why. Look at the vector field and its divergence":

In[11]:=
```
Clear[x,y,z,m,n,p,Field,divField]
{m[x_,y_,z_],n[x_,y_,z_],p[x_,y_,z_]} = {y - z, x + z, x - y};
Field[x_,y_,z_] = {m[x,y,z],n[x,y,z],p[x,y,z]};
divField[x_,y_,z_] = D[m[x,y,z],x] + D[n[x,y,z],y] + D[p[x,y,z],z]
```

Out[11]=
0

Calculus Cal went on to say: "It says right in B.1) that any vector field Field$[x, y, z]$ with divField$[x, y, z] = 0$ for all points $\{x, y, z\}$ has no sources or sinks. Therefore, its flow across any old surface is 0."

Say why Calculus Cal is wrong and then tell Cal where to go.

G.2.e) Here's a plot of the uv-paper ellipse

$$\left(\frac{u}{2}\right)^2 + \left(\frac{v}{4}\right)^2 = 1$$

on the plane

$$x + 2y + 4z = 0$$

under the parameterization:

$$x[u,v] = u,$$
$$y[u,v] = v$$

and

$$z[u,v] = \frac{-u - 2v}{4}.$$

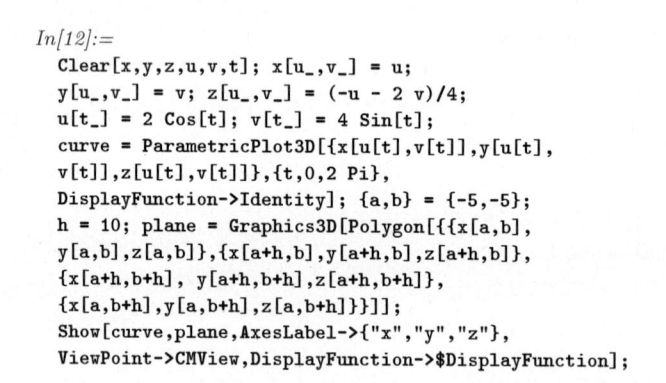

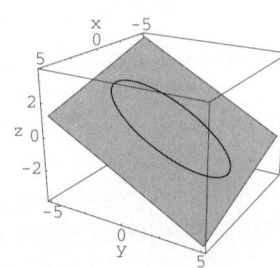

In[12]:=
```
Clear[x,y,z,u,v,t]; x[u_,v_] = u;
y[u_,v_] = v; z[u_,v_] = (-u - 2 v)/4;
u[t_] = 2 Cos[t]; v[t_] = 4 Sin[t];
curve = ParametricPlot3D[{x[u[t],v[t]],y[u[t],
v[t]],z[u[t],v[t]]},{t,0,2 Pi},
DisplayFunction->Identity]; {a,b} = {-5,-5};
h = 10; plane = Graphics3D[Polygon[{{x[a,b],
y[a,b],z[a,b]},{x[a+h,b],y[a+h,b],z[a+h,b]},
{x[a+h,b+h], y[a+h,b+h],z[a+h,b+h]},
{x[a,b+h],y[a,b+h],z[a,b+h]}}]];
Show[curve,plane,AxesLabel->{"x","y","z"},
ViewPoint->CMView,DisplayFunction->$DisplayFunction];
```

Come up with a measurement of the area enclosed by this curve as measured on the plotted plane.

■ G.3) Substitute surfaces for avoiding calculational nightmares*

G.3.a) Here is a surface called C_1:

In[13]:=
```
Clear[x1,y1,z1,s,t]
x1[s_,t_] = 2 + 2 Sin[s] Cos[t];
y1[s_,t_] = 1 + 2 Sin[s] Sin[t];
z1[s_,t_] = 1 + E^(Sin[y1[s,t] - 1]) Cos[s];
{{s1low,s1high},{t1low,t1high}} =
{{0, Pi/2},{0,2 Pi}}; C1plot =
ParametricPlot3D[Evaluate[{x1[s,t],y1[s,t],
z1[s,t]}],{s,s1low,s1high},{t,t1low,t1high},
Boxed->False,ViewPoint->CMView,
PlotRange->All,AxesLabel->{"x","y","z"}];
```

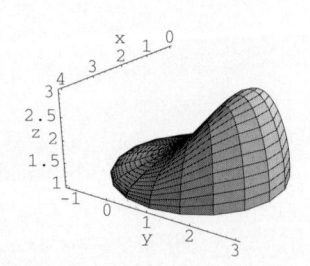

C_2 is a circular disk in the plane $z = 1$ centered at the point $\{2, 1, 1\}$:

In[14]:=
```
Clear[x2,y2,z2,s,t]; x2[s_,t_] = 2 + s Cos[t];
y2[s_,t_] = 1 + s Sin[t]; z2[s_,t_] = 1;
{{s2low,s2high},{t2low,t2high}} = {{0, 2},{0,2 Pi}};
C2plot = ParametricPlot3D[Evaluate[{x2[s,t],
y2[s,t],z2[s,t]}],{s,s2low,s2high},{t,t2low,t2high},
PlotPoints->{2,Automatic},Boxed->False,
ViewPoint->CMView,PlotRange->All,
AxesLabel->{"x","y","z"}];
```

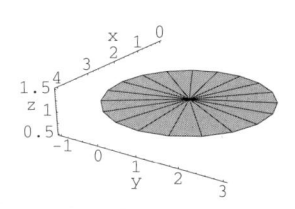

C_1 and C_2 share the same boundary curve, as you can see by studying the code or by looking from underneath:

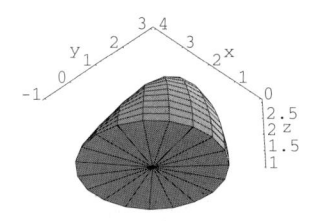

In[15]:=
```
Show[C1plot,C2plot,ViewPoint->{5,5,-5}];
```

Here's a vector field:

In[16]:=
```
Clear[x,y,z,Field]
Field[x_,y_,z_] = {z - y,x + z,y - x}
```

Out[16]=
{-y + z, x + z, -x + y}

And its divergence:

In[17]:=
```
Clear[m,n,p,divField]
{m[x_,y_,z_],n[x_,y_,z_],p[x_,y_,z_]} = Field[x,y,z];
divField[x_,y_,z_] = D[m[x,y,z],x] + D[n[x,y,z],y] + D[p[x,y,z],z]
```

Out[17]=
0

Calculate the net flow of this vector field across C_1,

$$\iint_{C_1} \text{Field} \cdot \text{topunitnormal} \, dA,$$

by calculating the net flow of this vector field across C_2,

$$\iint_{C_2} \text{Field} \cdot \text{topunitnormal} \, dA$$

and explain why this gives the correct answer.

G.3.b) Here is a new surface called C_1:

```
In[18]:=
  Clear[x1,y1,z1,r,s]; x1[r_,s_] = -3 + 2 r Cos[s];
  y1[r_,s_] = 1 + r Sin[s];
  z1[r_,s_] = 2 + E^(Sin[4(y1[r,s] - 1)]) (2 - r);
  {{r1low,r1high},{s1low,s1high}}= {{0,2},{0,2 Pi}};
  C1plot = ParametricPlot3D[Evaluate[
  {x1[r,s],y1[r,s],z1[r,s]}],
  {r,r1low,r1high},{s,s1low,s1high},
  Boxed->False,ViewPoint->CMView,
  PlotRange->All,PlotPoints->{30,30},
  AxesLabel->{"x","y","z"}];
```

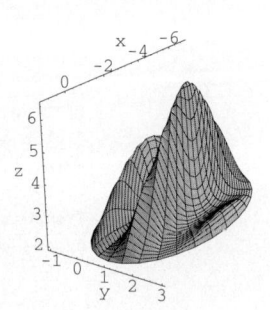

Here is a vector field:

```
In[19]:=
  Clear[x,y,z,Field]; F[x_,y_,z_] = {3 z - y,3 x + z,6 y - x}
Out[19]=
  {-y + 3 z, 3 x + z, -x + 6 y}
```

And its divergence:

```
In[20]:=
  Clear[m,n,p,divField]
  {m[x_,y_,z_],n[x_,y_,z_],p[x_,y_,z_]} = F[x,y,z];
  divField[x_,y_,z_] = D[m[x,y,z],x] +D[n[x,y,z],y] + D[p[x,y,z],z]
Out[20]=
  0
```

> Calculate the net flow of this vector field across C_1 by calculating
> $$\iint_{C_2} \text{Field} \bullet \text{topunitnormal} \, dA$$
> where C_2 is a substitute surface of your own choice.
>
> Explain why you believe in your answer.

■ G.4) Avoiding another calculational nightmare*

Here's a vector field:

```
In[21]:=
  Clear[Field,x,y,z,m,n,p,divField1]
  Field[x_,y_,z_] = {z,4 x,-3 y} + (3 ({x,y,z} - {0,0.5,0})/
  ((x - 0)^2 + (y - 0.5)^2 + (z - 0)^2)^(3/2));
  {m[x_,y_,z_],n[x_,y_,z_] ,p[x_,y_,z_]} = Field[x,y,z]
```

Note the big singularity at $\{0, 0.5, 0\}$. Here is the divergence of $\text{Field}[x, y, z]$:

In[22]:=
```
divField[x_,y_,z_] = Together[D[m[x,y,z],x] + D[n[x,y,z],y] + D[p[x,y,z],z]]
```
Out[22]=
0

Here comes a surface:

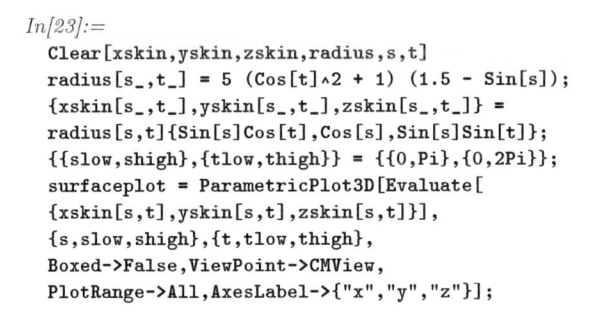

In[23]:=
```
Clear[xskin,yskin,zskin,radius,s,t]
radius[s_,t_] = 5 (Cos[t]^2 + 1) (1.5 - Sin[s]);
{xskin[s_,t_],yskin[s_,t_],zskin[s_,t_]} =
   radius[s,t]{Sin[s]Cos[t],Cos[s],Sin[s]Sin[t]};
{{slow,shigh},{tlow,thigh}} = {{0,Pi},{0,2Pi}};
surfaceplot = ParametricPlot3D[Evaluate[
   {xskin[s,t],yskin[s,t],zskin[s,t]}],
   {s,slow,shigh},{t,tlow,thigh},
   Boxed->False,ViewPoint->CMView,
   PlotRange->All,AxesLabel->{"x","y","z"}];
```

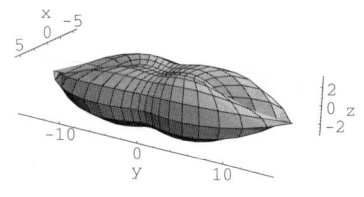

G.4.a.i) Is the singularity of Field[x, y, z] inside the surface plotted above?

G.4.a.ii) Why is it a bad idea to assume in advance that since divField[x, y, z] calculated out to 0, then the net flow of Field[x, y, z] across the surface plotted above is 0?

G.4.a.iii) Measure the flow of Field[x, y, z] across this surface.

Is the flow from outside to inside or inside to outside?

G.4.b.i) Here is a new vector field and its divergence:

In[24]:=
```
Clear[Field,x,y,z,m,n,p,divField]
Field[x_,y_,z_] = {z,4 x,-3 y} + (3({x,y,z} - {10,0,0})/
   ((x - 10)^2 + (y - 0)^2 + (z - 0)^2)^(3/2));
{m[x_,y_,z_],n[x_,y_,z_],p[x_,y_,z_]} = Field[x,y,z]
```

Out[24]=
$$\{z + \frac{3(-10 + x)}{((-10 + x)^2 + y^2 + z^2)^{3/2}},$$
$$4x + \frac{3y}{((-10 + x)^2 + y^2 + z^2)^{3/2}},$$
$$-3y + \frac{3z}{((-10 + x)^2 + y^2 + z^2)^{3/2}}\}$$

```
In[25]:=
  divField[x_,y_,z_] = Together[D[m[x,y,z],x] + D[n[x,y,z],y] + D[p[x,y,z],z]]
Out[25]=
  0
```

> Is the singularity of Field$[x, y, z]$ inside the surface plotted part G.4.a)?

G.4.b.ii) Why is it a good idea to assume in advance that since divField$[x, y, z]$ calculated out to 0, the net flow of Field$[x, y, z]$ across the surface plotted in part G.4.a) is 0?

■ G.5) Flux of the electric field and Gauss's law

This problem appears only in the electronic version.

■ G.6) Sources and sinks in the 3D gradient field, the Laplacian

$$\frac{\partial^2 f[x, y, z]}{\partial x^2} + \frac{\partial^2 f[x, y, z]}{\partial y^2} + \frac{\partial^2 f[x, y, z]}{\partial z^2},$$

and steady-state heat★

One way to come up with a 3D vector field is to take a function $f[x, y, z]$ and put

$$\text{Field}[x, y, z] = \text{grad} f[x, y, z].$$

Try it out:

```
In[26]:=
  Clear[f,gradf,Field,x,y]
  f[x_,y_,z_] = 1 - (x^2 + y^2 + z^2)/2;
  gradf[x_,y_,z_] = {D[f[x,y,z],x],D[f[x,y,z],y],D[f[x,y,z],z]};
  Field[x_,y_,z_] = gradf[x,y,z]
Out[26]=
  {-x, -y, -z}
```

This is the gradient field of

$$f[x, y] = 1 - \frac{x^2 + y^2 + z^2}{2}.$$

Note that $\{x, y, z\} = \{0, 0, 0\}$ maximizes $f[x, y, z]$. Here's a plot of the scaled gradient field of $f[x, y, z]$ shown with the maximizer at $\{0, 0, 0\}$:

```
In[27]:=
    maximizerplot = Graphics3D[{Red,PointSize[0.06],
    Point[{0,0,0}]}]; scalefactor = 0.4;
    gradfieldplot = Table[Arrow[scalefactor gradf[x,y,z],
    Tail->{x,y,z},Blue],{x,-1,1,2},{y,-1,1,1},{z,-1,1,1}];
    Show[maximizerplot,gradfieldplot,
    ViewPoint->CMView,Axes->Automatic,
    AxesLabel->{"x","y","z"}];
```

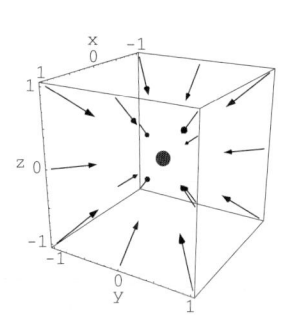

G.6.a.i) Why did this happen?

Where are the trajectories in this gradient field headed?

What do you try to learn about a function $f[x,y,z]$ by looking at a plot of its gradient field?

G.6.a.ii) You are given a certain function $f[x,y,z]$ and the information that

$$f[a,b,c] > f[x,y,z]$$

for all $\{x,y,z\}$ close to but not equal to $\{a,b,c\}$. You center a small sphere on $\{a,b,c\}$ and plot the gradient field of $f[x,y,z]$ on this sphere.

Is the net flow of the gradient field of $f[x,y,z]$ across this sphere from inside to outside or from outside to inside?

G.6.b.i) This time go with $f[x,y] = e^{(x^2+y^2+z^2)/16}$.

```
In[28]:=
    Clear[f,gradf,Field,x,y,z]
    f[x_,y_,z_] = E^((x^2 + y^2 + z^2)/16);
    gradf[x_,y_,z_] = {D[f[x,y,z],x],D[f[x,y,z],y],D[f[x,y,z],z]};
    Field[x_,y_,z_] = gradf[x,y,z]

Out[28]=
    { E^((x^2+y^2+z^2)/16) x / 8 ,  E^((x^2+y^2+z^2)/16) y / 8 ,  E^((x^2+y^2+z^2)/16) z / 8 }
```

Note that $\{x,y,z\} = \{0,0,0\}$ minimizes $f[x,y,z]$.

Here's a plot of the scaled gradient field of $f[x,y,z]$ shown with the minimizer at $\{0,0,0\}$:

Lesson 3.10 3D Surface Measurements

In[29]:=
```
minimizerplot = Graphics3D[{Red,PointSize[0.06],
Point[{0,0,0}]}]; scalefactor = 2.5;
gradfieldplot = Table[Arrow[scalefactor gradf[x,y,z],
Tail->{x,y,z},Blue],{x,-1,1,2},{y,-1,1,1},{z,-1,1,1}];
Show[minimizerplot,gradfieldplot,Axes->Automatic,
AxesLabel->{"x","y","z"}];
```

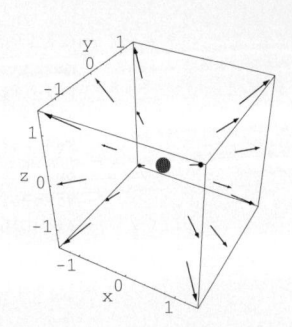

> Why did this happen?
>
> What are the trajectories in this gradient field trying to get away from?

G.6.b.ii) You are given a certain function $f[x, y, z]$ and the information that

$$f[a, b, c] < f[x, y, z]$$

for all $\{x, y, z\}$ close to but not equal to $\{a, b, c\}$. You center a small sphere at $\{a, b, c\}$ and plot the gradient field of $f[x, y, z]$ on this sphere.

> Is the net flow of the gradient field of $f[x, y, z]$ across this sphere from inside to outside or from outside to inside?

G.6.c.i) Here is a cleared function and its gradient field:

In[30]:=
```
Clear[x,y,z,f,gradf,m,n,p,Field,divField]
gradf[x_,y_,z_] = {D[f[x,y,z],x],D[f[x,y,z],y],D[f[x,y,z],z]};
{m[x_,y_,z_],n[x_,y_,z_],p[x_,y_,z_]} = gradf[x,y,z];
Field[x_,y_,z_] = {m[x,y,z],n[x,y,z],p[x,y,z]}
```

Out[30]=
$\{f^{(1,0,0)}[x, y, z], f^{(0,1,0)}[x, y, z], f^{(0,0,1)}[x, y, z]\}$

Here is the divergence of this gradient field:

In[31]:=
```
divField[x_,y_,z_] = D[m[x,y,z],x] + D[n[x,y,z],y] + D[p[x,y,z],z]
```

Out[31]=
$f^{(0,0,2)}[x, y, z] + f^{(0,2,0)}[x, y, z] + f^{(2,0,0)}[x, y, z]$

The Laplacian of $f[x, y, z]$ is given by

$$\frac{\partial^2 f[x, y, z]}{\partial x^2} + \frac{\partial^2 f[x, y, z]}{\partial y^2} + \frac{\partial^2 f[x, y, z]}{\partial z^2}.$$

> How does the divergence of the gradient field compare with the Laplacian?

G.6.c.ii) How do you check the Laplacian

$$\frac{\partial^2 f[x,y,z]}{\partial x^2} + \frac{\partial^2 f[x,y,z]}{\partial y^2} + \frac{\partial^2 f[x,y,z]}{\partial z}$$

to look for sources and sinks in the gradient field of a given function $f[x,y,z]$?

How do you check the Laplacian

$$\frac{\partial^2 f[x,y,z]}{\partial x^2} + \frac{\partial^2 f[x,y,z]}{\partial y^2} + \frac{\partial^2 f[x,y,z]}{\partial z}$$

to check whether the gradient field of a given function $f[x,y,z]$ is free of sources or sinks at points other than singularities?

G.6.c.iii) Explain this: If $f[x,y,z]$ is a function with no singularities and with

$$\frac{\partial^2 f[x,y,z]}{\partial x^2} + \frac{\partial^2 f[x,y,z]}{\partial y^2} + \frac{\partial^2 f[x,y,z]}{\partial z} = 0$$

for all points $\{x,y,z\}$, then there can be no point $\{a,b,c\}$ with $f[a,b,c] > f[x,y,z]$ for all $\{x,y,z\}$ close to but not equal to $\{a,b,c\}$.

And explain this: If $f[x,y,z]$ is a function with no singularities and with

$$\frac{\partial^2 f[x,y,z]}{\partial x^2} + \frac{\partial^2 f[x,y,z]}{\partial y^2} + \frac{\partial^2 f[x,y,z]}{\partial z} = 0$$

for all points $\{x,y,z\}$, then there can be no point $\{a,b,c\}$ with $f[a,b,c] < f[x,y,z]$ for all $\{x,y,z\}$ close to but not equal to $\{a,b,c\}$.

G.6.d.i) A large solid region in three dimensions represents a big rock. Part of the surface of the rock is kept at a prescribed temperature—maybe hotter at one point than at another. The remainder of the surface is perfectly insulated. You wait until the temperature inside the rock settles into its steady state condition. Say $\text{temp}[x,y,z]$ represents the steady state temperature at a position $\{x,y,z\}$ inside the rock.

In the steady state, no point inside the rock and not on the surface can be a source of new heat flow or a sink for old heat.

> Why does this tell you that if
>
> $$\text{Field}[x,y,z] = \text{gradtemp}[x,y,z],$$
>
> then

$$\text{divField}[x,y,z] = \frac{\partial^2 \text{temp}[x,y,z]}{\partial x^2} + \frac{\partial^2 \text{temp}[x,y,z]}{\partial y^2} + \frac{\partial^2 \text{temp}[x,y,z]}{\partial z^2} = 0$$

at each point $\{x,y,z\}$ inside but not on the surface of the rock?

G.6.d.ii) Explain why your answer to G.6.d.i) immediately above tells you that the hottest and the coldest locations of the rock must be on the outside skin of the rock and not inside the rock.

■ G.7) Morphing and Moebius strips

This problem appears only in the electronic version.

■ G.8) Bidding on rocket parts

You are an engineer working for the United Engineering Company, makers of specialty metal products for the industrial market. A call for bids comes in for some front sections of a rocket upon which ceramic nose cones will be mounted. The specifications say that the front section is to be made as follows: You start with the inner parabolic surface with all measurements in meters:

In[32]:=
```
Clear[innerx,innery,innerz,r,s];
innerx[r_,s_] = r Cos[s];
innery[r_,s_] = r Sin[s]; innerz[r_,s_] = 9 - r^2;
{innerrlow,innerrhigh} = {2,3};
{innerslow,innershigh} = {0,2 Pi};
innersurface = ParametricPlot3D[
{innerx[r,s],innery[r,s],innerz[r,s]},
{r,innerrlow,innerrhigh},
{s,innerslow,innershigh},ViewPoint->CMView,
PlotRange->All,AxesLabel->{"x","y","z"}];
```

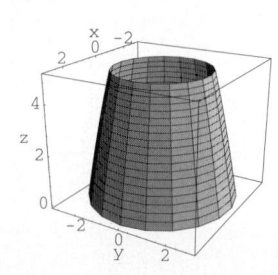

Then you have the outer parabolic surface with all measurements in meters:

In[33]:=
```
Clear[outerx,outery,outerz,r,s]
outerx[r_,s_] = 2 r Cos[s];
outery[r_,s_] = 2 r Sin[s];
outerz[r_,s_] = 36 - (2 r)^2;
{outerrlow,outerrhigh} = {2,3};
{outerslow,outershigh} = {0,2 Pi};
outersurface = ParametricPlot3D[
{outerx[r,s],outery[r,s],outerz[r,s]},
{r,outerrlow,outerrhigh},{s,outerslow,outershigh},
ViewPoint->CMView,PlotRange->All,AxesLabel->{"x","y","z"}];
```

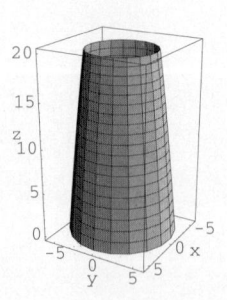

The specifications say that you get the top of the front section by running a straight line from each point on the top of the inner surface

$$\{\text{innerx}[2, s], \text{innery}[2, s], \text{innerz}[2, s]\}$$

to the corresponding point

$$\{\text{outerx}[2, s], \text{outery}[2, s], \text{outerz}[2, s]\}$$

on the top of the outer surface. You get the base, on which the section rests, by running a straight line from each point on the bottom of the inner surface

$$\{\text{innerx}[3, s], \text{innery}[3, s], \text{innerz}[3, s]\}$$

to the corresponding point

$$\{\text{outerx}[3, s], \text{outery}[3, s], \text{outerz}[3, s]\}$$

on the bottom of the outer surface.

G.8.a) Measure:

1) The surface area of the inner skin.
2) The surface area of the outer skin.
3) The surface area of the flat ring at the bottom.
4) The surface area of the conical ring at the top.
5) The volume enclosed by the outer skins composed of the outer shell, the flat ring at the bottom, the conical ring at the top, and the inner shell.

Report these measurements and throw in a plot of the finished product including the flat ring at the bottom and the conical ring at the top.

Show it from several advantageous viewpoints.

■ G.9) Volumes, flow, and bouyancy

This problem appears only in the electronic version.

■ G.10) Rotating and measuring: Formulas you may see elsewhere

This problem appears only in the electronic version.

LESSON 3.11

3D Flow Along

Basics

■ B.1) Measuring flow along a 3D curve

B.1.a) How do you measure the flow of a 3-dimensional vector field

$$\text{Field}[x, y, z] = \{m[x, y, z], n[x, y, z], p[x, y, z]\}$$

along a curve C given in parametric form

$$P[t] = \{x[t], y[t], z[t]\} \quad \text{with } a \leq t \leq b?$$

Answer: Just about the same way as you do it in two dimensions. You measure the net flow of Field$[x, y, z]$ along C by calculating the three-dimensional path integral

$$\int_C \text{Field} \bullet \text{unittan } ds = \int_C m[x, y, z] \, dx + n[x, y, z] \, dy + p[x, y, z] \, dz$$

$$= \int_a^b \left(m[x[t], y[t], z[t]] \, x'[t] + n[x[t], y[t], z[t]] \, y'[t] + p[x[t], y[t], z[t]] \, z'[t] \right) \, dt.$$

If this path integral calculates out as positive, then the net flow of Field$[x, y, z]$ along C is in the direction specified by the parameterization.

If this path integral calculates out as negative, then the net flow of Field$[x, y, z]$ along C is against the direction specified by the parameterization.

B.1.b) Here is a 3D vector field:

In[1]:=
```
Clear[Field,m,n,p,x,y,z]; m[x_,y_,z_] = y^2 z^3;
n[x_,y_,z_] = x y z^3/3; p[x_,y_,z_] = x y^2 z^2/2;
Field[x_,y_,z_] = {m[x,y,z],n[x,y,z],p[x,y,z]}
```

Out[1]=
$$\{y^2 z^3, \frac{x y z^3}{3}, \frac{x y^2 z^2}{2}\}$$

Here's a curve C in three dimensions:

In[2]:=
```
Clear[t]; {x[t_],y[t_],z[t_]} = {t, t + Sin[1.6 t], t - 1};
```

Out[2]=
{t, t + Sin[1.6 t], -1 + t}

This curve is given to start at:

In[3]:=
```
a = 0.6; start = {x[a],y[a],z[a]}
```

Out[3]=
{0.6, 1.41919, -0.4}

And this curve is given to end at:

In[4]:=
```
b = 2.8; end = {x[b],y[b],z[b]}
```

Out[4]=
{2.8, 1.82688, 1.8}

Here's a look at C:

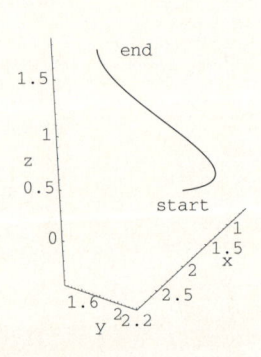

In[5]:=
```
Clear[P,t,Cplot,labels,setup]
P[t_] = {x[t],y[t],z[t]};
Cplot = ParametricPlot3D[Evaluate[P[t]],{t,a,b},
 DisplayFunction->Identity];
labels = {Graphics3D[Text["start",start]],
 Graphics3D[Text["end",end]]};
setup = Show[Cplot,labels,
 ViewPoint->CMView,Axes->Automatic,
 AxesLabel->{"x","y","z"},Boxed->False,
 DisplayFunction->$DisplayFunction];
```

Measure the net flow of Field[x, y, z] along C and interpret the result.

Answer: The measurement

$$\int_C \text{Field} \bullet \text{unittan} \, ds = \int_C m[x,y,z] \, dx + n[x,y,z] \, dy + p[x,y,z] \, dz$$

$$= \int_a^b m[x[t],y[t],z[t]] \, x'[t] + n[x[t],y[t],z[t]] \, y'[t] + p[x[t],y[t],z[t]] \, z'[t] dt$$

measures the flow of the vector field $\text{Field}[x,y,z] = \{m[x,y,z], n[x,y,z], p[x,y,z]\}$ along C:

In[6]:=
```
NIntegrate[m[x[t],y[t],z[t]] x'[t] +
  n[x[t],y[t],z[t]] y'[t] + p[x[t],y[t],z[t]] z'[t],{t,a,b}]
```

Out[6]=
16.2563

Positive.

This means that the net flow of $\text{Field}[x,y,z]$ along C is fairly strong in the direction of the parameterization of C. To see what this means, just plot C and some of its unit tangent vectors.

In[7]:=
```
Clear[unittan]
unittan[t_] = Expand[D[P[t],t]/Sqrt[D[P[t],t].D[P[t],t]],
  Trig->True]; jump = (b - a)/5;
Show[setup,Table[Arrow[unittan[t],Tail->P[t]],
  {t,a,b - jump,jump}]];
```

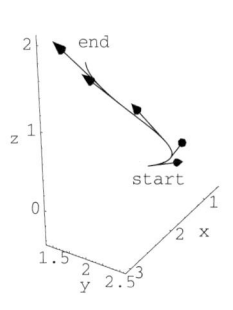

There you go with the flow.

■ B.2) The curl of a 3D vector field

Start with a cleared 3D vector field:

In[8]:=
```
Clear[Field,m,n,p,x,y,z]
Field[x_,y_,z_] = {m[x,y,z],n[x,y,z],p[x,y,z]}
```

Out[8]=
{m[x, y, z], n[x, y, z], p[x, y, z]}

The curl,

$\quad \text{curlField}[x,y,z]$,

of $\text{Field}[x,y,z]$ is given by:

Lesson 3.11 3D Flow Along

In[9]:=
```
Clear[curlField]
curlField[x_,y_,z_] = {D[p[x,y,z],y]-D[n[x,y,z],z],
 D[m[x,y,z],z]-D[p[x,y,z],x], D[n[x,y,z],x]-D[m[x,y,z],y]}
```

Out[9]=

$$\{-n^{(0,0,1)}[x,y,z] + p^{(0,1,0)}[x,y,z],\ m^{(0,0,1)}[x,y,z] - p^{(1,0,0)}[x,y,z],$$
$$-m^{(0,1,0)}[x,y,z] + n^{(1,0,0)}[x,y,z]\}$$

You can detect some cyclic patterns here. That's quite a pill to swallow, but the computer will swallow it for you. You will want to use it to help finger a 3D vector field.

B.2.a) Stick the tail of a unit vector V at a point $\{a, b, c\}$. Push your finger onto the point of V so that V spikes through the center of the bones of your finger, and the tip of your finger is at $\{a, b, c\}$.

> How do you use curlField$[a, b, c]$ to determine whether the tip of your finger feels a net clockwise or counterclockwise swirl resulting from the flow of the given 3D vector field Field$[x, y, z]$?

Answer: Here's how it works:

→ If curlField$[a, b, c] \bullet V > 0$, then you feel Field$[x, y, z]$ swirling in the counterclockwise way at the tip of your finger.

→ If curlField$[a, b, c] \bullet V < 0$, then you feel Field$[x, y, z]$ swirling in the clockwise way at the tip of your finger.

→ If curlField$[a, b, c] \bullet V = 0$, then you feel no net swirl at all.

Try it out for a given vector field, a given unit vector V, and a given point $\{a, b, c\}$:

In[10]:=
```
Clear[Field,curlField,unitvector,s,t,x,y,z,m,n,p];
unitvector[s_,t_] := N[{Sin[s] Cos[t], Sin[s] Sin[t], Cos[s]}]
m[x_,y_,z_] = x - 2 z; n[x_,y_,z_] = y - 2 x^2;
p[x_,y_,z_] = z - 2 y;
Field[x_,y_,z_] = {m[x,y,z],n[x,y,z],p[x,y,z]};
curlField[x_,y_,z_] = {D[p[x,y,z],y] - D[n[x,y,z],z],
 D[m[x,y,z],z] - D[p[x,y,z],x],D[n[x,y,z],x] - D[m[x,y,z],y]};
curlField[0,2,1].unitvector[Pi/3,Pi/4]
```

Out[10]=
-2.44949

Negative. This tells you that when you push your finger onto

$$V = \text{unitvector}\left[\frac{\pi}{3}, \frac{\pi}{4}\right]$$

so that V spikes through the center of your finger and the tip of your finger is at $\{0, 2, 1\}$, then your finger feels a net clockwise swirl.

Confirm this by plotting what's happening near $\{0, 2, 1\}$ using the plotting option ViewPoint->V to look from the tip of V to the tail of V at the point $\{0, 2, 1\}$:

In[11]:=
```
point = {0,2,1}; V = unitvector[Pi/3,Pi/4];
h = 0.3; scalefactor = 0.08;
fieldplot = Table[Arrow[
scalefactor Apply[Field,point +{x,y,z}],
Tail->N[point + {x,y,z}],Blue],
{x,-h,h,2 h},{y,-h,h,2 h},{z,-h,h,2 h}];
pointplot = Graphics3D[{PointSize[0.07],Point[point]}];
swirlplot = Show[pointplot,fieldplot,PlotRange->All,
Boxed->False,ViewPoint->V];
```

Put the tip of your finger at the plotted point on the paper and feel that net clockwise flow! Just as the calculation predicted! Try it for a different point and a different unit vector:

In[12]:=
```
curlField[0,0,0].unitvector[3 Pi/2,Pi/6]
```
Out[12]=
2.73205

Positive. This tells you that when you push your finger onto
$$V = \text{unitvector}\left[\frac{3\pi}{2}, \frac{\pi}{6}\right]$$
so that V spikes through the center of your finger and the tip of your finger is at $\{0, 0, 0\}$, then your finger feels a net counterclockwise swirl. Confirm with a plot:

In[13]:=
```
point = {0,0,0}; V = unitvector[3 Pi/2,Pi/6];
h = 0.2; scalefactor = 0.4;
fieldplot = Table[Arrow[
scalefactor Apply[Field,point +{x,y,z}],
Tail->N[point + {x,y,z}],Blue],
{x,-h,h,2 h},{y,-h,h,2 h},{z,-h,h,2 h}];
pointplot = Graphics3D[{PointSize[0.07],Point[point]}];
swirlplot = Show[pointplot,fieldplot,PlotRange->All,
Boxed->False,ViewPoint->V];
```

Put the tip of your finger at the plotted point on the paper and feel that net counterclockwise flow! Math happens again.

■ B.3) Stokes's formula for using the curl to measure the swirl of a vector field in 3D

In 2D, everyone likes the form of the Gauss-Green formula that tells you that if
$$\text{Field}[x, y] = \{m[x, y], n[x, y]\}$$

and R is a region in two dimensions with boundary curve C, then

$$\iint_R \text{rotField}\, dA = \oint_C \text{Field} \bullet \text{unittan}\, ds$$
$$= \oint_C m[x,y]\, dx + n[x,y]\, dy$$
$$= \text{net flow of Field}[x,y] \text{ along } C.$$

Stokes's formula is the 3D analogue of this formula. Stokes's formula says that if you go with a 3D vector field

$$\text{Field}[x,y,z] = \{m[x,y,z], n[x,y,z], p[x,y,z]\}$$

and R is a surface in three dimensions with boundary curve C, then

$$\iint_R \text{curlField} \bullet \text{topunitnormal}\, dA = \oint_C \text{Field} \bullet \text{unittan}\, ds$$
$$= \oint_C m[x,y,z]\, dx + n[x,y,z]\, dy + p[x,y,z]\, dz$$
$$= \text{net flow of Field}[x,y,z] \text{ along } C.$$

Before you can make complete sense of this formula, you have to deal with a few objections.

B.3.a.i) → Objection 1:

The integral $\oint_C \text{Field} \bullet \text{unittan}\, ds$ insists on a counterclockwise parameterization for the boundary curve C. But counterclockwise parameterization makes no sense in three dimensions, so Stokes's formula

$$\iint_R \text{curlField} \bullet \text{topunitnormal}\, dA = \oint_C \text{Field} \bullet \text{unittan}\, ds$$

makes no sense.

→ Objection 2:

Look at this piece of the yz-plane:

In[14]:=
```
yzplanepiece = Graphics3D[
  Polygon[{{0,0,0},{0,2,0},{0,2,1},{0,0,1}}]];
Show[yzplanepiece,Axes->True,
  AxesLabel->{"x","y","z"},ViewPoint->CMView,
  Boxed->False,PlotRange->All,
  DisplayFunction->$DisplayFunction];
```

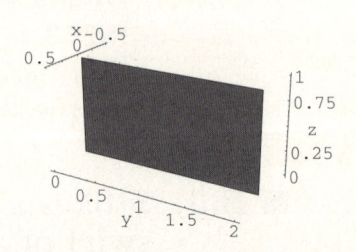

Since this surface sits in the yz-plane, you can't say whether the front side or the back side is the top side. Calculating $\iint_R \text{curlField} \bullet \text{topunitnormal}\, dA$ depends on knowing which side of the surface is the top side. So again Stokes's formula

$$\iint_R \text{curlField} \bullet \text{topunitnormal} \, dA = \oint_C \text{Field} \bullet \text{unittan} \, ds$$

makes no sense.

> How do you get around these legitimate objections?

Answer: Taken separately, each of these two objections ruins any possibility of making sense out of Stokes's formula, but if you deal with both of them simultaneously, then you can get a full understanding of what Stokes's formula says.

First, you pick one side of the surface at your own pleasure and call it the top side.

The side you pick for the top side may not be the same as the side designated the top side by the person sitting at the computer next to you.

Once you pick the top side of R, then you agree that you are going in the counterclockwise direction on C if, as you walk around C, R is on your left. With these two agreements carved in stone, Stokes's formula

$$\iint_R \text{curlField} \bullet \text{topunitnormal} \, dA = \oint_C \text{Field} \bullet \text{unittan} \, ds$$
$$= \oint_C m[x,y,z] \, dx + n[x,y,z] \, dy + p[x,y,z] \, dz$$

for a given 3D vector field $\text{Field}[x,y,z] = \{m[x,y,z], n[x,y,z], p[x,y,z]\}$ and a given surface R with boundary curve C makes perfect sense provided you agree to parameterize C in the counterclockwise direction dictated by choice of the top side of the surface according to the agreement above.

B.3.a.ii) Here is a surface R which is a cap of an ellipsoid:

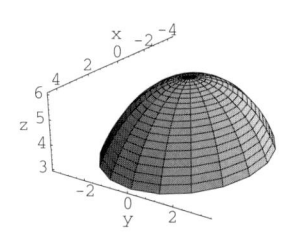

```
In[15]:=
  Clear[x,y,z,s,t]
  {x[s_,t_],y[s_,t_],z[s_,t_]} =
  {5 Sin[s] Cos[t],4 Sin[s] Sin[t],6 Cos[s]};
  {slow,shigh} = {0, Pi/3}; surfaceplot =
  ParametricPlot3D[{x[s,t],y[s,t],z[s,t]},
  {s,slow,shigh},{t,0,2 Pi},
  Boxed->False, PlotRange->All,
  ViewPoint->CMView,AxesLabel->{"x","y","z"}];
```

A parameterization of the boundary curve C is:

```
In[16]:=
  Clear[P]; P[t_] = {x[shigh,t],y[shigh,t],z[shigh,t]}

Out[16]=
  {5 Sqrt[3] Cos[t]
  {───────────────, 2 Sqrt[3] Sin[t], 3}
         2
```

Here are the surface R, its boundary curve C, and a few tangent vectors reflecting the direction of this parameterization of C:

In[17]:=
```
Cplot = ParametricPlot3D[Evaluate[P[t]],
{t,0,2 Pi},DisplayFunction->Identity];
tangents = Table[Arrow[P'[t],Tail->P[t]],
{t,0,2 Pi,Pi/3}];
Show[surfaceplot,Cplot,tangents,Boxed->False,
PlotRange->All,AxesLabel->{"x","y","z"},
ViewPoint->CMView,DisplayFunction->$DisplayFunction];
```

> If you want to call this parameterization of boundary curve C counterclockwise in accordance with the agreement above, then which side of the cap must you designate as the top side?

Answer: Walk around the boundary in the direction indicated by the tangents. As you walk, your left foot hits on the visible side (the high side). This is your designated top side.

B.3.a.iii) Take a look at this sorry excuse for a surface:

In[18]:=
```
Clear[P,x,y,z,t,unittan,mainunitnormal,
binormal,moebius]; P[t_] = 2 {-Cos[t],Sin[t],0};
unittan[t_] = Expand[P'[t]/
Sqrt[P'[t].P'[t]],Trig->True];
mainunitnormal[t_] = D[unittan[t],t]/
Sqrt[Expand[D[unittan[t],t].D[unittan[t],t],
Trig->True]]; binormal[t_] = Expand[Cross[unittan[t],
mainunitnormal[t]],Trig->True]; moebius[s_,t_] =
P[t] + s(Sin[t/2] mainunitnormal[t] -
Cos[t/2] binormal[t]);
ParametricPlot3D[Evaluate[moebius[s,t]],{s,-0.5,0.5},
{t,-Pi,Pi},PlotPoints->{2,40},ViewPoint->CMView,
PlotRange->All,Boxed->False,Axes->False];
```

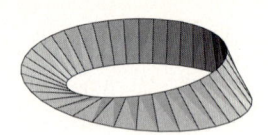

This surface, which folks like to call a "Moebius strip," is unfortunate enough to have only one side.

> How the heck are you going to designate a top side for a surface that has only one side?

Answer: You can't. It's impossible to choose a top side.

The upshot: It's impossible to use Stokes's formula for a surface that has only one side.

B.3.a.iv) | What else do you have to worry about?

Answer: Stokes's formula works only for surfaces that are not the complete boundary of a solid region.

For example, Stokes's formula is OK if R is a cap of a sphere, but Stokes's formula does not work if R is the whole sphere. This is the case simply because a whole sphere does not have a boundary curve, but any cap of a sphere does have a boundary curve.

B.3.b.i) Take a 3D vector field Field$[x, y, z]$. Stick the tail of a unit vector V at the point $\{x_0, y_0, z_0\}$. Push your finger onto V so that V spikes through the center of your finger and the tip of your finger is at $\{x_0, y_0, z_0\}$.

Why does the tip of your finger feel a counterclockwise swirl if

$$\text{curlField}[x_0, y_0, z_0] \bullet V > 0?$$

Why does the tip of your finger feel a clockwise swirl if

$$\text{curlField}[x_0, y_0, z_0] \bullet V < 0?$$

Answer: Put a plane with normal vector V through the point $\{x_0, y_0, z_0\}$. Put a small circle C with center at $\{x_0, y_0, z_0\}$ in this plane. Call R the part of this plane that is within the small circle. Designate the top side by taking the direction V points. This gives you

$$V = \text{topunitnormal}$$

for this little surface R. Stokes's formula tells you:

$$\text{the net flow of Field}[x, y, z] \text{ along } C = \oint_C m[x, y, z]\, dx + n[x, y, z]\, dy + p[x, y, z]\, dz$$

$$= \iint_R \text{curlField} \bullet \text{topunitnormal}\, dA$$

$$= \iint_R \text{curlField} \bullet V\, dA.$$

Here's the kicker: If

$$\text{curlField}[x_0, y_0, z_0] \bullet V > 0,$$

then curlField$[x, y, z] \bullet V > 0$ for all $\{x, y, z\}$'s close to $\{x_0, y_0, z_0\}$. So if C is small enough so that curlField$[x, y, z] \bullet V > 0$ at all $\{x, y, z\}$'s on R and inside C, then you see that

$$\text{the net flow of Field}[x, y, z] \text{ along } C = \iint_R \text{curlField} \bullet V\, dA > 0.$$

So the net flow of Field$[x, y, z]$ along such a small circle C is counterclockwise. This is why the tip of your finger feels a counterclockwise swirl when you stick the tail of V at the point $\{x_0, y_0, z_0\}$ and push your finger onto V so that V spikes through the center of your finger and the tip of your finger is at $\{x_0, y_0, z_0\}$.

Similarly, if curlField$[x_0, y_0, z_0] \bullet V < 0$ and you stick the tail of V at the point $\{x_0, y_0, z_0\}$ and push your finger onto V so that V spikes through the center of your finger and the tip of your finger is at $\{x_0, y_0, z_0\}$, then your fingertip feels a clockwise swirl.

B.3.b.ii) Given a 3D vector field Field$[x, y, z]$, when you stick the tail of a unit vector V at the point $\{x, y, z\}$ and push your finger onto V so that V spikes through the center of your finger and the tip of your finger is at $\{x, y, z\}$, then what unit vector V do you want to use so that your fingertip feels the greatest possible counterclockwise swirl?

What does the direction of curlField$[x, y, z]$ tell you?

What does the length of curlField$[x, y, z]$ tell you?

Answer: The swirl your fingertip feels is measured by

$$\text{curlField}[x, y, z] \bullet V = ||\text{curlField}[x, y, z]||\, ||V|| \cos[\text{angle between}]$$
$$= ||\text{curlField}[x, y, z]|| \cos[\text{angle between}]$$

because V is a unit vector. This is biggest when

$$\cos[\text{angle between}] = 1 = \cos[0].$$

So you feel the biggest counterclockwise swirl when you take the unit vector V to be in the same direction as curlField$[x, y, z]$. In other words, to feel the most counterclockwise swirl, you take

$$V = \frac{\text{curlField}[x, y, z]}{\sqrt{\text{curlField}[x, y, z] \bullet \text{curlField}[x, y, z]}}.$$

The upshots:

$\to$ The direction of curlField$[x, y, z]$ gives the direction of the axis of the actual swirl of Field$[x, y, z]$ at $\{x, y, z\}$.

$\to$ The length of curlField$[x, y, z]$ measures how vigorous the swirl at $\{x, y, z\}$ is.

Check this out:

In[19]:=
```
Clear[x,y,z,m,n,p,Field,curlField]
m[x_,y_,z_] = x + 2 z^2;  n[x_,y_,z_] = y - 2 x^2;
p[x_,y_,z_] = z + 4 y^2;
Field[x_,y_,z_] = {m[x,y,z],n[x,y,z],p[x,y,z]};
curlField[x_,y_,z_] = {D[p[x,y,z],y] - D[n[x,y,z],z],
  D[m[x,y,z],z] - D[p[x,y,z],x], D[n[x,y,z],x] - D[m[x,y,z],y]}
```

Out[19]=
{8 y, 4 z, -4 x}

To feel the greatest possible counterclockwise swirl of this vector field at $\{1,1,1\}$, you should use:

In[20]:=
```
point = {1,1,1};
V = curlField[1,1,1]/Sqrt[curlField[1,1,1].curlField[1,1,1]]
```

Out[20]=
$$\{\text{Sqrt}[\frac{2}{3}], \frac{1}{\text{Sqrt}[6]}, -(\frac{1}{\text{Sqrt}[6]})\}$$

Take a look down the vector V from its tip to its tail at $\{1,1,1\}$:

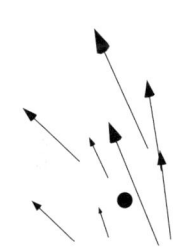

In[21]:=
```
h = 0.3; scalefactor = 0.1;
fieldplot = Table[Arrow[
  scalefactor N[Apply[Field,point +{x,y,z}]],
  Tail->N[point + {x,y,z}],Blue],
  {x,-h, h, 2 h },{y,- h, h,2 h},{z,-h,h,2 h}];
pointplot = Graphics3D[{PointSize[0.07],Point[point]}];
fingerplot = Show[pointplot,fieldplot,
Boxed-> False,ViewPoint->V];
```

Powerfully counterclockwise, just as you expected. Math really happens.

B.3.b.iii) Where did the name "curl" come from?

Answer: Take a look at the end of the answer to part B.3.b.ii) above.

B.3.c.i) Take a surface R with boundary curve C. Designate a top side of R. The path integral

$$\oint_C \text{Field} \bullet \text{unittan} \, ds = \oint_C m[x,y,z] \, dx + n[x,y,z] \, dy + p[x,y,z] \, dz$$

measures the net flow of a given 3D vector field

$$\text{Field}[x,y,z] = \{m[x,y,z], n[x,y,z], p[x,y,z]\}$$

along C.

If you know that

$$\text{curlField}[x,y,z] \bullet \text{topunitnormal}[x,y,z] > 0$$

at all points $\{x,y,z\}$ on the surface R, then how do you also know that the net flow of Field$[x,y,z]$ along the boundary curve C is counterclockwise?

Answer: You calculate the flow of Field$[x, y, z]$ along C by calculating

$$\oint_C m[x, y, z]\, dx + n[x, y, z]\, dy + p[x, y, z]\, dz.$$

Stokes's formula tells you that this is the same as calculating

$$\iint_R \text{curlField} \bullet \text{topunitnormal}\, dA.$$

Here's the kicker: If

$$\text{curlField}[x, y, z] \bullet \text{topunitnormal}[x, y, z] > 0$$

at all points of R, then the

$$\text{flow of Field}[x, y, z] \text{ along } C = \oint_C m[x, y, z]\, dx + n[x, y, z]\, dy + p[x, y, z]\, dz$$

$$= \iint_R \text{curlField} \bullet \text{topunitnormal}\, dA > 0,$$

because curlField$[x, y, z] \bullet$ topunitnormal$[x, y, z] > 0$ for all $\{x, y, z\}$'s on R.

So, if you know that curlField$[x, y, z] \bullet$ topunitnormal$[x, y, z] > 0$ for all $\{x, y, z\}$'s on R, then you know that the flow-along-C measurement is positive.

This means that the net flow of Field$[x, y, z]$ along the boundary curve C of R is counterclockwise.

B.3.c.ii) Take a surface R with boundary curve C. If you know that

$$\text{curlField}[x, y, z] = 0$$

at all points $\{x, y, z\}$ on the surface R, then how do you also know that the net flow of Field$[x, y, z]$ along C is 0?

Answer: You calculate the net flow of Field$[x, y, z]$ along C by calculating

$$\oint_C m[x, y, z]\, dx + n[x, y, z]\, dy + p[x, y, z]\, dz.$$

Stokes's formula tells you that this is the same as calculating

$$\iint_R \text{curlField} \bullet \text{topunitnormal}\, dA.$$

Here's the kicker: If curlField$[x, y, z] = 0$ at all points of R, then the

$$\text{flow of Field}[x, y, z] \text{ along } C = \oint_C m[x, y, z]\, dx + n[x, y, z]\, dy + p[x, y, z]\, dz$$

$$= \iint_R \text{curlField} \bullet \text{topunitnormal}\, dA$$

$$= \iint_R 0\, dA = 0,$$

because curlField$[x, y, z] = 0$ for all $\{x, y, z\}$'s on R. So curlField$[x, y, z] = 0$ for all $\{x, y, z\}$'s on R guarantees that the flow-along-C measurement is 0. This means that the net flow of Field$[x, y, z]$ along the boundary curve C of R is 0.

B.3.c.iii) Why do the fancy folks call a vector field Field$[x, y, z]$ with curlField$[x, y, z] = 0$ at all points $\{x, y, z\}$ irrotational?

Answer: As you saw above if curlField$[x, y, z] = 0$ for all points $\{x, y, z\}$, then the flow of Field$[x, y, z]$ along any closed curve is 0. This rules out any vortex or other swirl.

■ **B.4) An attempt to explain Stokes's formula as an outgrowth of the 2D Gauss-Green formula**

This problem appears only in the electronic version.

Tutorials

■ **T.1) Stokes's formula in theory and practice**

Stokes's formula says that if
$$\text{Field}[x, y, z] = \{m[x, y, z], n[x, y, z], p[x, y, z]\}$$
and if R is a surface in three dimensions with boundary curve C and you have made the right specifications of the top side and of what counterclockwise means, then you are guaranteed that
$$\iint_R \text{curlField} \bullet \text{topunitnormal } dA = \oint_C \text{Field} \bullet \text{unittan } ds$$
$$= \oint_C m[x, y, z]\, dx + n[x, y, z]\, dy + p[x, y, z]\, dz$$
$$= \text{net flow of Field}[x, y, z] \text{ along } C.$$

T.1.a) How good a theoretical tool is Stokes's formula?

Answer: It's a superb theoretical tool.

Check out B.3) to see how Stokes's formula confirms that the counterclockwise swirl of a 3D vector field Field$[x, y, z]$ in the direction of a unit vector V at a point $\{x_0, y_0, z_0\}$ is measured by
$$\text{curlField}[x_0, y_0, z_0] \bullet V.$$

This, in turn, tells you that if you stick the tail of curlField$[x_0, y_0, z_0]$ at the point $\{x_0, y_0, z_0\}$, then you get the axis of the direction of the greatest counterclockwise swirl of Field$[x, y, z]$ at the point $\{x_0, y_0, z_0\}$. This is the direction of the curl of the flow represented by Field$[x, y, z]$.

Stokes's formula is a great theoretical tool that sets up concrete calculations.

T.1.b.i) How good a calculational tool is Stokes's formula?

Answer: Look at Stokes's formula again.

$$\iint_R \text{curlField} \bullet \text{topunitnormal} \, dA = \oint_C \text{Field} \bullet \text{unittan} \, ds$$

$$= \oint_C m[x, y, z] \, dx + n[x, y, z] \, dy + p[x, y, z] \, dz$$

$$= \text{net flow of Field}[x, y, z] \text{ along } C.$$

If curlField$[x, y, z]$ calculates out to $\{0, 0, 0\}$, then you don't have to do any extra work to learn that the flow of Field$[x, y, z]$ along any closed curve is 0.

In fact, if you have a closed curve C, think of it as wire and throw a big bed sheet over the wire. The part of the sheet inside C defines a surface R with boundary curve C.

If curlField$[x, y, z]$ is always $\{0, 0, 0\}$, you get

$$0 = \iint_R 0 \, dA$$

$$= \iint_R \{0, 0, 0\} \bullet \text{topunitnormal} \, dA$$

$$= \iint_R \text{curlField} \bullet \text{topunitnormal} \, dA$$

$$= \oint_C \text{Field} \bullet \text{unittan} \, ds$$

$$= \oint_C m[x, y, z] \, dx + n[x, y, z] \, dy + p[x, y, z] \, dz$$

$$= \text{net flow of Field}[x, y, z] \text{ along } C.$$

This means that if curlField$[x, y, z]$ is always $\{0, 0, 0\}$, then the net flow of Field$[x, y, z]$ along any closed curve is 0.

And Stokes's formula gives you this juicy calculational fact with no work on your part. You've got to agree that if curlField$[x, y, z]$ is always 0, then Stokes's formula is an excellent calculational tool.

T.1.b.ii) What happens when curlField$[x, y, z]$ does not calculate out to $\{0, 0, 0\}$?

Answer: Take another look at Stokes's formula:

$$\iint_R \text{curlField} \bullet \text{topunitnormal}\, dA = \oint_C \text{Field} \bullet \text{unittan}\, ds$$
$$= \oint_C m[x,y,z]\, dx + n[x,y,z]\, dy + p[x,y,z]\, dz$$
$$= \text{flow of Field}[x,y,z] \text{ along } C.$$

If curlField$[x, y, z]$ is not always 0, then you are probably better off ignoring Stokes's formula and calculating the path integral

$$\oint_C m[x,y,z]\, dx + n[x,y,z]\, dy + p[x,y,z]\, dz$$

directly.

Reason: Most surface integrals are harder to calculate than most path integrals.

T.1.c) Here are two vector fields in three dimensions:

In[1]:=
```
Clear[x,y,z,m1,n1,p1,m2,n2,p2,Field1,Field2]
Field1[x_,y_,z_] = E^(-y) {2 x Sin[x^2 + z^2],
  Cos[x^2 + z^2],2 z Sin[x^2 + z^2]};
{m1[x_,y_,z_],n1[x_,y_,z_],p1[x_,y_,z_]} = Field1[x,y,z];
Field2[x_,y_,z_] = {(x - 1)^2,x y,(z - y)^2};
{m2[x_,y_,z_],n2[x_,y_,z_],p2[x_,y_,z_]} = Field2[x,y,z];
```

Here is a closed curve C in three dimensions:

In[2]:=
```
Clear[t,x,y,z]
{x[t_],y[t_],z[t_]} =
  {3 Sin[2 t],2 Cos[t],Sin[t] + 0.2 Cos[3 t]};
{tlow,thigh} = {0,2 Pi}; Clear[P]
P[t_] = {x[t],y[t],z[t]};
Cplot = ParametricPlot3D[
  Evaluate[P[t]],{t,tlow,thigh},
  ViewPoint->CMView,Boxed->False,
  Axes->Automatic,AxesLabel->{"x","y","z"}];
```

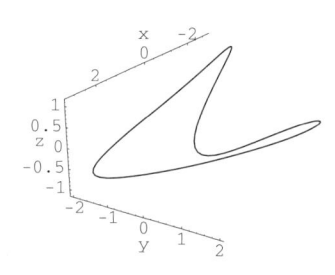

> Measure the flow of Field1$[x, y, z]$ along this curve, and then measure the flow of Field2$[x, y, z]$ along this curve.

Answer: Take a look at the curls:

In[3]:=
```
Clear[curlField1,x,y,z]
curlField1[x_,y_,z_] = {D[p1[x,y,z],y] - D[n1[x,y,z],z],
  D[m1[x,y,z],z] - D[p1[x,y,z],x],D[n1[x,y,z],x] - D[m1[x,y,z],y]}
```

Out[3]=
{0, 0, 0}

Ah-ha!

With no further work, you know that the flow of Field1$[x, y, z]$ along this curve is 0. In fact, thanks to Stokes's formula, you also know that the flow of Field1$[x, y, z]$ along any other closed curve is 0.

Now look at the second vector field and its curl:

In[4]:=
```
Clear[curlField2]; curlField2[x_,y_,z_] = {D[p2[x,y,z],y] - D[n2[x,y,z],z],
D[m2[x,y,z],z] - D[p2[x,y,z],x],D[n2[x,y,z],x] - D[m2[x,y,z],y]}
```
Out[4]=
{-2 (-y + z), 0, y}

No such luck this time; curlField2$[x, y, z]$ does not calculate out to $\{0, 0, 0\}$. To measure the flow of Field2$[x, y, z]$ along the given curve, calculate the path integral

$$\int_C \text{Field2} \bullet \text{unittan } ds = \int_C m_2[x, y, z] \, dx + n_2[x, y, z] \, dy + p_2[x, y, z] \, dz$$

$$= \int_{\text{tlow}}^{\text{thigh}} (m_2[x[t], y[t], z[t]] \, x'[t] + n_2[x[t], y[t], z[t]] \, y'[t] + p_2[x[t], y[t], z[t]] \, z'[t]) \, dt :$$

In[5]:=
```
{x[t_],y[t_],z[t_]}=P[t];
NIntegrate[m2[x[t],y[t],z[t]] x'[t] + n2[x[t],y[t],z[t]] y'[t] +
p2[x[t],y[t],z[t]] z'[t],{t,tlow,thigh},AccuracyGoal->2]
```
Out[5]=
-18.8496

Negative. This means the net flow of Field2$[x, y, z]$ along C is fairly strong in the direction opposite the direction of the parameterization of C. To see what this means, just plot C and some of its tangent vectors.

In[6]:=
```
scalefactor = 0.5;
Show[Cplot,
Table[Arrow[scalefactor P'[t],Tail->P[t],Red],
{t,tlow,thigh,(thigh - tlow)/8}],
ViewPoint->CMView,Boxed->False,
Axes->Automatic,AxesLabel->{"x","y","z"}];
```

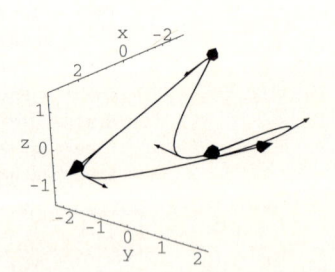

The net flow of Field2$[x, y, z]$ along C is in the direction opposite the direction indicated by the tangent vectors above.

■ T.2) Path independence

T.2.a) You've got a vector field

$$\text{Field}[x, y, z] = \{m[x, y, z], n[x, y, z], p[x, y, z]\}$$

and you are fortunate enough to find that curlField$[x, y, z]$ is always 0. You also have a curve C_1 with parameterization $P_1[t] = \{x_1[t], y_1[t], z_1[t]\}$ with $a_1 \le t \le b_1$ and you have a curve C_2 with parameterization $P_2[t] = \{x_2[t], y_2[t], z_2[t]\}$ with $a_2 \le t \le b_2$. In addition, you are given that $P_1[a_1] = P_2[a_2]$ and $P_1[b_1] = P_2[b_2]$; so that both curves start at the same point and end at the same point.

> Say how you know in advance with no calculation that
> $$\int_{C_1} m[x,y,z]\,dx + n[x,y,z]\,dy + p[x,y,z]\,dz = \int_{a_1}^{b_1} m[x_1[t], y_1[t], z_1[t]]\,x_1'[t]$$
> $$+ n[x_1[t], y_1[t], z_1[t]]\,y_1'[t] + p[x_1[t], y_1[t], z_1[t]]\,z_1'[t]\,dt$$
> is the same as
> $$\int_{C_2} m[x,y,z]\,dx + n[x,y,z]\,dy + p[x,y,z]\,dz = \int_{a_2}^{b_2} m[x_2[t], y_2[t], z_2[t]]\,x_2'[t]$$
> $$+ n[x_2[t], y_2[t], z_2[t]]\,y_2'[t] + p[x_2[t], y_2[t], z_2[t]]\,z_2'[t]\,dt.$$

Answer:

$$\int_{C_1} m[x,y,z]\,dx + n[x,y,z]\,dy + p[x,y,z]\,dz$$
$$- \int_{C_2} m[x,y,z]\,dx + n[x,y,z]\,dy + p[x,y,z]\,dz$$
$$= \int_C m[x,y,z]\,dx + n[x,y,z]\,dy + p[x,y,z]\,dz,$$

where C is the closed curve made by joining C_1 and C_2. Think of C as rigid wire, and hang a sheet over the wire, being careful that the sheet doesn't pass through any singularities. Call R the part of the sheet bounded by C. Remembering that curlField$[x, y, z] = \{0, 0, 0\}$ at all points on R, use Stokes's formula to see that

$$0 = \iint_R \{0, 0, 0\} \bullet \text{topunitnormal}\,dA$$
$$= \iint_R \text{curlField} \bullet \text{topunitnormal}\,dA$$
$$= \oint_C \text{Field} \bullet \text{unittan}\,ds$$
$$= \oint_C m[x,y,z]\,dx + n[x,y,z]\,dy + p[x,y,z]\,dz$$
$$= \int_{C_1} m[x,y,z]\,dx + n[x,y,z]\,dy + p[x,y,z]\,dz$$
$$- \int_{C_2} m[x,y,z]\,dx + n[x,y,z]\,dy + p[x,y,z]\,dz.$$

This tells you that

$$\int_{C_1} m[x,y,z]\,dx + n[x,y,z]\,dy + p[x,y,z]\,dz$$
$$= \int_{C_2} m[x,y,z]\,dx + n[x,y,z]\,dy + p[x,y,z]\,dz.$$

■ T.3) Work

T.3.a) | What is work?

Answer: What's work for some folks is fun for other folks. Trig identities come to mind; they always seem to be work for the math student but fun for the math teacher. Physicists have their own technical notion of work.

Just as in two dimensions, physicists envision a 3D vector field

$$\text{Field}[x,y,z] = \{m[x,y,z], n[x,y,z], p[x,y,z]\}$$

to represent the force (push) on an object positioned at $\{x,y,z\}$. In this interpretation, the vector field Field$[x,y,z]$ is called a force field. Next, the physicists say that if an object goes along a 3D curve C specified by

$$P[t] = \{x[t], y[t], z[t]\} \qquad \text{with } a \le t \le b,$$

then the work done by a force field Field$[x,y,z]$ on the object during the duration of the object's trip is measured by the path integral

$$\int_0^{\text{length}} \text{Field} \bullet \text{unittan } ds$$
$$= \int_C m[x,y,z]dx + n[x,y,z]dy + p[x,y,z]\,dz$$
$$= \int_a^b m[x[t],y[t],z[t]]\,x'[t] + n[x[t],y[t],z[t]]\,y'[t] + p[x[t],y[t],z[t]]\,z'[t]\,dt.$$

This might not be your own notion of work, but notice that it is not you who does this work. The physicists have a pretty good reason for using this word for this measurement. Think of it this way: If at a point on the trip,

→ Field • unittan > 0, then at this point the force field Field$[x,y,z]$ is working this much to push the object and the object does no work at all.

But if

→ Field • unittan < 0, then at this point the force field Field$[x,y,z]$ is against the object; the object is working this much and the force field Field$[x,y,z]$ does not work at all.

With this in mind, you can think of

$$\int_0^{\text{length}} \text{Field} \bullet \text{unittan} \, ds$$

as a measurement of: the force field's work − the object's work.

If

$$\int_0^{\text{length}} \text{Field} \bullet \text{unittan} \, ds > 0,$$

then the force field Field$[x, y, z]$ did most of the work during the object's trip. But if

$$\int_0^{\text{length}} \text{Field} \bullet \text{unittan} \, ds < 0,$$

then the object did most of the work during the object's trip. If you agree that the object's work represents negative work for the force field, then you'll probably agree that

$$\int_0^{\text{length}} \text{Field} \bullet \text{unittan} \, ds = \int_C m[x,y,z] \, dx + n[x,y,z] \, dy + p[x,y,z] \, dz$$
$$= \int_a^b (m[x[t], y[t], z[t]] \, x'[t] + n[x[t], y[t], z[t]] \, y'[t] + p[x[t], y[t], z[t]] \, z'[t]) \, dt$$

is a reasonable measurement of the work done by the force field during the duration of the trip.

T.3.b) Is there a difference between flow along the curve and work?

Answer: Mathematically there is no difference, because they are both measured by the same integral. The difference is the interpretation.

When you are talking about flow along the curve, then you envision Field$[x, y, z] = \{m[x, y, z], n[x, y, z], p[x, y, z]\}$ as the velocity vector at $\{x, y, z\}$ of a fluid flow. The fluid is flowing and the curve is just sitting there.

When you are talking about work, then you envision

$$\text{Field}[x, y, z] = \{m[x, y, z], n[x, y, z], p[x, y, z]\}$$

as the force on an object at $\{x, y, z\}$ moving on a curve. This time the force field is just sitting there and the object is moving on the curve.

T.3.c) Here's a force field:

In[7]:=
```
Clear[x,y,z,m,n,p,Field]
Field[x_,y_,z_] = E^(-y){1 - x,y,z - 2};
{m[x_,y_,z_],n[x_,y_,z_],p[x_,y_,z_]} = Field[x,y,z]
```

Lesson 3.11 3D Flow Along

Out[7]=
$$\{\frac{1-x}{E^y}, \frac{y}{E^y}, \frac{-2+z}{E^y}\}$$

Here's a space curve:

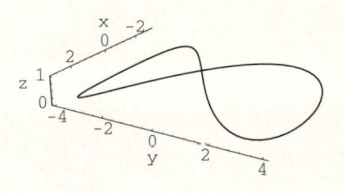

In[8]:=
```
Clear[t,P]; {x[t_],y[t_],z[t_]} =
  {3 Sin[2 t],4 Sin[t], Cos[t]^2};
P[t_] = {x[t],y[t],z[t]};
{tlow,thigh} = {0, 2 Pi};
Cplot = ParametricPlot3D[Evaluate[P[t]],
  {t,tlow,thigh},ViewPoint->CMView,Boxed->False,
  Axes->Automatic,AxesLabel->{"x","y","z"}];
```

An object starts at:

In[9]:=
```
P[tlow]
```
Out[9]=
{0, 0, 1}

and makes one trip around the curve, ending the trip at:

In[10]:=
```
P[thigh]
```
Out[10]=
{0, 0, 1}

Which way should you send the object around C to make the force field do most of the work?

Answer: Take a look at the curl:

In[11]:=
```
Clear[curlField]
curlField[x_,y_,z_] = {D[p[x,y,z],y] - D[n[x,y,z],z],
  D[m[x,y,z],z] - D[p[x,y,z],x], D[n[x,y,z],x] - D[m[x,y,z],y]}
```

Out[11]=
$$\{-(\frac{-2+z}{E^y}), 0, \frac{1-x}{E^y}\}$$

Too bad!

If curlField$[x, y, z]$ had calculated out to be $\{0, 0, 0\}$, the work measurement would have been 0; it wouldn't have mattered which way you sent the object. To measure the work done by the force field in moving the object along C in the given parameterization of C, calculate the path integral

$$\int_C \text{Field} \bullet \text{unittan} \, ds$$
$$= \int_C m[x,y,z] \, dx + n[x,y,z] \, dy + p[x,y,z] \, dz$$
$$= \int_{\text{tlow}}^{\text{thigh}} \left(m[x[t], y[t], z[t]] \, x'[t] + n[x[t], y[t], z[t]] \, y'[t] + p[x[t], y[t], z[t]] \, z'[t] \right) dt :$$

In[12]:=
```
NIntegrate[m[x[t],y[t],z[t]] x'[t] +
 n[x[t],y[t],z[t]] y'[t] + p[x[t],y[t],z[t]] z'[t],{t,tlow,thigh},AccuracyGoal->2]
```
Out[12]=
-242.111

Very negative. This means that to make the force field do the lion's share of the work, the object should move in the direction **opposite** to the parameterization of C. To see which direction that is, just plot C and a few of its **negative** tangent vectors.

In[13]:=
```
scalefactor = 0.5; Show[Cplot,
 Table[Arrow[scalefactor (-P'[t]),Tail->P[t],Red],
 {t,tlow,thigh,(thigh - tlow)/8}],
 ViewPoint->CMView,Boxed->False,
 Axes->Automatic,AxesLabel->{"x","y","z"}];
```

This is the way you should send the object around C to make the force field do most of the work. If you send the object the other way, then the force field will work against the object's motion.

■ T.4) The gradient test in 3D

T.4.a) The gradient test for a 2D vector field
$$\text{Field}[x,y] = \{m[x,y], n[x,y]\}$$
is run by checking whether
$$\text{rotField}[x,y] = D[n[x,y], x] - D[m[x,y], y] = 0.$$
If Field$[x,y]$ passes this test, then you know how to go about trying to find a function $f[x,y]$ with gradf$[x,y]$ = Field$[x,y]$.

> What is the three-dimensional version of the gradient test for a 3D vector field?

Answer: No surprise here; you just check whether
$$\text{curlField}[x,y,z] = \{0,0,0\}.$$

Lesson 3.11 3D Flow Along

To see why, check on what the curl of any gradient field is:

In[14]:=
```
Clear[f,gradf,Field,x,y,z,m,n,p,curlField]
gradf[x_,y_,z_] = {D[f[x,y,z],x],D[f[x,y,z],y],D[f[x,y,z],z]};
Field[x_,y_,z_] = gradf[x,y,z];
{m[x_,y_,z_],n[x_,y_,z_],p[x_,y_,z_]} = Field[x,y,z];
curlField[x_,y_,z_] = {D[p[x,y,z],y] - D[n[x,y,z],z],
D[m[x,y,z],z] - D[p[x,y,z],x], D[n[x,y,z],x] - D[m[x,y,z],y]}
```

Out[14]=
{0, 0, 0}

And that's that.

T.4.b) Here's a 3D vector field which passes the gradient test:

In[15]:=
```
Clear[Field,x,y,z,m,n,p,curlField]
Field[x_,y_,z_] = {-2 + y z,
x z - 5 z Sin[5 y z],4 E^(4 z) + x y - 5 y Sin[5 y z]};
{m[x_,y_,z_],n[x_,y_,z_],p[x_,y_,z_]} = Field[x,y,z];
curlField[x_,y_,z_] = {D[p[x,y,z],y] - D[n[x,y,z],z],
D[m[x,y,z],z] - D[p[x,y,z],x],D[n[x,y,z],x] - D[m[x,y,z],y]}
```

Out[15]=
{0, 0, 0}

> How do you try to come up with a function $f[x, y, z]$ with
>
> $\text{Field}[x, y, z] = \text{gradf}[x, y, z]$?

Answer: You just take the procedure that worked so well in two dimensions, and make appropriate adjustments. The goal is to come up with a function $f[x, y, z]$ with $\text{gradf}[x, y, z] = \text{Field}[x, y, z]$. To do this, fix any point $\{a, b, c\}$ you like, and parameterize a line C (or other curve) running from $\{a, b, c\}$ to the variable point $\{x, y, z\}$. The point $\{0, 0, 0\}$ is usually a good choice for $\{a, b, c\}$.

In[16]:=
```
Clear[t]; {a,b,c} = {0,0,0};
fixedpoint = {a,b,c}; variablepoint = {X,Y,Z};
tlow = 0; thigh = 1;
{x[t_],y[t_],z[t_]} = fixedpoint + t(variablepoint - fixedpoint)
```

Out[16]=
{t X, t Y, t Z}

To get a function $f[x, y, z]$ with $\text{gradf}[x, y, z] = \{m[x, y, z], n[x, y, z], p[x, y, z]\}$, all you gotta do is set

$$f[x, y, z] = \int_C m[x, y, z]\, dx + n[x, y, z]\, dy + p[x, y, z]\, dz,$$

where C is the line (or other curve) running from the fixed point $\{a, b, c\}$ to the variable point $\{x, y, z\}$:

In[17]:=
```
Clear[f]
f[X_,Y_,Z_] = Integrate[m[x[t],y[t],z[t]] x'[t] +
  n[x[t],y[t],z[t]] y'[t] + p[x[t],y[t],z[t]] z'[t],{t,tlow,thigh}]
```

Out[17]=
$$-2 + E^{4Z} - 2X + XYZ + \cos[5YZ]$$

Try it out:

In[18]:=
```
Clear[gradf]; gradf[x_,y_,z_] = {D[f[x,y,z],x],D[f[x,y,z],y],D[f[x,y,z],z]}
```

Out[18]=
$$\{-2 + y\,z,\ x\,z - 5\,z\,\sin[5\,y\,z],\ 4\,E^{4z} + x\,y - 5\,y\,\sin[5\,y\,z]\}$$

Compare:

In[19]:=
```
Field[x,y,z]
```

Out[19]=
$$\{-2 + y\,z,\ x\,z - 5\,z\,\sin[5\,y\,z],\ 4\,E^{4z} + x\,y - 5\,y\,\sin[5\,y\,z]\}$$

In[20]:=
```
gradf[x,y,z] == Field[x,y,z]
```

Out[20]=
True

Great. This tells you that $\text{gradf}[x, y, z] = \text{Field}[x, y, z]$, just as you wanted. See what happens when you go with a different fixed point:

In[21]:=
```
Clear[t]; {a,b,c} = {3,0,1}; fixedpoint = {a,b,c};
variablepoint = {X,Y,Z}; tlow = 0; thigh = 1;
{x[t_],y[t_],z[t_]} = fixedpoint + t(variablepoint - fixedpoint)
```

Out[21]=
$$\{3 + t\,(-3 + X),\ t\,Y,\ 1 + t\,(-1 + Z)\}$$

In[22]:=
```
Clear[f]; f[X_,Y_,Z_] = Integrate[m[x[t],y[t],z[t]] x'[t] +
  n[x[t],y[t],z[t]] y'[t] + p[x[t],y[t],z[t]] z'[t],{t,tlow,thigh}]
```

Out[22]=
$$5 - E^{4} + E^{4 + 4(-1 + Z)} - 2X + XYZ + \cos[5Y + 5Y(-1 + Z)]$$

Try it out:

In[23]:=
```
Clear[gradf]
gradf[x_,y_,z_] = {D[f[x,y,z],x],D[f[x,y,z],y],D[f[x,y,z],z]}
```

Lesson 3.11 3D Flow Along

Out[23]=
$\{-2 + y\,z,\ x\,z - (5 + 5\,(-1 + z))\,\text{Sin}[5\,y + 5\,y\,(-1 + z)],$
$4\,E^{4 + 4\,(-1 + z)} + x\,y - 5\,y\,\text{Sin}[5\,y + 5\,y\,(-1 + z)]\}$

Compare:

In[24]:=
```
Field[x,y,z]
```

Out[24]=
$\{-2 + y\,z,\ x\,z - 5\,z\,\text{Sin}[5\,y\,z],\ 4\,E^{4\,z} + x\,y - 5\,y\,\text{Sin}[5\,y\,z]\}$

Looks bad; check whether it feels good:

In[25]:=
```
ExpandAll[gradf[x,y,z] - Field[x,y,z],Trig->True]
```

Out[25]=
$\{0,\ 0,\ 0\}$

Yes! $\text{gradf}[x, y, z] = \text{Field}[x, y, z]$

It feels great!

Each time you change the fixed point $\{a, b, c\}$, you make a different function, $f[x, y, z]$, whose gradient is $\text{Field}[x, y, z]$.

Of course, this cannot be expected to work when you start with a 3D vector field that does not pass the 3D gradient test.

Give It a Try

■ G.1) Fingering a 3D vector field

G.1.a.i) Here's a 3D vector field:

In[1]:=
```
Clear[x,y,z,m,n,p,Field]
Field[x_,y_,z_] = {(y - z)^2,(z - x)^2,(x - y)^2};
{m[x_,y_,z_],n[x_,y_,z_],p[x_,y_,z_]} = Field[x,y,z]
```

Out[1]=
$\{(y - z)^2,\ (-x + z)^2,\ (x - y)^2\}$

> Stick the tail of the following unit vectors V at the indicated points $\{a, b, c\}$. Push your finger onto V so that V spikes through the center of your finger and the tip of your finger is at $\{a, b, c\}$.

> Use curlField[a, b, c] to help you report on whether the tip of your finger reveals a counterclockwise net swirl, a net clockwise swirl, or no swirl at all.
>
> Illustrate your results with a plot.

G.1.a.ii) Here is a point and a vector:

In[2]:=
```
{a,b,c} = {1,0,3};
V = {-3,2,1}/Sqrt[3^2 + 2^2 + 1^2];
```

G.1.a.iii) Now go with:

In[3]:=
```
{a,b,c} = {1.13,-2.67,-1.58};
V = {2,8,1}/Sqrt[2^2 + 8^2 + 1^2];
```

G.1.a.iv) And now try:

In[4]:=
```
{a,b,c} = {2.13,1.38,4.34};
V = {-6,4.5,1.5}/Sqrt[6^2 + 4.5^2 + 1.5^2];
```

G.1.a.v)
> For the vector field Field[x, y, z] as set above, there are many points at which your finger will feel no net clockwise or counterclockwise swirl no matter how you align your finger.
>
> Describe these points.

G.1.b) Here is another three-dimensional vector field:

In[5]:=
```
Clear[x,y,z,m,n,p,Field]
Field[x_,y_,z_] = {1.3 (y - z),0.2 (z - x),0.04 (x^2 - y)};
{m[x_,y_,z_],n[x_,y_,z_],p[x_,y_,z_]} = Field[x,y,z];
```

> Your job is to find the points at which this field rotates the very least.

G.1.c) Here is a new 3D vector field:

In[6]:=
```
Clear[x,y,z,m,n,p,Field]
Field[x_,y_,z_] = {z x^2,x (y - 1)^2,3 y z^2};
{m[x_,y_,z_],n[x_,y_,z_],p[x_,y_,z_]} = Field[x,y,z];
```

Take any three-dimensional unit vector

$$V = \{a, b, 0\} \qquad \text{with } a > 0 \text{ and } b > 0$$

and put the tail of V at any unspecified point $\{x,y,z\}$ with $x \neq 0$ and $z \neq 0$. Push your finger onto V so that V spikes through the center of your finger, and so the tip of your finger is at $\{x,y,z\}$.

> How do you know in advance that the tip of your finger will feel a net counter-clockwise swirl?

■ G.2) Flow-along measurements and estimates

G.2.a) Here are two 3D vector fields:

In[7]:=
```
Clear[x,y,z,m1,n1,p1,m2,n2,p2,Field1,Field2]
Field1[x_,y_,z_] = (E^(-y^2)){x,y,z};
{m1[x_,y_,z_],n1[x_,y_,z_],p1[x_,y_,z_]} = Field1[x,y,z];
Field2[x_,y_,z_] = {y z,x z,x y};
{m2[x_,y_,z_],n2[x_,y_,z_],p2[x_,y_,z_]} = Field2[x,y,z];
```

Here's a closed curve C in three dimensions:

In[8]:=
```
Clear[t,P]; {x[t_],y[t_],z[t_]} =
 {4 Cos[t],8 Sin[t],2 Sin[2 t] + 1};
P[t_] = {x[t],y[t],z[t]};
Cplot = ParametricPlot3D[
 Evaluate[P[t]],{t,0,2 Pi},
 ViewPoint->CMView,Boxed->False,
 Axes->Automatic,AxesLabel->{"x","y","z"}];
```

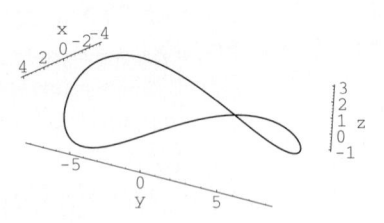

> Measure the flow of Field1$[x,y,z]$ along C and then measure the flow of Field2$[x,y,z]$ along C.
>
> Describe the direction of the net flow of each vector field along C.

G.2.b) Take a surface R with boundary curve C.

> If you are given a vector field Field$[x,y,z]$ with the extra property that curlField$[x,y,z]$ is tangent to the surface R at all points $\{x,y,z\}$ on the surface R, then how does Stokes's formula tell you that the net flow of Field$[x,y,z]$ along C is 0?

G.2.c.i) Here's a surface shown with a selection of normal vectors:

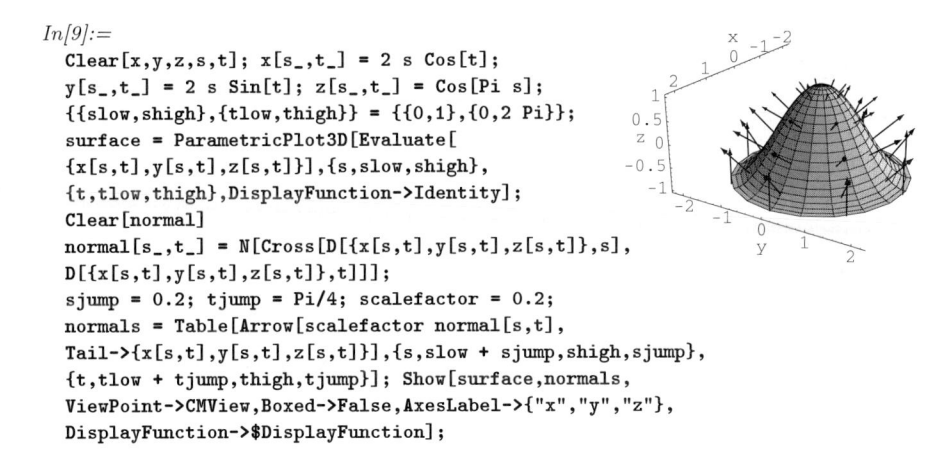

In[9]:=
```
Clear[x,y,z,s,t]; x[s_,t_] = 2 s Cos[t];
y[s_,t_] = 2 s Sin[t]; z[s_,t_] = Cos[Pi s];
{{slow,shigh},{tlow,thigh}} = {{0,1},{0,2 Pi}};
surface = ParametricPlot3D[Evaluate[
{x[s,t],y[s,t],z[s,t]}],{s,slow,shigh},
{t,tlow,thigh},DisplayFunction->Identity];
Clear[normal]
normal[s_,t_] = N[Cross[D[{x[s,t],y[s,t],z[s,t]},s],
D[{x[s,t],y[s,t],z[s,t]},t]]];
sjump = 0.2; tjump = Pi/4; scalefactor = 0.2;
normals = Table[Arrow[scalefactor normal[s,t],
Tail->{x[s,t],y[s,t],z[s,t]}],{s,slow + sjump,shigh,sjump},
{t,tlow + tjump,thigh,tjump}]; Show[surface,normals,
ViewPoint->CMView,Boxed->False,AxesLabel->{"x","y","z"},
DisplayFunction->$DisplayFunction];
```

You come across a 3D vector field, Field$[x,y,z]$, and happen to realize that

$$\text{curlField}[x[s,t], y[s,t], z[s,t]] \bullet \text{normal}[s,t] > 0$$

no matter what s and t you go with. This means that as you look down from the top of each normal, you see Field$[x,y,z]$ swirling around the tail of each normal vector in the counterclockwise way.

> Give a common-sense explanation of why this guarantees that the net flow of Field$[x,y,z]$ along the boundary curve of this surface is in the direction indicated by the following plot:

In[10]:=
```
Clear[P]
P[t_] = {x[shigh,t],y[shigh,t],z[shigh,t]};
Cplot = ParametricPlot3D[Evaluate[
P[t]],{t,tlow,thigh},DisplayFunction->Identity];
scalefactor = 1; tangents = Table[Arrow[
scalefactor P'[t],Tail->P[t],Red],{t,0,2 Pi,Pi/3}];
Show[surface,Cplot,tangents,Boxed->False,
PlotRange->All,ViewPoint->CMView,
AxesLabel->{"x","y","z"},
DisplayFunction->$DisplayFunction];
```

> Back up your common-sense explanation with an explanation based on Stokes's formula.

G.2.c.ii) Here's a surface shown with a selection of normal vectors:

In[11]:=
```
Clear[x,y,z,s,t,normal]; x[s_,t_] = 2 s Cos[t];
y[s_,t_] = 2 s Sin[t]; z[s_,t_] = 2 - x[s,t];
{{slow,shigh},{tlow,thigh}} = {{0,1},{0,2 Pi}};
surface = ParametricPlot3D[Evaluate[
 {x[s,t],y[s,t],z[s,t]}],{s,slow,shigh},{t,tlow,thigh},
PlotPoints->{2,Automatic},DisplayFunction->Identity];
normal[s_,t_] = N[Cross[D[{x[s,t],y[s,t],z[s,t]},s],
D[{x[s,t],y[s,t],z[s,t]},t]]];
sjump = 0.3; tjump = Pi/2; scalefactor = 0.5;
normals = Table[Arrow[scalefactor normal[s,t],
Tail->{x[s,t],y[s,t],z[s,t]}],
{s,slow + 0.1,shigh,sjump},{t,tlow + tjump,thigh,tjump}];
Show[surface,normals,ViewPoint->CMView,Boxed->False,
AxesLabel->{"x","y","z"},DisplayFunction->$DisplayFunction];
```

Here's a 3D vector field:

In[12]:=
```
Clear[Field,m,n,p,x,y,z,curlField]
{m[x_,y_,z_],n[x_,y_,z_],p[x_,y_,z_]} = { z , x^3 + 2 x + y,z^2 + x};
Field[x_,y_,z_] = {m[x,y,z],n[x,y,z],p[x,y,z]}
```

Out[12]=
$$\{z,\ 2x + x^3 + y,\ x + z^2\}$$

In[13]:=
```
curlField[x_,y_,z_] = {D[p[x,y,z],y] - D[n[x,y,z],z],
D[m[x,y,z],z] - D[p[x,y,z],x], D[n[x,y,z],x] - D[m[x,y,z],y]}
```

Out[13]=
$$\{0,\ 0,\ 2 + 3x^2\}$$

Now look at this:

In[14]:=
```
curlField[x[s,t],y[s,t],z[s,t]].normal[s,t]
```

Out[14]=
$$(4.\ s\ Cos[t]^2 + 4.\ s\ Sin[t]^2)\ (2 + 3\ x[s,\ t]^2)$$

> Remembering that $0 \leq s \leq 1$, use the information immediately above to say which direction is the direction of the net flow of the given vector field on the boundary curve of the surface.

■ G.3) Work

This problem appears only in the electronic version.

■ G.4) The gradient test in 3D

Here are two 3D vector fields. One of these vector fields passes the gradient test and the other does not.

In[15]:=
```
Clear[x,y,z,m1,n1,p1,m2,n2,p2,Field1,Field2]
Field1[x_,y_,z_] = {y + z,x - 2 z + 3,x - 2 y - 4 z + 3};
{m1[x_,y_,z_],n1[x_,y_,z_],p1[x_,y_,z_]} = Field1[x,y,z];
Field2[x_,y_,z_] = {y + 2 z,x - 2 z + 3,x - 2 y + 3 z + 3};
{m2[x_,y_,z_],n2[x_,y_,z_],p2[x_,y_,z_]} =Field2[x,y,z];
```

G.4.a.i) Which of these two vector fields passes the three-dimensional gradient test?

G.4.a.ii) Come up with function $f[x, y, z]$ whose gradient field is the same as the 3D vector field in part G.4.a.i) that passed the gradient test.

G.4.b) Here is the 3D electric field resulting from a single point charge of strength q placed at the point $\{a, b, c\}$

In[16]:=
```
Clear[Field,m,n,p,x,y,z,q,a,b,c]; point = {a,b,c};
Field[x_,y_,z_] = q ({x,y,z} - point)/
   ((x - a)^2 + (y - b)^2 + (z - c)^2)^(3/2);
{m[x_,y_,z_],n[x_,y_,z_],p[x_,y_,z_]} = Field[x,y,z]
```

Out[16]=
$$\left\{ \frac{q(-a + x)}{((-a + x)^2 + (-b + y)^2 + (-c + z)^2)^{3/2}}, \frac{q(-b + y)}{((-a + x)^2 + (-b + y)^2 + (-c + z)^2)^{3/2}}, \frac{q(-c + z)}{((-a + x)^2 + (-b + y)^2 + (-c + z)^2)^{3/2}} \right\}$$

Confirm that Field$[x, y, z]$ passes the gradient test and try to come up with a function $f[x, y, z]$ with

$$\text{gradf}[x, y, z] = \text{Field}[x, y, z].$$

Some folks call $f[x, y, z]$ a potential function for Field$[x, y, z]$.

The function $f[x, y, z]$ produces a measurement that is proportional to the voltage at $\{x, y, z\}$ resulting from placing a point charge of strength q at the point $\{a, b, c\}$ in the case that $\{a, b, c\} \neq \{0, 0, 0\}$.

■ G.5) Path independence and path dependence

G.5.a) The curve C_1 is parameterized to run on a straight line starting at $\{0,0,0\}$ and ending at $\{1,2,1\}$ as the parameter t runs from 0 to 1.

In[17]:=
```
Clear[x1,y1,z1,t]
{x1[t_],y1[t_],z1[t_]} = {0,0,0} + t {1,2,1};
startC1 = {x1[0],y1[0],z1[0]};
endC1 = {x1[1],y1[1],z1[1]};
```

The curve C_2 is also parameterized to run on a parabolic arc starting at $\{0,0,0\}$ and ending at $\{1,2,1\}$ as the parameter t runs from 0 to 1.

In[18]:=
```
Clear[x2,y2,z2,t]
{x2[t_],y2[t_],z2[t_]} = {6 t^2 - 5 t , 2 t^2,t^2};

startC2 = {x2[0],y2[0],z2[0]}

endC2 = {x2[1],y2[1],z2[1]};
```

The curve C_3 is also parameterized to run on another curve starting at $\{0,0,0\}$ and ending at $\{1,2,1\}$ as the parameter t runs from 0 to 1:

In[19]:=
```
Clear[x3,y3,z3,t]
{x3[t_],y3[t_],z3[t_]} = {t, 2 Sin[Pi t/2]^2,t^2};

startC3 = {x3[0],y3[0],z3[0]};

endC3 = {x3[1],y3[1],z3[1]};
```

Here is how they look:

In[20]:=
```
Clear[P1,P2,P3]
P1[t_] = {x1[t],y1[t],z1[t]};
C1plot = ParametricPlot3D[Evaluate[
P1[t]],{t,0,1},DisplayFunction->Identity];
P2[t_] = {x2[t],y2[t],z2[t]};
C2plot = ParametricPlot3D[Evaluate[
P2[t]],{t,0,1},DisplayFunction->Identity];
P3[t_] = {x3[t],y3[t],z3[t]};
C3plot = ParametricPlot3D[Evaluate[
P3[t]],{t,0,1},DisplayFunction->Identity];
points = {Graphics3D[{Red,PointSize[0.02],
Point[P1[0]]}],Graphics3D[{Red,PointSize[0.02],
Point[P1[1]]}]}; Show[C1plot,C2plot,C3plot,points,
ViewPoint->CMView,Boxed->False,
AxesLabel->{"x","y","z"},
DisplayFunction->$DisplayFunction];
```

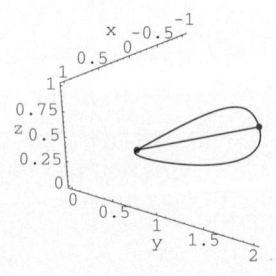

Here are calculations of the three path integrals

$$\int_{C_1} (6x - 4y + 3z)\,dx - (4x + 4z)\,dy + (3x - 4y)\,dz$$

$$\int_{C_2} (6x - 4y + 3z)\,dx - (4x + 4z)\,dy + (3x - 4y)\,dz$$

$$\int_{C_3} (6x - 4y + 3z)\,dx - (4x + 4z)\,dy + (3x - 4y)\,dz$$

in order:

In[21]:=
```
Clear[m,n,p,x,y,z]; m[x_,y_,z_] = 6 x - 4 y + 3 z;
n[x_,y_,z_] = -(4 x + 4 z); p[x_,y_,z_] = 3 x - 4 y;
NIntegrate[m[x1[t],y1[t],z1[t]] x1'[t] +
  n[x1[t],y1[t],z1[t]] y1'[t] + p[x1[t],y1[t],z1[t]] z1'[t],{t,0,1}]//Chop
```

Out[21]=
-10.

In[22]:=
```
NIntegrate[m[x2[t],y2[t],z2[t]] x2'[t] +
  n[x2[t],y2[t],z2[t]] y2'[t] + p[x2[t],y2[t],z2[t]] z2'[t],{t,0,1}]//Chop
```

Out[22]=
-10.

In[23]:=
```
NIntegrate[m[x3[t],y3[t],z3[t]] x3'[t] +
  n[x3[t],y3[t],z3[t]] y3'[t] + p[x3[t],y3[t],z3[t]] z3'[t],{t,0,1}]//Chop
```

Out[23]=
-10.

Keep the same curves, but tweak the integrals slightly by putting in a little minus sign on the middle term. Here are the resulting calculations

$$\int_{C_1} (6x - 4y + 3z)\,dx + (4x + 4z)\,dy + (3x - 4y)\,dz$$

$$\int_{C_2} (6x - 4y + 3z)\,dx + (4x + 4z)\,dy + (3x - 4y)\,dz$$

$$\int_{C_3} (6x - 4y + 3z)\,dx + (4x + 4z)\,dy + (3x - 4y)\,dz$$

in order:

In[24]:=
```
Clear[mm,nn,pp,x,y,z]; mm[x_,y_,z_] = 6 x - 4 y + 3 z;
nn[x_,y_,z_] = +(4 x + 4 z); pp[x_,y_,z_] = 3 x - 4 y;
NIntegrate[mm[x1[t],y1[t],z1[t]] x1'[t] +
  nn[x1[t],y1[t],z1[t]] y1'[t] + pp[x1[t],y1[t],z1[t]] z1'[t],{t,0,1}]
```

Out[24]=
6.

In[25]:=
```
NIntegrate[mm[x2[t],y2[t],z2[t]] x2'[t] +
   nn[x2[t],y2[t],z2[t]] y2'[t] + pp[x2[t],y2[t],z2[t]] z2'[t],{t,0,1}]
```
Out[25]=
-7.33333

In[26]:=
```
NIntegrate[mm[x3[t],y3[t],z3[t]] x3'[t] +
   nn[x3[t],y3[t],z3[t]] y3'[t] + pp[x3[t],y3[t],z3[t]] z3'[t],{t,0,1}]
```
Out[26]=
2.75772

> Explain how you could have predicted in advance that the first group of three integrals would have calculated out to the same value.
>
> Explain why it would have been quite a surprise if the second group of three path integrals had all calculated out to all be equal.

G.5.b) C_1 is a line segment parameterized to start at $\{1, -3, 5\}$ and to end at $\{-8, 7, 4\}$. C is any other curve parameterized to start at $\{1, -3, 5\}$ and to end at $\{-8, 7, 4\}$.

> Explain how you know that
> $$\int_C \left(-18\,x^2 + 24\,y + 2\,x\,z\right)\,dx + (24\,x - 4\,z)\,dy + \left(x^2 - 4\,y\right)\,dz$$
> $$= \int_{C_1} \left(-18\,x^2 + 24\,y + 2\,x\,z\right)\,dx + (24\,x - 4\,z)\,dy + \left(x^2 - 4\,y\right)\,dz.$$
> Then use this fact to help calculate
> $$\int_C \left(-18\,x^2 + 24\,y + 2\,x\,z\right)\,dx + (24\,x - 4\,z)\,dy + \left(x^2 - 4\,y\right)\,dz$$
> for any curve C parameterized to start at $\{1, -3, 5\}$ and end at $\{-8, 7, 4\}$.

G.5.c.i) You are given a 3D vector field

$$\text{Field}[x, y, z] = \{m[x, y, z], n[x, y, z], p[x, y, z]\}$$

and a closed curve C (like a deformed circle). You are also given the additional information that:

→ Field$[x, y, z]$ has only a couple of singularities and none of them lie on C and

→ curlField$[x, y, z] = \{0, 0, 0\}$ at all points $\{x, y, z\}$ that are not singularities.

Parameterize C and call the direction of your parameterization counterclockwise.

> Use Stokes's formula to explain why you are guaranteed that the path integral
> $$\oint_C m[x,y,z]\,dx + n[x,y,z]\,dy + p[x,y,z]\,dz = 0.$$

G.5.c.ii)
> Agree or disagree with each of the following statements:
> → In 2D, if you go with a vector field $\text{Field}[x,y] = \{m[x,y], n[x,y]\}$ with the property that
> $$\text{rotField}[x,y] = 0$$
> at all $\{x,y\}$'s that are not singularities, then it is possible that
> $$\oint_C m[x,y]\,dx + n[x,y]\,dy \neq 0$$
> for a closed curve C that does not pass through any of the singularities.
> → In 3D, if you go with a vector field
> $$\text{Field}[x,y,z] = \{m[x,y,z], n[x,y,z], p[x,y,z]\}$$
> with the property that
> $$\text{curlField}[x,y,z] = \{0,0,0\}$$
> at all $\{x,y,z\}$ that are not singularities, then it is possible that
> $$\oint_C m[x,y,z]\,dx + n[x,y,z]\,dy + p[x,y,z]\,dz \neq 0$$
> for a closed curve C that does not pass through any of the singularities.

■ G.6) Parallel flow and irrotational flow

This problem appears only in the electronic version.

■ G.7) Clockwise or counterclockwise?

This problem appears only in the electronic version.

■ G.8) The symbols ∇ (nabla) and △ (del)

This problem appears only in the electronic version.

■ G.9) Windpower

G.9.a) An electrical generator is driven by the wind. When you put the unit into the field, and the wind swirls in the clockwise way along the front of the unit, the generator charges a battery. If the wind swirls in the clockwise way, the generator discharges the battery. You are looking at a wind flow given by the vector field:

In[27]:=
```
Clear[Field,x,y,z,m,n,p]
Field[x_,y_,z_] = {2 x^2 - 3 y^2,4 y - 6 z,z - 2 y};
{m[x_,y_,z_],n[x_,y_,z_],p[x_,y_,z_]} = Field[x,y,z]
```

Out[27]=
$\{2 x^2 - 3 y^2, 4 y - 6 z, -2 y + z\}$

> If you stick the generator into this flow at the point $\{1,1,2\}$, then how should you align the unit (front to back) to make the generator charge the battery as much as possible?

■ G.10) How 2D rotField$[x,y]$ is related to 3D curlField$[x,y,z]$

G.10.a) Look at this cleared 2D vector field:

In[28]:=
```
Clear[m,n,x,y,Field,rotField]; Field[x_,y_] = {m[x,y],n[x,y]}
```

Out[28]=
`{m[x, y], n[x, y]}`

Here is rotField$[x,y]$:

In[29]:=
```
rotField[x_,y_] = D[n[x,y],x] - D[m[x,y],y]
```

Out[29]=
$-m^{(0,1)}[x, y] + n^{(1,0)}[x, y]$

Now make this 2D vector field into a 3D field as follows:

In[30]:=
```
Clear[Field3D,z]; Field3D = Append[Field[x,y],0]
```

Out[30]=
`{m[x, y], n[x, y], 0}`

> Calculate curlField3D$[x,y,z]$ and say how it is related to rotField$[x,y]$.

LITERACY SHEETS

Vector Calculus: Measuring in Two and Three Dimensions

3.01 Vectors Point the Way Literacy Sheet

L.1) When you plot a certain 3D vector X with its tail at $\{0, 0, 0\}$, it turns out that the tip of X is sitting on $\{9, -6, 4\}$.

When you plot the same vector with its tail at $\{-2, 7, 5\}$, where will the tip of X be?

L.2) Given two 2D vectors X and Y, what do you get when you plot Y and $X + Y$ with their tails at the origin, and you plot X with its tail at the tip of Y? Illustrate with a sketch.

What happens in 3D?

L.3) Given two 2-dimensional vectors X and Y, what do you get when you plot X and $X - Y$ with their tails at the origin, and you plot Y with its tail at the tip of $X - Y$? Illustrate with a sketch.

What happens in 3D?

L.4) Here is a the vector $X = \{5, 4\}$ shown with its tail at $\{0, 0\}$:

```
In[1]:=
  X = {5,4};
  Show[Arrow[X,Tail->{0,0}],Axes->True,
  AxesLabel->{"x","y"},
  PlotRange->{{-6,6},{-6,6}}];
```

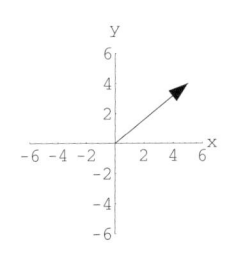

Add to this plot sketches of any squadron of its transplants you like, but make sure that the squadron contains some vectors with tails in each of the four quadrants.

L.5) Given $X = \{1, 0, 4\}$, calculate $3X$.

Can you change the direction of a vector by multiplying it by a positive number?

What happens when you multiply a vector by a negative number?

L.6) Calculate $X \bullet Y$ for $X = \{3, 8, 2\}$ and $Y = \{2, -2, 5\}$.

What does your result tell you about these two vectors?

L.7) Give the number t that makes $X = \{3, -1, 2\}$ and $Y = \{1, 1, t\}$ perpendicular.

L.8) Here are parametric formulas for two 3D lines:
$$L_1[t] = \{3, 0, 1\} + t\{-2, 1, 1\}$$
and
$$L_2[t] = \{3, 0, 1\} + t\{1, 1, 1\}.$$

Say how you can tell that these lines cross each other at right angles.

L.9) You are walking along a curve and at time t you are at the location
$$P[t] = \{x[t], y[t], z[t]\}.$$

You stop at a certain time t and plunk down the velocity vector
$$P'[t_0] = \{x'[t_0], y'[t_0], z'[t_0]\}$$
so that its tail is right on the point $P[t_0]$.

Does the resulting tangent vector point forward in the direction you are going, or does it point back against the direction you are walking?

L.10) The component of a vector X in the direction of another vector Y is given by
$$\left(\frac{X \bullet Y}{Y \bullet Y}\right) Y.$$

What does this formula reduce to if Y is a unit vector?

L.11) Put the tail of the vector
$$t\left\{\frac{1}{\sqrt{2}}, \frac{1}{\sqrt{2}}\right\} = \left\{\frac{t}{\sqrt{2}}, \frac{t}{\sqrt{2}}\right\}$$
at $\{0, 0\}$ and say what t must be to make the tip of this vector as close to $\{1, 2.5\}$ as it can be.

L.12) Here are the vectors $X = \{1/\sqrt{2}, 1/\sqrt{2}\}$ and $Y = \{1, 2.5\}$ with their tails at $\{0, 0\}$, shown in true scale:

Identify X and Y and then pencil in the component of Y in the direction of X.

Calculate the component of Y in the direction of X.

L.13) Here is a plot of an object's path shown with two acceleration vectors with their tails at the points on the curve at which they are calculated:

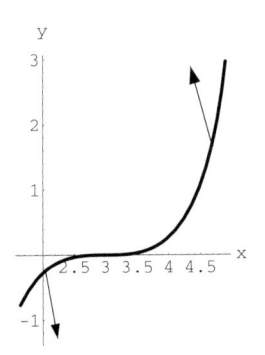

The object is moving on this curve from lower left to upper right. Pencil in the tangential and normal components of each of the two acceleration vectors in the plot.

At which of these points is the speed of the object increasing?

At which of these points is the speed of the object decreasing?

L.14) At time t with $0 \leq t$, an object is at the position

$$P[t] = \{\cos[2\,t], 2\,t, e^{-2t}\}.$$

Calculate its velocity, vel$[t]$, and its acceleration, accel$[t]$, as functions of t.

L.15) Ballistic projectiles (like cannonballs fired from a cannon) fired from the origin with muzzle velocity v_0 ft/sec and angle b with the horizontal are at the position

$$P[t] = \{v_0 \cos[b]\, t,\, v_0 \sin[b]\, t - 16\, t^2\}$$

t seconds after firing.

Calculate the acceleration vector as a function of t and explain the result.

L.16) If $X \bullet Y > 0$, then is the push of X in the direction of Y with Y or against Y?

If $X \bullet Y < 0$, then is the push of X in the direction of Y with Y or against Y?

L.17) Anywhere you happen to be, you feel a push whose direction is the same as the direction of the vector $\{2, 1\}$. Your velocity at time $t \geq 0$ is given by

$$\text{vel}[t] = \{-t, t^2\}.$$

For what times $t \geq 0$ are you being helped by the push?

At what times $t \geq 0$ are you being hindered by the push?

L.18) Write a parametric formula for the line that passes through the points $\{3, 1\}$ and $\{5, 11\}$.

Give a vector parallel to this line.

Give a vector perpendicular to this line.

L.19) Here is the line segment running from $\{-3, 2\}$ to $\{3, -4\}$:

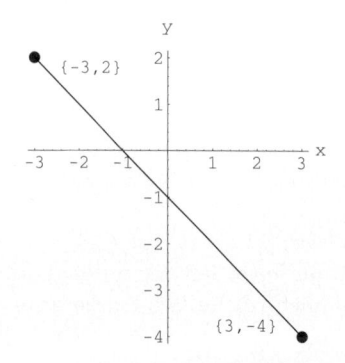

A parametric formula of the line through these two points is:

$$L[t] = \{-3, 2\} + t\, (\{3, -4\} - \{-3, 2\}).$$

→ What value of t makes $L[t]$ land on $\{-3, 2\}$?

→ What value of t makes $L[t]$ land on $\{3, -4\}$?

→ What value of t makes $L[t]$ land on the midpoint

$$\frac{\{\{-3, 2\} + \{3, -4\}\}}{2} = \{0, -1\}$$

of the indicated segment?
- → What t's do you use to plot the indicated segment?
- → If you take $t > 1$, where will $L[t]$ land?
- → If you take $t < 0$, where will $L[t]$ land?

L.20) Are the lines with parametric formulas
$$L_1[t] = \{2, 3\} + t\{-2, 4\}$$
and
$$L_2[t] = \{2, 3\} + t\{-4, 8\}$$
the same line or different lines?

Are the lines with parametric formulas
$$L_1[t] = \{2, 3\} + t\{-3, 5\}$$
and
$$L_2[t] = \{-4, 13\} + t\{-3, 5\}$$
the same line or different lines?

L.21) Give a parametric formula for the 3D-line through the points $\{2, 5, 0\}$ and $\{4, 1, 3\}$.

L.22) Here are two parallel circles:

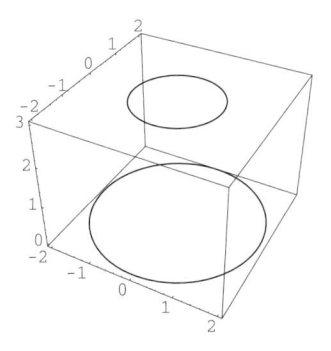

When you fix a given s with $0 \le s \le 2\pi$, then what happens to the points
$$\{2\cos[s], 2\sin[s], 0\} + t\,(\{\cos[s], \sin[s], 3\} - \{2\cos[s], 2\sin[s], 0\})$$
as you run t from 0 to 1? Use your answer to predict the output from:

In[2]:=
```
ParametricPlot3D[
 {2 Cos[s],2 Sin[s],0} + t ({Cos[s],Sin[s],3} - {2 Cos[s],2 Sin[s],0}),
 {s,0,2 Pi},{t,0,1}];
```

3.02 Perpendicularity Literacy Sheet

L.1) How do you know that the vector
$$X = \left\{t, -\frac{s}{2}, -\frac{s}{2}\right\}$$
is perpendicular to the vector $Y = \{s, t, t\}$?

L.2) Given $X = \{1, 1, 1\}$ and $Y = \{2, 0, 0\}$, calculate $X \bullet Y$ and $X \times Y$.

L.3) Given any two 3D vectors X and Y, say why you are sure that
$$X \bullet (X \times Y) = Y \bullet (X \times Y) = 0.$$

L.4) Come up with a unit vector that points in the same direction as the vector
$$X = \{3, 0, 4\}.$$

L.5) If X is a non-zero vector, then what is the length of
$$\frac{X}{\sqrt{X \bullet X}} = \frac{1}{\sqrt{X \bullet X}} X?$$
What is the relation between the direction of X and the direction of
$$\frac{X}{\sqrt{X \bullet X}}?$$

L.6) A plane has xyz-equation
$$x - y + 5z = 10.$$
Give a vector perpendicular to the plane, and give a point on the plane.

L.7) Here are xyz-equations for two planes:
$$x + y + z = 9$$
and
$$2x - y - z = 3.$$
Say how you can tell that these planes cut each other at right angles.

L.8) Put
$$\text{normal1} = \{1, 1, 1\}$$
and
$$\text{normal2} = \{2, -1, -1\}.$$
Explain how you know that the cross product, normal1 × normal2, is parallel to the line in which the planes
$$x + y + z = 9$$
and
$$2x - y - z = 3$$
intersect each other.

L.9) In an effort to find a formula for the line in which the planes
$$x + y + z = 9$$
and
$$2x - y - z = 3$$
intersect, the lab pest Calculus Cal added the two equations, and announced that a formula for the line of intersection of these planes is
$$3x = 12 \quad (\text{or } x = 4).$$
Say why Cal's answer is a hundred yards short of the green.

L.10) How do you know that the planes
$$x - 2y + 4z = 12$$
and
$$3x - 6y + 12z = 4$$
are parallel?

L.11) Given that X and Y are perpendicular unit vectors in 2D, describe the curve traced out by
$$P[t] = \cos[t]\, X + \sin[t]\, Y$$
as t runs from 0 to 2π.

L.12) Given that X and Y are perpendicular unit vectors in 3D, describe the curve traced out by
$$P[t] = 2\cos[t]\, X + 4\sin[t]\, Y$$
as t runs from 0 to 2π.

L.13) Given a fixed point $\{a, b, c\}$ and fixed, perpendicular unit vectors X and Y in 3D, describe the curve traced out by
$$P[t] = \{a, b, c\} + 2\cos[t]\, X + 2\sin[t]\, Y$$
as t runs from 0 to 2π.

L.14) Given a fixed point $\{a, b, c\}$ and given fixed, perpendicular unit vectors X and Y in 3D, describe the curve traced out by
$$P[t] = \{a, b, c\} + 2\cos[t]\, X + 3\sin[t]\, Y$$
as t runs from 0 to 2π.

L.15) You are walking around a closed curve with no loops (a curve like a distorted circle) in the counterclockwise way, and at time t, you are at the point $\{x[t], y[t]\}$. Does the unit normal vector
$$\frac{\{y'[t], -x'[t]\}}{\sqrt{x'[t]^2 + y'[t]^2}}$$
transplanted with its tail at $\{x[t], y[t]\}$ point out away from the curve toward your right foot, or does it point inside the curve toward your left foot?

L.16) Here is a curve parameterized by a certain formula $P[t] = \{x[t], y[t]\}$ for $0 \le t \le 2\pi$:

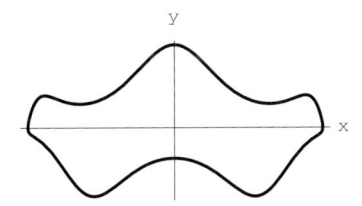

As t advances from 0 to 2π, you are given that $P[t]$ moves along this curve in the counterclockwise direction. Pencil in some of the unit normals
$$\frac{\{y'[t], -x'[t]\}}{\sqrt{x'[t]^2 + y'[t]^2}}$$
with their tails at $\{x[t], y[t]\}$.

L.17) Here is a curve parameterized by a certain formula $P[t] = \{x[t], y[t]\}$ for $0 \le t \le 4$:

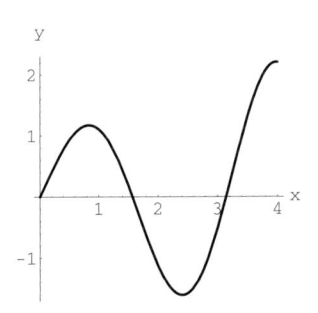

As t advances from 0 to 4, you are given that $P[t]$ moves from left to right along this curve. Pencil in some of the main unit normals
$$\text{mainunitnormal}[t] = \frac{D[\text{unittan}[t], t]}{\sqrt{D[\text{unittan}[t], t] \bullet D[\text{unittan}[t], t]}}$$
with tail at $\{x[t], y[t]\}$. (In order to be able to answer this question, experience with the Give It a Try problem called "Kissing circles and curvature" would be helpful.)

L.18) What is curvature?

L.19) Describe how you can use unit tangents, main unit normals, and binormals to make a plot like this:

Describe how you can use unit tangents, main unit normals, and binormals to make a tube like this:

L.20) You are going to cut an elliptical hole out of a piece of plywood with a drill bit whose radius is 1.05 inches. Here is the shape of the hole with all units in inches:

In[1]:=
```
Clear[x,y,P,t]
{x[t_],y[t_]} = {4 Cos[t],2 Sin[t]};
curve = ParametricPlot[{x[t],y[t]},
  {t,0,2 Pi},AxesLabel->{"x","y"},
  AspectRatio->Automatic];
```

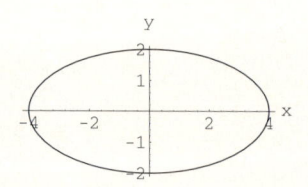

Why should the plot of the following curve coincide with the plot of the center of the drill bit as you cut the hole out of the plywood?

In[2]:=
```
Clear[P]; P[t_] = {x[t],y[t]} -
1.05 {y'[t],-x'[t]}/Sqrt[x'[t]^2 + y'[t]^2];
drillpath = ParametricPlot[P[t],
  {t,0,2 Pi},DisplayFunction->Identity];
Show[curve,drillpath,AspectRatio->Automatic,
  DisplayFunction->$DisplayFunction];
```

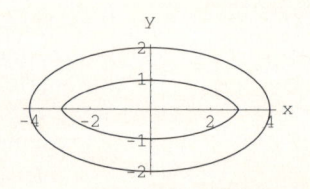

Is the path of the drill bit an ellipse?

3.03 Gradient Literacy Sheet

L.1) Calculate the gradient, $\nabla f[x,y] = \text{gradf}[x,y]$, of $f[x,y] = e^{-(x^2/2 + y^2/4)}$.

Calculate the gradient, $\nabla f[x,y] = \text{gradf}[x,y]$, of $f[x,y] = e^{-3xy}$.

Calculate the gradient, $\nabla f[x,y,z] = \text{gradf}[x,y,z]$, of $f[x,y,z] = x\,\sin[x\,y\,z]$.

L.2) Go with
$$f[x,y] = \frac{0.6}{1 + 1.2\,x^2 + 1.4\,y^2}$$
and look at a plot of gradf$[x,y]$ with tails at $\{x,y\}$ for some selected points $\{x,y\}$ in the vicinity of $\{0,0\}$:

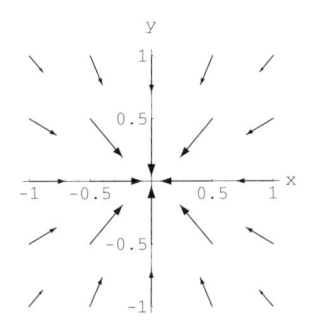

Why are these gradient vectors so attracted to the point $\{0,0\}$?

L.3) Go with
$$f[x,y] = 0.6\,(1 + 1.2\,x^2 + 1.4\,y^2)$$
and look at a plot of gradf$[x,y]$ with tails at $\{x,y\}$ for some selected points $\{x,y\}$ in the vicinity of $\{0,0\}$:

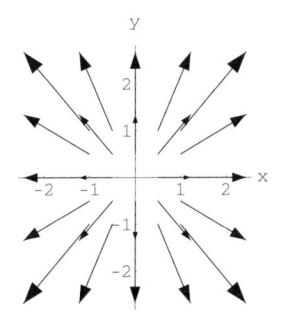

Why are these gradient vectors so repelled by the point $\{0,0\}$?

L.4) Given a certain function $f[x,y]$, you find that the plot of gradf$[x,y]$ with tails at $\{x,y\}$ for some selected points $\{x,y\}$ in the vicinity of $\{0,0\}$ looks like this:

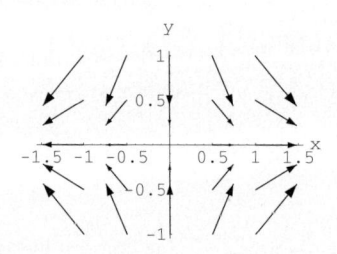

Explain why this plot suggests that when you go with a point $\{x,y\}$ with

$-1 < x < 1$ and $-1 < y < 1$,

then you can expect that:

→ If $x > 0$, then increasing x results in increasing $f[x,y]$.
→ If $x < 0$, then increasing x results in decreasing $f[x,y]$.
→ If $y > 0$, then increasing y results in decreasing $f[x,y]$.
→ If $y < 0$, then increasing y results in increasing $f[x,y]$.

L.5) You are at the point $\{1,0\}$. In the direction of what vector should you step off $\{1,0\}$ in order to get the greatest initial increase in the function

$$f[x,y] = x^2 + 7\,x\,e^y?$$

In the direction of what vector should you step off $\{1,0\}$ in order to get the greatest initial decrease in the function

$$f[x,y] = x^2 + 7\,x\,e^y?$$

L.6) Discuss the general idea behind *Mathematica*'s FindMinimum instruction.

L.7) Go with

$$f[x,y,z] = x\,y\,\sin[z].$$

In which direction should you leave the point $\{1,2,0\}$ to get the greatest possible initial increase of $f[x,y,z]$?

In which direction should you leave the point $\{1,2,0\}$ to get the greatest possible initial decrease of $f[x,y,z]$?

L.8) Here are side-by-side plots of the negative gradient field of a certain function $f[x,y]$ and a contour plot of the same function.

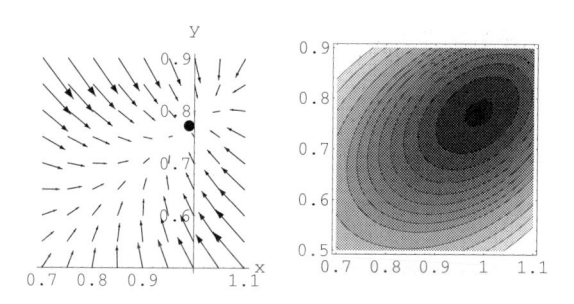

And here's a plot of the surface $z = f[x,y]$ (for the same function $f[x,y]$) with a pole planted at the deepest dip:

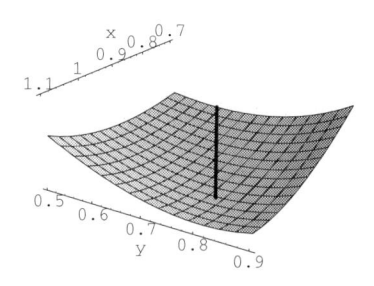

Explain why a literate person will say that all three plots are in total harmony.

L.9) The point $\{1,1\}$ is on the level curve
$$x^2 y^3 e^{x+y} = 7.38906.$$

Here's a plot of part of this level curve with axes going through the point $\{1,1\}$:

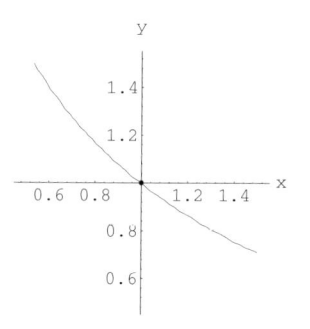

Pencil in a vector with its tail at $\{1,1\}$ that points in the same direction as gradf$[1,1]$.

L.10) Here is the level surface
$$f[x,y,z] = x^2 + \left(\frac{y}{2}\right)^2 + z^2 = 1$$
together with a selection of gradient vectors of $f[x,y,z]$ with tails at the points at which they are calculated:

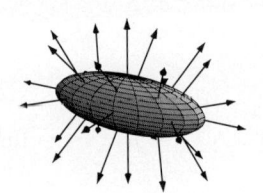

Explain:

→ Why you are not surprised by the fact that the plotted gradient vectors are perpendicular to the level surface.

→ Why these gradient vectors point out away from the surface and not in toward the interior of the surface.

L.11) Does
$$f[x,y] = x^6 - 3x^2y + y^4 - 6x + 5y$$
have a maximizer or minimizer? How do you know?

Does
$$f[x,y] = e^{(-x^2/2 - y^2/2)}\left(x^6 + 7y^3 + 4\right)$$
have a maximizer or a minimizer? How do you know?

Does
$$f[x,y] = \frac{x^5 + 3y^2}{1 + x^4 + y^6}$$
have a maximizer or a minimizer? How do you know?

L.12) You are sitting at a given point $\{x_0, y_0, z_0\}$ and are trying to study a function $f[x,y,z]$. As you leave this point in the direction of a vector U, is it true that the function $f[x,y,z]$ initially goes up if
$$U \bullet \text{gradf}[x_0, y_0, z_0] > 0$$
and intially goes down if
$$U \bullet \text{gradf}[x_0, y_0, z_0] < 0?$$

L.13) You are trying to minimize a function $f[x,y,z]$ subject to the constraint that $\{x,y,z\}$ must lie on a given line in 3D. Explain why you want to become very

interested in points on the line at which
$$\nabla f[x,y,z] = \text{gradf}[x,y,z]$$
is perpendicular to the line.

L.14) When you are trying to find the highest crest or the deepest dip on the plot of a surface $z = f[x,y]$, why do you become interested in solving
$$\text{gradf}[x,y] = \{0,0\}$$
for x and y?

L.15) Use the total differential of
$$f[x,y] = x\,y$$
to give a quick hand derivation of the "product rule" formula
$$D[x[t]\,y[t], t] = x'[t]\,y[t] + x[t]\,y'[t].$$
Use the total differential of
$$f[x,y] = \frac{x}{y}$$
to give a quick hand derivation of the "quotient rule" formula
$$D\left[\frac{x[t]}{y[t]}, t\right] = \frac{x'[t]}{y[t]} - \frac{x[t]\,y'[t]}{y[t]^2} = \frac{x'[t]\,y[t] - x[t]\,y'[t]}{y[t]^2}.$$

L.16) How is the total differential related to this:
$$\nabla f[x,y] \bullet \{dx, dy\} = \left\{\frac{\partial f[x,y]}{\partial x}, \frac{\partial f[x,y]}{\partial y}\right\} \bullet \{dx, dy\}?$$

L.17) The chain rule formulas say:
→ $D[f[x[t]], t] = f'[x[t]]\,x'[t].$
→ $D[f[x[t], y[t]], t] = \text{gradf}[x[t], y[t]] \bullet \{x'[t], y'[t]\}.$
→ $D[f[x[t], y[t], z[t]], t] = \text{gradf}[x[t], y[t], z[t]] \bullet \{x'[t], y'[t], z'[t]\}.$

The fundamental formula of calculus and the chain rule for functions of one variable give you the clean formula
$$\int_a^b g'[x[t]]\,x'[t]\,dt = g[x[b]] - g[x[a]].$$
Use the fundamental formula of calculus and the chain rule
$$D[f[x[t], y[t]], t] = \text{gradf}[x[t], y[t]] \bullet \{x'[t], y'[t]\}$$
to give a clean formula for
$$\int_a^b \text{gradf}[x[t], y[t]] \bullet \{x'[t], y'[t]\}\,dt.$$

If you are given that $a = 0$ and $b = 2\pi$, and $x[t] = \cos[t]$ and $y[t] = \sin[t]$, explain why you also know

$$\int_a^b \text{gradf}[x[t], y[t]] \bullet \{x'[t], y'[t]\} \, dt = 0.$$

L.18) Lots of folks like to use the notations

$$f^{(1,0)}[x, y] = \frac{\partial f[x, y]}{\partial x} = D[f[x, y], x]$$

and

$$f^{(0,1)}[x, y] = \frac{\partial f[x, y]}{\partial y} = D[f[x, y], y].$$

Calculate

$$\frac{\partial f[x, y]}{\partial x} = f^{(1,0)}[x, y]$$

and calculate

$$\frac{\partial f[x, y]}{\partial y} = f^{(0,1)}[x, y]$$

for

$$f[x, y] = e^{x^2/2 - 3y^2}.$$

L.19) Calculate

$$\frac{\partial f[x, y, z]}{\partial z} = f^{(0,0,1)}[x, y, z]$$

by hand for $f[x, y, z] = \sin[x^2 y^3 z^4]$.

L.20) Here is

$$\frac{\partial f[x, y]}{\partial x} = f^{(1,0)}[x, y]$$

for

$$f[x, y] = e^{\sin[xy - x]}:$$

In[1]:=
```
Clear[f,x,y]; f[x_,y_] = E^Sin[x y - x]; D[f[x,y],x]
```

Out[1]=

$$-\left(\frac{(1 - y) \cos[x - x y]}{E^{\sin[x - x y]}}\right)$$

And here is $\partial f[2.3, 8.4]/\partial x = f^{(1,0)}[2.3, 8.4]$ for the same function:

In[2]:=
```
D[f[x,y],x]/.{x->2.3,y->8.4}
```

Out[2]=
-0.72017

What does this calculation tell you about what happens to $f[x,y]$ when you hold y at 8.4, but make x just a teensy-weensy bit bigger than 2.3?

L.21) Put $f[x,y] = \sin[2x - 3y]$ and say what
$$\frac{3\,\partial f[x,y]}{\partial x} + \frac{2\,\partial f[x,y]}{\partial y}$$
calculates out to.

When you go with any function $g[x]$ and you put
$$f[x,y] = g[2x - 3y],$$
then what does
$$\frac{3\,\partial f[x,y]}{\partial x} + \frac{2\,\partial f[x,y]}{\partial y}$$
calculate out to?

3.04 2D Vector Fields and Their Trajectories Literacy Sheet

L.1) Here's a rather detailed plot of a certain vector field:

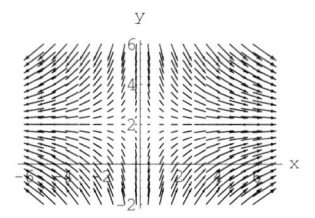

Visible in this plot are four families of trajectories in this vector field. Pencil in one trajectory of each type.

L.2) Here's a plot of a certain vector field shown with the plots of two curves:

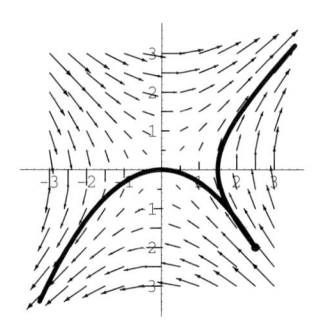

One of these curves is a genuine trajectory in the vector field. The other is bogus. Identify the trajectory and discuss the information that led to your choice.

L.3) You are given that the point $\{2, 1\}$ is on a certain trajectory in the vector field
$$\text{Field}[x, y] = \{x + y, x - y\}.$$
Write down a vector that is tangent to this trajectory at the point $\{2, 1\}$.

L.4) How do you know that when you run a trajectory in the vector field
$$\text{Field}[x, y] = \{y^2, x^2\},$$
starting at any point you like, you are guaranteed that the trajectory is the graph of a function whose derivative is positive?

L.5) Here's an ellipse C and the plot of a certain vector field Field$[x, y]$ plotted with tail at $\{x, y\}$ for a fairly generous selection of points $\{x, y\}$ on the ellipse.

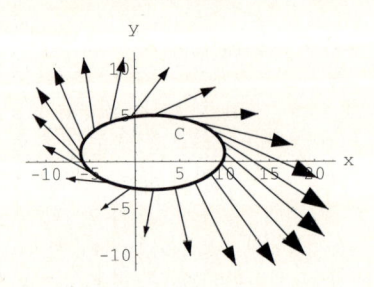

Look at this plot, and then estimate whether the net flow of this vector field across C is from inside to outside, or is from outside to inside, and estimate whether the net flow of this vector field along this curve is clockwise or counterclockwise.

L.6) Here's a circle and the plot of a certain vector field Field$[x, y]$ plotted with tails at $\{x, y\}$ for a generous selection of points $\{x, y\}$ on the circle.

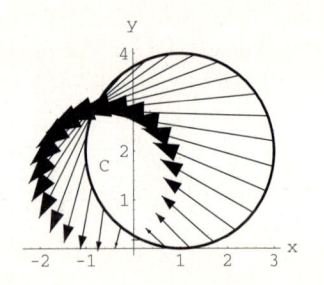

Here are tangential and normal components of what you see above:

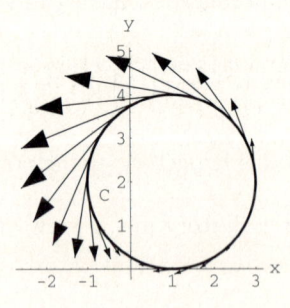

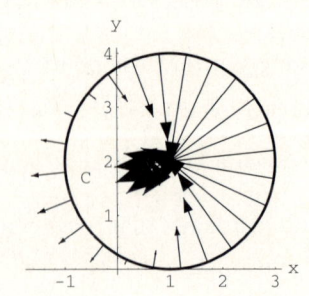

Look at these plots, and then estimate whether the net flow of this vector field across the circle is from inside to outside, or is from outside to inside, and estimate whether the net flow of this vector field along this circle is clockwise or counterclockwise.

L.7) Here's a circle and the plot of a different vector field Field$[x, y]$ plotted with tails at $\{x, y\}$ for a generous selection of points $\{x, y\}$ on the circle.

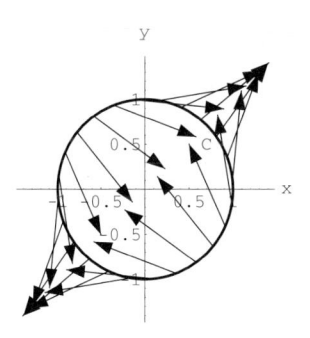

Here are tangential and normal components of what you see above:

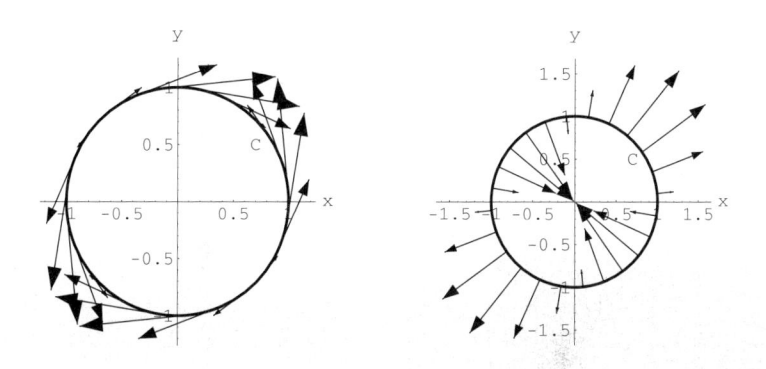

Look at these plots and then estimate whether the net flow of this vector field across the circle is from inside to outside, from outside to inside, or zero, and estimate whether the net flow of this vector field along this circle is clockwise or counterclockwise or zero.

L.8) True or false: When you plot a given vector field at points $\{x, y\}$ on a given curve, then all the field vectors you plot turn out to be perpendicular to the curve.

L.9) Here is a rectangle C with eight labeled points:

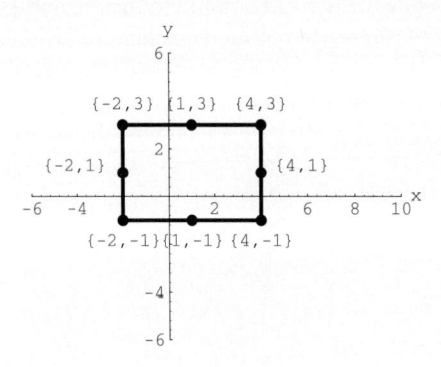

Go with the vector field

$$\text{Field}[x, y] = \{y + x, -x\}.$$

For each labeled point $\{x, y\}$, pencil in the field vector $\text{Field}[x, y]$ with tail at $\{x, y\}$.

On the basis of your plot, estimate whether the net flow of this vector field across the edge of this rectangle is from inside to outside, or is from outside to inside, and estimate whether the net flow of this vector field along this curve is clockwise or counterclockwise.

L.10) Your boss gives you a differential equation

$$y'[x] = f[x, y[x]] \quad \text{with } y[0] = 2$$

and asks you to give an approximate plot of its solution. As a literate calculus person, you know that when you start the trajectory of a certain vector field at $\{0, 2\}$, then the plot of this trajectory will be the same as the plot of the solution you want. The question here is: What vector field do you go with?

L.11) Can two trajectories of a vector field ever cross over each other like this:

Why or why not?

L.12) You have a differential equation

$$y'[x] = f[x, y[x]].$$

If you have a formula for $f[x, y]$ and you start with $y[0] = 1$, then you can use NDSolve to plot out a function $y[x]$ with

$$y[0] = 1 \quad \text{and} \quad y'[x] = f[x, y[x]].$$

On the other hand if you start with $y[0] = 2$, then you can use NDSolve to plot out a function $y[x]$ with

$$y[0] = 2 \quad \text{and} \quad y'[x] = f[x, y[x]].$$

The question here is: Can the plots of these two solutions ever cross over each other like this:

Why or why not?

L.13) Here is a plot of the gradient field of a certain function near the maximizer of the function, which is plotted as a heavy dot:

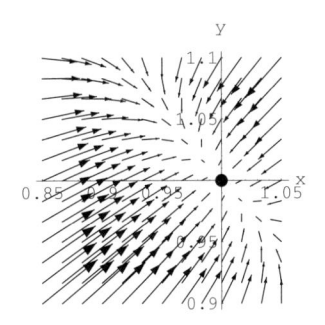

Pencil in a couple of trajectories.

Where are the trajectories in this gradient field trying to go?

Why are they trying to go there?

L.14) Here is a plot of the negative gradient field of a certain function near the minimizer of the function, which is plotted as a heavy dot:

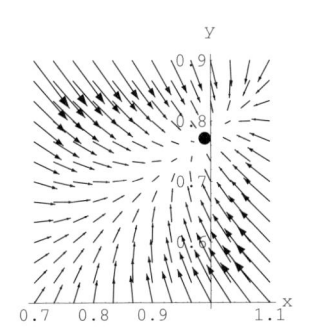

Pencil in a couple of trajectories.

Where are the trajectories in this negative gradient field trying to go?

Why are they trying to go there?

L.15) Suppose $f[x_0, y_0] > f[x, y]$ for all other $\{x, y\}$ near $\{x_0, y_0\}$.

If you center a small circle C at $\{x_0, y_0\}$, why do you expect that the net flow of gradf$[x, y]$ across C is from outside to inside?

Suppose $f[x_0, y_0] < f[x, y]$ for all other $\{x, y\}$ near $\{x_0, y_0\}$.

If you center a small circle C at $\{x_0, y_0\}$, why do you expect that the net flow of gradf$[x, y]$ across C is from inside to outside?

L.16) True or false: Trajectories through the gradient field of $f[x, y]$ try to seek out maximizers of $f[x, y]$, but they sometimes get sidetracked by a pretender. Trajectories through the negative gradient field of $f[x, y]$ try to find minimizers of $f[x, y]$ but they sometimes get sidetracked by a pretender.

3.05 Flow Measurements by Integrals Literacy Sheet

L.1) You have a given vector field
$$\text{Field}[x, y] = \{m[x, y], n[x, y]\}.$$
You have parameterized the ellipse
$$\left(\frac{x}{2}\right)^2 + \left(\frac{y}{3}\right)^2 = 1$$
in the counterclockwise way with
$$\{x[t], y[t]\} = \{2\cos[t], 3\sin[t]\} \qquad \text{with } 0 \le t \le 2\pi.$$
You calculate
$$\oint_C m[x, y]\, dx + n[x, y]\, dy = \int_0^{2\pi} (m[x[t], y[t]]\, x'[t] + n[x[t], y[t]]\, y'[t])\, dt,$$
and learn that
$$\oint_C m[x, y]\, dx + n[x, y]\, dy = 9.01.$$
Then you calculate
$$\oint_C -n[x, y]\, dx + m[x, y]\, dy = \int_0^{2\pi} (-n[x[t], y[t]]\, x'[t] + m[x[t], y[t]]\, y'[t])\, dt,$$
and learn that
$$\oint_C -n[x, y]\, dx + m[x, y]\, dy = -3.52.$$
Is the net flow of Field$[x, y]$ along C clockwise or counterclockwise?

Is the net flow of Field$[x, y]$ across C from outside to inside, or is it from inside to outside?

L.2) You are walking around a closed curve C with no loops (a curve like a distorted circle) in the counterclockwise way, and at time t, you are at the point $\{x[t], y[t]\}$. The unit tangent vector
$$\text{unittan}[t] = \frac{\{x'[t], y'[t]\}}{\sqrt{x'[t]^2 + y'[t]}}$$
with tail at $\{x[t], y[t]\}$ points out away from your belly in the direction you are walking. When you use this unit tangent to measure the flow of a vector field Field$[x, y] = \{m[x, y], n[x, y]\}$ along C, you calculate
$$\oint_C m[x, y]\, dx + n[x, y]\, dy = \oint_C \text{Field} \bullet \text{unittan}\, ds.$$

→ How do you interpret the result if it turns out that
$$\oint_C m[x,y]\,dx + n[x,y]\,dy > 0?$$

→ How do you interpret the result if it turns out that
$$\oint_C m[x,y]\,dx + n[x,y]\,dy < 0?$$

→ How do you interpret the result if it turns out that
$$\oint_C m[x,y]\,dx + n[x,y]\,dy = 0?$$

L.3) You are walking around a closed curve C with no loops (a curve like a distorted circle) in the counterclockwise way, and at time t, you are at the point $\{x[t], y[t]\}$. The unit normal vector
$$\text{outerunitnormal}[t] = \frac{\{y'[t], -x'[t]\}}{\sqrt{x'[t]^2 + y'[t]}}$$
with tail at $\{x[t], y[t]\}$ points out away from the curve toward your right foot. When you use this unit normal to measure the flow of a vector field
$$\text{Field}[x, y] = \{m[x, y], n[x, y]\}$$
across C, you calculate
$$\oint_C -n[x,y]\,dx + m[x,y]\,dy = \oint_C \text{Field} \bullet \text{outerunitnormal}\,ds.$$

→ How do you interpret the result if
$$\oint_C -n[x,y]\,dx + m[x,y]\,dy > 0?$$

→ How do you interpret the result if
$$\oint_C -n[x,y]\,dx + m[x,y]\,dy < 0?$$

→ How do you interpret the result if
$$\oint_C -n[x,y]\,dx + m[x,y]\,dy = 0?$$

L.4) Measure, by hand calculation, the net flow across the circle
$$x^2 + y^2 = 4$$
of a fluid whose velocity is given by the vector field
$$\text{Field}[x, y] = \{x, y\}.$$
(Don't forget that $\sin[t]^2 + \cos[t]^2 = 1$.)

Is the net flow across this curve from inside to outside or outside to inside?

L.5) Measure, by hand calculation, the net flow along the circle
$$x^2 + y^2 = 9$$
of a fluid whose velocity is given by the vector field
$$\text{Field}[x, y] = \{-y, x\}.$$
Is the net flow of Field$[x, y]$ along this curve clockwise or counterclockwise?

L.6) When you go with a given vector field
$$\text{Field}[x, y] = \{m[x, y], n[x, y]\}$$
and a closed curve C, you measure the flow of this vector field across C by calculating the path integral
$$\oint_C -n[x, y]\, dx + m[x, y]\, dy.$$
To calculate
$$\oint_C -n[x, y]\, dx + m[x, y]\, dy,$$
you go with two functions $x[t]$ and $y[t]$ and two numbers, tlow and thigh. And then you calculate
$$\oint_C -n[x, y]\, dx + m[x, y]\, dy = \int_{\text{tlow}}^{\text{thigh}} -n[x[t], y[t]]\, x'[t] + m[x[t], y[t]]\, y'[t]\, dt.$$
What conditions must the two functions, $x[t]$ and $y[t]$, and the two numbers, tlow and thigh, satisfy for you to be sure that your calculation is correct?

L.7) Suppose C_1 and C_2 are physically the same curve, but they are parameterized so that the starting point of C_1 is the ending point of C_2, and the ending point of C_1 is the starting point of C_2. Express
$$\int_{C_2} m[x, y]\, dx + n[x, y]\, dy$$
in terms of
$$\int_{C_1} m[x, y]\, dx + n[x, y]\, dy.$$

L.8) Take a given vector field
$$\text{Field}[x, y] = \{m[x, y], n[x, y]\}.$$
If C_1 and C_2 are two curves that are parameterized to start at a given point $\{a, b\}$ and end at another given point $\{c, d\}$, but are not the same physical curve, what do you look for to determine whether you are guaranteed that
$$\int_{C_1} m[x, y]\, dx + n[x, y]\, dy = \int_{C_2} m[x, y]\, dx + n[x, y]\, dy?$$

L.9) Go with a given vector field

$$\text{Field}[x, y] = \{m[x, y], n[x, y]\}$$

with the extra information that $\text{Field}[x, y]$ is a gradient field. If C is a closed curve, then what do you expect the value of

$$\oint_C m[x, y]\, dx + n[x, y]\, dy$$

to be?

L.10) What do you mean when you say that a given vector field

$$\text{Field}[x, y] = \{m[x, y], n[x, y]\}$$

is a gradient field?

L.11) Given that

$$\text{Field}[x, y] = \{m[x, y], n[x, y]\}$$

is the gradient field of the function

$$f[x, y] = e^{-x^2/2 - y^2/4},$$

exactly what are the formulas for $m[x, y]$ and $n[x, y]$?

L.12) Here are four vector fields:

$$\text{Field1}[x, y] = \{y\, e^{xy}, x\, e^{xy}\},$$
$$\text{Field2}[x, y] = \{y, -x\},$$
$$\text{Field3}[x, y] = \{\sin[2\, x], \cos[5\, y]\}$$

and

$$\text{Field4}[x, y] = \{3\, x - 2\, y, -2\, x + 5\, y\}.$$

All but one of these vector fields are gradient fields. Identify the oddball.

L.13) If $\text{Field}[x, y]$ is a gradient field and C is a closed curve, then what is the net flow of $\text{Field}[x, y]$ along C?

L.14) Two legitimate parameterizations of the circle C with equation $x^2 + y^2 = 16$ are:

Parameterization 1: $\{x[t], y[t]\} = \{4\sin[t], 4\cos[t]\}$ with $0 \le t \le 2\pi$.

Parameterization 2: $\{x[t], y[t]\} = \{4\cos[t], 4\sin[t]\}$ with $0 \le t \le 2\pi$.

Which of these two parameterizations do you use to calculate

$$\oint_C -y\, dx + x\, dy?$$

Calculate this integral by hand.

What does the result tell you about the net flow of Field1$[x, y] = \{-y, x\}$ along this circle?

What does the result tell you about the net flow Field2$[x, y] = \{x, y\}$ across this circle?

L.15) Here are three curves all parameterized to start at $\{0, 0\}$ and end at $\{3, 9\}$:

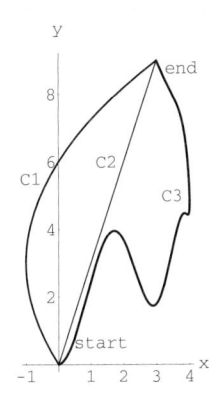

Explain how you know, before calculations begin, that the following three path integrals all calculate out to the same value:

$$\int_{C_1} 3x^2 y^4 \, dx + 4x^3 y^3 \, dy$$

$$\int_{C_2} 3x^2 y^4 \, dx + 4x^3 y^3 \, dy$$

$$\int_{C_3} 3x^2 y^4 \, dx + 4x^3 y^3 \, dy.$$

L.16) Does

$$\text{Field}[x, y] = \{y \sin[x\, y], x \sin[x, y]\}$$

pass the gradient test?

Does

$$\text{Field}[x, y] = \{x \sin[x\, y], y \sin[x, y]\}$$

pass the gradient test?

L.17) Does this vector field

$$\text{Field}[x, y] = \{1 + y, 3 + x\}$$

pass the gradient test?

Come up with a function $f[x, y]$ whose gradient is
$$\text{Field}[x, y] = \{1 + y, 3 + x\}.$$

L.18) You are given:

→ A vector field $\text{Field}[x, y] = \{m[x, y], n[x, y]\}$

→ a closed curve C (like a deformed circle) and

→ a counterclockwise parameterization of C, $P[t] = \{x[t], y[t]\}$ with $P[t]$ sweeping out C exactly one time as t advances from tlow to thigh.

With these ingredients, you can make the following seven measurements:

i) $\oint_C \text{Field} \bullet \text{outerunitnormal} \, ds$

ii) $\oint_C m[x, y] \, dx + n[x, y] \, dy$

iii) $\int_{\text{tlow}}^{\text{thigh}} \text{Field}[x[t], y[t]] \bullet \{x'[t], y'[t]\} \, dt$

iv) $\oint_C \text{Field} \bullet dP$

v) $\oint_C -n[x, y] \, dx + m[x, y] \, dy$

vi) $\oint_C \text{Field} \bullet \text{unittan} \, ds$

vii) $\int_{\text{tlow}}^{\text{thigh}} \text{Field}[x[t], y[t]] \bullet \{y'[t], -x'[t]\} \, dt.$

Some of these measurements measure the net flow of $\text{Field}[x, y]$ along C and some measure the net flow of $\text{Field}[x, y]$ across C.

Which are which?

3.06 Sources, Sinks, Swirls, and Singularities Literacy Sheet

L.1) Go with
$$\text{Field}[x, y] = \{e^x \sin[y], e^x \cos[y]\}$$
and calculate divField$[x, y]$. What does your result tell you about the net flow of Field$[x, y]$ across the ellipse
$$\left(\frac{x}{3}\right)^2 + \left(\frac{y}{7}\right)^2 = 1?$$

L.2) Go with
$$\text{Field}[x, y] = \{-e^x \cos[y], e^x \sin[y]\}$$
and calculate rotField$[x, y]$. What does your result tell you about the net flow of Field$[x, y]$ along the circle
$$(x - 1)^2 + (y - 6)^2 = 49?$$

L.3) Given a point $\{x, y\}$, how does the sign of x tell you whether the given point $\{x, y\}$ is a source or is a sink of the vector field Field$[x, y] = \{x^2 + \sin[y], e^x\}$?

L.4) Is the flow of the vector field
$$\text{Field}[x, y] = \{x^4 - 6x^2 y^2 + 5y^4, -4x^3 y + 4xy^3\}$$
suitable for modelling the flow of an incompressible fluid such as water?

Why or why not?

L.5) Here's the rectangle R with corners at $\{-2, 0\}$, $\{3, 0\}$, $\{3, \pi\}$, and $\{-2, \pi\}$:

Use a 2D integral to measure the net flow of the vector field
$$\text{Field}[x, y] = \{x \cos[y], e^{-x^2/2} + y^2\}$$
across the boundary curve C of this rectangle.

Say why you are happy to make this measurement by calculating your 2D-integral instead of making this measurement by calculating the path integral
$$\oint_C -\left(e^{-x^2/2} + y^2\right) dx + x \cos[y] \, dy.$$

L.6) Take a look at divField$[x, y]$ for the vector field
$$\text{Field}[x, y] = \{\sin[y^2] + x, e^{\sin(x)} + y\}:$$

In[1]:=
```
Clear[x,y,m,n,Field,divField]
{m[x_,y_],n[x_,y_]} = {Sin[y^2] + x,E^(Sin[x]) + y};
Field[x_,y_] = {m[x,y],n[x,y]};
divField[x_,y_] = D[m[x,y],x] + D[n[x,y],y]
```
Out[1]=
2

You look at this and note that divField$[x, y] > 0$ no matter what $\{x, y\}$ is. And then you say to yourself: "Good, this tells me that the flow of this vector field across any closed curve is from inside to outside."

Are you right? Explain your response.

L.7) Here's the rectangle R with corners at $\{-2, -\pi\}$, $\{2, -\pi\}$, $\{2, \pi\}$, and $\{-2, \pi\}$:

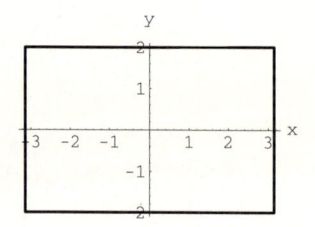

Use a 2D integral to measure the net flow of the vector field
$$\text{Field}[x, y] = \{y \sin[2x], x^2 - e^{-y}\}$$
along the boundary curve C of this rectangle.

Say why you are happy to make this measurement by calculating your 2D-integral instead of making this measuremnt by calculating the path integral
$$\oint_C y \sin[2x]\, dx + \left(x^2 - e^{-y}\right)\, dy.$$

L.8) Take a look at rotField$[x, y]$ for the vector field
$$\text{Field}[x, y] = \{\sin[x] - y, \cos[y] - e^{-2x}\}$$

In[2]:=
```
Clear[x,y,m,n,Field,rotField]
{m[x_,y_],n[x_,y_]} = {Sin[x] - y,Cos[y] - E^(-2 x)};
Field[x_,y_] = {m[x,y],n[x,y]};
rotField[x_,y_] = D[n[x,y],x] - D[m[x,y],y]
```

Out[2]=
$$1 + \frac{2}{E^{2x}}$$

Remembering that $2/e^{2x} = 2e^{-2x}$ can never be negative, you look at this and note that rotField$[x, y] > 0$ no matter what $\{x, y\}$ is. And then you say to yourself: "Good, this tells me that the flow of this vector field along any closed curve is counterclockwise."

Are you right? Explain your response.

L.9) Here's a closed curve:

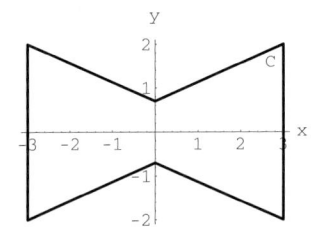

Measure the flow of
$$\text{Field}[x, y] = \{e^x \cos[y], -e^x \sin[y]\}$$
across this curve.

L.10) Here's a curve:

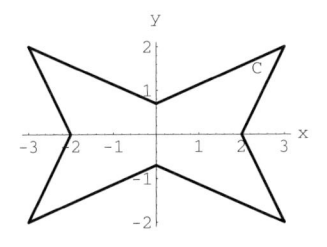

Here's a vector field:

In[3]:=
```
Clear[Field,x,y,m,n]
{m[x_,y_],n[x_,y_]} = {x/(x^2 + y^2),y/(x^2 + y^2)};
Field[x_,y_] = {m[x,y],n[x,y]}
```

Out[3]=
$$\{\frac{x}{x^2+y^2}, \frac{y}{x^2+y^2}\}$$

Here is its divergence, divField$[x,y]$:

In[4]:=
```
Together[D[m[x,y],x] + D[n[x,y],y]]
```

Out[4]=
0

When Calculus Cal saw this, he said, "Good, this tells me that the net flow of Field$[x,y]$ across this curve is 0."

Tell Cal where he made his mistake and show him how to make the calculation correctly.

L.11) Here's a closed curve:

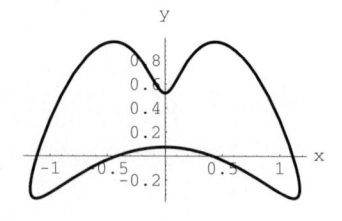

Here is a vector field:

In[5]:=
```
Clear[Field,x,y,m,n]
{m[x_,y_],n[x_,y_]} = {x/(x^2 + y^2),y/(x^2 + y^2)};
Field[x_,y_] = {m[x,y],n[x,y]}
```

Out[5]=
$$\{\frac{x}{x^2+y^2}, \frac{y}{x^2+y^2}\}$$

Here is its divergence, divField$[x,y]$:

In[6]:=
```
Together[D[m[x,y],x] + D[n[x,y],y]]
```

Out[6]=
0

When Calculus Cal saw this, he said, "Good, this tells me that the net flow of Field$[x,y]$ across this curve is 0."

For once, Cal is right. Why is he right?

L.12) Here are two curves:

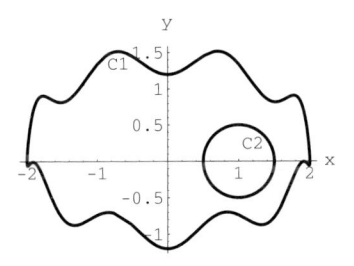

Given functions $m[x, y]$ and $n[x, y]$, what do you check to be sure that
$$\oint_{C_1} m[x, y]\, dx + n[x, y]\, dy = \oint_{C_2} m[x, y]\, dx + n[x, y]\, dy$$
without going to all the trouble to make the individual calculations?

L.13) Here are the same two curves:

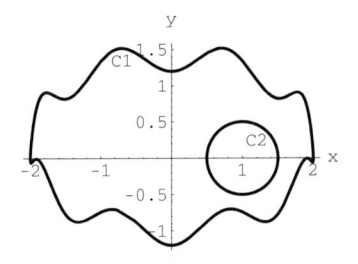

Given a vector field
$$\text{Field}[x, y] = \{m[x, y], n[x, y]\}$$
what do you check to be sure that the net flow of $\text{Field}[x, y]$ across C_1 is the same as the net flow of $\text{Field}[x, y]$ across C_2 without going to all the trouble of making the individual calculations?

L.14) Identify the points $\{x, y\}$ of the vector field
$$\text{Field}[x, y] = \{-x^2 + y, y + 2\sin[\pi x]\}$$
that are sources of new fluid.

Identify the points $\{x, y\}$ of the vector field
$$\text{Field}[x, y] = \{-x^2 + y, y + 2\sin[\pi x]\}$$
that are sinks for old fluid.

L.15) Here's a new curve:

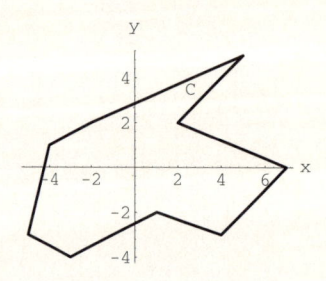

Here are two new functions $m[x, y]$ and $n[x, y]$:

In[7]:=
```
Clear[m,n,x,y]
m[x_,y_] = 3(1 - y)/((x + 2)^2 + (y - 1)^2) + (y + 1)/((x + 2)^2 + (y + 1)^2)
```
Out[7]=
$$\frac{3(1-y)}{(2+x)^2 + (-1+y)^2} + \frac{1+y}{(2+x)^2 + (1+y)^2}$$

In[8]:=
```
n[x_,y_] = 3 (x + 2)/((x + 2)^2 + (y - 1)^2) - (x + 2)/((x + 2)^2 + (y + 1)^2)
```
Out[8]=
$$\frac{3(2+x)}{(2+x)^2 + (-1+y)^2} - \frac{2+x}{(2+x)^2 + (1+y)^2}$$

Note the singularities at $\{-2, 1\}$ and $\{-2, -1\}$ and plot them:

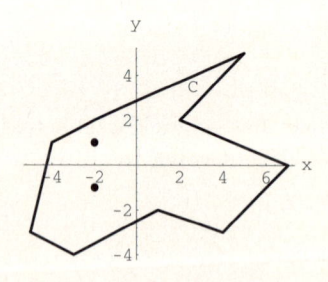

The singularities are inside the region enclosed by C. Now check

$$D[n[x, y], x] - D[m[x, y], y] :$$

In[9]:=
```
Together[D[n[x,y],x] - D[m[x,y],y]]
```

Out[9]=
0

If there were no singularities inside C, then this information would be enough to tell you that

$$\oint_C m[x,y]\,dx + n[x,y]\,dy$$

is 0, but this does not tell you that

$$\oint_C m[x,y]\,dx + n[x,y]\,dy = 0$$

because of the nasty singularities inside the curve C.

But this does tell you that you can express

$$\oint_C m[x,y]\,dx + n[x,y]\,dy$$

in terms of

$$\oint_{C_1} m[x,y]\,dx + n[x,y]\,dy$$

and

$$\oint_{C_2} m[x,y]\,dx + n[x,y]\,dy$$

where C_1 and C_2 are the little circles plotted below:

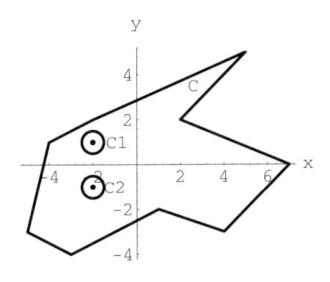

Just how do you express $\oint_C m[x,y]\,dx + n[x,y]\,dy$ in terms of

$$\oint_{C_1} m[x,y]\,dx + n[x,y]\,dy \quad \text{and} \quad \oint_{C_2} m[x,y]\,dx + n[x,y]\,dy?$$

No calculation of any of these three integrals is asked for here.

L.16) Gauss's law in physics says that if you take n different points

$$\{a_1, b_1\}, \{a_2, b_2\}, \ldots, \{a_n, b_n\}$$

inside a closed curve C and you place electrical charges of strength q_1 at $\{a_1, b_1\}$, q_2 at $\{a_2, b_2\}, \ldots, q_n$ at $\{a_n, b_n\}$, then the flux (= flow) of the resulting electric field across C is simply $2\pi(q_1 + q_2 + q_3 + \cdots + q_n)$.

Explain where Gauss's law comes from.

L.17) Suppose Field$[x,y]$ is a gradient field of a function $f[x,y]$. Say what rotField$[x,y]$ is and then say what the net flow of any gradient field along any closed curve is.

L.18) Here's the gradient field of a cleared function $f[x,y]$:

In[10]:=
```
Clear[f,x,y,m,n]
```

In[11]:=
```
Clear[gradField,x,y,m,n]
{m[x_,y_],n[x_,y_]} = {D[f[x,y],x],D[f[x,y],y]};
gradf[x_,y_] = {m[x,y],n[x,y]}
```

Out[11]=
$\{f^{(1,0)}[x, y], f^{(0,1)}[x, y]\}$

The gradient field is such a big deal in science that there are many commonly used notations for it. Here are some:

$$\nabla f[x,y] = \text{gradf}[x,y]$$
$$= \{D[f[x,y],x], D[f[x,y],y]\}$$
$$= \{f^{(1,0)}[x,y], f^{(0,1)}[x,y]\}$$
$$= \left\{\frac{\partial f[x,y]}{\partial x}, \frac{\partial f[x,y]}{\partial y}\right\}.$$

Folks use these notations interchangeably. The Laplacian $\triangle f[x,y]$ of $f[x,y]$ is defined by

$$\triangle f[x,y] = \frac{\partial^2 f[x,y]}{\partial x^2} + \frac{\partial^2 f[x,y]}{\partial y^2}.$$

The Laplacian of $f[x,y]$ is also a big deal in science. Here is *Mathematica*'s calculation of the Laplacian, $\triangle f[x,y]$, of a cleared function $f[x,y]$:

In[12]:=
```
Clear[laplacianf]
laplacianf[x_,y_] = D[f[x,y],{x,2}] + D[f[x,y],{y,2}]
```

Out[12]=
$f^{(0,2)}[x, y] + f^{(2,0)}[x, y]$

One of the reasons that the Laplacian is a big deal is that the Laplacian is related to the gradient field.

Just what is the exact nature of the relationship between the divergence of the gradient field of $f[x,y]$ and the Laplacian of $f[x,y]$?

L.19) Suppose you are going with a function $f[x,y]$ and you know that its Laplacian
$$\triangle f[x,y] = \frac{\partial^2 f[x,y]}{\partial x^2} + \frac{\partial^2 f[x,y]}{\partial y^2} = 0$$
no matter what x and y you go with. Suppose also that $f[x,y]$ has no singularities. Say why the net flow of gradf$[x,y]$ across any closed curve C is 0.

L.20) Suppose you are going with a function $f[x,y]$ and you know that its Laplacian
$$\triangle f[x,y] = \frac{\partial^2 f[x,y]}{\partial x^2} + \frac{\partial^2 f[x,y]}{\partial y^2} = 0$$
no matter what x and y you go with. Suppose also that $f[x,y]$ has no singularities. Explain why it is impossible to find a point $\{x_0, y_0\}$ with the property that
$$f[x_0, y_0] > f[x,y]$$
for all points $\{x,y\}$ near $\{x_0, y_0\}$ but not the same as $\{x_0, y_0\}$. Think of it this way: If there is a point $\{x_0, y_0\}$ with the property that $f[x_0, y_0] > f[x,y]$ for all points $\{x,y\}$ near but not the same as $\{x_0, y_0\}$, then what is the net flow of gradf$[x,y]$ across small circles centered at $\{x_0, y_0\}$?

L.21) What is it about each of the following plots that tells you that none of the plots are plots of functions $f[x,y]$ with the extra property that
$$\triangle f[x,y] = \frac{\partial^2 f[x,y]}{\partial x^2} + \frac{\partial^2 f[x,y]}{\partial y^2} = 0$$
no matter what x and y you go with?

Plot 1:   Plot 2:   Plot 3:

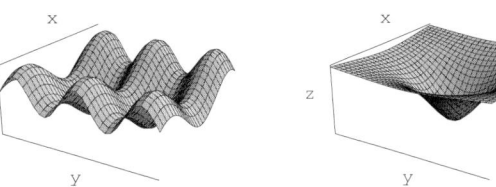

L.22) Go with $f[x,y] = e^{-x}\sin[y]$ and note that its Laplacian,
$$\triangle f[x,y] = \frac{\partial^2 f[x,y]}{\partial x^2} + \frac{\partial^2 f[x,y]}{\partial y^2} = 0,$$
no matter what $\{x,y\}$ you go with.

In[13]:=
```
Clear[f,x,y]
f[x_,y_] = 1 + 0.4 E^(-x) (Cos[y] + Sin[y]);
Expand[D[f[x,y],{x,2}] + D[f[x,y],{y,2}],Trig->True]
```

Out[13]=
0

Here's a plot of the surface $z = f[x, y]$ over (or under) the region R in the xy-plane consisting of all points $\{x, y, 0\}$ with $\{x, y\}$ inside the 2D circle $x^2 + y^2 = 1$:

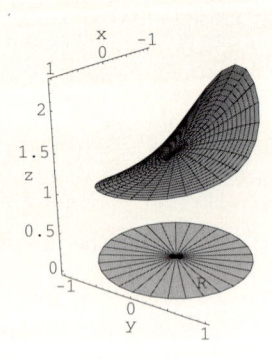

Notice that the highest and lowest plotted points on the surface $z = f[x, y]$ sit directly above the boundary curve of R.

When you take any other function $f[x, y]$ with the extra property that its Laplacian

$$\triangle f[x, y] = \frac{\partial^2 f[x, y]}{\partial x^2} + \frac{\partial^2 f[x, y]}{\partial y^2} = 0$$

no matter what $\{x, y\}$ you go with, then why do you also know in advance that the highest and lowest plotted points on the surface $z = f[x, y]$ will sit directly above the boundary curve of R?

3.07 Transforming 2D Integrals Literacy Sheet

L.1) You are faced with a hand calculation of
$$\iint_R x^2 + y^2 \, dx \, dy,$$
where R is the two-dimensional region consisting of everything inside and on the circle
$$x^2 + y^2 = 9.$$
Switch paper by using polar coordinates
$$x = r\cos[t] \quad \text{and} \quad y = r\sin[t]$$
and then calculate the integral.

L.2) You are faced with a hand calculation of
$$\iint_R x \, dx \, dy,$$
where R is the two-dimensional region consisting of everything bounded by the lines
$$x+y=0, \ x+y=1, \ x-y=0, \text{ and } x-y=1.$$
This is just a little nasty, because doing it without going to better paper involves solving some equations, and then doing two separate integrals. Switch to paper on which you can evaluate this with one sweet integral and then calculate its value.

L.3) Here's a plot of the part of the surface
$$z = e^{-(x^2 + (y/2)^2)}$$
above everything inside and on the ellipse
$$x^2 + \left(\frac{y}{2}\right)^2 = 1$$
in the xy-plane:

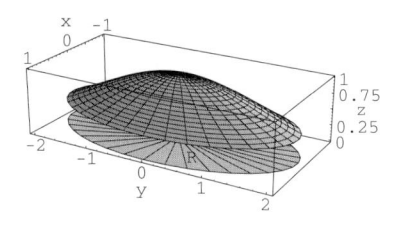

By hand calculation, measure the volume,
$$\iint_R e^{-(x^2 + (y/2)^2)} \, dx \, dy,$$

VECTOR CALCULUS: Measuring in Two and Three Dimensions

of the solid whose top skin is the surface plotted above, and whose base is everything on the xy-plane directly below this surface, by transforming to more favorable paper. (Remember that $\sin[t]^2 + \cos[t]^2 = 1$.)

L.4) The area conversion factor $A_{(x,y)}[u,v]$ is the absolute value of the determinant of a matrix composed of what two gradient vectors?

L.5) All the plots below give information about the same phenomenon, namely about what happens when you plot the region within the uv-paper rectangle consisting of all points $\{u,v\}$ with

$$-2 \leq u \leq 2 \quad \text{and} \quad \frac{1}{2} \leq v \leq 4$$

on xy-paper for

$$x[u,v] = \arctan[u] \quad \text{and} \quad y[u,v] = \log[u^2 + v].$$

Interpret the information conveyed by each plot.

In[1]:=
```
Clear[x,y,u,v,gradx,grady,Axy]
x[u_,v_] = ArcTan[u]; y[u_,v_] = Log[u^2 + v];
gradx[u_,v_] = {D[x[u,v],u],D[x[u,v],v]};
grady[u_,v_] = {D[y[u,v],u],D[y[u,v],v]};
Axy[u_,v_] = Det[{gradx[u,v],grady[u,v]}]
```

Out[1]=

$$\frac{1}{(1+u^2)(u^2+v)}$$

In[2]:=
```
ContourPlot[Axy[u,v],{u,-2,2},{v,1/2,4}];
```

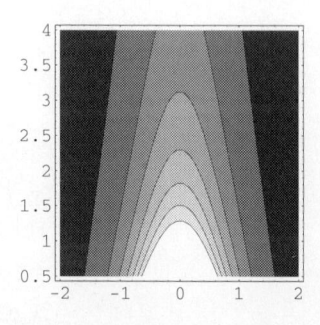

In[3]:=
```
Clear[sizer]; Clear[k]
scalefactor = 0.1;
sizer[u_,v_] = scalefactor Sqrt[Axy[u,v]];
uvpoints = Table[{Random[Real,{-2,2}],
Random[Real,{1/2,4}]},{k,1,300}];
Show[Table[Graphics[{PointSize[
Apply[sizer,uvpoints[[k]]]],
Point[uvpoints[[k]]]}],{k,1, 300}],
Axes->True,AxesLabel->{"u","v"}];
```

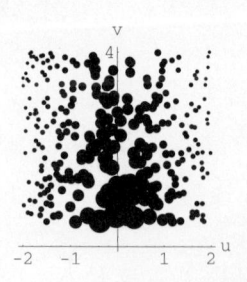

```
In[4]:=
  Clear[gradAxy]
  gradAxy[u_,v_] = {D[Axy[u,v],u],D[Axy[u,v],v]};
  scalefactor = 0.4;
  Show[Table[Arrow[scalefactor gradAxy[u,v],Tail->{u,v}],
  {u,-2,2,1/4},{v,1/2,4,1/4}],Axes->Automatic];
```

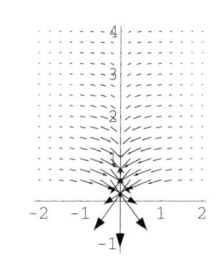

```
In[5]:=
  Plot3D[Axy[u,v],{u,-2,2},{v,1/2,4},
  PlotPoints->30,ViewPoint->CMView,
  AxesLabel->{"u","v","Axy[u,v]"},
  PlotRange->All];
```

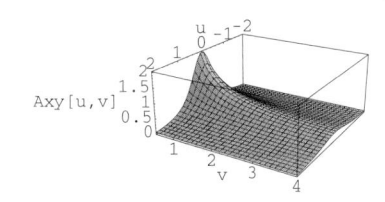

L.6) Say how to linearize a function $f[x,y]$ at a point $\{a,b\}$ and discuss how well the linearized version of $f[x,y]$ at $\{a,b\}$ approximates $f[x,y]$ near $\{a,b\}$.

Does the quality of the approximation improve or deteriorate as you go closer and closer to $\{a,b\}$?

L.7) Briefly explain how linearizations can be used to come up with the formula for the area conversion factor.

L.8) Look at:

```
In[6]:=
  Clear[u,v,x,y]
  {x[u_,v_],y[u_,v_]} = {3 u^2,u^2 + v^2};
  Clear[gradx,grady,Axy]
  gradx[u_,v_] = {D[x[u,v],u],D[x[u,v],v]};
  grady[u_,v_] = {D[y[u,v],u],D[y[u,v],v]};
  Axy[u_,v_] = Det[{gradx[u,v],grady[u,v]}]
Out[6]=
  12 u v
```

The uv-paper circle
$$(u-1)^2 + (v-1)^2 = 2$$
goes right through $\{0,0\}$:

```
In[7]:=
  Clear[t]
  ParametricPlot[
  {1 + Sqrt[2] Cos[t],1 + Sqrt[2] Sin[t]},{t,0,2 Pi},
  PlotStyle->{{Thickness[0.01]}},
  AspectRatio->Automatic,AxesLabel->{"u","v"}];
```

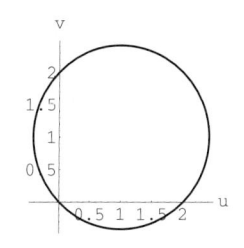

Now look at the plot of this circle on xy-paper with $x[u,v]$ and $y[u,v]$ as given above:

In[8]:=
```
Clear[t]
ParametricPlot[
 {x[1 + Sqrt[2] Cos[t],1 + Sqrt[2] Sin[t]],
  y[1 + Sqrt[2] Cos[t],1 + Sqrt[2] Sin[t]]},
 {t,0,2 Pi},PlotStyle->{{Thickness[0.01]}},
 PlotRange->All,AxesLabel->{"x","y"}];
```

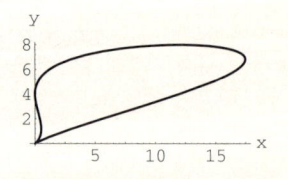

Why do you think the corner popped up when you went from xy-paper to uv-paper?

L.9) To calculate the net flow of a vector field

$$\text{Field}[x,y] = \{m[x,y], n[x,y]\}$$

across the boundary C of a region R, you have your choice:

$\rightarrow$ You can go to the labor of parameterizing C and then calculate

$$\oint_C -n[x,y]\, dx + m[x,y]\, dy.$$

$\rightarrow$ Or, if the field has no singularities inside R, you can put

$$\text{divField}[x,y] = D[m[x,y],x] + D[n[x,y],y]$$

and calculate the 2D-integral

$$\iint_R \text{divField}[x,y]\, dx\, dy.$$

Here's a vector field:

In[9]:=
```
Clear[x,y,m,n,Field]
{m[x_,y_],n[x_,y_]} = {-x^2 + Cos[y] ,E^x + y^2};
```

Agree that R is everything inside and on the parallelogram you see below:

In[10]:=
```
Plot[{-x ,-x + 3,
 0.5 x,0.5 x + 4}, {x,-2.67,2.0},
 PlotStyle->{{Thickness[0.01]}},
 PlotRange->{0,3.68},AspectRatio->Automatic,
 AxesLabel->{"x","y"}];
```

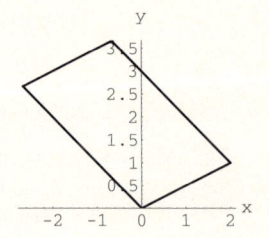

Transform the 2D-integral

$$\iint_R \text{divField}[x,y]\, dx\, dy$$

to favorable uv-paper to measure the net flow of this vector field across the parallelogram.

Is the net flow of this vector field across the boundary of this parallelogram from outside to inside, or inside to outside?

L.10) To calculate the net flow of a vector field
$$\text{Field}[x,y] = \{m[x,y], n[x,y]\}$$
along the boundary C of a region R, you have your choice:

→ You can go to the labor of parameterizing C and then calculate
$$\oint_C m[x,y]\,dx + n[x,y]\,dy$$

→ or, if the field has no singularities inside R, you can put
$$\text{rotField}[x,y] = D[n[x,y], x] - D[m[x,y], y]$$
and calculate the 2D-integral
$$\iint_R \text{rotField}[x,y]\,dx\,dy.$$

Here's a vector field:

In[11]:=
```
Clear[x,y,m,n,Field]
{m[x_,y_],n[x_,y_]} = {(y - 2)^2, x - y};
```

Agree that R is everything inside the parallelogram plotted in the last problem. Transform the 2D integral
$$\iint_R \text{rotField}[x,y]\,dx\,dy$$
to favorable uv-paper to measure the net flow of this vector field along the parallelogram.

Is the net flow of this vector field along this parallelogram clockwise or counterclockwise?

L.11) When you start with the ellipse
$$\left(\frac{x}{a}\right)^2 + \left(\frac{y}{b}\right)^2 = r^2$$
on xy-paper and you go to uv-paper using
$$u[x,y] = \frac{x}{a} \quad \text{and} \quad v[x,y] = \frac{y}{b},$$
then the xy-ellipse plots out as the circle $u^2 + v^2 = r^2$. On uv-paper, the area enclosed by this circle measures out to πr^2 square units.

Calculate the area conversion factor $A_{(u,v)}[x,y]$ and use the result to explain why the xy-area measurement of the region enclosed by the ellipse

$$\left(\frac{x}{a}\right)^2 + \left(\frac{y}{b}\right)^2 = r^2$$

is $ab\pi r^2$ square units.

3.08 Transforming 3D Integrals Literacy Sheet

L.1) R is the cube with sides of length 2 centered at $\{1, 0, 2\}$. It looks like this:

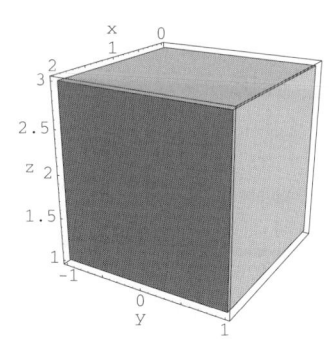

Calculate $\iiint_R x \, dx \, dy \, dz$.

L.2) R is the solid whose boundary surfaces are the xy-plane, the yz-plane, the xz-plane, and the plane
$$2x + 2y + 3z = 12.$$
Here are two looks at R:

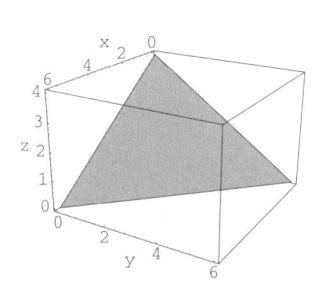

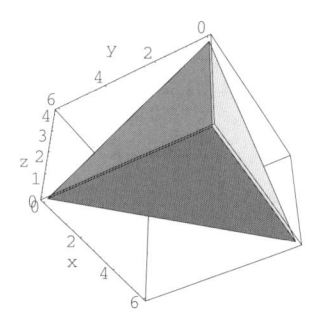

The viewpoint on the left is the usual CMView; the viewpoint on the right is a look from underneath.

Calculate $\iiint_R (x + z) \, dx \, dy \, dz$.

L.3) The volume of a certain solid R is measured by calculating
$$\iiint_R dx \, dy \, dz = \int_0^1 \int_0^2 \int_0^3 dx \, dy \, dz.$$
Describe the solid and measure its volume in cubic units.

L.4) All the linear dimensions of a certain solid are tripled (multiplied by 3). What is the ratio of the new volume to the original volume?

L.5) When you start with the ellipsoid (egg)
$$\left(\frac{x}{a}\right)^2 + \left(\frac{y}{b}\right)^2 + \left(\frac{z}{c}\right)^2 = r^2$$
in xyz-space and you go to uvw-space using:
$$u[x,y,z] = \frac{x}{a}, v[x,y,z] = \frac{y}{b} \quad \text{and} \quad w[x,y,z] = \frac{z}{c},$$
then the xyz-ellipsoid plots out as the sphere $u^2 + v^2 + w^2 = r^2$.

In uvw-space, the volume enclosed by this sphere measures out to $4\pi r^3/3$ cubic units.

Calculate the volume conversion factor $V_{(u,v,w)}[x,y,z]$ and use the result to explain why the xyz-space volume-measurement of the region enclosed by the ellipsoid
$$\left(\frac{x}{a}\right)^2 + \left(\frac{y}{b}\right)^2 + \left(\frac{z}{c}\right)^2 = r^2$$
is $4\,a\,b\,c\,\pi\,r^3/3$ cubic units.

L.6) Look at this:

In[1]:=
```
Clear[x,y,z,r,s,t]
{x[r_,s_,t_],y[r_,s_,t_],z[r_,s_,t_]} =
{2 r Cos[t],s (2 - r) (2 + 0.5 Sin[t]),3 r Sin[t]};
rightskin = ParametricPlot3D[{x[r,1,t],y[r,1,t],
z[r,1,t]},{r,0,1},{t,0,2 Pi},DisplayFunction->Identity];
leftskin = ParametricPlot3D[{x[r,-1,t],y[r,-1,t],
z[r,-1,t]},{r,0,1},{t,0,2 Pi},DisplayFunction->Identity];
middleskin = ParametricPlot3D[{x[1,s,t],y[1,s,t],
z[1,s,t]},{s,-1,1},{t,0,2 Pi},DisplayFunction->Identity];
wholeskin = Show[leftskin,rightskin,middleskin,
PlotRange->All,ViewPoint->CMView,AxesLabel->{"x","y","z"},
DisplayFunction->$DisplayFunction];
```

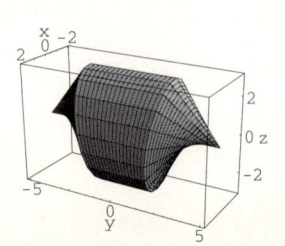

And a look from behind:

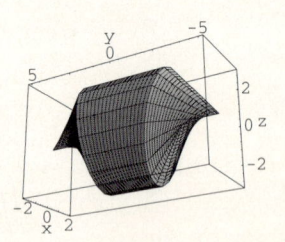

In[2]:=
```
Show[wholeskin,ViewPoint->-CMView];
```

Now look at this:

In[3]:=
```
Clear[gradx,grady,gradz,Vxyz]
gradx[r_,s_,t_] = {D[x[r,s,t],r],D[x[r,s,t],s],D[x[r,s,t],t]};
grady[r_,s_,t_] = {D[y[r,s,t],r],D[y[r,s,t],s],D[y[r,s,t],t]};
gradz[r_,s_,t_] = {D[z[r,s,t],r],D[z[r,s,t],s],D[z[r,s,t],t]};
Vxyz[r_,s_,t_] = Expand[Det[{gradx[r,s,t],grady[r,s,t],
  gradz[r,s,t]}],Trig->True]
```

Out[3]=
$$24.\, r - 12.\, r^2 + 6.\, r\, \mathrm{Sin}[t] - 3.\, r^2\, \mathrm{Sin}[t]$$

Now look at this:

In[4]:=
```
Plot3D[Vxyz[r,s,t],{r,0,1},{t,0,2 Pi},
ViewPoint->CMView,
AxesLabel->{"r","t","Vxyz[r,s,t]"}];
```

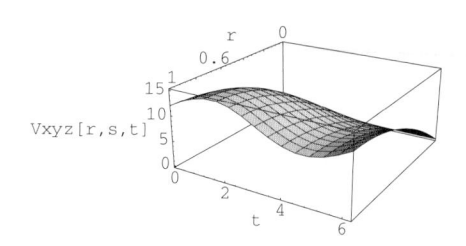

The plot tells you that $V_{(x,y,z)}[r,s,t] \geq 0$ no matter what r, s, and t are.

Now you are in the catbird seat because you have enough information to give a short, sweet, hand calculation of the volume enclosed by the skin plotted above. Do it.

L.7) Given a transformation coming from
$$x = x[u, v, w],$$
$$y = y[u, v, w],$$
and
$$z = z[u, v, w]$$
say how you use
$$\mathrm{gradx}[u, v, w],$$
$$\mathrm{grady}[u, v, w],$$
and
$$\mathrm{gradz}[u, v, w]$$
to calculate $V_{(x,y,z)}[u, v, w]$.

Then discuss the meaning of $V_{(x,y,z)}[u, v, w]$.

L.8) Give a hand calculation of
$$\iiint_{R_{xyz}} y\, dx\, dy\, dz$$

where R_{xyz} is the region inside and on the "box" bounded by the planes:

$z = 2\,x$ (bottom),
$z = 2\,x + 3$ (top),
$y = x$ (side),
$y = x + 2$ (side),
$y = -2\,x$ (side), and
$y = -2\,x + 4$ (side).

3.09 Spherical Coordinates Literacy Sheet

L.1) You can specify a point in three dimensions as usual by its three coordinates, $\{x, y, z\}$.

Another way to specify a point is to plant the x-, y- and z-axes at $\{0, 0, 0\}$ and then run a stick from $\{0, 0, 0\}$ to the point:

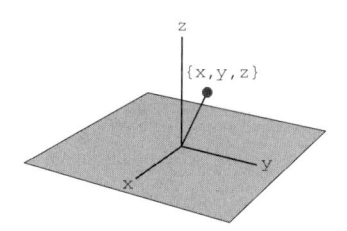

The spherical coordinates $\{r, s, t\}$ of the plotted point are specified by the equations:
$$x = r \sin[s] \cos[t],$$
$$y = r \sin[s] \sin[t],$$
and
$$z = r \cos[s].$$

Take out a pencil and label the physical meaning of the measurements r, s, and t on the plot above.

L.2) Set numbers r^*, slow, shigh, tlow, and thigh so that the xyz-points
$$\{x[s, t], y[s, t], z[s, t]\} = \{r^* \sin[s] \cos[t], r^* \sin[s] \sin[t], r^* \cos[s]\}$$
with
$$\text{slow} \le s \le \text{shigh} \quad \text{and} \quad \text{tlow} \le t \le \text{thigh}$$
describe the whole sphere
$$x^2 + y^2 + z^2 = 16^2$$
of radius r centered at $\{0, 0, 0\}$.

L.3) Set numbers r^*, slow, shigh, tlow, and thigh so that the xyz-points
$$\{x[s, t], y[s, t], z[s, t]\} = \{r^* \sin[s] \cos[t], r^* \sin[s] \sin[t], r^* \cos[s]\}$$
with
$$\text{slow} \le s \le \text{shigh} \quad \text{and} \quad \text{tlow} \le t \le \text{thigh}$$

describe the part of the sphere
$$x^2 + y^2 + z^2 = 9$$
consisting of those points with $z \geq 0$.

L.4) Set numbers r^*, slow, shigh, tlow, and thigh so that the xyz-points
$$\{x[s,t], y[s,t], z[s,t]\} = \{r^* \sin[s] \cos[t], r^* \sin[s] \sin[t], r^* \cos[s]\}$$
with
$$\text{slow} \leq s \leq \text{shigh} \quad \text{and} \quad \text{tlow} \leq t \leq \text{thigh}$$
describe the part of the sphere
$$x^2 + y^2 + z^2 = 25$$
consisting of those points with $x \geq 0$.

L.5) Set numbers slow, shigh, tlow, and thigh so that the xyz-points
$$\{x[s,t], y[s,t], z[s,t]\} = \{2\sin[s] \cos[t], 2\sin[s] \sin[t], 2\cos[s]\}$$
with
$$\text{slow} \leq s \leq \text{shigh} \quad \text{and} \quad \text{tlow} \leq t \leq \text{thigh}$$
describe something like this:

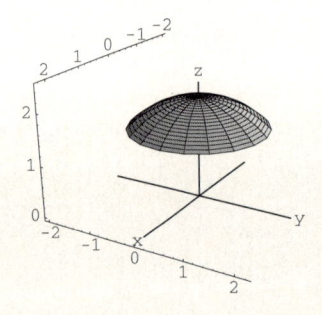

L.6) Set numbers slow, shigh, tlow, and thigh so that the xyz-points
$$\{x[s,t], y[s,t], z[s,t]\} = \{2\sin[s] \cos[t], 2\sin[s] \sin[t], 2\cos[s]\}$$
with
$$\text{slow} \leq s \leq \text{shigh} \quad \text{and} \quad \text{tlow} \leq t \leq \text{thigh}$$
describe something like this:

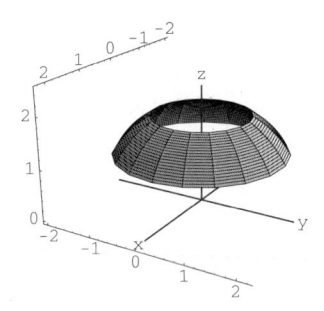

L.7) Set numbers s^*, rlow, rhigh, tlow and thigh so that the xyz-points
$$\{x[r,t], y[r,t], z[r,t]\} = \{r\sin[s^*]\cos[t], r\sin[s^*]\sin[t], r\cos[s^*]\}$$
with
$$\text{rlow} \leq r \leq \text{rhigh} \quad \text{and} \quad \text{tlow} \leq t \leq \text{thigh}$$
describe something like this:

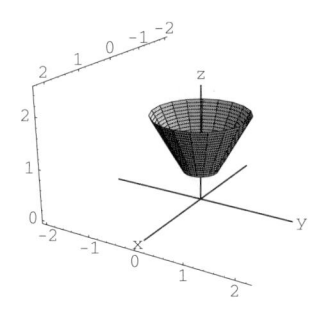

L.8) Set numbers s^*, rlow, rhigh, tlow and thigh so that the xyz-points
$$\{x[r,t], y[r,t], z[r,t]\} = \{r\sin[s^*]\cos[t], r\sin[s^*]\sin[t], r\cos[s^*]\}$$
with
$$\text{rlow} \leq r \leq \text{rhigh} \quad \text{and} \quad \text{tlow} \leq t \leq \text{thigh}$$
describe something like this:

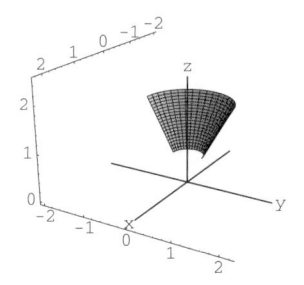

L.9) Set numbers t^*, rlow, rhigh, slow and shigh so that the xyz-points
$$\{x[r,t], y[r,t], z[r,t]\} = \{r\sin[s]\ \cos[t^*], r\sin[s]\ \sin[t^*], r\cos[s]\}$$
with
$$\text{rlow} \leq r \leq \text{rhigh} \quad \text{and} \quad \text{slow} \leq s \leq \text{shigh}$$
describe something like this:

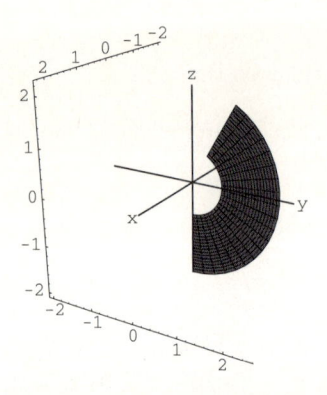

L.10) You go to a library reference and learn that the volume of a sphere of radius r measures out to $4\pi r^3/3$ cubic units. You think about this for a second and then you type and execute:

In[1]:=
```
Clear[x,y,z,r,s,t]
x[r_,s_,t_] = r Sin[s] Cos[t]; y[r_,s_,t_] = r Sin[s] Sin[t];
z[r_,s_,t_] = r Cos[s]; Clear[gradx,grady,gradz,Vxyz]
gradx[r_,s_,t_] = {D[x[r,s,t],r],D[x[r,s,t],s],D[x[r,s,t],t]};
grady[r_,s_,t_] = {D[y[r,s,t],r],D[y[r,s,t],s],D[y[r,s,t],t]};
gradz[r_,s_,t_] = {D[z[r,s,t],r],D[z[r,s,t],s],D[z[r,s,t],t]};
Vxyz[r_,s_,t_] = Expand[Det[{gradx[r,s,t],grady[r,s,t],gradz[r,s,t]}],Trig->True]
```
Out[1]=
$$r^2\ \text{Sin}[s]$$

And then you say that you can give your own hand calculation that explains where the formula $4\pi r^3/3$ comes from.

Do it.

L.11) Given a radius r, you can plot the sphere
$$x^2 + y^2 + z^2 = 1$$
by going with spherical coordinates, and plotting
$$\{\sin[s]\ \cos[t], \sin[s]\ \sin[t], \cos[s]\},$$
running s from 0 to π and running t from 0 to 2π.

Given positive numbers a, b, and c, say how you go about plotting the ellipsoid (egg)
$$\left(\frac{x}{a}\right)^2 + \left(\frac{y}{b}\right)^2 + \left(\frac{z}{c}\right)^2 = 1.$$

L.12) Comment on the following statement: "Spherical coordinates have to be in the bag of tricks of any serious plotting artist."

L.13) Use spherical coordinates to pull off a very quick hand calculation of
$$\iiint_{R_{xyz}} (x^2 + y^2 + z^2) \, dx \, dy \, dz$$
where R_{xyz} is everything inside and on the sphere
$$x^2 + y^2 + z^2 = a^2.$$

3.10 3D Surface Measurements Literacy Sheet

L.1) Give the formula for the divergence, divField$[x, y, z]$, of a three-dimensional vector field
$$\text{Field}[x, y, z] = \{m[x, y, z], n[x, y, z], p[x, y, z]\}.$$
Say how the sign (positive or negative) of divField$[x, y, z]$ tells you whether $\{x, y, z\}$ is a source or a sink of the three-dimensional flow represented by $F[x, y, z]$.

L.2) How do you know that
$$\text{Field}[x, y, z] = \{y\,z - x, \sin[z] - 2\,y, \cos[x\,y] - 5\,z^3\}$$
has sinks at every point $\{x, y, z\}$?

How do you know that the net flow of this vector field across the skin of any solid surface is from outside to inside?

L.3) Determine the sources and sinks of
$$\text{Field}[x, y, z] = \{3\,z - 3\,x^2, 3\,z + 2\,y, 4\,z\}.$$
Eyeball your answer and then respond quickly to the following questions.

If C_1 is the sphere of radius 2 centered at $\{4, 0, 0\}$, is the net flow of this vector field across C_1 from outside to inside, or is it from inside to outside?

If C_2 is the sphere of radius 2 centered at $\{-3, 0, 0\}$, is the net flow of this vector field across C_2 from outside to inside, or is it from inside to outside?

L.4) Here's a vector field:

In[1]:=
```
Clear[x,y,z,m,n,p,Field]
{m[x_,y_,z_],n[x_,y_,z_],p[x_,y_,z_]} = {x^2 + E^(-y),y^2,x - 2 y z^2};
Field[x_,y_,z_] = {m[x,y,z],n[x,y,z],p[x,y,z]}
```

Out[1]=
$$\{E^{-y} + x^2,\ y^2,\ x - 2\,y\,z^2\}$$

Use Gauss's formula to help come up with a measurement of the flow of this vector field across the surface of the three dimensional box consisting of all points $\{x, y, z\}$ with
$$0 \leq x \leq 2, \quad -1 \leq y \leq 3, \quad \text{and} \quad 0 \leq z \leq 3.$$
Is the net flow of this vector field across this skin from inside to outside, from outside to inside, or 0?

L.5) Suppose C is the skin of a solid region in three dimensions. Suppose Field$[x, y, z]$ is a given vector field with divField$[x, y, z] = 0$ at all points $\{x, y, z\}$ and suppose that this vector field has no singularities inside C.

Explain how you know that the net flow of this vector field across C is 0.

L.6) Suppose C is the skin of a solid region in three dimensions. Suppose Field$[x, y, z]$ is a given vector field with divField$[x, y, z] > 0$ at all points $\{x, y, z\}$ inside and on C and suppose that this vector field has no singularities inside C.

Explain how you know that the net flow of this vector field across C is from inside to outside.

L.7) Here's the gradient field of a cleared function $f[x, y, z]$:

In[2]:=
```
Clear[f,gradf,x,y,z,m,n,p]
{m[x_,y_,z_],n[x_,y_,z_],p[x_,y_,z_]} = {D[f[x,y,z],x],D[f[x,y,z],y],D[f[x,y,z],z]};
gradf[x_,y_,z_] = {m[x,y,z],n[x,y,z],p[x,y,z]}
```

Out[2]=
$$\{f^{(1,0,0)}[x, y, z], f^{(0,1,0)}[x, y, z], f^{(0,0,1)}[x, y, z]\}$$

The gradient field is such a big deal in science that there are many commonly used notations for it. Here are some:

$$\nabla f[x, y, z] = \text{gradf}[x, y, z]$$
$$= \{D[f[x, y, z], x], D[f[x, y, z], y], D[f[x, y, z], z]\}$$
$$= \{f^{(1,0,0)}[x, y, z], f^{(0,1,0)}[x, y, z], f^{(0,0,1)}[x, y, z]\}$$
$$= \left\{ \frac{\partial f[x, y, z]}{\partial x}, \frac{\partial f[x, y, z]}{\partial y}, \frac{\partial f[x, y, z]}{\partial z} \right\}.$$

Folks use these notations interchangeably.

The Laplacian $\triangle f[x, y, z]$ of $f[x, y, z]$ is defined by

$$\triangle f[x, y] = \frac{\partial^2 f[x, y, z]}{\partial x^2} + \frac{\partial^2 f[x, y, z]}{\partial y^2} + \frac{\partial^2 f[x, y, z]}{\partial z^2}.$$

The Laplacian of $f[x, y, z]$ is also a big deal in science. Here is *Mathematica*'s calculation of the Laplacian, $\triangle f[x, y, z]$, of a cleared function $f[x, y, z]$:

In[3]:=
```
Clear[laplacianf]
laplacianf[x_,y_,z_] = D[f[x,y,z],{x,2}] + D[f[x,y,z],{y,2}] + D[f[x,y,z],{z,2}]
```

Out[3]=
$$f^{(0,0,2)}[x, y, z] + f^{(0,2,0)}[x, y, z] + f^{(2,0,0)}[x, y, z]$$

One of the reasons that the Laplacian is a big deal is that it is related to the gradient field.

Just what is the exact nature of the relationship between the divergence of the gradient field of $f[x, y, z]$ and the Laplacian of $f[x, y, z]$?

L.8) Suppose you are going with a function $f[x, y, z]$ and you know that its Laplacian

$$\triangle f[x, y, z] = \frac{\partial^2 f[x, y, z]}{\partial x^2} + \frac{\partial^2 f[x, y, z]}{\partial y^2} + \frac{\partial^2 f[x, y, z]}{\partial z^2} = 0$$

no matter what x, y and z you go with. Suppose also that $f[x, y, z]$ has no singularities.

Say how you know that the net flow of gradf$[x, y, z]$ across the skin C of the surface of any given 3D solid is 0.

L.9) Suppose you are going with a function $f[x, y, z]$ and you know that its Laplacian

$$\Delta f[x, y, z] = \frac{\partial^2 f[x, y, z]}{\partial x^2} + \frac{\partial^2 f[x, y, z]}{\partial y^2} + \frac{\partial^2 f[x, y, z]}{\partial z^2} = 0$$

no matter what x, y and z you go with. Suppose also that $f[x, y, z]$ has no singularities.

Explain why it is impossible to find a point $\{x, y, z\}$ with the property that

$$f[x_0, y_0, z_0] > f[x, y, z]$$

for all points $\{x, y, z\}$ near $\{x_0, y_0, z_0\}$, but not the same as $\{x_0, y_0, z_0\}$.

Think of it this way: If there is a point $\{x_0, y_0, z_o\}$ with the property that

$$f[x_0, y_0, z_0] > f[x, y, z]$$

for all points $\{x, y, z\}$ near $\{x_0, y_0, z_0\}$ but not the same as $\{x_0, y_0, z_0\}$, then what is the net flow of gradf$[x, y, z]$ across small spheres centered at $\{x_0, y_0, z_0\}$?

L.10) A large solid region in three dimensions represents a big rock. Part of the surface of the rock is kept at a prescribed temperature—hotter at one point than at another. The remainder of the surface is perfectly insulated.

You wait until the temperature inside the rock settles into its steady state condition. Say that temp$[x, y, z]$ represents the steady state temperature at a position $\{x, y, z\}$ inside the rock. In the steady state, no point inside the rock and not on the surface can be a source of new heat flow or a sink for old heat.

Why does this tell you that

$$\Delta \text{temp}[x, y, z] = \frac{\partial^2 \text{temp}[x, y, z]}{\partial x^2} + \frac{\partial^2 \text{temp}[x, y, z]}{\partial y^2} + \frac{\partial^2 \text{temp}[x, y, z]}{\partial z^2} = 0$$

at each point $\{x, y, z\}$ inside but not on the surface of the rock?

Explain why the hottest and the coldest locations of the rock must be on the outside skin of the rock, and not inside the rock.

L.11) C_1 is the top half of a distorted ellipsoid centered at the origin. It is parameterized as indicated:

In[4]:=
```
Clear[x1,y1,z1,s,t]
x1[s_,t_] = 3 Sin[s] Cos[t];
y1[s_,t_] = 2 Sin[s] Sin[t];
z1[s_,t_] = (2 - Sin[6 s^2]) Cos[s];
C1plot = ParametricPlot3D[{x1[s,t],y1[s,t],z1[s,t]},
  {s,0,Pi/2},{t,0,2 Pi},PlotPoints->{30,30},
  AxesLabel->{"x","y","z"},ViewPoint->CMView,
  Boxed->False];
```

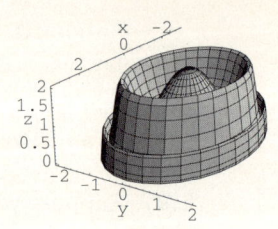

C_2 is just a plain elliptical disk in the xy-plane parameterized as indicated:

In[5]:=
```
Clear[x2,y2,z2,r,t]
x2[r_,t_] = 3 r Cos[t];
y2[r_,t_] = 2 r Sin[t];
z2[r_,t_] = 0;
C2plot = ParametricPlot3D[{x2[r,t],y2[r,t],z2[r,t]},
  {r,0,1},{t,0,2 Pi},PlotPoints->{2,30},
  AxesLabel->{"x","y","z"},ViewPoint->CMView,
  Boxed->False];
```

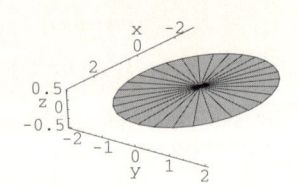

Note that C_1 and C_2 have the same boundary curve:

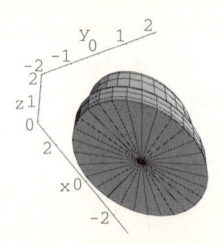

In[6]:=
```
Show[C1plot,C2plot,ViewPoint->{4,2,-4}];
```

Explain how you know that when you go with a vector field Field[x, y, z] with divField[x, y, z] = 0 throughout the solid region whose top skin is C_1 and whose bottom skin is C_2, then

$$\iint_{C_1} \text{Field} \bullet \text{topunitnormal} \, dA = \iint_{C_2} \text{Field} \bullet \text{topunitnormal} \, dA;$$

so that the flow of this vector field across both surfaces is the same.

What calculational nightmare does this help you avoid?

L.12) You are given the formula for a vector field Field[x, y, z]. The first thing you notice is that this vector field has a singularity at $\{0, 0, 0\}$. The second thing you do is calculate divField[x, y, z] and you are happy to learn that Field[x, y, z] has no sources or sinks other than at the singularity at $\{0, 0, 0\}$.

Your ultimate goal is to measure the flow of this vector field across the skin plotted below:

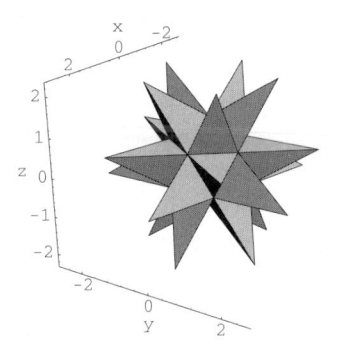

The point $\{0, 0, 0\}$ is inside this skin, but the exact parameterization of this skin is disgustingly tedious.

Outline the steps you would take to measure the flow of the given vector field across this skin.

L.13) You are given the formula for a vector field $\text{Field}[x, y, z]$. The first thing you notice is that this vector field has a singularity at $\{0.8, -0.5, 1.0\}$. The second thing you do is calculate $\text{divField}[x, y, z]$ and you are happy to learn that $\text{Field}[x, y, z]$ has no sources or sinks other than at the singularity at $\{0.8, -0.5, 1.0\}$.

Your ultimate goal is to measure the flow of this vector field across the skin plotted below:

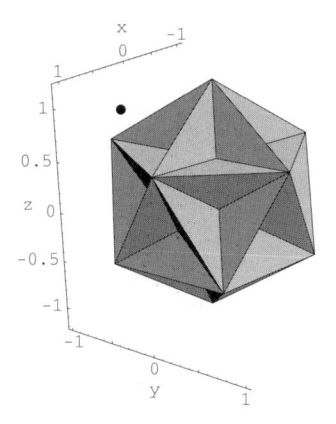

This is the skin of a solid region and the plotted point, $\{0.8, -0.5, 1.0\}$, is outside this skin. The exact parameterization of this skin is disgustingly tedious.

Outline the steps you would take to measure the flow of the given vector field across this skin.

L.14) Discuss the statement: "Part of mathematics is the art of replacing heavy calculations by simple calculations."

L.15) When you place an electrical charge of strength q at a point $\{a, b, c\}$ and you center a small sphere at $\{a, b, c\}$, then the flux (= flow) of the resulting electric field across the skin of the sphere is $4\pi q$. Look at this skin of a solid region:

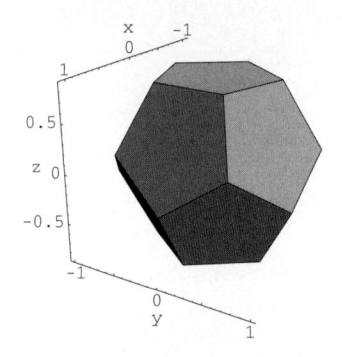

Place an electrical charge of strength q at a point $\{a, b, c\}$ inside this skin and measure the flux (= flow) of the resulting electric field across this skin.

L.16) When you place an electrical charge of strength q at a point $\{a, b, c\}$ and you center a small sphere at $\{a, b, c\}$, then the flux (= flow) of the resulting electric field across this small sphere is $4\pi q$.

Gauss's law in physics says that if you have a surface C that is the boundary surface of a solid in three dimensions, and if you take k different points:

$$\{a_1, b_1, c_1\}, \{a_2, b_2, c_2\}, \ldots, \{a_k, b_k, c_k\}$$

inside C and you place electrical charges of strength

$$q_1 \text{ at } \{a_1, b_1, c_1\}, q_2 \text{ at } \{a_2, b_2, c_2\}, \ldots, q_k \text{ at } \{a_k, b_k, c_k\},$$

then the flux (= flow) of the resulting electric field across C is simply

$$4\pi (q_1 + q_2 + q_3 + \cdots + q_k).$$

Explain how Gauss's law is based on Gauss's formula.

L.17) Given a surface in parametric form $\{x[u, v], y[u, v], z[u, v]\}$, the two vectors

$$\tan1[u, v] = D[\{x[u, v], y[u, v], z[u, v]\}, u] = \left\{\frac{\partial x[u, v]}{\partial u}, \frac{\partial y[u, v]}{\partial u}, \frac{\partial z[u, v]}{\partial u}\right\}$$

and
$$\tan 2[u,v] = D[\{x[u,v], y[u,v], z[u,v]\}, v] = \left\{\frac{\partial x[u,v]}{\partial v}, \frac{\partial y[u,v]}{\partial v}, \frac{\partial z[u,v]}{\partial v}\right\}$$

are both tangent to the surface when their tails are put at the point

$$\{x[u,v], y[u,v], z[u,v]\}.$$

As a result, the cross product

$$\text{normal}[u,v] = \tan 1[u,v] \times \tan 2[u,v]$$

is perpendicular to the surface when its tail is put at the point $\{x[u,v], y[u,v], z[u,v]\}$.

What does the length of normal$[u,v]$ measure?

L.18) Given numbers a, b, and c, you can parameterize the plane

$$z = ax + by + c$$

by going with

$$\{x[u,v], y[u,v], z[u,v]\} = \{u, v, au + bv + c\}.$$

Calculate the area conversion factor

$$SA_{(x,y,z)}[u,v]$$

by hand and use it to give a formula for the area of that part of the plane

$$z = ax + by + c$$

that sits directly above (or below) the disk

$$x^2 + y^2 \leq r^2$$

in the xy-plane.

3.11 3D Flow Along Literacy Sheet

L.1) A 3D vector field $\text{Field}[x, y, z] = \{m[x, y, z], n[x, y, z], p[x, y, z]\}$ comes into your hands. You calculate $\text{curlField}[1, 0, 0]$ and learn

$$\text{curlField}[1, 0, 0] = \{-2, 0, 1\}.$$

You stick the tail of the unit vector

$$V = \left\{\frac{1}{\sqrt{2}}, \frac{1}{\sqrt{2}}, 0\right\}$$

at the point $\{1, 0, 0\}$ and push your finger onto V so that V spikes through the center of your finger, and the tip of your finger is at $\{1, 0, 0\}$. Does the tip of your finger feel a net clockwise or counterclockwise swirl?

What happens when you take $V = \{0, 1/\sqrt{2}, 1/\sqrt{2}\}$?

What unit vector V do you take if you want to feel the biggest possible net counterclockwise swirl at the point $\{1, 0, 0\}$?

What unit vector V do you take if you want to feel the biggest possible net clockwise swirl at the point $\{1, 0, 0\}$?

L.2) You are given a certain 3D vector field $\text{Field}[x, y, z]$. You calculate $\text{curlField}[1, 2, 1]$ and show it with a little disk plotted in the plane through $\{1, 2, 1\}$ that is perpendicular to $\text{curlField}[1, 2, 1]$. It turns out like this:

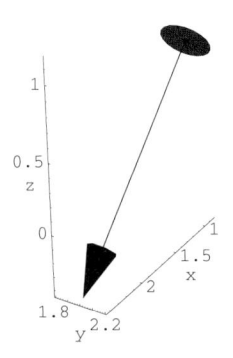

(From this viewpoint, the disk does not block your view of any part of the vector.) The question is, which of the following plots do you expect to indicate the direction of the net flow of the given vector field along the boundary curve of the disk?

L.3) Here is a calculation of

$$\text{curlField}[0.3, 0.9, 0.2] \bullet V$$

for a given 3D vector field and a given unit vector V:

In[1]:=
```
V = {0.205092,0.489097,0.847774};
Clear[Field,x,y,z,m,n,p,curlField]
Field[x_,y_,z_] = {x,-y,z}/(x^2 + y^2 + z^2);
{m[x_,y_,z_],n[x_,y_,z_],p[x_,y_,z_]} = Field[x,y,z];
curlField[x_,y_,z_] = {D[p[x,y,z],y] - D[n[x,y,z],z],
D[m[x,y,z],z] - D[p[x,y,z],x],D[n[x,y,z],x] - D[m[x,y,z],y]};
curlField[0.3,0.9,0.2].V
```
Out[1]=
0.869092

When you push your finger onto V so that V spikes through the center of your finger, and the tip of your finger is at $\{0.3, 0.9, 0.2\}$, does the tip of your finger feel a net clockwise or net counterclockwise swirl?

L.4) Here's a surface shown with a selection of normal vectors with tails planted at the points at which the normals are calculated:

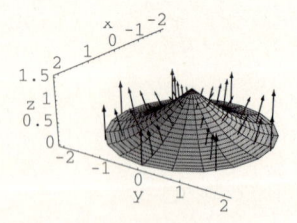

Here's the same surface shown with a selection of curlField$[x, y, z]$ vectors for a certain 3D vector field, Field$[x, y, z]$, with tails planted at the same place the tails of the normal vectors are planted above:

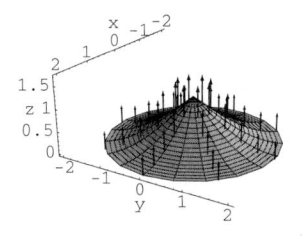

Here they are together:

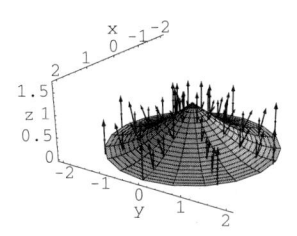

On the basis of these two plots, which of the following plots do you expect to indicate the direction of the net flow of the given 3D vector field along the boundary curve of the surface?

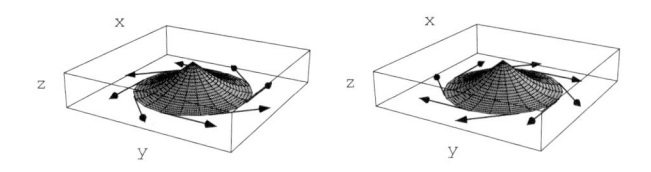

L.5) Here's a calculation of curlField$[x, y, z]$ for a given 3D vector field:

In[2]:=

```
Clear[Field,x,y,z,m,n,p,curlField]
Field[x_,y_,z_] = {x^2,y^2,z^2};
{m[x_,y_,z_],n[x_,y_,z_],p[x_,y_,z_]} = Field[x,y,z];
curlField[x_,y_,z_] = {D[p[x,y,z],y] - D[n[x,y,z],z],
    D[m[x,y,z],z] - D[p[x,y,z],x], D[n[x,y,z],x] - D[m[x,y,z],y]}
```

Out[2]=
{0, 0, 0}

Is there any way to stick your finger into this vector field so that the tip of your finger feels a net clockwise or net counterclockwise swirl?

How do you know?

L.6) You are given a certain 3D vector field

$$\text{Field}[x, y, z] = \{m[x, y, z], n[x, y, z], p[x, y, z]\}$$

and learn that no matter what point $\{x, y, z\}$ you go with, curlField$[x, y, z]$ points in the same direction as $\{0, 0, 1\}$. You drop some paddle wheels into the flow represented by this vector field so that the axis of each paddle wheel is parallel to the vector $\{0, 0, 1\}$ like this:

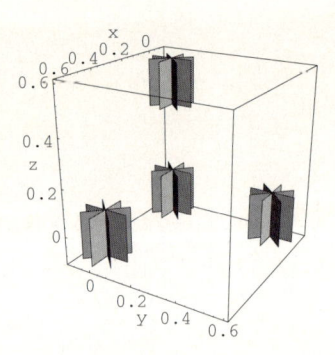

How do you know in advance that the flow represented by the given vector field will make these paddle wheels rotate?

Which way will they rotate?

How about in the case that curlField$[x, y, z]$ points in the same direction as $\{0, 0, -1\}$ at all points $\{x, y, z\}$?

L.7) For a given 3D vector field

$$\text{Field}[x, y, z] = \{m[x, y, z], n[x, y, z], p[x, y, z]\},$$

you calculate

$$\text{curlField}[x, y, z] =$$
$$\{D[p[x, y, z], y] - D[n[x, y, z], z],$$
$$D[m[x, y, z], z] - D[p[x, y, z], x],$$
$$D[n[x, y, z], x] - D[m[x, y, z], y]\}.$$

You are asked to manufacture your own choice of special functions $m[x, y, z], n[x, y, z]$ and $p[x, y, z]$ so that no matter what point $\{x, y, z\}$ you go with,

$$\text{curlField}[x, y, z] = \{0, 0, \text{positive}\}.$$

In other words, no matter what point $\{x, y, z\}$ you go with, curlField$[x, y, z]$ points in the same direction as $\{0, 0, 1\}$.

How would you align paddle wheels to be driven by the flow?

L.8) What is the gradient test for a 3D vector field?

L.9) You are given a certain 3D vector field
$$\text{Field}[x, y, z] = \{m[x, y, z], n[x, y, z], p[x, y, z]\},$$
with the extra information that it is a gradient field.

Is there any way to drop a small paddle wheel into the flow represented by this vector field and align it so that the given vector field will make the paddle rotate? (Assume that there are no singularities.)

L.10) Is there any way to stick your finger into the flow described by a 3D gradient field so that the tip of your finger feels a net clockwise or net counterclockwise swirl?

How do you know? (Assume that there are no singularities.)

L.11) Take a surface R with boundary curve C. If you are given a 3D vector field $\text{Field}[x, y, z]$ with the extra property that $\text{curlField}[x, y, z]$ is tangent to the surface R at all points $\{x, y, z\}$ on the surface R, then how do you use Stokes's formula to tell you once and for all that the net flow of $\text{Field}[x, y, z]$ along C is 0?

L.12) The curve C is parameterized by
$$\{x[t], y[t], z[t]\} = \{2\,t^2, 2 + t^2, 1 - t^2\} \qquad \text{with } 0 \le t \le 2.$$
This tells you that C starts at $\{0, 2, 1\}$ and ends at $\{8, 6, -3\}$.

Measure the flow of the 3D vector field
$$\text{Field}[x, y, z] = \{z, y, x\}$$
along C and interpret the result.

Use a hand calculation to determine whether the net flow of $\text{Field}[x, y, z]$ along C is in the direction of the parameterization, or against the direction of the parameterization.

L.13) You plot a surface R and you get:

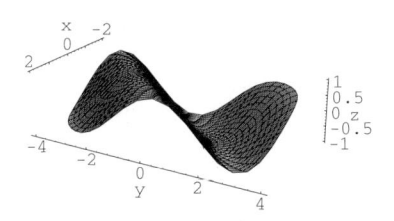

Then you parameterize the boundary curve of R, plot a few tangents to see the direction your parameterization takes, and you get:

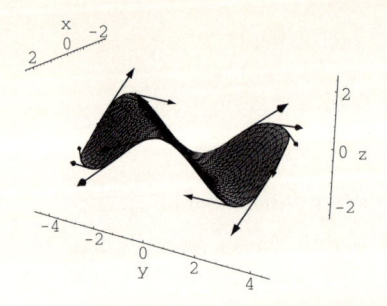

If you want to call your parameterization of the boundary curve counterclockwise in accordance with the agreement made to understand the meaning of Stokes's formula, then which side of the surface must you designate as the top side?

L.14) A 3D vector field

$$\text{Field}[x, y, z] = \{m[x, y, z], n[x, y, z], p[x, y, z]\}$$

comes into your hands. You are given two curves C_1 and C_2, both parameterized to start at a given point $\{a_s, b_s, c_s\}$ and end at a given point $\{a_e, b_e, c_e\}$.

You calculate curlField$[x, y, z]$ and are pleased to find that

$$\text{curlField}[x, y, z] = \{0, 0, 0\}$$

at all points $\{x, y, z\}$. What is the relationship between the flow along the curve measurements

$$\int_{C_1} m[x, y, z]\, dx + n[x, y, z]\, dy + p[x, y, z]\, dz$$

and

$$\int_{C_2} m[x, y, z]\, dx + n[x, y, z]\, dy + p[x, y, z]\, dz?$$

Do you expect this relationship to hold up in the case that curlField$[x, y, z]$ is not $\{0, 0, 0\}$ at all points $\{x, y, z\}$?

L.15) Here is the 3D electric field resulting from an electrical charge of strength q placed at a point $\{a, b, c\}$:

In[3]:=
```
Clear[x,y,z,m,n,p,a,b,c,Field]
{m[x_,y_,z_], n[x_,y_,z_],p[x_,y_,z_]} = q({x,y,z} - {a,b,c})/
  ((x - a)^2 + (y - b)^2 + (z - c)^2)^(3/2)
```

Out[3]=
$$\left\{\frac{q\,(-a+x)}{((-a+x)^2+(-b+y)^2+(-c+z)^2)^{3/2}},\right.$$
$$\frac{q\,(-b+y)}{((-a+x)^2+(-b+y)^2+(-c+z)^2)^{3/2}},$$
$$\left.\frac{q\,(-c+z)}{((-a+x)^2+(-b+y)^2+(-c+z)^2)^{3/2}}\right\}$$

Here is a calculation of DivField$[x,y,z]$:

In[4]:=
```
Together[D[m[x,y,z],x] + D[n[x,y,z],y] + D[p[x,y,z],z]]
```
Out[4]=
0

And here is a calculation of curlField$[x,y,z]$:

In[5]:=
```
{D[p[x,y,z],y] - D[n[x,y,z],z],
 D[m[x,y,z],z] - D[p[x,y,z],x],
 D[n[x,y,z],x] - D[m[x,y,z],y]}
```
Out[5]=
{0, 0, 0}

Discuss what information you get from these two calculations.

L.16) Write down Stokes's formula and say how the formula is to be interpreted.

L.17) If C is a closed curve in three dimensions and

$$\text{Field}[x,y,z] = \{m[x,y,z], n[x,y,z], p[x,y,z]\}$$

with curlField$[x,y,z] = \{0,0,0\}$ at all points, then what is the flow along C measurement

$$\oint_C m[x,y,z]\,dx + n[x,y,z]\,dy + p[x,y,z]\,dz?$$

How does Stokes's formula help you to back up your answer?

L.18) A force field

$$\text{Field}[x,y,z] = \{m[x,y,z], n[x,y,z], p[x,y,z]\}$$

comes into your hands. You are required to send an object through this force field starting at a given point $\{a_s, b_s, c_s\}$ and ending at a given point $\{a_e, b_e, c_e\}$. You can send the object along any curve C you desire, provided it is parameterized to start at $\{a_s, b_s, c_s\}$ and to end at $\{a_e, b_e, c_e\}$. Naturally you want to pick the curve C that makes the force field do as much of the work as possible.

You evaluate curlField$[x, y, z]$ and are pleased to find that

$$\text{curlField}[x, y, z] = \{0, 0, 0\}$$

at all points $\{x, y, z\}$. Armed with this information, you say, "What the hell, I'll send it on a straight line segment running from $\{a_s, b_s, c_s\}$ to $\{a_e, b_e, c_e\}$."

Why are you right?

Why does picking the right curve become much more complicated if curlField$[x, y, z]$ is not $\{0, 0, 0\}$ at all points $\{x, y, z\}$?

L.19) Lots of folks say that a three-dimensional vector field

$$\text{Field}[x, y, z] = \{m[x, y, z], n[x, y, z], p[x, y, z]\}$$

represents ideal fluid flow if

$$\text{divField}[x, y, z] = 0 \quad \text{and} \quad \text{curlField}[x, y, z] = \{0, 0, 0\}$$

at all points $\{x, y, z\}$.

What do these folks have in mind?

L.20) Give a hand calculation of curlField$[x, y, z]$ by calculating

$$\nabla \times \text{Field}[x, y, z] \quad \text{for} \quad \text{Field}[x, y, z] = \{x, y, z\}.$$

L.21) Calculate

$$\nabla \bullet \text{Field}[x, y, z] \quad \text{for} \quad \text{Field}[x, y, z] = \{x, y, z\}$$

by hand and interpret the result in terms of divField$[x, y, z]$.

L.22) Calculate the Laplacian

$$\triangle f[x, y, z] = \nabla \bullet \nabla f[x, y, z] \quad \text{for} \quad f[x, y, z] = e^x \cos[y] + z^2$$

by hand, and interpret the result in terms of

$$\frac{\partial^2 f[x, y, z]}{\partial x^2} + \frac{\partial^2 f[x, y, z]}{\partial y^2} + \frac{\partial^2 f[x, y, z]}{\partial z^2},$$

and the divergence of the gradient field of $f[x, y, z]$.

Index

Entries are listed by lesson number and problem number.

Acceleration, 3.01: T.1, G.4
    breaking acceleration vectors into normal and tangential components and, perpendicularity and, 3.02: G.6
Area, on surfaces, 3D surface measurements of, 3.10: B.2
Area conversion factors, transforming 2D integrals and, 3.07: T.3, G.10
Average values, transforming 3D integrals and, 3.08: B.4, G.6

Centers of mass, transforming 3D integrals and, 3.08: G.6
Centroids, transforming 3D integrals and, 3.08: G.6
Chain rule, gradients and
    linearizations to explain, 3.03: B.4
    uses of, 3.03: T.2
Circles, kissing, perpendicularity and, 3.02: G.8
Cobb-Douglas manufacturing model, for industrial engineering, 3.03: G.7
Coordinate axes, in three dimensions, 3.01: G.5
Coordinate planes, in three dimensions, 3.01: G.5
Cross product, measurements with, 3.02: G.9, G.10
Curly field, 3.11: B.2
Curvature, perpendicularity and, 3.02: G.8
Curved surfaces, in 3D, normal vectors for, 3.02: B.3

Cylinders, plotting and integrating on, transforming 3D integrals and, 3.08: T.1

Data fit, in two variables, 3.03: G.8
Density measurement
    transforming 3D integrals and, 3.08: B.3
    transforming 2D integrals and, 3.07: T.4, G.7
Differential equations, 2D vector fields and, 3.04: T.2, G.5
Dipole fields, 3.06: G.5
Directed curves, flow measurements with integrals and, 3.05: B.4
Distance, measuring, 3.01: B.3, G.2
Divergence, 3.06: B.2, 3.10: B.1
Dot plots, 2D vector fields and, 3.04: G.3
Dot products, 3.01: B.3, G.2
    sign of, flow along curves and, 3.04: T.3

Earth, size of, spherical coordinates and, 3.09: G.6
Eigenvalues
    spherical coordinates and, 3.09: G.11
    transforming 2D integrals and, 3.07: G.11
Eigenvectors
    spherical coordinates and, 3.09: G.11
    transforming 2D integrals and, 3.07: G.11
Ellipsoids, spherical coordinates and, 3.09: T.1
Elliptical reflectors, 3.01: G.9
Estimates, 3D flow along and, 3.11: G.2

Index

FindMinimum instruction, 3.03: T.3
Flatness, plotting and, 3.02: T.2
Flowers, and spherical coordinates, 3.09: T.3
Flow measurements
   with integrals. *See* Integrals
   singularities and, 3.06: G.4
      flow-across-the-curve measurements and, 3.06: B.3
   3D flow along and. *See* 3D flow along
   3D surface measurements and. *See* 3D surface measurements
   2D vector fields and. *See* 2D vector fields
Force fields, trajectories in, 3.05: G.10

Gauss's formula, 3.06: G.5. *See also* 3D surface measurements
Gradient(s), 3.03: B.1
   chain rule and
      linearizations to explain, 3.03: B.4
      uses of, 3.03: T.2
   Cobb-Douglas manufacturing model for industrial engineering, 3.03: G.7
   data fit and, 3.03: T.5
      in two variables, 3.03: G.8
   direction of, 3.03: B.3, G.1
   level curves, level surfaces, and gradient as normal vector, 3.03: B.2, G.2
   linearizations and, 3.03: G.9
      total differentials and, 3.03: G.9
   maximization and minimization and, 3.03: T.3, T.4, G.3, G.5
      constrained, Lagrange's method for, 3.03: T.6, G.6
   solving problems by hand and, 3.03: G.4
   total differentials and, 3.03: T.1, G.9
Gradient fields, 3.04: G.4
   flow measurements with integrals and, 3.05: B.4, G.6
   gradient test and, 3.05: T.3
Gradient test, in 2D, 3.05: T.3
Gradient test, in 3D, 3D flow along and, 3.11: T.4, G.4
Grids, linearizing, transforming 2D integrals and, 3.07: B.2

Integrals
   flow measurements with
      calculations and interpretation and, 3.05: G.3
      directed curves and, 3.05: B.4
      flow across curves, 3.05: B.1; G.1
      flow along curves, 3.05: B.2, G.1
      force fields and trajectories and, 3.05: G.10
      gradient fields and, 3.05: B.4, T.3, G.6
      line integrals and, 3.05: T.4
      path independence and, 3.05: B.4
      path integrals and, 3.05: B.3, B.4, T.2
      reversing direction and, 3.05: T.1, G.2
      sources and sinks and, 3.05: G.5
      spin fields and, 3.05: G.8
      summary of, 3.05: T.5
      work measurement and, 3.05: G.7
   line, flow measurements with, 3.05: T.4
   path
      calculating, 3.06: T.1
      flow measurements with, 3.05: B.3, B.4, T.2
   surface, 3D surface measurements and, 3.10: B.3
   3D. *See* 3D integrals
   2D. *See* 2D integrals
Integration, in 2D, 3.07, in 3D, 3.08
   with spherical coordinates, 3.09: B.2, G.2
   switching order of, transforming 3D integrals and, 3.08: T.4, G.7
Irrotational flow, 3D flow along and, 3.11: G.6

Lagrange's method, for constrained maximization and minimization, 3.03: T.6, G.6
Laplacian
   as divergence of the gradient field, 3.06: G.6, 3.10: G.6
   steady-state heat and, 3.06: G.6
   3D surface measurements, 3.10: G.6
Line(s), 3.01: G.7
   fundamentals of, 3.01: G.1
   specifying using vectors, 3.01: T.3
   tangent, 3.01: B.2
Linear equations, transforming 2D integrals and, 3.07: G.10
Linearizations
   gradients and, 3.03: G.9
   perpendicularity and, 3.02: G.4
Linearizing grids, transforming 2D integrals and, 3.07: B.2
Line integrals. *See* path integrals
Log-log paper, transforming 2D integrals and, 3.07: G.5

Mass measurement
   centers of, transforming 3D integrals and, 3.08: G.6
   transforming 3D integrals and, 3.08: B.3
   transforming 2D integrals and, 3.07: T.4, G.7
Maximization, gradients and, 3.03: T.3, G.3, G.5

Maximization (*continued*)
  constrained, Lagrange's method for, 3.03: T.6, G.6
Minimization, gradients and, 3.03: G.3, G.5
  constrained, Lagrange's method for, 3.03: G.6
Moebius strips, 3D surface measurements and, 3.10: G.7
Morphing, 3D surface measurements and, 3.10: G.7

Parabolic reflectors, 3.01: G.9
Parallelepipeds, transforming 3D integrals and, 3.08: T.3
Parallel flow
  rotational nature of, 3.06: G.9
  3D flow along and, 3.11: G.6
Parametric formulas
  transforming 2D integrals with boundary given by, 3.07: T.1
  transforming 2D integrals with boundary not given by, 3.07: T.2
Parametric plots, 3.01: G.6, G.7
Path independence
  flow measurements with integrals and, 3.05: B.4
  3D flow along and, 3.11: T.2, G.5
Path integrals
  calculating, 3.06: T.1
  flow measurements with, 3.05: B.3, B.4, T.2
Perpendicularity
  borings and, 3.02: G.5
  bouncing light beams off surfaces using normal vector, 3.02: G.7
  breaking acceleration vectors into normal and tangential components and, 3.02: G.6
  cross product $X \times Y$ of two 3D vectors, 3.02: B.1
  flatness and plotting and, 3.02: T.2
  of gradient to level curves and surfaces, 3.03: G.2
  kissing circles and curvature and, 3.02: G.8
  linearizations and, 3.02: G.4
  normal vectors for curved surfaces in 3D, 3.02: B.3
  planes and
    fundamentals of, 3.02: G.1
    plotting on, 3.02: G.2
    in 3D, 3.02: B.2
  3D curves and, 3.02: G.3
  true scale plots, 3.02: T.1
  unit vectors and, 3.02: T.3, T.4

Planes
  perpendicularity and, 3.02: G.1
    plotting on planes and, 3.02: G.2
    plotting curves on, perpendicularity and, 3.02: T.3
  between spheres, spherical coordinates and, 3.09: G.8
  in 3D, 3.02: B.2
Plotting
  of cylinders, transforming 3D integrals and, 3.08: T.1
  flatness and, 3.02: T.2
  planes and, perpendicularity and, 3.02: T.3, G.2
  with spherical coordinates, 3.09: G.1
  of trajectories and, 2D vector fields, 3.04: G.7, G.8
  transforming 2D integrals and, 3.07: G.8

Rotation
  3D surface measurements and, 3.10: G.10
  of vector fields, 3.06: T.3
rotField, 3.06: T.3

Semi-log paper, transforming 2D integrals and, 3.07: G.5
Singularities, 3.06: G.7
  flow-across-the-curve measurements and, 3.06: B.3
  flow calculations and, 3.06: G.4
  sources, sinks, and swirls and, 3.06: G.2
  summary of, 3.06: T.4
  3D surface measurements and, 3.10: T.4
Sinks, 3.06: G.1
  divergence of vector fields and, 3.06: B.2
  Gauss's formula in 3D and, 3.10: B.1
  singularities and, 3.06: G.2
  summary of, 3.06: T.4
Solids bounded by sets of surfaces, integrating on, transforming 3D integrals and, 3.08: T.2
Sources, 3.06: G.1
  divergence of vector fields and, 3.06: B.2
  Gauss's formula in 3D and, 3.10: B.1
  singularities and, 3.06: G.2
  summary of, 3.06: T.4
Spheres
  drilling and slicing, transforming 3D integrals and, 3.08: G.3
  planes between, spherical coordinates and, 3.09: G.8
  plotting and integrating on, transforming 3D integrals and, 3.08: T.1
  spherical coordinates and, 3.09: T.1

Index

Spherical coordinates, 3.09: B.1
  art and, 3.09: G.9
  eigenvalues and eigenvectors and, 3.09: G.11
  four dimensions and, 3.09: G.10
  integrating with, 3.09: B.2, G.2
  planes between spheres and, 3.09: G.8
  plotting with, 3.09: G.1
  skins and, 3.09: G.4
  spheres and ellipsoids and, 3.09: T.1
  star wars and, 3.09: G.7
Spherical reflectors, 3.01: G.9
Spin fields, 3.06: G.10
  flow measurements with integrals and, 3.05: G.8
Steady-state heat, Laplacian and, 3.06: G.6
Stokes's formula. *See* 3D flow along, Stokes's formula and
Surface integrals, 3D surface measurements and, 3.10: B.3
Surface packaging, 3D surface measurements and, 3.10: T.5
Swirls, 3.06: G.1
  singularities and, 3.06: G.2
  summary of, 3.06: T.4

Tangent(s)
  spying along, 3.01: T.5
  2D vector fields and, 3.04: G.3
Tangent lines, 3.01: B.2
Tangent vectors, 3.01: B.2
3D
  cross product $X \times Y$ of two 3D vectors, perpendicularity of, 3.02: B.1
  normal vectors for curved surfaces in, 3.02: B.3
  planes in, 3.02: B.2
3D curves, perpendicularity and, 3.02: G.3
3D flow along
  clockwise and counterclockwise, 3.11: G.7
  curl of 3D vector field and, 3.11: B.2
  gradient test in 3D and, 3.11: T.4, G.4
  measurements and estimates and, 3.11: G.2
  measuring flow along 3D curves and, 3.11: B.1
  parallel flow and irrotational flow and, 3.11: G.6
  path independence and, 3.11: T.2, G.5
  relationship of 2D rotField$[x, y]$ to 3D curlField$[x, y, z]$ and, 3.11: G.10
  Stokes's formula and
    as outgrowth of 2D Gauss-Green formula and, 3.11: B.4
    in theory and practice, 3.11: T.1
  for using curl of 3D vector field to measure swirl of vector field in 3D, 3.11: B.3
  symbols ∇ and $\triangle$ and, 3.11: G.8
  vector fields and, 3.11: G.1
  work and, 3.11: T.3, G.3
3D integrals, transforming, 3.08: B.1, B.2, G.1
  average values and, 3.08: B.4
  centroids and centers of mass and, 3.08: G.6
  drilling and slicing spheres and, 3.08: G.3
  integrating on solids bounded by sets of surfaces and, 3.08: T.2
  mass and density and, 3.08: B.3
  measuring when you can't see and, 3.08: G.5
  parallelepipeds and, 3.08: T.3
  plotting and integrating on cylinders, spheres, and tubes and, 3.08: T.1
  switching order of integration and, 3.08: T.4, 3.08: G.7
  tubes, horns, and squashed doughnuts and, 3.08: G.4
  volume and, 3.08: G.2, G.8
  volume conversion factor and, 3.08: G.9
3D surface measurements, 3.10: G.1
  flux of electric field and, 3.10: B.5
  Gauss's law and, 3.10: G.5
  Laplacian and, 3.10: G.6
  measuring area on surfaces and, 3.10: B.2
  measuring flow across surfaces using, 3.10: B.4
  calculating flow by calculating flow across substitute sphere using Gauss's formula, 3.10: T.3
  calculating flow by calculating flow across substitute surfaces using Gauss's formula, 3.10: T.2, G.3, G.4
  Gauss's 3D formula versus calculation by surface integrals, 3.10: T.1
  morphing and Moebius strips and, 3.10: G.7
  rotating and measuring and, 3.10: G.10
  sources, sinks, and Gauss's formula in 3D and, 3.10: B.1
  surface integrals and, 3.10: B.3
  volumes, flow, and buoyancy and, 3.10: G.9
Total differentials, gradients and, 3.03: T.1, G.9
Trajectories. *See* 2D vector fields
True scale plots, 3.02: T.1
Tubes, plotting and integrating on, transforming 3D integrals and, 3.08: T.1, G.4

Index

2D curves, bouncing light beams off, using normal vector, 3.01: T.2

2D integrals
measuring flow across closed curves using, 3.06: B.1, T.2
transforming, 3.07: G.1
area conversion factor and, 3.07: T.3
with boundary given by parametric formulas, 3.07: T.1
with boundary not given by parametric formulas, 3.07: T.2
flow measurements and, 3.07: G.3
interpreting plots and, 3.07: G.4
linear equations and area conversion factors and, 3.07: G.10
linearizing grids and, 3.07: B.2
plotting and, 3.07: G.8
problems with, 3.07: G.9
semi-log paper and log-log paper and, 3.07: G.5
uv-paper and xy-paper and, 3.07: B.1
volume, mass, and density measurements and, 3.07: T.4, G.7
when $A_{(x,y)}[u,v]$ is 0, 3.07: G.6
xy-paper and, 3.07: G.2

2D vector fields, 3.04: B.1
crossing of trajectories and, 3.04: G.6
differential equations and, 3.04: T.2, G.5
electrical fields and, 3.04: T.4, G.10, G.5
flow across curves, 3.04: B.2, B.3, T.1, T.3, G.2
flow along curves, 3.04: B.2, B.3, T.1, T.3, G.1, G.2
sign of dot product and, 3.04: T.3
gradient field, 3.04: G.4
normals, tangents, and dot plots and, 3.04: G.3
plots of, troubleshooting, 3.04: T.5
plotting trajectories, 3.04: G.7, G.8
trajectories of, 3.04: B.1
trajectories of force field, 3.05: G.10
water flow and, 3.04: G.9

Unit vectors, perpendicularity and, 3.02: T.3, T.4

Vector(s), 3.01: G.3. *See also* 2D vector fields
acceleration and, 3.01: T.1, G.4
adding, 3.01: B.1
bouncing light beams off surfaces using normal vector, 3.02: G.7
two-dimensional curves, 3.01: T.2
coordinate axes and planes in three dimensions, 3.01: G.5
cross product $X \times Y$ of two 3D vectors, perpendicularity of, 3.02: B.1
fundamentals of, 3.01: G.1
length of, 3.01: B.3
moving, 3.01: B.1
multiplying, 3.01: B.1
normal
for curved surfaces in 3D, 3.02: B.3
gradient as, 3.03: B.2
parabolic, spherical, and elliptical reflectors, 3.01: G.9
push of one vector in direction of another, 3.01: B.4
specifying lines using, 3.01: T.3
stealth technology, 3.01: G.11
subtracting, 3.01: B.1
tangent, 3.01: B.2
unit, perpendicularity and, 3.02: T.3, T.4
velocity and, 3.01: B.2, T.1, G.4, G.10

Vector fields
curl of, 3.11: B.2
divergence of, sources and sinks and, 3.06: B.2, 3.10: B.1
rotation of, 3.06: T.3
3D flow along and, 3.11: G.1

Velocity measurement, 3.01: T.1, G.4, G.10

Velocity vectors, 3.01: B.2

Volume measurement
3D surface measurements and, 3.10: G.9
transforming 3D integrals and, 3.08: G.2, G.8
volume conversion factor and, 3.08: G.9
transforming 2D integrals and, 3.07: T.4, G.7

Work measurement
flow measurements with integrals and, 3.05: G.7
3D flow along and, 3.11: T.3, G.3

$X \bullet Y = 0$, 3.01: B.5
xy-paper and, transforming 2D integrals and, 3.07: G.2
$X \bullet Y = \|X\| \|Y\| \cos[b]$, 3.01: B.4